Ram Accelerators

Springer
Berlin
Heidelberg
New York
Barcelona
Budapest
Hong Kong
London
Milan
Paris
Singapore
Tokyo

K. Takayama · A. Sasoh (Eds.)

Ram Accelerators

Proceedings
of the Third International Workshop
on Ram Accelerators
Held in Sendai, Japan, 16-18 July 1997

With 305 Figures and 24 Tables

Springer

Professor K. Takayama
Associate Professor A. Sasoh

Shock Wave Research Center
Institute of Fluid Science
Tohoku University
2-1-1 Katahira, Aoba-ku
Sendai 980-8577, Japan

Library of Congress Cataloging-in-Publication Data.

International Workshop on Ram Accelerators (3rd: 1997: Sendai-han, Japan) Ram accelerators: proceedings of the third International Workshop on Ram Accelerators, held in Sendai, Japan, 16-18 July 1997 / K. Takayama, A. Sasoh (eds.) p. cm. Includes bibliographical references (p.). ISBN-13: 978-3-642-46878-0 (hardcover: alk. paper) 1. Hypervelocity guns--Congresses. 2. shock tubes--Congresses. 3. Projectiles--Congresses. I. Takayama, K. (Kazuyashi), 1940- . II. Sasoh, A. III. Title.
TL567.S4I53 1997 629.132'306--dc21 98-26286 CIP

ISBN-13: 978-3-642-46878-0 e-ISBN-13: 978-3-642-46876-6
DOI: 10.1007/978-3-642-46876-6

Softcover reprint of the hardcover 1st edition 1998

Typesetting: Camera-ready by editors
Cover design: *design & production* GmbH, Heidelberg

SPIN 10683444 55/3144 – 5 4 3 2 1 0 – Printed on acid-free paper

Foreword

The Ram Accelerator is one of the most challenging projects that I have ever undertaken. I was fortunate in having a group of very imaginative colleagues working with me at the time. Together we developed a working system that we call the "Ram Accelerator." This system and its variations have been discussed at International Workshops on Ram Accelerators (In Saint Louis 1993, Seattle 1995 and Sendai 1997).

The Ram Accelerator remains a subtle and demanding project involving the combined action of combustion, wave systems, and turbulence. While I am very pleased with what scientists and engineers have done out over the years since the concept was introduced, there still remains much to do.

The quality of the work in this book shows that we are now acquiring much of the basic information and, indeed, some of the practical engineering knowledge that is necessary for us to move forward in this area. I hope that I shall be able to attend the next meetings and enjoy the opportunity to meet with colleagues.

The basic concept of the Ram Accelerator offers a new approach to mass launches, and still remains one of our most interesting tools for the generation of basic fluid dynamics data in supersonic reacting systems. Of all of the ground-based mass launchers that may ultimately be able to reach orbital velocity, the Ram Accelerator remains the only scalable system and hence might be useful to impulsively lift masses into space in a cost-effective manner. It is a worthwhile, intellectual challenge and a fascinating experimental tool.

Seattle, Washington
September 1997

Abraham Hertzberg

Preface

The Third International Workshop on Ram Accelerators was held at the Shock Wave Research Center, Institute of Fluid Science, Tohoku University, Sendai, Japan from July 16 to 18, 1997. Scientists and engineers from twelve nations, USA, France, Germany, Korea, Canada, Israel, Australia, Russia, Iran, China, Brazil, and Japan, participated in the workshop or submitted papers. The first success in the operation of this noble device at the University of Washington motivated other researchers to join the venture of developing the device. A micro-workshop, which initiated the international collaboration in ram accelerator activities, was held during the Eighteenth International Symposium on Shock Waves, Sendai, July 1991. The First International Workshop on Ram Accelerators was organized by the French-German Research Institute of Saint-Louis (ISL), France, in 1993; the second one by the University of Washington, WA, USA in 1995. In each international workshop, various aspects of ram accelerators were intensely discussed. The presented material was bound and served as proceedings at the venues.

In spite of the previous progress in this technology, it still warrants further intense investigation for its practical application. So far, knowledge on ram accelerators has spread to only a limited number of scientists and engineers. It might not be easy for people who have not been involved in related projects to participate in further research adventure on this device. Based on this background, this book was edited not only as the proceedings of the workshop but also as an introductory textbook on ram accelerators. This book includes three lecture notes on important aspects – overview of research activities, performance modeling and high-pressure detonation dynamics. Readers will also benefit from the reviews or summaries of the related topics as given in other contributed papers. We hope that this book will act both as a bound volume on current technologies and as an introduction for many scientists and researchers.

We would like to express our gratitude to all the contributors and participants in the workshop, including Dr. Tsutomu Saito, Dr. Osamu Onodera, Messrs Hidenori Ojima, Toshihiro Ogawa, Ms Aki Tanimoto, Ms Naoko Yagi, Ms Takako Iijima, and students of the Shock Wave Research Center for their valuable help in organizing the workshop. Also, we thank Ms Etsuko Hagita for her assistance with this edition and Ms Michiyo Sasoh for designing the cover figures from our experimental visualization. The organizer wishes to acknowledge the support provided by Japan Society for Promotion of Science (JSPS).

Shock Wave Research Center
Institute of Fluid Science, Tohoku University
Sendai, Japan, May 1998

Kazuyoshi Takayama
Akihiro Sasoh

Contents

Lecture Notes

Facilities and Experiments

Performance Prediction

Starting Processes

Hypersonic Blunt Body in Chemically Reacting Flows

Diagnostics

Computational Fluid Dynamics

Lecture notes

The ram accelerator: overview and state of the art

A. P. Bruckner
Aerospace and Energetics Research Program, University of Washington, Seattle, WA, 98195-2250, U.S.A.

Abstract. A new launcher technology called the ram accelerator has been under development since 1983 for applications as a scalable hypervelocity accelerator capable, in principle, of launching projectiles at velocities of 8 km/s or greater. The device is based on an in-bore ramjet concept in which a subcaliber projectile, shaped like the centerbody of a supersonic ramjet, is propelled down the center of a stationary tube filled with a pressurized mixture of gaseous fuel and oxidizer. This propellant burns near the base of the moving projectile, generating thrust. The highest pressure in the system is always in the vicinity of the projectile base, rather than at the breech as in a gun. The chemical energy density and speed of sound of the propellant can be adjusted (via pressure and composition) to control the Mach number and acceleration history experienced by the projectile. Three different propulsive modes, centered on the Chapman-Jouguet (CJ) detonation speed of the propellant, have been experimentally observed. Projectiles have been accelerated smoothly from velocities below to above the CJ speed within a single propellant. During this process the nature and location of the combustion changes from thermally choked subsonic combustion behind the projectile to shock-induced supersonic combustion in the region between the projectile and the tube wall. The ram accelerator is easily scalable for a variety of interesting applications ranging from hypersonics research to direct launch of acceleration-insensitive payloads to orbit. Growing interest in this technology has motivated the establishment of several ram accelerator research facilities throughout the world. This paper presents an overview of the technology and the current state of the art.

Key words: Ramjet-in-tube, Combustion, Detonation, Hypervelocity launcher, Hypersonic test facility, Mass launcher, Space launcher

1. Introduction

The ram accelerator is a novel chemically propelled mass launcher conceived at the University of Washington (UW) in 1983 by Hertzberg et al. (1986, 1988).* The principle of operation of the ram accelerator is similar to that of a supersonic airbreathing ramjet engine. A stationary tube, analogous to the cylindrical outer cowling of a ramjet engine (Fig. 1), is filled with a premixed gaseous propellant consisting of fuel, oxidizer, and an inert diluent (typically methane, oxygen, and nitrogen, respectively) at a pressure of 0.5-20 MPa. Thin diaphragms close off each end of the tube to contain the propellant. The projectile is similar in shape to the centerbody of a ramjet and has a diameter smaller than the bore of the accelerator tube.

The operational sequence of the ram accelerator (Fig. 2) is initiated by injecting the projectile into the ram accelerator tube at speeds greater than 700 m/sec by means of a conventional powder gun or light gas gun. A lightweight obturator, or piston, in contact with the base of the projectile seals the gun bore during this initial impulse. The acceleration from rest of the projectile/obturator combination compresses residual air in the gun's launch tube via a series of shock waves that reflect back and forth between the moving projectile and the entrance diaphragm of

* Unknown to the UW group at the time, a number of prior concepts having similarities to the ram accelerator had been proposed by others, one dating as far back as 1962. None of these, however, were ever reduced to practice. See Hertzberg et al. (1988) for references to several of these early concepts.

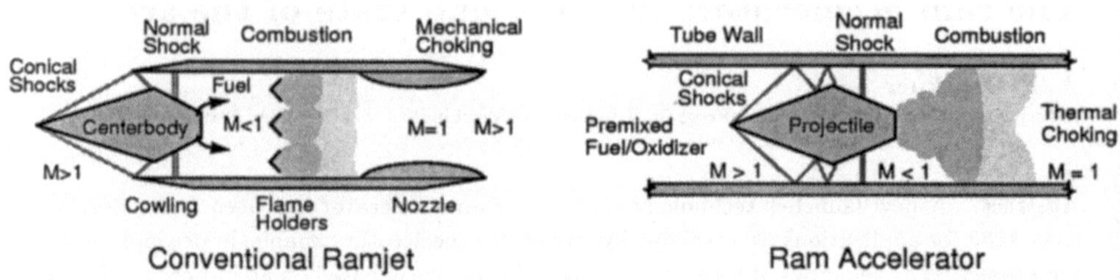

Fig. 1. Comparison of ram accelerator and conventional ramjet.

the ram accelerator. When the projectile punctures the diaphragm and enters the ram accelerator, the shock-heated air contacts the propellant, igniting it near the base of the projectile. The combustion stabilizes behind the projectile, giving rise to a wave of high pressure that drives the projectile forward, in a manner analogous to an ocean wave propelling a surfboard (Fig. 3). The obturator rapidly decelerates following ignition and does not participate in the subsequent acceleration process. To keep the projectile centered in the tube, either the projectile is fabricated with fins that span the bore (Fig. 4a), or the tube is equipped with several internal rails that run its length (Fig. 4b).

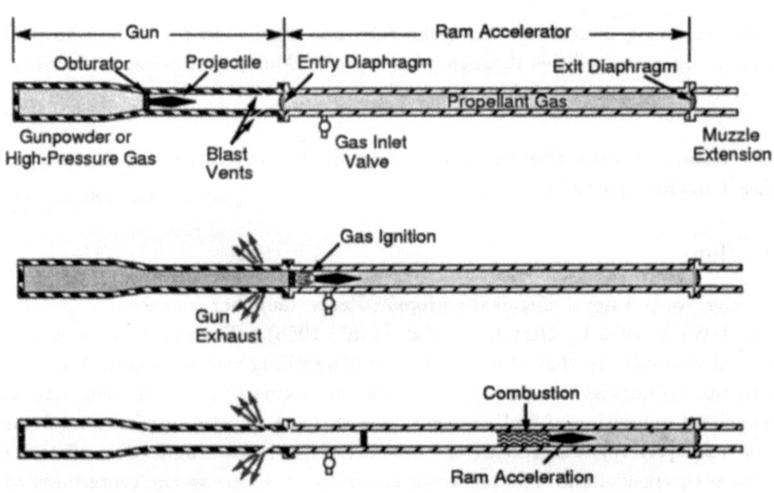

Fig. 2. Operational sequence of ram accelerator. a) Gun is loaded with projectile and obturator, and charge of gunpowder or high pressure gas. Ram accelerator is loaded with combustible gas mixture at 0.5-20 MPa. b) Gun fires obturator/projectile combination into ram accelerator. c) Combustion is initiated and moves with projectile, sustaining traveling pressure wave that accelerates projectile to high velocity.

What distinguishes the ram accelerator from a gun is that its source of energy (the combustible gas mixture) is uniformly distributed throughout the entire length of the accelerator tube, whereas in a gun the energy source is concentrated at the breech as either a charge of gunpowder or high pressure gas. During the ram acceleration process the highest pressure in the tube is always at the projectile's base, rather than at the breech as in a gun (Fig. 3), and the bulk of the combustion products moves in a rearward direction. Only a small volume of high pressure gas exits the tube with the projectile. These characteristics of the ram accelerator result in much more uniform acceleration of the projectile, very high velocity capability, and very little muzzle blast and recoil. Furthermore, the acceleration and muzzle velocity of the ram accelerator can be easily tailored to specific needs by adjusting the propellant composition and fill pressure. Potential applications of the ram accelerator include hypersonic aerodynamic testing (Witcofski et al. 1991, Bruckner et al. 1992a, Naumann and Bruckner 1994), scramjet simulation (Bruckner et al. 1991a), and direct launch to orbit (Bruckner and Hertzberg 1987, Kaloupis and Bruckner 1988, Bogdanoff 1992).

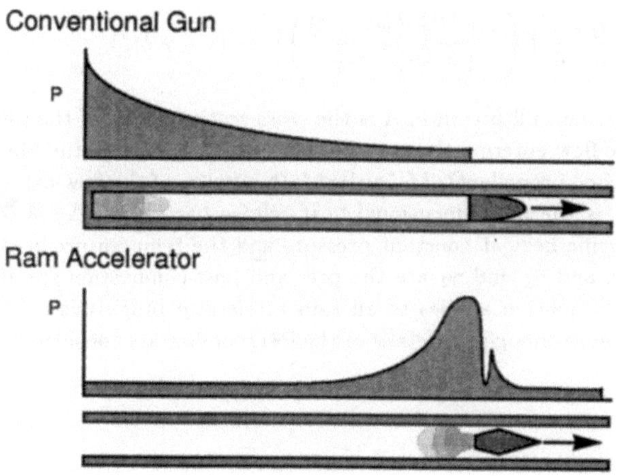

Fig. 3. Pressure distributions in conventional gun and ram accelerator.

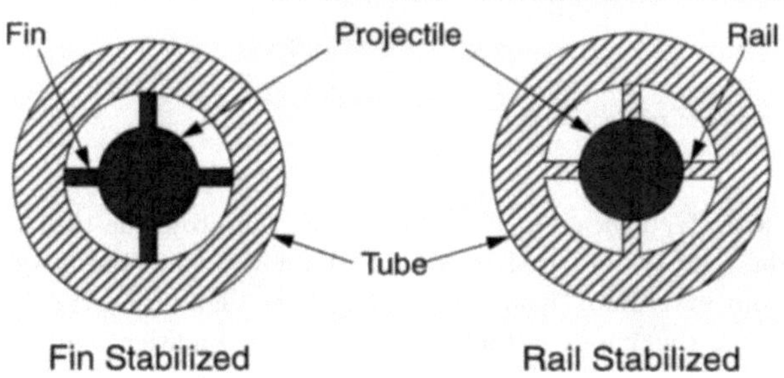

Fig. 4. Methods for centering projectile in tube. a) Fins on projectile. b) Rails in tube.

The propulsive cycle illustrated in Fig. 1 is the thermally choked ram accelerator mode, which operates with in-tube projectile Mach numbers typically ranging from 2.5 to 4 and at velocities below the Chapman-Jouguet (CJ) detonation speed of the propellant, i.e., at subdetonative velocities (Hertzberg et al. 1988, 1991, Bruckner et al. 1991b). In this mode the thrust is provided by the high projectile base pressure generated by the normal shock system that is stabilized on the body by thermally choked subsonic combustion behind the projectile.

The ram accelerator can be modeled analytically using a simple one-dimensional control volume approach originally formulated by the author (Hertzberg et al. 1988, Bruckner et al. 1991b) and later further developed into a generalized Hugoniot theory of the ram accelerator by Knowlen and Bruckner (1992). By applying the steady flow gasdynamic conservation equations to a control volume that contains the projectile (Fig. 5), and assuming that the propellant and combustion products behave as ideal, calorically perfect gases, the following expression for the non-dimensional thrust, τ, on the projectile can be derived:

$$\tau = \frac{F}{p_1 A} = \frac{\gamma_1 M_1}{\gamma_2 M_2}(1+\gamma_2 M_2^2)\left[\left(\frac{\gamma_2-1}{\gamma_1-1}\right)\left(\frac{1+\frac{(\gamma_1-1)}{2}M_1^2+Q}{1+\frac{(\gamma_2-1)}{2}M_2^2}\right)\right]^{\frac{1}{2}} - (1+\gamma_1 M_1^2) \quad (1)$$

where F is the thrust, P_1 is the propellant fill pressure, A is the cross-sectional area of the tube bore, M_1 is the Mach number of the flow entering the control volume (i.e., the projectile Mach number with respect to the undisturbed propellant), M_2 is the Mach number of the flow exiting the control volume, $Q = \Delta q / c_{p1} T_1$ is the non-dimensional heat release parameter, Δq is the heat release, c_{p1} and T_1 are the specific heat at constant pressure and the temperature of the undisturbed propellant, respectively, and γ_1 and γ_2 are the pre- and post-combustion specific heat ratios, respectively. This thrust equation applies to all ram accelerator propulsive modes operating in a quasi-steady manner, even though no details of the internal flow are considered in its derivation.

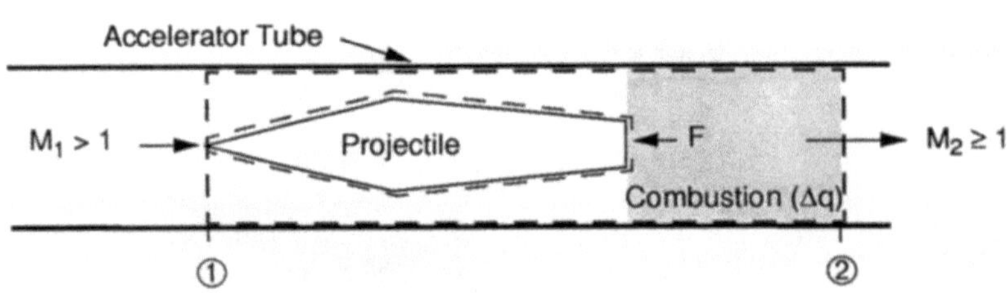

Fig. 5. One-dimensional control volume model of ram accelerator.

If one knows how M_2 varies with M_1 in a given propellant, then the projectile thrust can be readily computed for any flight velocity. Thermal choking of the flow behind the projectile ($M_2 = 1$) corresponds to an entropy extremum (Knowlen and Bruckner 1992); thus, the details of the process which brings the flow to choking do not affect the end state conditions and do not have to be known to predict the thrust. For propulsive cycles that do not involve thermal choking, such as the transdetonative and superdetonative modes discussed later in this paper, the details of the flow field around the projectile must be considered to accurately predict M_2 (Knowlen and Bruckner 1992, Knowlen and Sasoh 1998).

Figure 6 shows a plot of the predicted non-dimensional thrust in the thermally choked mode, as a function of M_1 and Q, for a case in which $\gamma_1 = \gamma_2 = 1.35$. It can be seen that thrust increases with increasing values of the heat release. The model also predicts that the thrust goes through a maximum and decreases with increasing Mach number, reaching zero when the projectile velocity is equal to the CJ detonation speed, V_{cj}, of the propellant (Hertzberg et al. 1988, Bruckner et al. 1991b). In order to achieve velocities higher than the V_{cj} of a particular propellant, the ram accelerator tube can be subdivided into several sections, called stages, each separated from its neighbor by a thin diaphragm and filled with a different propellant (Fig. 7)

(Bruckner et al. 1991b). By selecting the sequence of propellants in such a manner that the speed of sound and detonation speed increase toward the exit of the ram accelerator, the projectile Mach number can be kept within limits that maximize thrust and efficiency, resulting in high average acceleration and a higher final velocity than is achievable with a single propellant stage. Figure 8 shows a velocity-distance plot for a four-stage configuration of the UW 38-mm ram accelerator. Note that the experimental data correlate very well with the velocity profiles predicted by the one-dimensional theoretical model.

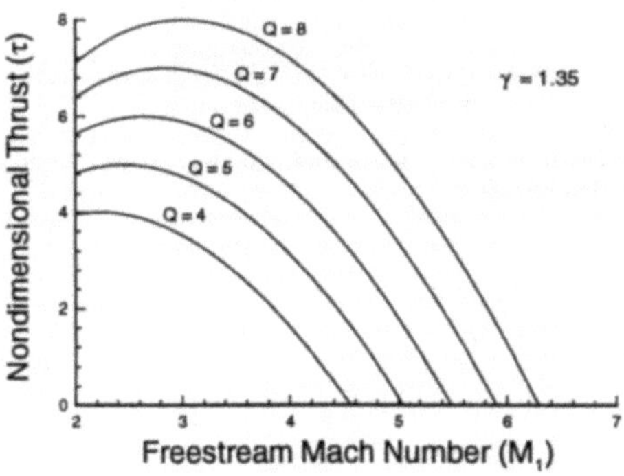

Fig. 6. Non-dimensional thrust as a function of projectile Mach number and non-dimensional heat release.

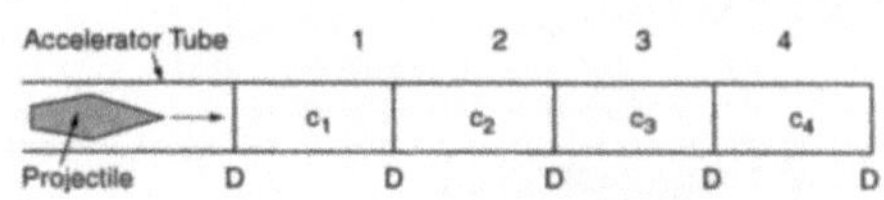

Fig. 7. Schematic of multi-stage ram accelerator. D denotes diaphragms separating propellant mixtures. Pressure is same in all stages. Note that $c_4 > c_3 > c_2 > c_1$, where c_n is the speed of sound in Stage n.

It has been observed that acceleration is also possible when the projectile is traveling above the CJ speed of the propellant - this is called the superdetonative velocity regime (Fig. 9a) (Kull et al. 1989). The transition from subdetonative to superdetonative operation occurs smoothly, through the transdetonative velocity regime (Fig. 9b) (Burnham et al. 1990). As the projectile approaches V_{cj} of the propellant in the thermally choked subdetonative mode, the combustion begins to move forward relative to the projectile, so that some of it takes place in the space between the projectile and the tube wall (Hertzberg et al. 1991, Knowlen and Sasoh 1998). As the projectile continues to accelerate to velocities above about $1.1V_{cj}$, i.e., into the superdetonative regime, the combustion appears to move almost entirely forward of the projectile's base. It is postulated that during this transition from subdetonative to superdetonative operation the combustion changes from purely subsonic to purely supersonic, and may even stabilize into an oblique detonation wave (Hertzberg et al. 1991). During operation in the transdetonative velocity regime, between approximately $0.9V_{cj}$ and $1.1V_{cj}$, it is believed that regions of both subsonic and supersonic combustion coexist (Burnham et al. 1990). The experimentally observed variation in the thrust as a function of the velocity ratio V/V_{cj}, as the propulsive mode makes the transition from subdetonative to superdetonative, is shown in Fig. 10 for several propellants. The thrust reaches a minimum in the transdetonative velocity regime and then increases in the superdetonative regime, reaching values exceeding the maximum thrust observed in the subdetonative mode.

To date, maximum superdetonative velocity ratios (V/V_{cj}) in the range of 1.5-1.57 have been observed (Kull et al. 1989, Hertzberg et al. 1991, Seiler et al. 1995a). It has been suggested that a maximum velocity in the range of approximately $1.5 - 2V_{cj}$, depending on projectile geometry, may exist due to energy balance considerations, i.e., the thrust equals drag limit (Rom 1997), but

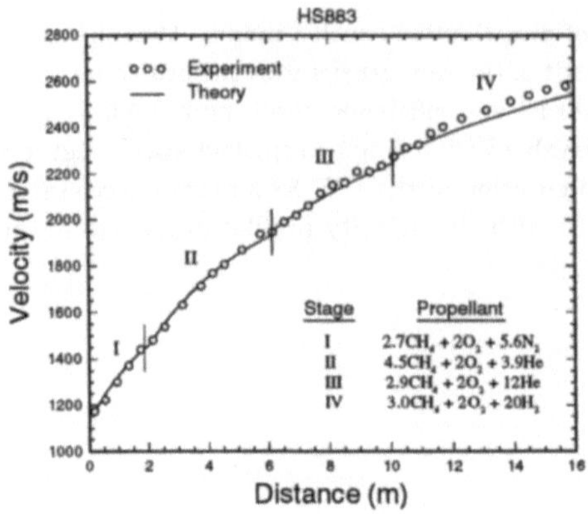

Fig. 8. Velocity-distance profile in four-stage ram accelerator (UW).

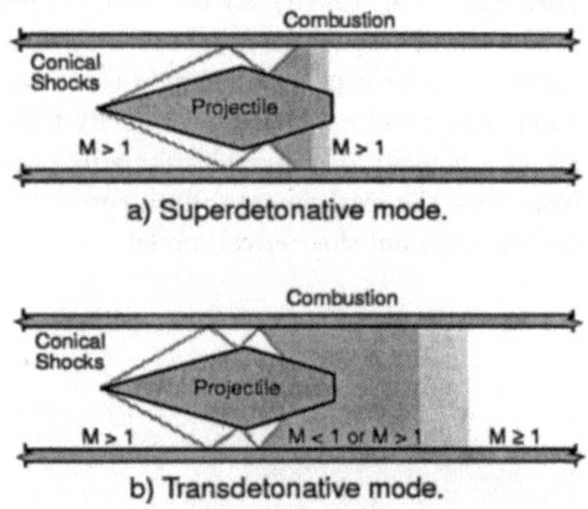

Fig. 9. Superdetonative and transdetonative propulsion modes.

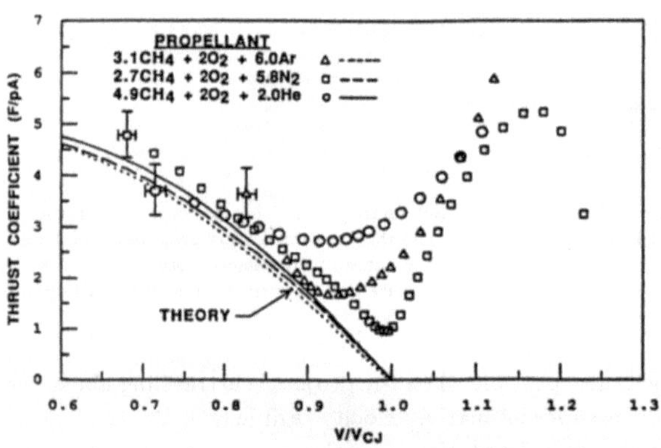

Fig. 10. Dependence of thrust on velocity ratio for various propellants (Hertzberg et al. 1991).

this limit has yet to be confirmed experimentally. In other work the maximum velocity of the ram accelerator has also been predicted to be about $2V_{cj}$. For example, in the superdetonative regime maximum velocities in the range of 7-9 km/s for operation in hydrogen-based propellants with CJ speeds of 3-4 km/s have been predicted (Yungster et al. 1991, Yungster and Bruckner 1992). Another velocity limiting mechanism that has been explored is the so-called "doomed propellant fraction," which refers to the possibility that at sufficiently high Mach numbers the bow shock standing off the finite radius of the projectile's nose tip may pre-ignite a sufficient fraction of the propellant to cause thermal choking of the flow at or ahead of the projectile throat (Ghorbanian and Pratt 1993). This limit has not yet been observed experimentally, either, and in any case was predicted to occur at velocities above $2V_{cj}$.

2. State of the art and recent work

A 38-mm-caliber ram accelerator (the first in the world) has been in operation at the UW since 1985 (Fig. 11) and has propelled 50-120 gm projectiles (Fig. 12) to velocities up to 2.7 km/s. Ram accelerator facilities have also been constructed at several other laboratories, both in the U.S.A. and in other countries. They include a 120-mm-caliber ram accelerator (currently the world's largest) at the U.S. Army Research Laboratory (ARL) at Aberdeen Proving Ground, Maryland (Fig. 13) (Kruczynski and Nusca 1992); ram accelerators of 90 mm and 30 mm caliber at the

French-German Research Institute (ISL) in Saint-Louis, France (Giraud et al. 1992, Seiler et al. 1993); a 25-mm-bore installation at Tohoku University in Sendai, Japan (Sasoh et al 1996a) and a 15×20 mm rectangular bore facility at Hiroshima University, also in Japan (Chang et al. 1995). In addition, a 37 mm bore ram accelerator was recently built and successfully tested at the Central Aerodynamics Research and Development Center (CARDC) in Mianyang, China (Sen et al. 1998). Projectile masses from 5 gm to 5 kg have been launched to velocities up to 2 km/s in these facilities. Research at all ram accelerator facilities has focused on improving the understanding of the physical principles of ram acceleration, achieving higher velocities, developing robust projectile designs, and studying various near- and long-term applications.

Fig. 11. Ram accelerator facility at University of Washington, Seattle. Bore diameter: 38 mm; Length of test section: 16 m.

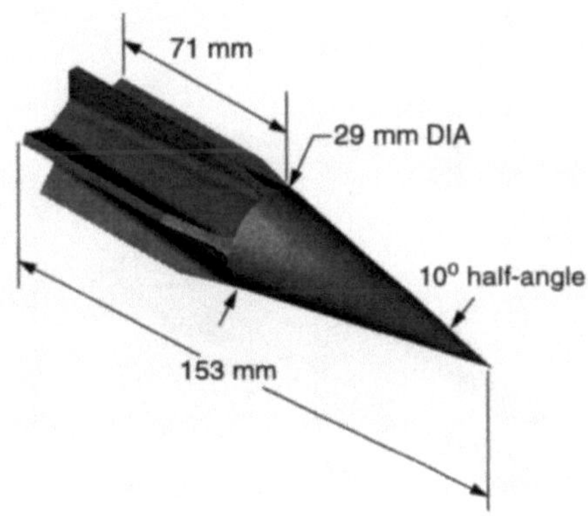

Fig. 12. Standard, finned ram accelerator projectile geometry used at UW.

2.1 Highlights of experimental results

During the nearly 12 years since the first experimental proof-of-concept was demonstrated at the UW (Hertzberg et al. 1986), much has been learned about the phenomena that govern the ram accelerator. Here only some of the salient results are summarized; the interested reader is directed to the references and to the other papers in this volume of Proceedings for further details.

University of Washington At the UW high spatial resolution pressure measurements of the flow around the projectile have revealed a complex three-dimensional flow structure associated with the centering fins (Fig. 14) (Hinkey et al. 1992). These observations have been corroborated by high-speed in-bore photography of projectiles through transparent polycarbonate tube sections (Knowlen et al. 1995a). Canting of the projectile in the tube has also been detected on a number of occasions (Hinkey et al. 1993), and is likely due to lateral forces and pitching moments generated by non-uniform pressure distributions around the nose and body of the projectile, coupled with erosion or bending of the projectile's centering fins. This problem can be mitigated through the use of titanium alloy as the projectile material, which is significantly stronger and more heat-resistant than the magnesium and aluminum alloys commonly used (Knowlen et al. 1995b).

Fig. 13. 120-mm ram accelerator facility at ARL.

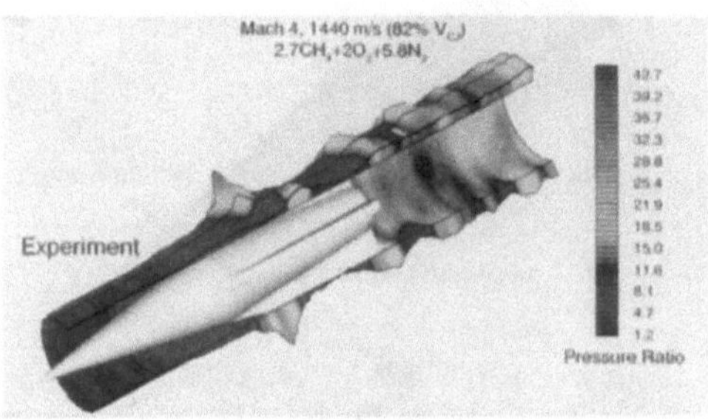

Fig. 14. 3-D pressure distribution around projectile in subdetonative mode (Hinkey et al. 1992).

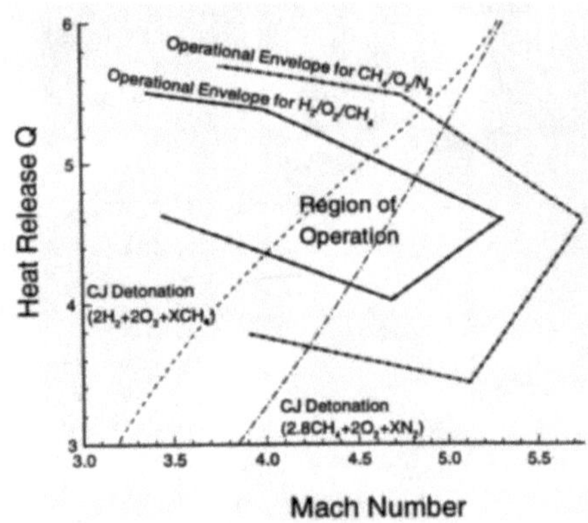

Fig. 15. Operational envelopes of ram accelerator for two classes of mixtures (Knowlen et al. 1995b).

Experiments performed with a variety of propellants have demonstrated the existence of operational limits which are governed by the heat release of the combustion, the Mach number of the projectile, and the projectile material (Fig. 15) (Higgins et al. 1993, Knowlen et al. 1994). If the heat release is too small, the driving pressure wave is unable to remain coupled to the projectile and falls behind, resulting in a cessation of thrust. If the heat release is too high, the driving pressure wave surges ahead of the projectile, causing a sudden deceleration (this is called an "unstart"). Hence, selection of the appropriate propellant composition is crucial to successful operation. The velocity limits, on the other hand, are believed to be related to projectile structural integrity and to the thrust equals drag limit. As the velocity increases, the pressure and aerodynamic heating increase markedly, and are capable of causing structural failure of the projectile. Computations of heat transfer to the nose cone, and to the leading edges and lateral surfaces of the centering fins, performed at the UW (Chew and Bruckner 1994, 1995) and ISL (Naumann 1993, 1996, Seiler and Naumann 1995), have shown that magnesium and aluminum alloys reach their melting points rapidly at these locations, resulting in potentially severe erosion by ablation, and loss of structural strength. Projectiles made of titanium alloy do not suffer these deleterious effects and have been found to attain higher velocities (Knowlen et al. 1995b).

Studies of the starting dynamics of the ram accelerator were first carried out several years ago at the UW (Bruckner et al. 1992b, Burnham 1993), but more recently this topic has received broader attention (Giraud et al. 1995, Sasoh et al. 1996a, Schultz 1997, Schultz et al. 1997, Henner et al. 1997, Stewart et al. 1997, 1998, Chang et al. 1998). Research on low velocity starting dynamics in propellants with low acoustic speeds carried out at the UW (Knowlen et al. 1997, Schultz et al. 1998) has shown that the starting process is very complicated and highly dependent on initial conditions, such as propellant composition, fill pressure, projectile and obturator mass, entrance diaphragm thickness, and projectile velocity.

Until very recently, UW experiments were conducted at ram accelerator fill pressures up to 5 MPa. By judicious choice of propellant chemistry and projectile geometry and mass, sustained accelerations of nearly 40,000 g over 2 m and 34,000 g over 6 m were achieved, attaining a velocity of 2.4 km/s in only 6 m of tube (Elvander et al. 1998). At the time of writing, experiments at 7.5 MPa fill pressure were in progress, with successful results. These are precursors to planned experiments at up to 20 MPa in an upgraded facility (Knowlen and Bruckner 1997).

U.S. Army Research Laboratory (ARL) At ARL the main thrust of the work since the installation of the 120-mm ram accelerator in 1991 has been to demonstrate its operation and then to increase its velocity capability (Kruczynski 1993, 1995, 1996a, 1998) and to develop computational fluid dynamics (CFD) codes for improved predictive capabilities (Nusca 1993, 1994, 1996a, 1996b, 1997). Projectiles of 5-kg-mass have been accelerated to velocities up to 2 km/s in a two-stage configuration. The experimental research to date has also included flow visualization of the subdetonative mode in sacrificial 2-m-long transparent acrylic tubes (Kruczynski, 1996b). The results were recorded using high-speed cinematography and showed that in the high subdetonative velocity regime the combustion zone enveloped the afterbody of the projectile well behind the throat (Fig. 16). Recently, the ARL facility was upgraded to permit operation at pressures as high as 20 MPa (Kruczynski 1998). The computational work has developed both steady and unsteady numerical codes that model various aspects of the subdetonative regime quite well.

Institut Saint-Louis (ISL) The 90 mm ram accelerator at ISL became operational in 1992, and has been used primarily in the subdetonative and transdetonative regimes, with the goal of attaining 3 km/s in a multi-stage configuration (Giraud et al. 1992, 1995, 1998). This facility uses a powder gun as the initial launcher, as does every other facility, except at the UW and Hiroshima University. Various projectile geometries, having masses in the vicinity of 1 kg, have been tested to velocities up to 2 km/s. Transdetonative performance has been observed, multi-stage experiments have been conducted, and projectile erosion has been monitored via flash X-ray photography (Giraud et al. 1995). The latter has shown that fin erosion can be severe in certain cases. The ram accelerator section was recently extended to a length of 21.2 m (Giraud et al. 1998), which makes it currently the longest in the world. The obturator used in this facility, as well as the transition section that joins the powder gun to the ram accelerator test section, are of a design different from those used elsewhere (Giraud et al. 1995). There is also a 30-mm-bore ram accelerator that has been used to study the effects of the projectile's fins on the initiation of combustion (Henner et al. 1997). This facility is similar to the smooth-bore ram accelerator used for superdetonative studies by another group at ISL (see below).

Experimental investigations of the superdetonative mode have been under way at ISL in the 30 mm railed ram accelerator, which also became operational in 1992 (Seiler et al. 1993, 1995, 1996). The test section is equipped with rails to center the axisymmetric projectiles. It was found that rail erosion due to combustion of the projectile material itself (magnesium or aluminum) was considerable, and the test section was recently replaced with one having five rails instead of four (Seiler et al. 1998). A smooth-bore 30 mm accelerator tube was also tested with finned projectiles having a straight mid-section, in order to be able to compare the results with those obtained in the railed tube (Patz et al. 1995, 1998). Flash X-ray imagery at the

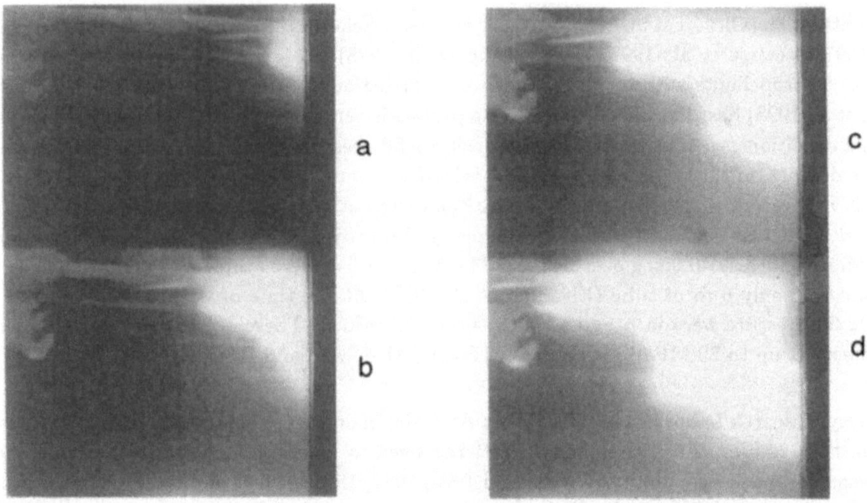

Fig. 16. Frames from high speed movie of projectile accelerating through transparent tube; $3CH_4 + 2O_2 + 10N_2$, 5.1 MPa, 1480 m/s (Kruczynski 1996b).

muzzles of both the railed and smooth-bore tubes has shown that projectile erosion is severe for magnesium and aluminum projectiles. Other materials tested have included refractory-coated aluminum, steel, and titanium in an effort to identify the best material for successful acceleration under the extreme pressures and temperatures characteristic of superdetonative operation. The highest velocity ratio, V/V_{cj}, reported to date by this group is approximately 1.57 in a propellant consisting of $2H_2 + O_2 + 6.8CO_2$, while the highest absolute velocity reported is nearly 2 km/s in $2H_2 + O_2 + 3.8CO_2$ (Seiler et al 1998).

Tohoku and Hiroshima Universities In Japan, as noted earlier, experimental ram accelerator research has been performed at Tohoku University in a 25-mm-bore facility (Sasoh et al. 1996a, 1997, 1998) and at Hiroshima University (Chang et al. 1995, 1996, 1998). At Tohoku University, Sasoh et al. (1996a) have performed flow visualization studies of the transition of the projectile from the initial launcher to the ram accelerator section. This work has confirmed the existence of a shock-heated, high-temperature slug of gas in front of the obturator when the projectile reaches the test section's entrance diaphragm. As noted earlier, it is believed that it is this hot gas that ignites the ram accelerator propellant. Both experimental and computational investigations of the diaphragm rupturing process itself have also been done (Maemura et al. 1998). In addition to firing standard projectiles, the Tohoku group has also collaborated with the UW in the study of hollow cylindrical projectiles (Sasoh et al. 1996b), which have internal flow similar to ramjets. Although successful operation with this projectile geometry was demonstrated, its performance was significantly below that of the standard geometry, most likely because of the very high sliding friction between the projectile and tube wall.

The facility at Hiroshima University is unusual in that the cross-section of the tube bore is rectangular (15×20 mm) and was designed to permit unobstructed optical diagnostics for investigations of the two-dimensional supersonic reacting flow phenomena attendant to the ram accelerator. Schlieren photographs of the flow around projectiles have been obtained in both inert and combustible propellant mixtures (Chang et al. 1995). In a collaborative effort with the UW the Hiroshima group has also successfully tested quasi-two-dimensional projectiles in the UW's 38-mm ram accelerator (Chang et al. 1997).

2.3 Computational modeling

In addition to the experimental work outlined here, considerable activity has also occurred in computational modeling, beginning shortly after the ram accelerator was first conceived (Brackett and Bogdanoff 1989, Yungster 1991, Yungster et al. 1991, Yungster and Bruckner 1992, Kruczynski and Nusca 1992, Nusca 1993, 1994, 1996, 1997, Soetrisno and Imlay 1991, Soetrisno et al. 1992, 1993, Li et al. 1993, 1995, Burnham 1993, Hinkey et al. 1992, 1993, Choi et al. 1997, Leblanc 1997, Taki et al. 1998). It is beyond the scope of this paper to review in any detail the numerous contributions in this area of ram accelerator research; the interested reader is directed to the references and the several papers on numerical modeling published in this volume of Proceedings for recent developments and current trends. Here only a very brief general overview is provided.

Computational fluid dynamics codes have been developed to support experimental work in order to provide detailed information on the flowfield surrounding the projectile. Most of these codes have had success in matching a number of the major flow features observed, however, they have not yet become able to predict performance and flow features with sufficient accuracy to direct the experiments. Computational schemes and strategies have been used which vary in complexity; ranging from simple control volume methods using equilibrium chemistry for determining thrust performance to 3-D unsteady calculations with finite-rate chemistry to investigate flow phenomena. Time and machine limitations have typically led to the use of approximate schemes which depend on the specific problem being studied. For example, the use of axisymmetric rather than 3-D modeling is common but neglects the fin effects which may play a major role in igniting the propellant. Other common simplifications include the assumption of inviscid flow and the use of reduced chemistry algorithms. New chemistry algorithms (Petersen et al. 1997) have been a welcome addition, since until recently, reduced chemistry sets had not been validated for the high pressures that exist in the ram accelerator. Finally, there has been a trend toward time accurate CFD codes because of the high acceleration levels experienced by the ram accelerator and the intrinsically unsteady nature of various phenomena, such as the starting process, unstart phenomena, and mode transition from subdetonative to superdetonative (Li et al. 1995, 1996, Nusca 1997, Maemura et al. 1998, Taki et al. 1998).

2.4 Related studies

Related experimental and theoretical investigations on detonation initiation phenomena in various propellants at pressures up to several tens of atmospheres have been carried out recently at the UW (Schultz 1997, Schultz et al. 1997), and at the Ecole Nationale Supérieure de Mécanique et d'Aerotechnique (ENSMA) and at ISL in France (Legendre 1996, Bauer et al. 1998). At the UW the conditions under which a supersonic blunt body will initiate a detonation wave in a combustible gas at pressures of 0.5-1.0 MPa have also been investigated experimentally (Higgins and Bruckner 1996, Higgins 1996, 1997). This work is relevant to the ram accelerator because of the shock-induced combustion processes that may occur in some of the propulsive modes and which may define the velocity limit in some cases (Ghorbanian and Pratt 1993). Computational modeling of blunt bodies in detonable mixtures has been performed by others (Wilson and MacCormack 1992, Matsuo et al. 1995, Ju et al. 1998).

Shock and expansion tube experiments using methane-based propellants and stationary models or components of projectiles have been carried out at ISL (Srulijes et al. 1992, 1995). Diagnostics have included high-speed multi-frame Schlieren photography and planar laser-induced fluorescence (PLIF). More recently a new technique that allows the use of hydrogen without the problem of pre-ignition was successfully tested (Srulijes et al. 1998). Researchers at Stanford University have used a high pressure shock tube facility to experimentally investigate the ignition kinetics of high pressure ram accelerator propellants in an effort to obtain chemical reaction data at conditions that have not been heretofore explored (Petersen et al. 1996). Also at Stanford, expansion tube experiments have been performed to study the details of the reacting flow around stationary ram accelerator projectiles using diagnostic techniques such as PLIF and Schlieren photography (Morris et al. 1995, 1998).

3. Applications

Because the ram accelerator's operating principle is essentially independent of dimension, the device can be scaled up or down over a broad range of sizes, as has already been experimentally demonstrated. Consequently, the ram accelerator is readily amenable to a variety of important applications, such as hypersonic aerodynamic testing (Bruckner et al. 1992a) and low-cost direct launch of payloads into low Earth orbit (Bruckner and Hertzberg 1987). Other applications of the ram accelerator, such as hypervelocity impact studies, testing of re-entry thermal protection materials, and even anti-asteroid defense (Kryukov 1995) have also been suggested. Here only the hypersonics and space launch applications are discussed.

3.1 Hypersonic test facility applications The easily controllable acceleration level of the ram accelerator gives it the potential to "soft" launch, i.e., gently accelerate large (0.3-0.6 m) instrumented models of hypersonic aircraft or atmospheric re-entry vehicles to high Mach numbers for aerodynamic testing (Bruckner et al. 1992a, Naumann and Bruckner 1994). Since the propulsion principles of the ram accelerator are similar to those of ramjets and scramjets, the device is also an excellent testbed for research on the reacting flow phenomena characteristic of these propulsive devices (Bruckner et al. 1991a).

Preliminary calculations have been made for large scale ram accelerators based on experimental results (Bruckner et al. 1992a). These hypervelocity launchers are constrained only by the maximum allowable internal pressure of the accelerator tube and the acceleration limits imposed by the test models. The parameters of ram accelerator launchers for 0.3 m and 0.6 m bore facilities, designed to accelerate projectiles having an average density of 1.5×10^3 kg/m^3 to a velocity of 6 km/s, are shown in Table 1. These facilities would be capable of launching highly instrumented projectiles having a total mass of 49 kg and 390 kg, respectively, with a propellant fill pressure of 14 MPa. The projectiles, with an external geometry designed for optimal ram acceleration, would act as sabots for the enclosed test models which would be released prior to entering the test section. The estimates for the barrel masses are based on steel walls having the thickness required to keep their internal stress below 450 MPa for a peak static pressure load of 225 MPa. This results in a ratio of outer-to-inner diameter of 1.41.

A schematic of an aeroballistic facility using a ram accelerator launcher is shown in Fig. 17. The pre-launcher, ram accelerator section, muzzle vent chamber and sabot stripping sections are shown, along with a free-flight test section and projectile decelerator. The ram accelerator is assumed to be partitioned into four different sections: the initial launcher, thermally choked stages, transdetonative stages, and superdetonative stages. For these scaling examples a conventional gas gun is assumed to bring the projectiles up to an initial operating speed of 0.7 km/s. Each of the subsequent phases of ram acceleration involves staging propellants whose composition is selected to maintain the desired in-tube Mach number and thrust level as the projectile accelerates. If all of the stages are operated at constant pressure then the partitions between stages could be opened just before launching to allow some mixing at the stage transitions to minimize sudden in-tube Mach number changes and the corresponding acceleration jumps, while also eliminating unnecessary diaphragm impacts. The entrance and exit seals can be burst just before projectile impact to reduce potential projectile and test model damage (Naumann and Bruckner 1994).

An inherent benefit of the ram accelerator launcher is that its muzzle blast is very small compared to that issuing from an equivalent gas gun launch. This is a consequence of the fact that in a ram accelerator the bulk of the burnt propellant moves in the rearward direction and is

Table 1. Physical parameters for 6 km/s ram accelerator launchers

Bore diameter	0.3 m	0.6 m
Projectile mass	49 kg	390 kg
Average acceleration	6780 g	3400 g
Barrel length	270 m	540 m
Barrel Diameter	0.42 m	0.85 m
Barrel mass	150 tons	1200 tons
Fill pressure	14 MPa	14 MPa
Peak pressure	225 MPa	225 MPa

vented at the coupling between the initial launcher and the ram accelerator. Thus, a short muzzle vent chamber filled with an inert gas would be sufficient to inhibit the remaining muzzle gases from preceding the projectile into the test section. An additional advantage of the ram accelerator as a ballistic launcher is associated with adjustments in operating pressure. Experiments at the UW have demonstrated that transitions through relatively large pressure differentials can be sustained, allowing projectiles to be ram accelerated up to test speed and then injected into a low pressure test section (Knowlen et al. 1995a). The low pressure propellant mixtures result in low accelerations, which lengthen the test time and improve data resolution from a test section having a fixed instrument density. Data obtained in such a facility would be directly applicable to current ramjet and scramjet research.

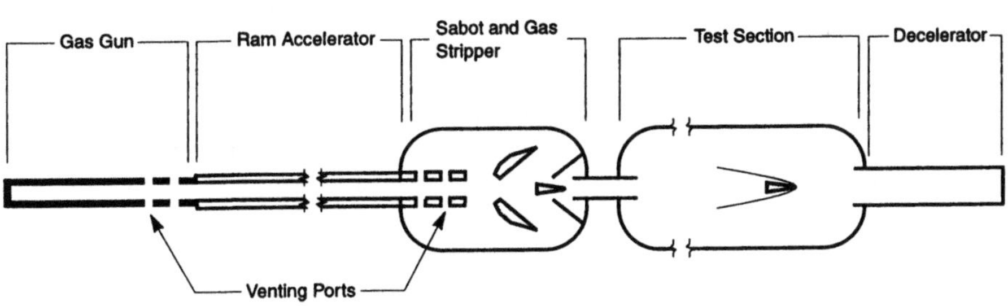

Fig. 17. Schematic of ram accelerator aeroballistic facility.

The ballistic coefficient of a large test model is sufficiently high that the flight through low density test gases (simulating upper atmospheres of various planets) would take place at essentially constant speed. The test duration would be governed by the length of the test section. For example, to obtain a 50 msec test duration at 6 km/s would require a test section 300 m long. Data collected from onboard sensors could be broadcast during transit of the test section or else saved and transmitted before the deceleration phase. Alternatively, the hypersonic test model could eject a data recorder which would be stopped in a less severe manner than the model and interrogated after recovery (Witcofski et al. 1991).

3.2 Space launch

The potential use of the ram accelerator as a space launcher is perhaps one of the most intriguing. The high cost of existing rocket launch systems is a significant barrier to developing any large scale permanent space infrastructure designed for space manufacturing or for the exploration and colonization of our solar system. Components of large space structures, the various raw materials required for space manufacturing, and hydrogen, oxygen, water, and other consumables are capable of withstanding high accelerations. This acceleration insensitivity has attracted the attention of various investigators over the years, and a number of mass launch systems to carry out the function of direct launch have been proposed. These have ranged from hypervelocity guns (Bruckner 1965, Murphy and Bull 1967) to various electromagnetic accelerators (O'Neill 1974, Chilton et al. 1977, Hawke et al. 1982) and beamed energy concepts (Glumb and Krier 1984, Powers et al. 1986). The relatively low efficiencies of the gun concepts, the formidable problem of the large instantaneous electric loads in the electromagnetic systems, the problems of atmospheric beam propagation in the beamed energy systems, and the general difficulty of scaling these concepts to useful payloads have been serious impediments to their implementation. The ram accelerator, on the other hand, is scalable for vehicle masses ranging from grams to tons, and thus offers the potential for impulsive launch of acceleration-insensitive payloads to low Earth orbit (Bruckner and Hertzberg 1987, Kaloupis and Bruckner 1988, Bogdanoff 1992).

A number of possible operating configurations for the ram accelerator mass launch system have been examined at the UW (Bruckner and Hertzberg 1987, Kaloupis and Bruckner 1988). A system capable of delivering a 2000 kg vehicle to a 500 km circular orbit was selected as a representative baseline case. Launch velocities in the range of 8 to 10 km/s and various orbital mechanics scenarios were considered. The peak permissible acceleration was limited to the range of 1000-1500 g as a compromise between limiting vehicle structural mass and limiting the length of the launch tube. A launch tube inside diameter of 1.0 m and a vehicle diameter of 0.76 m were selected. As in the case of the hypersonic test facility discussed above, acceleration to the desired launch velocity was assumed to be accomplished using the three observed ram accelerator propulsion modes in sequence (subdetonative, transdetonative, and superdetonative). A structural configuration that integrates the payload vessel with the vehicle design was chosen and a graphite/epoxy composite was selected in order to withstand the high pressures and accelerations attendant to the launch process (Fig. 18). A carbon-carbon ablating nosecone was incorporated to protect the vehicle from the aerodynamic heating which occurs in the accelerator tube and during atmospheric transit. The ablative mass loss in the atmosphere was computed to be very small, of the order of 1-2% of the vehicle mass. More recent studies have shown that the in-tube ablation losses would be greater (Bogdanoff 1992).

For a muzzle velocity of 8 km/s, a ram accelerator with the above requirements would have a length of approximately 3.3 km. The optimum launch angle would be about 15° which could be met by installing the launch facility up the side of a suitable mountain (Fig. 19). The projectile would lose less than 20% of its initial velocity to aerodynamic drag. An on-board rocket ignited at the peak of the projectile's trajectory would provide the necessary additional impulse to circularize the projectile's orbit at 500 km. Because the payload mass fraction of the projectile would be nearly 50% (compared to 3-4% for conventional rocket launch vehicles) and because multiple launches per day would be possible, a ram accelerator would be capable of very low launch costs, approximately $200-600 per kilogram of payload to orbit, nearly two orders of magnitude less than current costs (Hertzberg and Bruckner 1988). Additional studies of the ram accelerator as a space cargo launcher, with specific attention to projectile survival in the ram accelerator environment have also been done (Bogdanoff 1992), as well as studies of the stability and control of the vehicle through the Earth's atmosphere (Kaloupis 1989), and other issues (Messersmith et al. 1995).

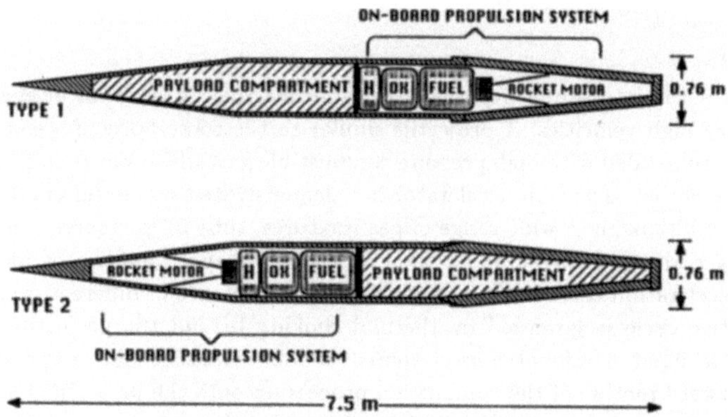

Fig. 18. Possible ram accelerator projectile geometries for space launch.

Fig. 19. Artist's conception of ram accelerator space launch facility.

4. Future work

Current areas of interest that will continue to be investigated in the future include operation of the ram accelerator at elevated pressures, up to 20 MPa (Knowlen and Bruckner 1997, Kruczynski 1998), investigation of projectile geometry effects (Imrich 1995), modeling of the ram accelerator propulsive modes with the inclusion of real gas effects (Buckwalter et al. 1998, Bauer et al. in press); further development of expansion tube techniques for the simulation of ram accelerator flowfields (Srulijes et al. 1998), studies of the starting dynamics, especially at low initial projectile velocities (Knowlen et al. 1997, Schultz et al. 1998, Stewart et al. 1997, 1998, Maemura et al. 1998), investigations of the superdetonative mode (Knowlen et al. 1996, Seiler et al. 1998), and improved CFD modeling of all the propulsive modes of the ram accelerator (Choi et al. 1997, Nusca 1997, Leblanc 1997, Taki et al. 1998). The ultimate aim of all ram accelerator research is to attain the theoretical velocity capability of this novel launcher technology, namely 6-8 km/s, at which the most interesting applications, such as hypersonic testing and direct space launch become practicable.

5. Conclusions

The ram accelerator is a ramjet-in-tube hypervelocity launcher that uses chemical energy to propel projectiles to very high velocities. A projectile similar to the centerbody of a supersonic ramjet travels through a tube filled with high pressure combustible gas, which burns on or behind the projectile to provide thrust. The ram accelerator has demonstrated successful operation at a variety of operating conditions, in a wide range of gas mixtures, tube fill pressures, and Mach numbers. Three velocity regimes, centered about the CJ detonation speed, have been identified that exhibit different acceleration characteristics, indicating the existence of different propulsive cycles. The subdetonative cycle is governed by thermal choking behind the projectile and is predicted very well by a one-dimensional control volume model. Transdetonative operation is characterized by the forward motion of the combustion process up onto the projectile body and the existence of regions of mixed supersonic and subsonic combustion. In the superdetonative mode the combustion is supersonic and occurs almost exclusively in the annular region between the projectile body and the tube wall. Ram accelerator facilities are currently operational in the U.S.A., France, Japan, and China. Velocities up to 2.7 km/s have been attained with 38-mm, 75-gm projectiles at the University of Washington, 2 km/s with 90-mm, 1-kg projectiles at ISL, and 2 km/s with 120-mm, 5-kg projectiles at the U.S. Army Research Laboratory. The ease with which the ram accelerator can be scaled up in size offers unique opportunities for its use as a hypersonic research tool, a potentially low-cost space launcher, and other applications. Progress in these areas is predicated on the further development of the velocity capability of this innovative launcher concept.

References

Bauer P, Legendre JF, Henner M, Giraud M (1998) Real gas effects in ram accelerator propellant mixtures: theoretical concepts and applied thermochemical codes. In: Takayama K, Sasoh A (eds) Ram Accelerators. Springer-Verlag, Heidelberg, pp 39-52

Bauer P, Knowlen C, Bruckner AP (1998) Real gas effects on the prediction of ram accelerator performance. Shock Waves, in press

Brackett DC, Bogdanoff DW (1989) Computational investigation of oblique detonation ramjet-in-tube concepts. J Prop Power 5:276-281

Bogdanoff DW (1992) Ram accelerator direct space launch system: new concepts. J Prop Power 8:481-490

Bruckner AP (1965) A gun-launched satellite. Spaceflight 7:118-121

Bruckner AP, Hertzberg A (1987) Ram accelerator direct launch system for space cargo. IAF paper 87-211

Bruckner AP, Chew G, Auzias de Turenne J, Dunmire B (1991a) Investigation of hypersonic ramjet propulsion cycles using a ram accelerator test facility. IAF paper 91-275

Bruckner AP, Knowlen C, Hertzberg A, Bogdanoff DW (1991b) Operational characteristics of the thermally choked ram accelerator. J Prop Power 7:828-836

Bruckner AP, Knowlen C, Hertzberg A (1992a) Applications of the ram accelerator to hypersonic aerothermodynamic testing. AIAA paper 92-3949

Bruckner AP, Burnham EA, Knowlen C, Hertzberg A, Bogdanoff DW (1992b) Initiation of combustion in the thermally choked ram accelerator. In: Takayama K (ed) Shock Waves, Springer-Verlag, Berlin, pp 623-630

Buckwalter D, Knowlen C, Bruckner AP (1998) Real gas effects on thermally choked ram accelerator performance. In: Takayama K, Sasoh A (eds) Ram Accelerators. Springer-Verlag, Heidelberg, pp 125-134

Burnham EA, Kull AE, Knowlen C, Bruckner AP, Hertzberg A (1990) Operation of the ram accelerator in the transdetonative velocity regime. AIAA paper 90-1985

Burnham EA (1993) Investigation of starting and ignition transients in the thermally choked ram accelerator. Ph.D. Thesis, Dept Aero Astro, Univ Washington, Seattle, WA

Chang X et al (1995) A ram accelerator with rectangular bore is working at Hiroshima University. AIAA paper 95-2496

Chang X, Higgins AJ, Schultz E, Bruckner AP (1997) Operation of quasi-two-dimensional projectiles in a ram accelerator. J. Prop Power 13:802-804

Chang X, Matsuoka S, Watanabe T, Taki S (1998) Ignition study for low pressure combustible mixture in a ram accelerator. In: Takayama K, Sasoh A (eds) Ram Accelerators. Springer-Verlag, Heidelberg, pp 105-109

Chew G, Bruckner AP (1994) A computational study of nose heating in the ram accelerator. AIAA paper 94-2964

Chew G, Bruckner AP (1995) Computational study of projectile nose heating in a ram accelerator. Proc RAMAC II, Second Int Workshop on Ram Accelerators, Univ Washington, Seattle, WA, July 17-20

Chilton F, Hibbs B, Kolm H, O'Neill GK, Phillips J (1977) Electromagnetic mass drivers. Progress in Aeronautics and Astronautics 57:37-61 AIAA, Reston, VA, USA

Choi J, Jeung I, Yoon Y (1997) Numerical Study of Scram Accelerator Starting Characteristics. AIAA paper 97-0915

Elvander JE, Knowlen C, Bruckner AP (1998) High acceleration experiments using a multi-stage ram accelerator. In: Takayama K, Sasoh A (eds) Ram Accelerators. Springer-Verlag, Heidelberg, pp 55-64

Ghorbanian K, Pratt DT (1993) Doomed propellant fraction for a superdetonative ram accelerator. AIAA paper 93-0359

Giraud M, Legendre JF, Simon G, Catoire L. (1992) Ram accelerator in 90 mm caliber: first results concerning the scale effect in thermally choked propulsion mode. 13th Int Symp on Ballistics, Stockholm, Sweden, June 1-3

Giraud M, Legendre JF, Simon G, Henner M, Voisin D (1995) RAMAC in 90 mm caliber or RAMAC90. Starting process, control of the ignition location and performance in the thermally choked propulsion mode. Proc RAMAC II, Second Int Workshop on Ram Accelerators, Univ Washington, Seattle, WA, USA, July 17-20

Giraud M, Legendre, JF, Henner M (1998) RAMAC in subdetonative propulsion mode - State of the ISL studies. In: Takayama K, Sasoh A (eds) Ram Accelerators. Springer-Verlag, Heidelberg, pp 65-77

Glumb RJ, Krier H (1984) Concepts and status of laser supported rocket propulsion. J Spacecraft Rockets 21:70-79

Hawke RS, Brooks AL, Fowler CA, Peterson DR (1982) Electromagnetic railgun launchers: direct launch feasibility. AIAA J 20:978-985

Henner M, Legendre J-F, Giraud M (1997) Initiation of reactive mixtures in a ram accelerator. AIAA paper 97-3173

Henner M, Giraud M, Legendre JF, Berner C (1998) CFD computations of steady and non-reactive flow around fin-guided RAM projectiles. In: Takayama K, Sasoh A (eds) Ram Accelerators. Springer-Verlag, Heidelberg, pp 325-332

Hertzberg A, Bruckner AP, Bogdanoff DW (1986) The ram accelerator: A new chemical method of achieving ultrahigh velocities. Proc 37th Meeting of Aeroballistic Range Assoc, Quebec, Canada, September 9-12

Hertzberg A, Bruckner AP (1988) Study of a ram cannon for a space cargo launcher. Final Report, NASA Grant No. NAG 1-746, Univ Washington, Seattle, WA

Hertzberg A, Bruckner AP, Bogdanoff, DW (1988) Ram accelerator: A new chemical method for accelerating projectiles to ultrahigh velocities. AIAA J 26:195-203

Hertzberg A, Bruckner AP, Knowlen C (1991) Experimental investigation of ram accelerator propulsion modes. Shock Waves 1:17-25

Higgins AJ, Knowlen C, Bruckner AP (1993) An investigation of ram accelerator gas dynamic limits. AIAA paper 93-2181

Higgins AJ, Bruckner AP (1996) Experimental investigation of detonation initiation by hypervelocity blunt projectiles. AIAA paper 96-0342

Higgins AJ (1996) Investigation of detonation initiation by supersonic blunt bodies. PhD Thesis, Dept Aero Astro, Univ Washington, Seattle, WA, USA

Higgins AJ (1997) The effect of confinement on detonation initiation by blunt projectiles. AIAA paper 97-3179

Hinkey JB, Burnham EA, Bruckner AP (1992) High spatial resolution measurements of ram accelerator gas dynamic phenomena. AIAA paper 92-3244

Hinkey JB, Burnham EA, Bruckner AP (1993) Investigation of ram accelerator flow fields induced by canted projectiles. AIAA paper 93-2186

Imrich TS (1995) The impact of projectile geometry on ram accelerator performance, MSAA Thesis, Dept Aero Astro, Univ Washington, Seattle, WA, USA

Ju Y, Sasoh A, Masuya G (1998) On the detonation initiation by a supersonic sphere. In: Takayama K, Sasoh A (eds) Ram Accelerators. Springer-Verlag, Heidelberg, pp 255-262

Kaloupis P, Bruckner AP (1988) The ram accelerator: A chemically driven mass launcher. AIAA paper 88-2968

Kaloupis P (1989) Atmospheric transit of a hypersonic, axisymmetric spin-stabilized vehicle. MSAA Thesis, Dept Aero Astro, Univ Washington, Seattle, WA

Knowlen C, Bruckner AP (1992) A Hugoniot analysis of the ram accelerator. In: Takayama K (ed) Shock Waves, Springer-Verlag, Berlin, pp 617-622

Knowlen C, Higgins AJ, Bruckner AP (1994) Investigation of operational limits to the ram accelerator. AIAA paper 94-2967

Knowlen C, Higgins AJ, Bruckner AP, Hertzberg A (1995a) In-tube photography of ram accelerator projectiles. In: Brun R, Dumitrescu LZ (eds) Shock Waves @Marseille, Springer-Verlag, Berlin, Vol 1, pp 189-194

Knowlen C, Higgins AJ, Bruckner AP, (1995b) Aerothermodynamics of the ram accelerator. AIAA paper 95-0289

Knowlen C, Higgins AJ, Bruckner AP, Bauer P (1996) Ram accelerator operation in the superdetonative velocity regime. AIAA paper 96-0098

Knowlen C, Schultz E, Bruckner AP (1997) Investigation of low velocity starting techniques for the ram accelerator. AIAA paper 97-3174

Knowlen C, Bruckner AP (1997) Facility upgrade for high pressure ram accelerator experiments. Paper No 9, Proc 3rd Int Workshop on Ram Accelerators, Tohoku University, Sendai, Japan, July 16-18

Knowlen C, Sasoh A (1998) Ram accelerator performance modeling. In: Takayama K, Sasoh A (eds) Ram Accelerators. Springer-Verlag, Heidelberg, pp 25-37

Kruczynski DL, Nusca MJ (1992) Experimental and computational investigation of scaling phenomena in a large caliber ram accelerator. AIAA paper 92-3245

Kruczynski D (1993) New experiments in a 120-mm ram accelerator at high pressures. AIAA paper 93-2589

Kruczynski D and Liberatore F (1995) Ram accelerator experiments with unique projectile geometries. AIAA paper 95-2490

Kruczynski D (1996a) Experimental investigation of high pressure/performance ram accelerator operation. AIAA paper 96-2676

Kruczynski D (1996b) Flow visualization of steady and transient combustion in a 120-mm ram accelerator. Report ARL-TR-1059, US Army Res Lab, Aberdeen, MD, USA

Kruczynski DL (1998) High performance ram accelerator research. In: Takayama K, Sasoh A (eds) Ram Accelerators, Springer-Verlag, Heidelberg, pp 97-104

Kryukov PV (1995) BALSAD - ballistic system for antiasteroid defense. Proc RAMAC II, Second Int Workshop on Ram Accelerators, Univ Washington, Seattle, WA, USA, July 17-20

Kull AE, Burnham EA, Knowlen C, Bruckner AP, Hertzberg A (1989) Experimental studies of superdetonative ram accelerator mode. AIAA paper 89-2632

Leblanc JE, Fujiwara T (1997) Numerical simulation of the RAMAC benchmark test configuration. In: Takayama K (eds) Proc 3d Int Workshop on Ram Accelerators, Sendai, July 16-18

Legendre JF (1996) Contribution a l'étude de la sensibilité et des caractéristiques de détonation de mélanges explosifs gazeux denses à base de méthane utilisés pour la propulsion dans les accélérateurs à effet stato. PhD Thesis, Univ Poitiers, Poitiers, France

Li C, Kailasanath K, Oran ES, Boris JP (1993) Numerical simulations of unsteady reactive flows in ram accelerators. Proc First Int Workshop on Ram Accelerators, Paper No 29, French-German Research Institute (ISL), Saint Louis, France, September 7-10

Li C, Kailasanath K, Oran ES, Boris JP (1995) Pressure distributions on ram accelerator projectiles. Proc RAMAC II, Second Int Workshop on Ram Accelerators, University of Washington, Seattle, WA, USA, July 17-20

Li C, Kailasanath K, Oran ES (1996) Stability of projectiles in thermally choked ram accelerators. J Prop Power 12:807-809

Maemura J, Hirakata S, Sasoh A, Takayama K (1998) Diaphragm rupturing processes by a ram accelerator projectile. In: Takayama K, Sasoh A (eds) Ram Accelerators. Springer-Verlag, Heidelberg, pp 205-213

Messersmith NL, Burke MJ (1995) Ram accelerator ballistic launch requirements. AIAA paper 95-2495

Matsuo A, Fujii K, Fujiwara T (1995) Flow features of shock-induced combustion around projectiles traveling at hypervelocities. AIAA J 33:1056-1863

Morris CI, Kamel MR, Thurber MC, Wehe SD, Hanson RK (1995) Development of an expansion tube for investigation of combustion in supersonic projectile flowfields. AIAA paper 95-2717

Morris CI, Kamel MR, Ben-Yakar A, Hanson RK (1998) Combined Schlieren and OH PLIF imaging study of ram accelerator flowfields. AIAA paper 98-0244

Murphy CH, Bull GV (1967) Aerospace application of gun launched projectiles and rockets. American Astronautical Society Paper No 67-38

Naumann KW (1993) Heating and ablation of projectiles during acceleration in a ram accelerator tube. AIAA paper 93-2184

Naumann KW, Bruckner AP (1994) Ram accelerator ballistic range concept for softly accelerating hypersonic free-flying models. J Aircraft 31:1310-1316

Naumann KW (1996) Thermomechanical constraints on ram accelerator projectile design. AIAA paper 96-2678

Nusca MJ (1993) Numerical simulation of fluid dynamics with finite rate and equilibrium combustion kinetics for the 120-mm ram accelerator. AIAA paper 93-2182

Nusca MJ (1995) Reacting flow simulation of transient, multi-stage ram accelerator operation and design studies. Proc RAMAC II, Second Int Workshop on Ram Accelerators, Seattle, WA USA, July 17-20

Nusca MJ (1996a) Reacting flow simulation for a large scale ram accelerator. J Prop Power 12:61-69

Nusca MJ (1996b) Investigation of ram accelerator flows for high pressure mixtures of various chemical compositions. AIAA paper 96-2946

Nusca MJ (1997) Computational simulation of the ram accelerator using a coupled CFD/interior ballistic approach. AIAA paper 97-2653

O'Neill GK (1974) The colonization of space. Physics Today 27:32-40

Patz G, Seiler F, Smeets G, Srulijes J (1995) Status of ISL's RAMAC 30 with fin guided projectiles accelerated in a smooth bore. Proc RAMAC II, Second Int Workshop on Ram Accelerators, Univ Washington, Seattle, WA, July 17-20

Patz G, Seiler F, Smeets G, Srulijes J (1998) The behaviour of fin-guided projectiles superdetonatively accelerated in ISL's RAMAC 30. In: Takayama K, Sasoh A (eds) Ram Accelerators. Springer-Verlag, Heidelberg, pp 89-95

Petersen EL, Davidson DF, Hanson RK (1996) Ignition delay times of ram accelerator CH_4/O_2 mixtures. AIAA paper 96-2681

Petersen EL, Davidson DF, Hanson RK (1997) Reduced Kinetics mechanisms for ram accelerator combustion. AIAA paper 97-2892

Powers MV, Zaretsky C, Myrabo LN (1986) Analysis of beamed energy ramjet/scramjet performance. AIAA paper 86-1761

Rom J (1997) Performance limits for projectile flight in the ram and external propulsion accelerators. J Prop Power 13:583-591

Sasoh A, Hirakata S, Ujigawa Y, Takayama, K (1996a) Operation tests of a 25 mm bore ram accelerator. AIAA paper 96-2677

Sasoh A, Higgins AJ, Knowlen C, Bruckner AP (1996b) Hollow projectile operation in the ram accelerator. J Prop Power 12:1183-1186

Sasoh A, Hirakata S, Maemura J, Takayama K (1997) Experimental studies of 25-mm-bore ram accelerator at the Shock Wave Research Center. AIAA paper 97-2652

Sasoh A, Hirakata S, Maemura J, Hamate Y, Takayama K (1998) Thermally choked operation in the 25-mm-bore ram accelerator. In: Takayama K, Sasoh A (eds) Ram Accelerators, Springer-Verlag, Heidelberg, pp 111-118

Schultz E (1997) The Subdetonative Ram Accelerator Starting Process. MSAA Thesis, Dept Aero Astro, Univ Washington, Seattle, WA

Schultz E, Knowlen C, Bruckner AP (1997) Detonation limits applied to the subdetonative ram accelerator starting process. AIAA paper 97-0807

Schultz E, Knowlen C, Bruckner AP (1998) Overview of the subdetonative ram accelerator starting process. In: Takayama K, Sasoh A (eds) Ram Accelerators. Springer-Verlag, Heidelberg, pp 189-203

Seiler F, Patz G, Smeets G, Srulijes J (1993) Status of ISL's RAMAC 30 with rail stabilized projectiles. Proc First Int Workshop on Ram Accelerators, Saint Louis, France, Sept 7-10

Seiler F, Patz G, Smeets G, Srulijes J (1995) The rail tube in ram acceleration: feasibility study with ISL's RAMAC 30. Proc 2nd Int Workshop on Ram Accelerators, Univ Washington, Seattle, WA, USA, July 17-20

Seiler F, Naumann KW (1995) Bow shock wave heating and ablation of a sharp-nosed projectile flying inside a ram accelerator. In: Brun R, Dumitrescu LZ (eds) Shock Waves @Marseille, Springer-Verlag, Berlin, Vol 1, pp 183-188

Seiler F, Patz G, Smeets G, Srulijes J (1996) Gasdynamic limits of ignition and combustion of a gas mixture in ISL's RAMAC30 scram accelerator. In: Sturtevant B, Shepherd JE, Hornung HG (eds) Proc 20th Int Symp Shock Waves, World Scientific, Singapore, Vol II, pp 1057-1062

Seiler F, Patz G, Smeets G, Srulijes J (1998) Presentation of the rail tube version II of ISL's RAMAC 30. In: Takayama K, Sasoh A (eds) Ram Accelerators. Springer-Verlag, Heidelberg, pp 79-87

Liu S, Bai ZY, Jian HX, Ping XH, Bu SQ (1998) 37-mm bore ram accelerator of CARDC. In: Takayama K, Sasoh A (eds) Ram Accelerators. Springer-Verlag, Heidelberg, pp 119-122

Soetrisno M, Imlay ST (1991) The flow field of a ram accelerator. AIAA paper 91-1915

Soetrisno M, Imlay ST, Roberts D (1992) Numerical simulations of the transdetonative ram accelerator combusting flow field on a parallel computer. AIAA paper 92-3249

Soetrisno M, Imlay ST, Roberts D (1993) Numerical simulations of the superdetonative ram accelerator combusting flow field. AIAA paper 93-2185

Srulijes J, Smeets G, Seiler F, George A, Mathieu G, Resweber R (1992a) Shock tube validation experiments for the simulation of ram-accelerator-related combustion and gasdynamic problems. In: Takayama K (ed) Shock Waves, Springer-Verlag, Heidelberg, Vol 1, pp 611-616

Srulijes J, Smeets, G, Seiler F (1992b) Expansion tube experiments for the investigation of ram-accelerator-related combustion and gasdynamic problems. AIAA paper 92-3246

Srulijes J, Eichhorn A, Smeets G, Seiler F (1995) Shock tube experiments for modeling ram-accelerator-related phenomena. Proc RAMAC II, Second Int Workshop on Ram Accelerators, Univ Washington, Seattle, WA, USA, July 17-20

Srulijes J, Smeets G, Patz G, Seiler F (1998) Accelerating hydrogen/air mixtures to superdetonative speeds using an expansion tube. In: Takayama K, Sasoh A (eds) Ram Accelerators. Springer-Verlag, Heidelberg, pp 295-302

Stewart JF, Knowlen C, Bruckner AP (1997) Effects of launch tube gases on starting of the ram accelerator. AIAA paper 97-3175

Stewart JF, Bruckner AP, Knowlen C (1998) Effects of launch tube shock dynamics on initiation of ram accelerator operation. In: Takayama K, Sasoh A (eds) Ram Accelerators. Springer-Verlag, Heidelberg, pp 181-188

Wilson GJ and MacCormack RW (1992) Modeling supersonic combustion using a fully implicit numerical method. AIAA J 30:1008-1015

Witcofski RD, Scallion WI, Carter DJ, Courter RW (1991) An advanced hypervelocity aerophysics facility: a ground-based flight test range. AIAA paper 91-0296

Yungster S (1991) Navier-Stokes simulation of the supersonic combustion flowfield in a ram accelerator. AIAA paper 91-1916

Yungster S, Eberhardt S, and Bruckner AP (1991) Numerical simulation of hypervelocity projectiles in detonable gases. AIAA J 29:187-199

Yungster S, Bruckner AP (1992) Computational studies of a superdetonative ram accelerator mode. J Prop Power 8:457-463

Ram accelerator performance modeling

C. Knowlen[1] and A. Sasoh[2]
[1]Aerospace and Energetics Research Program, University of Washington, Seattle, WA 98195-2250, U.S.A.
[2]Shock Wave Research Center, Institute of Fluid Science, Tohoku University, 2-1-1 Katahira, Aoba-ku, Sendai 980-8577, Japan

Abstract Relatively simple flowfield models are useful for examining the effects of Mach number, propellant composition, and projectile geometry on the thrust generated by the ram accelerator propulsive modes. Experimental results are predicted very well by assuming that the flow is thermally choked behind the projectile while its velocity is below about 90% Chapman-Jouguet detonation speed. The subsequent upsweep in thrust observed in experiments is anticipated by a flowfield model that incorporates a shock-induced combustion process which governs the distribution of heat release on the projectile. Ram accelerator operating characteristics at superdetonative velocities are estimated by assuming that supersonic combustion occurs at the projectile throat and that the aerodynamic drag can be determined independently. These propulsive mode models enable straight-forward evaluation of the performance potential of many different ram accelerator configurations.

Key words: Nondimensional thrust, Propulsive modes, Thermal choking, Generalized choking, Supersonic combustion

1. Introduction

The ram accelerator is a chemically powered mass launcher which has the potential to reach muzzle velocities in excess of 8 km/s (Hertzberg et al. 1988). It consists of an initial launcher, a tube pressurized with gaseous propellant, and a carefully contoured subcaliber projectile. The passage of the projectile at supersonic velocity initiates combustion of propellant which, in turn, continuously accelerates the projectile through the tube. This device can be operated over a wide range of Mach numbers using several aerothermodynamic cycles. The projectile geometry, propellant composition, and projectile Mach number are the primary factors determining which ram accelerator propulsive mode will ensue. The high ballistic efficiency and ease of scaling of the ram accelerator have generated much interest in developing this mass launcher concept for a wide range of hypervelocity applications (Bruckner 1998).

The ram accelerator propulsive modes are generally classified as subdetonative, transdetonative, and superdetonative (Hertzberg et al. 1991); based on their operational velocity range relative to the Chapman-Jouguet (CJ) detonation speed of the propellant, as illustrated in Fig. 1. For projectile velocities (V_p) below the CJ speed (V_{cj}), the thermally choked ram accelerator propulsive mode with subsonic combustion is routinely used. At superdetonative velocity, shock induced combustion processes result in all heat release occurring in the flow passage between the projectile and tube wall, and the exhaust products expand supersonically back to full tube area. In the transdetonative velocity regime ($0.9V_{cj} < V_p < 1.1V_{cj}$) the combustion is assumed to be relatively spread out, resulting in supersonic heat release occurring on and behind the projectile body. As the projectile Mach number increases, the region of heat release moves completely onto the projectile body and the flowfield can make a transition to a superdetonative propulsive mode.

One-dimensional models of the ram accelerator flowfields associated with each velocity regime (Fig. 1) are presented here and their general operating characteristics are determined. The simplified flowfield models are useful for assessing the performance potential of a wide range of ram accelerator design options; i.e., tube diameter and stage length, projectile geometry and mass, propellant composition and fill pressure, entrance Mach number, etc. The derivation of a thrust

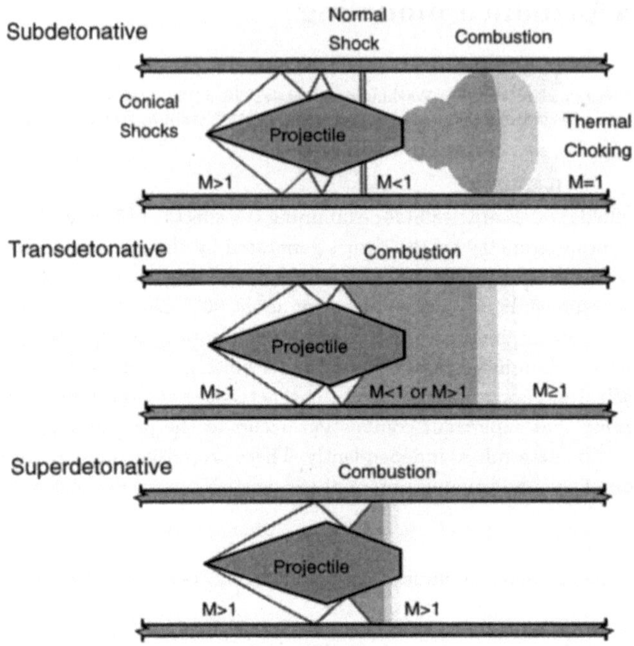

Fig. 1. Ram accelerator propulsive modes

equation, which can be used to predict the effects of Mach number and propellant heat release on all ram accelerator propulsive modes, is provided in the next section. This is followed by the presentation of three idealized flowfield models which incorporate sufficient physics to account for the thrust characteristics observed in many ram accelerator experiments.

2. Theoretical modeling

Relatively simple flowfield models can be used to determine the effects of Mach number, propellant composition, and projectile geometry on the thrust generated by the ram accelerator propulsive modes. Even though this mass launcher concept inherently involves unsteady gasdynamic processes (Brouillette et al. 1993), reasonable estimates of projectile acceleration can be made using steady flow assumptions (Bruckner et al. 1991, Sasoh and Knowlen 1997). In addition, viscous drag effects can generally be ignored for first order performance calculations, although they do become more influential at superdetonative velocities (Knowlen et al. 1996). Another factor to consider is that the ram compression of propellant often results in combustion taking place at such high pressure that real gas effects become significant, resulting in higher peak cycle pressure and more thrust than predicted when using the ideal gas equation of state (Bauer and Knowlen 1995, Buckwalter et al. 1996, 1998, Nusca and Kruczynski 1996). For purposes of illustrating the performance characteristics of various ram accelerator propulsive modes, however, the real gas effects will not be included in this theoretical modeling, and the resulting thrust predictions can be considered conservative.

Propulsive cycles for inviscid, steady flow are presented for ram accelerator operation in all velocity regimes (Fig. 1). A few of the terms that are commonly used in descriptions of ram accelerator operating characteristics are defined here. The "projectile throat" refers to the point of maximum projectile diameter and, hence, minimum flow passage area between the projectile and tube wall. "Staging" refers to the use of different propellants that are loaded into adjacent segments of the ram accelerator tube to keep the projectile operating within a desired Mach number range as its velocity increases (Bruckner et al. 1991). Another term in general use is "unstart;" this refers to a flowfield feature that often ceases ram accelerator operation. This phenomenon is characterized by a strong shock or overdriven detonation wave that is driven through the accelerator tube by the projectile (Bruckner et al. 1987). The unstart shock wave results in the choking of the flow at the throat, exposing the projectile to extremely high temperatures and pressures which often leads to its destruction. There are many mechanisms for generating an unstart, only a few of which will be discussed in this lecture note.

2.1 Ram accelerator Hugoniot analysis All of the flowfields are analyzed in the reference frame of the projectile which, in turn, is enclosed by a control volume as shown in Fig. 2. The boundaries of the control volume are denoted by dashed lines and the flow properties are assumed uniform at any axial location. Heat release from chemical reactions is added to the flow and total enthalpy is conserved. Detailed flowfield modeling for the ram accelerator propulsive modes that operate in the transdetonative and superdetonative velocity regimes is necessary to determine the state of the propellant as it exits the control volume, whereas the assumption of a sonic end state is sufficient to predict the performance of the thermally choked propulsive mode. However the end state properties of the propellant are determined, the difference between the stream thrust (i.e., the sum of the propellant momentum flux and static pressure) of the propellant flow as it enters and exits the control volume is the net propulsive force applied to the projectile (Knowlen and Bruckner 1992). Thus the general thrust characteristics can be determined by assuming that the ram accelerator flowfield induces a "jump" in the propellant properties, which can be evaluated with the same approach used in a Rankine-Hugoniot analysis for shocks and detonation waves.

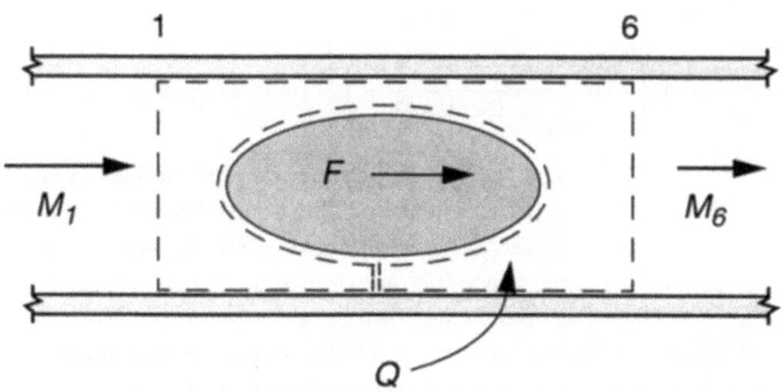

Fig. 2. Control volume for ram accelerator Hugoniot analysis

Applying the mass, momentum, and energy conservation equations to the propellant flowing through the control volume results in the following:

$$\rho_1 u_1 = \rho_6 u_6 \tag{1}$$

$$\rho_1 u_1^2 + p_1 + \frac{F}{A} = \rho_6 u_6^2 + p_6 \tag{2}$$

$$h_1 + \frac{1}{2}u_1^2 = h_6 + \frac{1}{2}u_6^2 \tag{3}$$

where ρ, u, p, F, A, and h are the density, mass-averaged particle velocity, pressure, thrust, tube cross-sectional area, and specific enthalpy, respectively. The subscripts "1" and "6" refer to the entrance and exit planes of the control volume, respectively. This system of equations is independent of equation of state and is applicable to all ram accelerator propulsive modes that are modeled with steady flow.

The enthalpy of the mixture is given by:

$$h = \sum_{i=1}^{N} Y_i h_i = \sum_{i=1}^{N} Y_i (h_{f,i} + \int c_{p,i} dT) \tag{4}$$

where the subscript i ($i = 1,N$) represents the index number of a particular chemical specie, Y denotes its mass fraction, $h_{f,i}$ is its enthalpy of formation at 0 K, and $c_{p,i}$ is the corresponding constant pressure specific heat capacity at temperature T.

The mixture specific heat capacity (c_p), ideal gas equation of state, and the mixture molecular weight (W) can be expressed as follows:

$$c_p = \sum_{i=1}^{N} Y_i c_{p,i} \tag{5}$$

$$p = \sum_{i=1}^{N} p_i = \sum_{i=1}^{N} \rho_i \frac{\Re}{W_i} T = \rho \frac{\Re}{W} T \tag{6}$$

$$\frac{1}{W} = \sum_{i=1}^{N} \frac{Y_i}{W_i} \tag{7}$$

where $\Re$ is the universal gas constant.

The second Damköhler number is used for the nondimensional heat release Q which is defined as:

$$Q = \frac{\sum_{i=1}^{N} [(Y_i h_{f,i})_1 - (Y_i h_{f,i})_6]}{c_{p1} T_1} \tag{8}$$

Solving this set of equations yields the following expression for the nondimensional thrust:

$$\frac{F}{p_1 A} = \frac{\gamma_1 M_1}{\gamma_6 M_6} (1 + \gamma_6 M_6^2) \left[\left(\frac{\gamma_6 - 1}{\gamma_1 - 1} \right) \left(\frac{\frac{h_1}{c_{p1} T_1} + \frac{(\gamma_1 - 1)}{2} M_1^2 + Q}{\frac{h_6}{c_{p6} T_6} + \frac{(\gamma_6 - 1)}{2} M_6^2} \right) \right]^{\frac{1}{2}} - (1 + \gamma_1 M_1^2) \tag{9}$$

where γ is the specific heat capacity ratio. To account for the temperature dependent heat capacity effects, the general expression $h/c_p T$ is used in the thrust equation. This term deviates from unity by 20-30% for many propellants of interest; however, it does not have a strong Mach number dependence and can generally be assumed to equal the value determined at the CJ speed.

This expression for nondimensional thrust applies for all ram accelerator propulsive modes. Once one has determined the exit Mach number as function of M_1 for any propulsive mode, the thrust is readily calculated. In the case of the thermally choked ram accelerator, $M_6 = 1$ for all projectile velocities below the CJ speed. More detailed flowfield modeling is required to predict end state Mach and net thrust in the transdetonative and superdetonative velocity regimes.

2.2 Thermally choked propulsive mode The one-dimensional flowfield model for the thermally choked ram accelerator propulsive mode is shown in Fig. 3. This propulsive mode is initiated by a starting process that establishes a normal shock on the rear of the projectile body, which is then stabilized by thermal choking of the flow at full tube area behind the projectile (Bruckner et al. 1992, Schultz et al. 1998). As the projectile accelerates, the normal shock recedes and eventually falls off the body. The flowfield described here presents three limits on the Mach number range of ram accelerator operation; a minimum Mach corresponding to that at which the incoming flow will just reach sonic velocity at the throat, a minimum Mach which is greater than the geometric constraint, yet is just at the point where the combustion heat release drives the normal shock up to the throat, and a maximum Mach which corresponds to the point where the shock falls off the projectile body, or recedes completely behind the projectile in the case when it has a pointed tail (Hertzberg et al. 1988, Higgins et al. 1993).

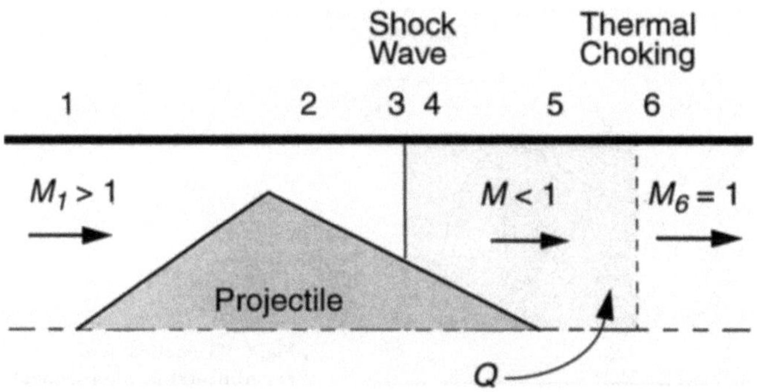

Fig. 3. Flowfield model for thermally choked propulsive mode

Standard gasdynamic relations can be used to predict the position of the normal shock at any M_1, and to determine the theoretical bounds of thermally choked ram accelerator operation (Shapiro 1953, Sasoh and Knowlen 1997). The thrust, however, can be determined directly from Eq. (9) without regards to the shock position. The theoretical basis for this was developed by Knowlen and Bruckner (1992), and has been experimentally verified for a wide range of projectile geometries and tube diameters (Hertzberg et al. 1991, Giraud et al. 1993, Kruczynski and Liberatore 1996, Sasoh et al. 1997, Chang et al. 1997). The independence of thrust on projectile geometry is due to the constraints imposed on the flowfield by the thermal choking of the flow. Since the thermal choke point corresponds to an entropy extremum, the flowfield must adjust to produce the equilibrium end state for any given in-tube Mach number. Under ideal flow conditions, the only increases in entropy come from the normal shock wave and heat addition at finite Mach numbers. The primary mechanism for adjusting the flow under non-ideal conditions is the positioning of the normal shock wave; e.g., it moves forward when the entropy rise due to reflected shock waves is accounted for and the normal shock moves backward when the flow is closer to ideal. Since the equilibrium flowfield properties at the thermal choking point are independent of the history of the propellant, the net axial force on the projectile is not altered by the placement of the normal shock. Thus, even though geometric features of the projectile govern its operating limits, the thrust generated once the thermally choked ram accelerator has been started is readily computed.

Theoretical and experimental nondimensional thrust profiles for the thermally choked ram accelerator are shown in Fig. 4. The effects of changing Q are investigated by varying the nitrogen dilution level ($4.0N_2$ - $8.2N_2$) of the propellant while keeping the methane-oxygen equivalence ratio constant. It can be seen that the theoretical thrust goes through a maximum and then steadily decreases as M_1 increases. Eventually the thrust goes to zero at the CJ speed of the propellant. Increasing Q increases the thrust, increases the propellant's CJ speed, and raises the Mach number for maximum thrust generation. Thus the highest projectile acceleration will result from operating the ram accelerator with the most energetic propellants feasible, at the maximum thrust Mach number. Multiple propellant stages can be used to maintain peak thrust performance over a wide velocity range (Elvander et al. 1998).

The theoretical nondimensional thrust profiles are in reasonably good agreement with those of typical ram accelerator experiments (Fig. 4). The experimental data begin at the entrance M_1 to the stage containing the test propellant and are terminated at the point where the projectile unstarted. It can be seen that the experimental thrust does indeed decrease with increasing

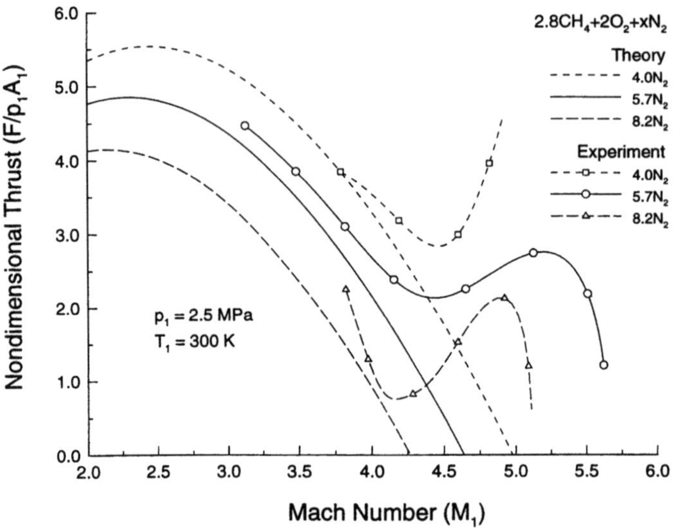

Fig. 4. Theoretical and experimental nondimensional thrust-Mach profiles for thermally choked propulsive mode

M_1 in the manner predicted for the thermally choked propulsive mode. As the projectile reaches $\approx 90\% V_{cj}$, however, the thrust deviates from theory and tends to increase with increasing M_1. This "upsweep" in thrust with respect to M_1 is the main characteristic of ram accelerator operation in the transdetonative velocity regime; and it suggests that the flow is no longer thermally choked behind the projectile and that the combustion process may have moved, in part, up onto the body. The thrust in experiments with 5.7 and $8.2N_2$ molar content ultimately reaches a relative maximum at $\approx 110\% V_{cj}$ and then begins to decrease with increasing M_1, until structural failure of the projectile causes an unstart (Naumann 1993, Chew and Bruckner 1994, Knowlen et al. 1995). When the nitrogen dilution is reduced to $4.0N_2$ an unstart occurs at the CJ speed of the propellant, presumably because of excessive heat release behind the projectile throat (Higgins et al. 1993).

2.3 Transdetonative propulsive mode Even though the transition process from the thermally choked propulsive mode is inherently unsteady, it is still useful to consider steady flowfield models which contain the principal physics that govern the outcome of ram accelerator experiments. Two gasdynamic phenomena that are assumed to be primary mechanisms governing the thrust characteristics in this velocity regime are the ram compression of the propellant and the system of intersecting shock waves that exists in the flow passages between the projectile and tube wall. The ram compression ahead of the projectile throat, increases the temperature and pressure in the flowfield around the projectile as M_1 increases, thus the chemical reaction rates increase and the length of the combustion zone decreases. The strength of the shock wave system also increases with increasing M_1, which further elevates the propellant temperature and generates more total pressure loss. Sophisticated computational fluid dynamic techniques, which require substantial computer resources for timely results, are needed to analyze the three-dimensional, viscous, time dependent ram accelerator flowfields with reasonable accuracy (Li et al. 1993, Soetrisno et al. 1993, Nusca 1995, Nusca and Kruczynski 1996). Thus a simplified one-dimensional model is proposed here which can be used to predict at what M_1 the onset of propulsive mode transition will occur and the corresponding thrust profile.

The one-dimensional transdetonative flowfield is assumed to be similar to that of the thermally choked ram accelerator, except that the combustion is now induced to occur on the rear of the projectile by the normal shock, as shown in Fig. 5, and the distribution of heat release in the control volume is governed by the finite chemistry rates of the propellant (Sasoh et al. 1995). This results in subsonic combustion thermally choking the flow behind the projectile at low Mach numbers ($2.5 < M_1 < 4$). Since the normal shock wave gets stronger as the projectile accelerates, the thermal choke point moves closer to the base and, under appropriate conditions, eventually moves up onto the body as the projectile velocity approaches the CJ speed. If choking of the flow can take place on the body, then the combustion begins behind the normal shock wave and continues to progress as the subsonic flow passes through the choke point. Chemical reactions are still proceeding as the propellant is subsequently expanded back to full tube area at supersonic velocity. This propulsive model accounts for the effects of ram compression and the shock wave system on the body as M_1 increases by incorporating the temperature dependent chemical reaction rates in the heat release calculations. The properties of the propellant as it exits the control volume are then used in Eq. (9) to estimate thrust in the transdetonative velocity regime.

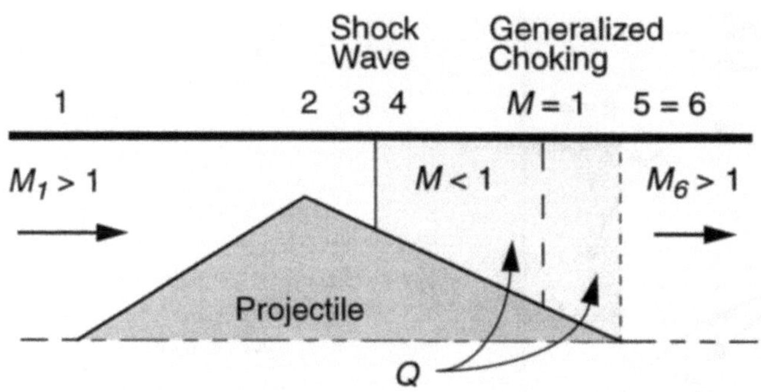

Fig. 5. Flowfield model for transdetonative propulsive mode

The flowfield proposed here for the transdetonative ram accelerator propulsive mode can be calculated with readily available ordinary-differential-equation solvers (Kee et al. 1993), or by algebraic methods applied to finite volumes (Sasoh et al. 1997). The key feature to determine is the flow-to-tube area ratio at which choking would occur for a given projectile area profile, and the corresponding position for the normal shock. In order to obtain a continuous variation of flow properties, the following condition must be satisfied at the choking point:

$$\frac{dq}{c_p T} - \frac{dA}{A} - \frac{dW}{W} = 0 \tag{10}$$

where

$$dq = -\sum_{i=1}^{N} h_{f,i}\, dY_i \tag{11}$$

Equation (10) implies that, neglecting the last term (which is usually very small relative to the others), the heat release (the first term on left) needs to be balanced with the effect of area change (middle term). This constraint will hereafter be referred to as the "generalized choking" condition. A unique solution for a given M_1 has been found once the shock location which results in the occurrence of a generalized choking condition has been determined (Sasoh and Knowlen 1997).

This propulsive model can be used to examine the effects of projectile length and combustion zone length on the thrust generated in the transdetonative ram accelerator. For example, the nondimensional thrust-Mach number profiles determined with a one-step reaction mechanism for the $2.8CH_4+2O_2+5.7N_2$ propellant are shown in Fig. 6, where β is the ratio of projectile length to combustion zone length. At low Mach numbers the combustion zone is relatively large and the predicted thrust performance is that of the thermally choked ram accelerator, regardless of body length. When the projectile length is large relative to the shock induced combustion zone, the choke point is predicted to move up onto the body as the projectile approaches the CJ speed. The M_1 at which this occurs for a given β is indicted by the point where the thrust profile begins to deviate from the thermally choked performance curve. As M_1 increases and the choke point moves farther up on the body, the normal shock moves farther toward the throat and more of the heat release occurs on the projectile. This results in higher thrust levels at velocities near CJ speed than the thermally choked propulsive mode would produce. Eventually, if the heat release does not first force the normal shock beyond the projectile throat, both the choke point and the normal shock will cease to move forward and begin to recede, causing the thrust to decrease with increasing M_1.

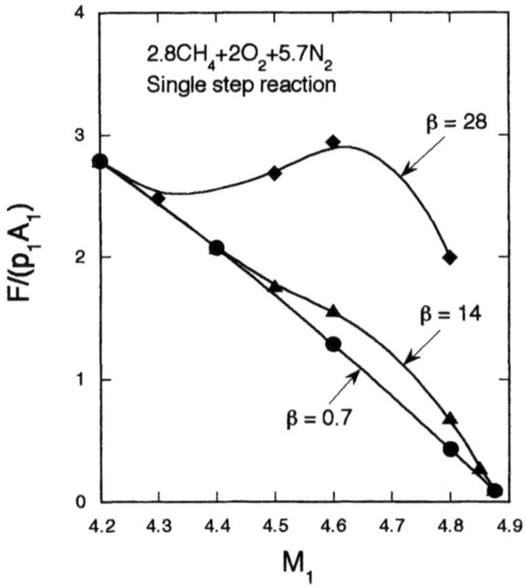

Fig. 6. Nondimensional thrust in the transdetonative velocity regime

In situations where the length of the combustion zone is very short relative to the projectile ($\beta >> 1$), an upsweep in the nondimensional thrust-Mach profile, followed by a relative maximum, is predicted in the transdetonative velocity regime (Fig. 6). The relative maximum in thrust occurs when the normal shock and the choke point are at their farthest forward positions. Under these conditions almost all of the heat release is added ahead of the choke point, resulting in a shock-induced combustion wave that is very much like a CJ detonation wave. Since a CJ wave cannot be stabilized in a diverging duct under steady flow conditions, this transdetonative propulsion model is not relevant once the normal shock begins to recede. Thus more detailed flowfield modeling is necessary to account for ram accelerator operation at and beyond the CJ speed. Even though the maximum operating Mach number cannot be predicted with this propulsive model, it is still useful for the insight it provides on how variable area heat release can affect the thrust characteristics as the projectile approaches CJ speed.

2.4 Supersonic combustion propulsive mode

The highest velocity potential for the ram accelerator is with the superdetonative propulsive mode. Of immediate interest is the thrust performance of simplified projectile geometries and the maximum velocity they can attain in a given propellant mixture. Modeling superdetonative ram accelerator flowfields with computational fluid dynamic techniques has been done with high levels of sophistication (Yungster and Bruckner 1992, Li et al. 1993, Soetrisno et al. 1993). Results have been encouraging, but very computationally intensive, and their accuracy is questionable due to real gas effects and the lack of appropriate kinetic rate data at the conditions existing in the combustion zone at hypersonic Mach numbers. Since detailed flowfield information is not always necessary to qualitatively determine the effects of propellant heat release on thrust performance, more simplified approaches are useful. Gatau et al. (1995) have applied a quasi-steady, quasi-one-dimensional flowfield model to the axisymmetric projectile experiments with good results; however, they did not address the issue of drag. A similar propulsive cycle analysis was carried out with the inclusion of drag effects by Knowlen et al. (1996). This theoretical model can readily be applied to examine the effects of geometry variations and propellant chemistries on the thrust-Mach number performance of a superdetonative ram accelerator.

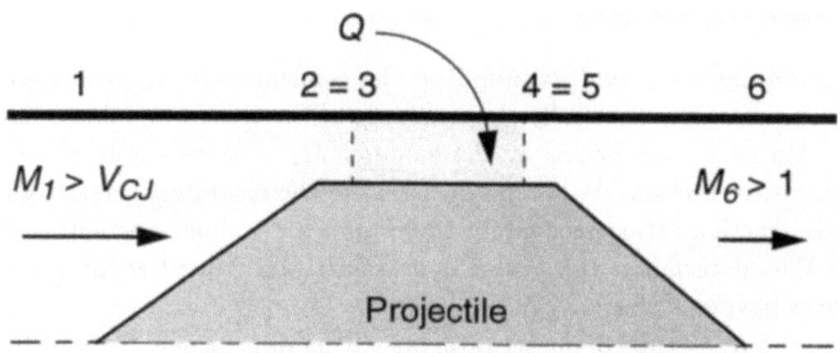

Fig. 7. Flowfield model for supersonic combustion propulsive mode

This analysis of the superdetonative propulsive mode is conducted in the projectile frame of reference using the control volume shown in Fig. 7. The ideal gas equation of state is used and the steady flow is considered both adiabatic and inviscid. The standard channel flow equations are applied to determine flow properties at each station. Ram compression over the conical nose cone occurs between the entrance to the control volume (station 1) and the projectile throat (station 2). The entrance to the combustion zone (station 3) has the same flow area as the projectile throat. For this part of the analysis, the ram compression process between stations 1 and 3 is considered isentropic. The chemical heat release is assumed to occur supersonically in the constant area section between stations 3 and 4, with the combustion products determined by computing the chemical equilibria based on the initial flow properties at station 3. The final expansion process between stations 4 and 6 is assumed to be isentropic. When the supersonic combustion process occurs at the point of maximum flow contraction, the exit Mach number, M_6, is primarily affected by the projectile throat-to-tube diameter ratio, d_2/d_1, the nondimensional heat release parameter Q, and the entrance Mach number, M_1. Once the exit Mach M_6 has been determined, the thrust is readily computed from Eq. (9).

Nondimensional thrust versus Mach number profiles in $CH_4/O_2/CO_2$ based propellants, determined for both the supersonic combustion and the thermally choked ram accelerator propulsive modes, are shown in Fig. 8. The ideal thrust profiles of the supersonic combustion ram accelerator were determined for $d_2/d_1 = 0.76$ and a fill pressure of 2.5 MPa. The mole number of CO_2 diluent was varied between 3 and 8 ($9.8 > Q_{cj} > 5.4$) to illustrate the effects of heat release on both of these propulsive modes. The nondimensional thrust of the supersonic combustion mode

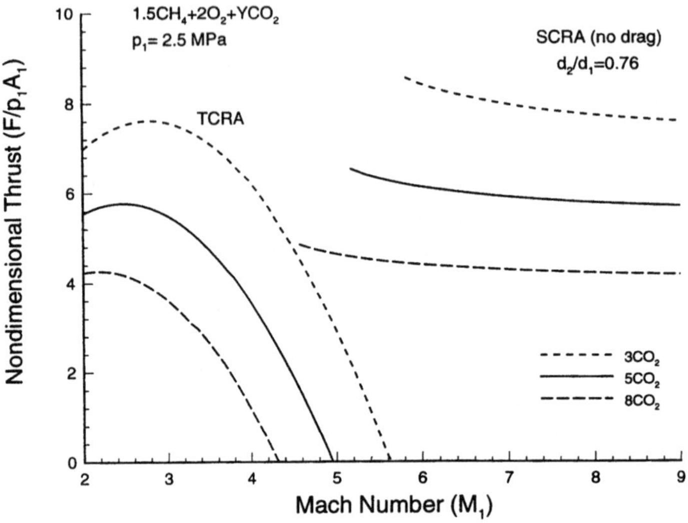

Fig. 8. Nondimensional thrust for thermally choked (TCRA) and supersonic combustion ram accelerator (SCRA) with CO_2 dilution variations

remains relatively constant as M_1 increases and its amplitude is, coincidentally, approximately the same as that of the peak thrust of the thermally choked mode in the same propellant. Each of the supersonic combustion thrust curves begins at the value of M_1 that results in thermal choking of the flow in the combustor section. At lower velocities the chemical heat release, Q_{cj}, cannot be added at the flow contraction ratio specified for the combustor without unstarting the projectile. This phenomenon thus determines the lowest operational Mach number limit for a superdetonative ram accelerator having a given d_2/d_1.

The effects of the projectile throat-to-tube diameter ratio on the nondimensional thrust-Mach number profiles of the thermally choked and supersonic combustion propulsive modes are shown in Fig. 9. The propellant composition assumed was $1.5CH_4 + 2O_2 + 5CO_2$ and its fill pressure was 2.5 MPa. The thrust of the thermally choked mode is independent of throat diameter, thus one curve represents the performance of all projectiles. The thrust profiles of the supersonic combustion mode are plotted from the minimum operating M_1 for each d_2/d_1. The curves indicate the maximum thrust available in the superdetonative velocity regime for these projectiles. The ideal nondimensional thrust more than doubles as d_2/d_1 is increased from 0.60 to 0.85. This is a consequence of the higher thermal efficiency that results from increasing the compression ratio of the propulsive cycle.

The projectile drag must be assessed in order to predict the effects of d_2/d_1 on its upper velocity limit in a particular propellant. Assuming that the net drag can be approximated by the sum of the skin friction, conical shocks, fin blockage, and base expansion losses results in a drag profile that is nearly linear in the Mach number range of $5 < M_1 < 8$ (Knowlen et al. 1996). Subtracting the net drag of a ram accelerator projectile from the ideal superdetonative thrust (Fig. 9) yields the nondimensional thrust profiles shown in Fig. 10. In this plot the projectile velocity is normalized by the propellant CJ speed. The thrust generated by the smallest of the projectiles ($d_2/d_1 = 0.60$) could not overcome its drag penalty at minimum M_1; thus this projectile is not expected to operate at all in the superdetonative velocity regime. Increasing d_2/d_1 results in higher thrust and higher peak velocity, indicating that projectile throat diameters should be as large as feasible to maximize the muzzle velocity of the ram accelerator. Minimizing the drag arising from nose cone shocks, fin bow shocks, and rear nozzle expansion losses will also increase peak velocity. Estimates for peak ram accelerator velocity based on this supersonic combustion model indicate that speeds up to twice V_{cj} are feasible for a projectile having $d_2/d_1 = 0.85$, which

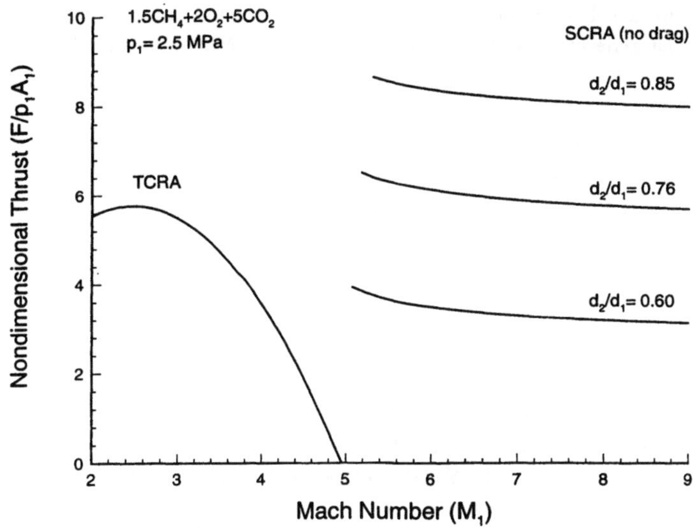

Fig. 9. Nondimensional thrust for thermally choked (TCRA) and supersonic combustion ram accelerator (SCRA) for various projectile diameters

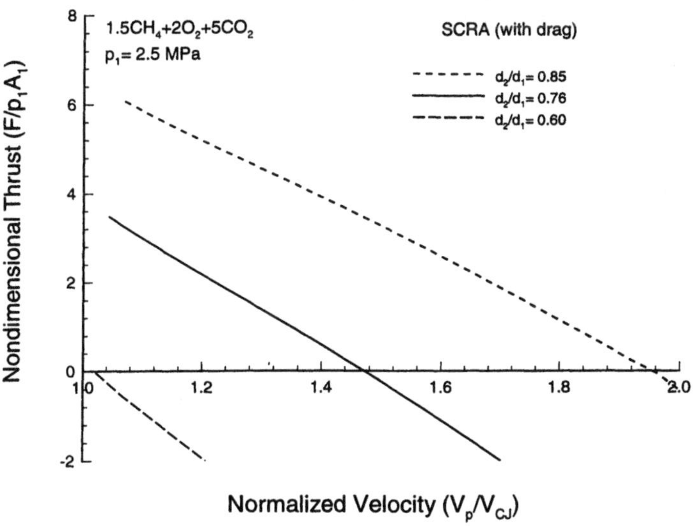

Fig. 10. Nondimensional thrust vs. normalized velocity for supersonic combustion ram accelerator (SCRA) with inclusion of drag

is consistent with the predictions of other computational analyses (Brackett and Bogdanoff 1989, Yungster and Bruckner 1992).

3. Summary

The aerothermodynamics of the ram accelerator propulsive modes are difficult to accurately model with reasonable computational expense. Thus, it is useful to evaluate the thrust potential of various ram accelerator configurations with one-dimensional flowfield models to estimate the end state Mach number. Experimental results are predicted very well by the model for the thermally choked propulsive mode, when the projectile velocity is below about 90% CJ speed. The subsequent upsweep in thrust is anticipated with a flowfield model that incorporates a shock-induced combustion process that governs the distribution of heat release. Superdetonative performance is predicted with a flowfield model that assumes supersonic combustion at the projectile throat and an appropriate hypersonic drag model. The combination of these three propulsive models provides reasonable estimates of the thrust generated by the ram accelerator at any Mach number.

References

Bauer P, Knowlen C (1995) Real gas effects on the operating envelope of the ram accelerator. Proc 2nd Int Workshop on Ram Accelerators, Univ Washington, Seattle, WA USA, July 17-20

Brackett DC, Bogdanoff DW (1989) Computational investigation of oblique detonation ramjet-in-tube concepts. J Prop Power 5:276-281

Brouillette M, Zhang F, Frost DL, Chue RC, Lee JHS, Thibault P and Yee C (1993) One-dimensional analysis of the ram accelerator. Proc 1st Int Workshop on Ram Accelerators, paper No 13, French-German Research Institute of Saint Louis, France, September 7-10

Bruckner AP (1998) The ram accelerator: overview and state of the art. In: Takayama K, Sasoh A (eds) Ram Accelerators, Springer-Verlag, Heidelberg, pp 3-23

Bruckner AP, Bogdanoff DW, Knowlen C, Hertzberg A (1987) Investigation of gasdynamic phenomena associated with the ram accelerator concept. AIAA paper 87-1327

Bruckner AP, Knowlen C, Hertzberg A, Bogdanoff DW (1991) Operational characteristics of the thermally choked ram accelerator. J Prop Power 7:828-836

Bruckner AP, Burnham EA, Knowlen C, Hertzberg A, Bogdanoff DW (1992) Initiation of combustion in the thermally choked ram accelerator. In: Takayama K (ed) Shock Waves, Springer-Verlag, Heidelberg, pp 623-630

Buckwalter DL, Knowlen C, Bruckner AP (1996) Ram accelerator performance analysis code incorporating real gas effects. AIAA paper 96-2945

Buckwalter DL, Knowlen C, Bruckner AP (1998) Real gas effects on thermally choked ram accelerator performance. In: Takayama K, Sasoh A (eds) Ram Accelerators, Springer-Verlag, Heidelberg, pp 125-134

Chang X, Matsuoka S, Watanabe T, Taki S (1998) Ignition study for low pressure combustible mixture in a ram accelerator. In: Takayama K, Sasoh A (eds) Ram Accelerators, Springer-Verlag, Heidelberg, pp 105-109

Chew G, Bruckner AP (1994) A computational study of projectile nose heating in the ram accelerator. AIAA paper 94-2964

Elvander JE, Knowlen C, Bruckner AP (1998) High acceleration experiments using a multi-stage ram accelerator. In: Takayama K, Sasoh A (eds) Ram Accelerators, Springer-Verlag, Heidelberg, pp 55-64

Gatau F, Smeets G, Srulijes J (1995) Theoretical model for calculating the projectile acceleration in a ram accelerator. Proc 2nd Int Workshop on Ram Accelerators, Univ Washington, Seattle, WA USA, July 17-20

Giraud M, Legendre JF, Simon G (1993) RAMAC 90: Experimental studies and results in 90mm caliber, length 108 calibers. Proc 1st Int Workshop on Ram Accelerators, paper No 3, French-German Research Institute of Saint Louis, France, September 7-10

Hertzberg A, Bruckner AP, Bogdanoff DW (1988) Ram accelerator: A new chemical method for accelerating projectiles to ultrahigh velocities. AIAA J 26:195-203

Hertzberg A, Bruckner AP, Knowlen C (1991) Experimental investigation of ram accelerator propulsion modes. Shock Waves 1:17-25

Higgins AJ, Knowlen C, Bruckner AP (1993) An investigation of ram accelerator gas dynamic limits. AIAA paper 93-2181

Kee RJ, Rupley FM, Miller JA (1993) Chemkin-II: A FORTRAN chemical kinetics package for the analysis of gas phase chemical kinetics. Sandia Lab Rep SAND89-8009B

Knowlen C, Bruckner AP (1992) A Hugoniot analysis of the ram accelerator. In: Takayama K (ed) Shock Waves, Springer-Verlag, Heidelberg, pp 617-622

Knowlen C, Higgins AJ, Bruckner AP (1995) Aerothermodynamics of the ram accelerator. AIAA paper 95-0289

Knowlen C, Higgins AJ, Bruckner AP, Bauer P (1996) Ram accelerator operation in the superdetonative velocity regime. AIAA paper 96-0098

Kruczynski D, Liberatore F (1996) Experimental investigation of high pressure/performance ram accelerator operation. AIAA paper 96-2676

Li C, Kailasanath K, Oran ES, Boris JP (1993) Numerical simulations of unsteady reactive flows in ram accelerators. Proc 1st Int Workshop on Ram Accelerators, paper No 29, French-German Research Institute of Saint Louis, France, September 7-10

Naumann KW (1993) Heating and ablation of projectiles during acceleration in a ram accelerator tube. AIAA paper 93-2184

Nusca MJ (1995) Reacting flow simulation of transient, multi-stage ram accelerator operation and design studies. Proc 2nd Int Workshop on Ram Accelerators, Univ Washington, Seattle, WA USA, July 17-20

Nusca MJ, Kruczynski DL (1996) Reacting flow simulation for a large-scale ram accelerator. J Prop Power 12:61-69

Sasoh A, Hirakata S, Maemura J, Takayama K (1997) Experimental studies of 25-mm-bore ram accelerator at the Shock Wave Research Center. AIAA paper 97-2652

Sasoh A, Knowlen C, Bruckner AP (1995) Effect of finite rate chemical reactions on ram accelerator thrust characteristics. AIAA paper 95-2492

Sasoh A, Knowlen C (1997) Ram accelerator operation analysis in thermally choked and transdetonative propulsive modes. Trans Japan Soc Aeronautical and Space Sciences 40:130-148

Schultz E, Knowlen C, Bruckner AP (1998) Overview of the subdetonative ram accelerator starting process. In: Takayama K, Sasoh A (eds) Ram Accelerators, Springer-Verlag, Heidelberg, pp 189-203

Shapiro AH (1953) The dynamics and thermodynamics of compressible fluid flow. John Wiley & Sons, New York, Vol I, pp 219-261

Soetrisno M, Imlay ST, Roberts DW (1993) Numerical simulations of the superdetonative ram accelerator combusting flow field. AIAA paper 93-2185

Yungster S, Bruckner AP (1992) Computational studies of a superdetonative ram accelerator mode. J Prop Power 8:457-463

Real gas effects in ram accelerator propellant mixtures: theoretical concepts and applied thermochemical codes

P. Bauer[1], J.F. Legendre[2], M. Henner[2], M. Giraud[2]
[1]Laboratoire de Combustion et de Détonique (UPR 9028 du CNRS), BP 109, 86960 Futuroscope Cdx, France
[2]ISL, French-German Research Institute of Saint Louis, BP 34, 68301 Saint Louis Cdx, France

Abstract. An a priori determination of the characteristics of detonation or combustion of combustible mixtures is often required to predict the performance of a ram accelerator. The difficulty any user is facing is that related to the high equivalence ratio and high pressure of the propellant mixtures that are used in the ram accelerators. These mixtures are mainly composed of methane, oxygen, and a diluent that can be nitrogen, carbon dioxide, or helium. The main aspects that should be addressed in a particular thermochemical code is the appropriate choice of the equation of state and, moreover, the reliability of the thermophysical data that are used. To date, a series of thermochemical codes are available and most of them run on a PC which gives a greater capability and flexibility for current use. The Quartet computer program has been widely used since it has been updated with the most recent data provided through extensive ram accelerator experiments. It offers a large frame of calculations, based on several equations of state. Other codes were also used for this purpose, including simple methods of calculation. These codes are currently dealing with a variety of equations of state. The virial form turns out to be the most pertinent way for a rapid and reliable calculation. Moreover, although they are more appropriate for condensed explosives, other equations of state, namely JCZ3 or BKW, for instance, are also available. A recent version of Cheetah thermochemical computer code was used, as well, and results were compared with those derived from Quartet. A slight disagreement was first observed since a number of default data included in the library of Cheetah are aimed at high explosive calculations and, therefore, do not fit the ram accelerator combustion products. A series of updated data were included in order to build a new library, allowing a more specific use in the case of ram accelerator calculation. The aim of the present paper is to describe the theoretical key elements of a thermochemical computer program and to improve the knowledge on its design and operating procedure. Moreover, additional information on the optimized version of Cheetah and on the appropriate library of thermophysical data is provided.

Key words: Equation of state, Equilibrium composition, Detonation, Real gas, High pressure

1. Introduction

The propellant mixtures used in the ram accelerator are fuel rich and they are based on hydrocarbon, namely methane, more or less diluted in nitrogen or eventually in another diluent. In most cases, the initial pressure of the mixtures that are used in the facilities throughout the world is of the order of 2.5 to 5.0 MPa (Legendre 1996, Giraud et al. 1992, Knowlen 1991). In some cases, the pressure is considerably greater, i.e., of the order of 10 MPa or higher (Kruczynski 1993, Knowlen and Bruckner 1997).

When the projectile travels in the ram accelerator tube, the mixture undergoes a complex thermodynamic cycle (Knowlen and Bruckner 1991) and the calculation of the various thermodynamic steps can be obtained by solving the set of conservation equations. A thorough description of the propulsion mode (Hertzberg et al. 1991) shows that, depending on whether the regime is sub-detonative (Giraud et al. 1997) or super-detonative (Knowlen et al. 1996), the final pressure of the products may reach a value of the order of 8 to 15 times the initial pressure, respectively. The mode termed super-detonative is related to the speed of the projectile, compared with the theoretical Chapman-Jouguet (CJ) detonation velocity of the mixture it travels in. Anyhow, according to the range of pressure of the combustion products, the conservation equations must be implemented with a real gas equation of state. This set of equations can be readily solved and it provides the final state of the combustion products. This was exemplified in the particular case of a sub-detonative regime (Bauer et al. 1998).

Most of the thermochemical codes that are available provide a series of data, including detonation characteristics, for two different types of reactive mixtures:
(i) gaseous explosives at atmospheric or sub-atmospheric initial pressures, and
(ii) high explosives.

Over the last decades, this was appropriate for most industrial situations. However, presently, as far as the ram accelerator technology is concerned, the gap between these two distinct fields must be filled. In other words, the gaseous combustion or detonation products require the use of an equation of state (e.o.s.) derived from that of liquid or solid explosive detonation products (Bauer 1985). This raises the main question on which e.o.s. to choose (Bauer et al. 1985, Heuzé et al. 1986). At this level of complexity, although some attempts have been made in this direction (Heuzé and Bauer 1994), the calculation can no longer be conducted by hand, and a thermochemical code running on a fast computer is required. An elaborate code must be designed, and, moreover, the appropriate thermophysical parameters must be included. The aim of the present paper is to highlight the main requirements a thermochemical computer code should meet in order to become a reliable tool for the purpose of ram accelerator studies, including CFD calculations that often require the input of thermodynamic data (Henner et al. 1997). Moreover, the results derived from calculation with two specific codes, namely Cheetah and Quartet are presented and compared with experimental data. Quartet code is derived from an earlier version, named Quatuor (Heuzé et al. 1987a) and it has been extensively used. It remains suitable in most circumstances. Moreover, the ability of Cheetah in this specific field of research is emphasized.

2. Main calculation steps

2.1 Calculation of equilibrium composition The calculation proceeds through a series of steps which, at all times involve many iterations. The equilibrium composition of the combustion products is calculated for each of these iterations and, therefore, it is one of the key operations of the calculation procedure. The calculation is based on the minimization of Gibbs free energy, but the iterative technique may differ from one case to another. A new technique based on matrix algebra was developed (Heuzé et al. 1985). It can eventually be reliably applied, even for hand calculation, to a system that contains as much as 10 different species in the combustion products. This point raises a major difficulty in the codes that are currently used. The more species are included, the better the calculation of thermodynamic properties. Therefore, a calculation that deals with a limited number of species, which, in turn, skips some important dissociation processes, may yield over-estimated data. On the other hand, the use of a too large set of components slows down the calculation and requires a much greater computer capacity. There is a very sharp domain where the species should be carefully chosen. On the basis of an extensive series of calculations, Greenlee and Butler (1993) showed the importance of selecting appropriate species. Their discussion is based on the equilibrium composition calculation derived from various codes running with different e.o.s.

The technique used in the Quartet or Quatuor codes is based on the matrix solver method and is extremely fast and reliable. One of the advantages of this method is that a precise initial guess on the nature or possible amount of the components of the combustion products helps, but it is not strictly required. This calculation procedure was validated by the numerous experiments made with various explosive mixtures, at all levels of density.

In the Cheetah code, one can also choose an initial set of species in the combustion products that should be accounted for during the calculation. In our present concern, the main lack of the default library of this code is that some dissociated species, which play a prominent role, are actually missing. This yields over-estimated values (Bauer 1995), but it has been corrected by the addition of an adapted library.

2.2 Thermodynamic properties The thermodynamic parameters such as the heat capacities, enthalpy, Gibbs free energy are readily accessible through polynomial developments such as those updated by Kee et al. (1990) which are available in the literature. They yield values that are totally consistent with the JANAF thermodata (Stull and Prophet 1971). Moreover, the pressure influence can now be accounted for (Heuzé et al. 1987b, Byers Brown and Amaee 1992). This correction may be conducted either analytically or numerically (Bauer et al. 1998, Byers Brown 1987). For this purpose, an update of the computer code designed by Knowlen (1991), aimed at the calculation of ram accelerator performance, is in progress (Buckwalter et al. 1997).

In some cases where a high initial pressure of the combustible mixture is involved, the input data, i.e., mainly the initial density should be corrected. The Redlich and Kwong (1949) (R-K) e.o.s. (Kemp et al. 1975, Austing and Selman 1988) turned out to be very suitable in this case.

2.3 Equation of state of combustion/detonation products The gas–liquid critical density, i.e., $1/v_c$, of each substance is often referred to, in order to define higher density ranges for that substance. The high to very high density domain corresponds to $v_c/6 < v < v_c$ (Byers Brown and Amaee 1992).

An equation of state can be expressed in the following explicit form:

$$\sigma = \frac{pv}{RT} = f(v, T, x_i) \tag{1}$$

or in an implicit form:

$$\Phi(p, v, T, x_i) = 0 \tag{2}$$

where v, T, x_i, and R are the specific volume, temperature, number of moles of specie i, and the universal gas constant, respectively. The term σ is the compressibility factor that can take any value around 1 (Kemp et al. 1975).

Fluids that are correctly described as gases, i.e., at a low enough density, obey the ideal gas (IG) e.o.s., which states that the compressibility factor takes the value $\sigma = 1$. At elevated initial pressure, it turns out that this e.o.s. which does not take into account the molecular interactions, is no longer suitable to predict the thermodynamic characteristics of the combustion products. It always yields under-evaluated results. Instead, a virial development is more appropriate (Bauer et al. 1994a). This is often expressed as:

$$f(v, T, x_i) = 1 + g(\frac{b}{v}) \tag{3}$$

b is the covolume that can take various analytical forms, whereas $g(b/v)$ is in a polynomial form. Sometimes, it is expressed as an analytical summation of all the terms in the development. An extensive study of the molecular interactions is proposed by Byers Brown and Amaee (1992). Another description of the hard sphere modeling is that given by Heuzé (1986).

Although the Carnahan and Starling (1969) (C-S) development is a better hard sphere modeling, the experience acquired over the last decades has shown the reliability of the development involved in the Boltzmann (B) e.o.s., and a better flexibility for current calculations (Bauer et al. 1983). In this e.o.s., molecules are modeled as rigid spheres, and it is assumed that the interactions are limited to molecules of the same type. The compressibility factor, based on the virial development that includes five terms, is written in the form:

$$\sigma = \frac{pv}{RT} = 1 + x + 0.625x^2 + 0.287x^3 + 0.193x^4 \tag{4}$$

with $x = b/v$, b being the covolume of the rigid spheres. A simple description of the mixing law for gases has the following form :

$$b = \sum x_i b_i \tag{5}$$

where x_i and b_i are respectively the number of moles and covolume of specie i. The covolumes that are used for computation of detonation products are derived from the literature. They are presented and discussed elsewhere (Legendre 1996).

The validity of this e.o.s. was based on a variety of experimental data (Bauer et al. 1988). In the specific case of ram accelerator propellants, a large series of experiments were conducted at ISL in the 90L35 detonation tube with stoichiometric and fuel rich mixtures at T_0=298K and p_0=2.5 and 3.5 MPa, including different diluents. These were mono-, di-, or tri-atomic, i.e., helium, nitrogen, or carbon dioxide, respectively (Legendre 1996, Legendre et al. 1993, 1995, 1997). As already published for experiments conducted in the 90L15 detonation tube, at p_0=2.5 MPa (Bauer and Legendre 1993), these stoichiometric mixtures could develop a stable self-sustained detonation wave up to a nitrogen dilution ratio of $\beta=[N_2]/[O_2]=2.65$. This conclusion was based on a deviation of less than 2% of the detonation velocities over three sets of measurement. Since other measurements, such as pressure or temperature, are much less accurate, the velocity of a stable detonation has always been regarded as the best criterion to check the validity of a given e.o.s. (Bauer 1985).

A similar type of development based on different values of the constants was proposed by Heuzé (1986).

A better agreement was observed at a higher level of pressure or with another diluent, i.e., helium, with the Percus and Yevick (1958) (P-Y) e.o.s. This e.o.s. has been extensively used in the past for condensed explosives. It is also based on hard spheres modeling (Lebovitz 1964), but it takes into account interactions between molecules of different natures. Some of its parameters have already been experimentally adjusted for a wide variety of both gaseous and condensed explosives (Edwards and Chaiken 1974).

Basically the P–Y e.o.s. is expressed as :

$$\sigma = \frac{pv}{RT} = \frac{1 + Z + Z^2 + Z^3}{(1 - Z)^3} \tag{6}$$

with:

$$Z = \frac{2\pi r^3}{v} \tag{7}$$

where r is the radius of the rigid sphere.

This form is just an analytical summation of all terms of the virial development.

The mixing law for gases is in the form:

$$r^3 = \lambda \sum \sum x_i x_j r_{ij}^3 \tag{8}$$

where x_i, (x_j) is the number of moles of specie i, (j), and r_{ij} is the interaction distance between molecules i and j:

$$r_{ij} = \frac{r_i + r_j}{2} \tag{9}$$

λ is a constant but its value can be adjusted. Calculations were conducted with values of r_i, i.e., the molecular radius of specie i, presented in Table 1 (Edwards and Chaiken 1974) and, for fuel-rich methane-based mixtures, the adjustable parameter was set at λ = 0.5, 0.3, 0.475, and 0.3, for nitrogen, carbon dioxide, argon, and helium, respectively (Legendre 1996).

On the basis of these interaction parameters, an excellent agreement between experimental and calculated values was observed, i.e., a deviation less than 1% . Data related to the detonation characteristics of more energetic systems, namely non-diluted methane and oxygen mixtures, in a pressure range of 1.5 to 4 MPa, showed that a correction of the P–Y e.o.s. was required, keeping identical both the development and thermodynamic data but adjusting the constant at a value: λ =0.5 (Bauer et al. 1994b). To some extent, this is merely a correction of the hard sphere model.

At extremely higher initial pressures, i.e., beyond p_0=20 MPa, the JCZ3 e.o.s. which is derived from studies involving high explosives, turned out to be very reliable (Bauer et al. 1991). Initially introduced by Jacobs (1969) for pure compounds, it was further developed and extended to the general case of mixtures by Cowperthwaite and Zwisler (1976). The equation which has a rather complicated formulation includes two sets of terms. One of them is a virial development, similar to those presented earlier in this paper. This semi empirical e.o.s. is based on an exponential potential and the Helmoltz free energy F is expressed as:

$$F = F^{id} + E_0 + RT \ln f(v, T) \tag{10}$$

where F^{id} is the ideal gas contribution to this energy, E_0 is a volume potential for a solid lattice, and f is the sum of a gaseous and solid contribution. This e.o.s. has been successful in the prediction of detonation characteristics of C, H, N, O high explosives. It can also be used at a significantly lesser initial density, as it will be shown in the following part of this paper.

A similar comment can be made for the Becker–Kistiakowsky–Wilson (Kistiakowsky and Wilson 1941)(BKW) e.o.s. (Mader 1979):

$$\sigma = \frac{pv}{RT} = 1 + X \exp(\beta X) \tag{11}$$

with:

$$X = \kappa \frac{\sum x_i B_i}{v(T+\theta)^\alpha} \tag{12}$$

β, κ, θ, and α are semi-empirical constants that must be adjusted, whereas B_i is the covolume of specie i. One should note that this covolume is not identical to that involved in the virial developments. Again, this form of e.o.s. can be merely regarded as an analytical summation of all the terms included in the polynomial form of a virial development. Ultimately, the appropriate covolumes used in the virial e.o.s. can be used in the case of gaseous detonations. This e.o.s. is currently used for high explosives engineering. However, this paper shows that, as previously demonstrated (Bauer et al. 1985, Heuzé et al. 1986), it can also be used in the present case as long as the parameters are carefully adjusted.

Moreover, the validity of these e.o.s., among others, is thoroughly discussed by Braithwaite et al. (1996a, 1996b) in an attempt to improve the reliability of the prediction of detonation characteristics of explosives. These authors also provide an extensive description and comparison of the main codes where these e.o.s. are used.

Although the Kihira–Hikita–Tanaka (KHT) e.o.s. (Kihira and Hikita 1953) has been also extensively used in the past for high explosives and, to a lesser extent, for dense gaseous ones (Bauer 1985, Heuzé 1985), it will not be presented here.

Schmitt and Butler (1995) performed a thorough study of real gas e.o.s. parameters to upgrade the Chemkin code (Schmitt et al. 1993). It provides a new set of valuable relations applicable to high pressure systems. This important aspect will not be presented in this paper.

The different e.o.s. together with their corresponding range of validity are summarized in Table 2.

Table 1. Molecular diameter of species

Specie	CO_2	H_2	H_2O	He	CO	O_2	N_2	OH	H	O
r_i (Å)	4.076	2.870	3.258	2.500	3.788	3.500	3.715	4.214	3.500	3.500

Table 2. Initial pressure domain of the equations of state

e.o.s.	IG	B	P-Y	C-S	BKW	JCZ3
p_0 (MPa)	0–1	1–6	6–10	6–10	5–100	10–100

2.4 Corrected thermodynamic equation of state Although it does not appear explicitely in the general form of its thermal e.o.s., namely $pv = nRT$, the IG law states that the main thermodynamic parameters are solely dependent on temperature. This property refers to the thermodynamic e.o.s. In the range of density involved in the present study, this statement is no longer appropriate. All thermodynamic parameters must be corrected. This is the generalized form of the thermodynamic equation of state. The most convenient procedure is to split a given parameter Ψ into two parts: a perfect or ideal term and an imperfect, residual or excess term (Byers Brown and Amaee 1992):

$$\Psi = \Psi^{id} + \Psi^{ex} \tag{13}$$

The first term corresponds to non–interacting molecules. The excess term entirely describes the molecule interactions. A general description of this method applied to the ram accelerator analysis is given by Bauer et al. (1998). It is actually correctly suited to most analytical studies involving real gas aspects.

2.5 Analytical and numerical tools for the correction terms in the thermodynamic equation of state The ideal term of the internal energy and enthalpy can be expressed respectively in the form:

$$U^{id} = \int c_v dT \tag{14}$$

$$H^{id} = \int c_p dT \tag{15}$$

The imperfection terms have the following form:

$$U^{ex} = \int \left[T \left(\frac{\partial p}{\partial T} \right)_v - p \right] dv \tag{16}$$

$$H^{ex} = \int \left[v - T \left(\frac{\partial v}{\partial T} \right)_p \right] dp \tag{17}$$

On the basis of the e.o.s. general formulation : $\sigma = pv/nRT$, a series of operators is presented (Heuzé et al. 1986):

$$\sigma_v = v \left(\frac{\partial \sigma}{\partial v} \right)_T \tag{18}$$

$$\sigma_T = T\left(\frac{\partial\sigma}{\partial T}\right)_v \tag{19}$$

$$\sigma_i = \left(\frac{\partial\sigma}{\partial x_i}\right)_{v,T} \tag{20}$$

This convenient formulation yields a simple analytical expression of the respective correction terms of internal energy, enthalpy, entropy, Gibbs free energy, and chemical potential:

$$\frac{U^{ex}}{RT} = \int \sigma_T \frac{dv}{v} \tag{21}$$

$$\frac{H^{ex}}{RT} = \int \sigma_T \frac{dv}{v} + (\sigma - 1) \tag{22}$$

$$\frac{S^{ex}}{R} = \int (\sigma - 1)\frac{dv}{v} + \int \sigma_T \frac{dv}{v} - \ln\sigma \tag{23}$$

$$\frac{G^{ex}}{RT} = -\Phi(v,T) + (\sigma - 1) - \ln\sigma \tag{24}$$

$$\frac{\mu_i^{ex}}{RT} = -\Phi(v,T) + \int \sigma_i \frac{dv}{v} - \ln\sigma \tag{25}$$

where:

$$\Phi(v,T) = -\int (\sigma - 1)\frac{dv}{v} \tag{26}$$

In the specific case where $\sigma_T = f(T)\sigma_v$, one can express:

$$\frac{U^{ex}}{RT} = f(T)(\sigma - 1) \tag{27}$$

$$\frac{H^{ex}}{RT} = [1 + f(T)](\sigma - 1) \tag{28}$$

$$\frac{S^{ex}}{R} = -\Phi(v,T) + f(T)(\sigma - 1) + \ln\sigma \tag{29}$$

where $f(T)$ can be readily derived from the molecular interaction potential.

Here again, it shows that despite the universal formulation involving just a series of mathematical operators which can be applied to any e.o.s., a further analytical calculation still requires the knowledge of the thermal e.o.s. of the substance. This is a fairly straightforward calculation when a simple e.o.s., namely a virial type, is used. It becomes more complex in the case where a complicated form like JCZ3, for instance, is involved.

Furthermore, another series of operators is introduced (Heuzé et al. 1987b):
$\left(\frac{\partial}{\partial v}\right)_T$, $\left(\frac{\partial}{\partial T}\right)_v$, $\left(\frac{\delta}{\delta p}\right)_T$, $\left(\frac{\delta}{\delta T}\right)_p$
where ∂ and δ respectively denote a partial derivative and a finite difference of the variable.

This method allows to perform simultaneously an analytical and a numerical calculation which yields a lesser number of numerical iterations. These operators are related to the dissociation rate of the species in the mixture (Bauer et al. 1998):

$$n_v = \frac{v}{n}\left(\frac{\partial n}{\partial v}\right)_T \tag{30}$$

$$n_T = \frac{T}{n}\left(\frac{\partial n}{\partial T}\right)_v \tag{31}$$

$$N_p = \frac{p}{n}\left(\frac{\delta n}{\delta p}\right)_T \tag{32}$$

$$N_T = \frac{T}{n}\left(\frac{\delta n}{\delta T}\right)_p \tag{33}$$

where n is the total number of moles in the mixture.

This allows the calculation of the calorimetric and thermo-mechanical coefficients. These latter, especially the heat capacities, play a major role in the determination of the thermodynamic states of the combustion products. Furthermore, the simplified ideal gas formulation of Mayer's law: $c_p - c_v = R$, which is no longer valid at a high pressure can be appropriately corrected to account for real gas effects. This point is addressed in a more extensive description given by Bauer et al. (1998). It is beyond the scope of the present paper.

2.6 Calculation of detonation properties The main detonation properties, i.e., detonation velocity and sound speed for instance are based on the parameters that have just been presented, including their appropriate correction.

The analytical calculation of detonation properties proceeds through the classical solution of the conservation equations. This process is very similar to that required for the determination of the ram accelerator performance and, basically, the same equations are used, except for the thrust term which is missing in the momentum equation related to a detonation calculation. At this point, it is convenient to define the following dimensionless parameters:

$$Q = \frac{h_{F1} - h_{F2}}{c_{p1}T_1} \tag{34}$$

$$P = \frac{p_2}{p_1} \tag{35}$$

$$V = \frac{v_2}{v_1} \tag{36}$$

which are the non-dimensional heat release, pressure and specific volume ratios between the initial and final state, respectively.

The parameter η denotes a measure of the calorific imperfection gas behavior, while σ keeps the same meaning as before: $\eta = h/c_pT$; $\sigma = pv/RT$. An algebraic combination of the relationships yields the general form of the Hugoniot (Bauer et al. 1998):

$$P = \frac{2c_{p1}R_1(Q+\eta_1) - (V+1)}{V(\frac{2c_{p2}\eta_2}{\sigma R_2} - 1) - 1} \tag{37}$$

R_1 and R_2 denote the specific values of the gas constant in the initial and final state, respectively.

At this stage, unless otherwise specified, the ideal gas e.o.s. is used to describe the initial properties of the mixtures, thus $\sigma_1 = 1$

The sound velocity is expressed as follows:

$$a^2 = \Gamma RT \tag{38}$$

where Γ, R, and T are the "adiabatic" heat capacity ratio, also named "adiabatic gamma" (Byers Brown and Amaee 1982), the universal gas constant, and temperature, respectively. One may express Γ in the form:

$$\Gamma = \left(\frac{\partial h}{\partial u}\right)_s = -\left(\frac{\partial \ln p}{\partial \ln v}\right)_s \tag{39}$$

where u, and h are the specific internal energy and enthalpy, respectively. This Γ, specifically used in the case of combustion products undergoing a rapid phase or composition change through compression or expansion processes is readily correlated to the usual γ (Heuzé et al. 1986). A complete description of these corrected parameters is given by Bauer et al. (1998) for the calculation of ram accelerator performance.

The rest of the detonation parameters are obtained through the usual relations, and the thermodynamic state variables are accounted for real gas effects like it has just been shown.

3. Thermochemical computer codes

3.1 Available codes Following the early computer program designed by Gordon and Mac Bride (1971), that was aimed at the calculation of the properties of propulsion systems, as well as Chapman-Jouguet (CJ) detonation characteristics, it was shown that an appropriate thermochemical code (Heuzé et al. 1987a) is a valuable tool to compute the state of equilibrium reached for different types of combustion including the CJ detonation. Several equations of state of the combustion products, such as ideal gas and, more importantly, a virial type, must be included, as well as others, especially fit for studies dealing with high explosives. Depending on the level of pressure, i.e., from gaseous explosives at any initial pressure to condensed explosives, and on the type of combustion, i.e., constant volume, constant pressure, or CJ detonation, the most suitable one can be chosen. Either Quartet or Quatuor computer codes, meet these requirements. Calculations have been made for methane, oxygen and nitrogen, at different equivalence ratios, including stoichiometric, as well as mixtures with various amounts of nitrogen and heavier hydrocarbon additives (Bauer et al. 1986). Mixtures such as those currently used in the 90-mm ram accelerator studies at ISL were especially emphasized. The experimental data were used to validate all the parameters of these codes.

The pressure range of validity of the different codes described here is summarized in Table 3. The early code of NASA (Gordon and Mac Bride 1971) is quoted as a reference, but it is based on the IG e.o.s., which does not make it suitable to our present concern. Quatuor, is better suited to extremely high initial pressures (Bauer et al. 1991) since the initial state of the combustible mixture is corrected by means of the R-K e.o.s. Moreover, should this correction be required, the use of Quatuor becomes mandatory to prepare the input for any other code.

Table 3. Which code for which range of initial pressure

Code	NASA	QUARTET	QUATUOR	CHEETAH
p_0 (MPa)	0– 1	0 – 10	0 – 100	0 – 100

3.2 Cheetah: general description and adaptability Among the various codes that are available in the different laboratories, a special emphasize is given in this paper to the Cheetah thermochemical computer code that can be obtained at the Lawrence Livermore National Laboratory (LLNL), University of California (Fried 1996). This code is very flexible and offers a wide variety of calculations, e.g., the Hugoniot, adiabat, CJ state, constant volume explosions. The latest version, namely 1.49 has an extremely user-friendly form. Another very important point is the fact that, as expected, several equations of state are available in the code: IG, BKW, JCZ3, and a virial form, namely, Blake e.o.s. However, according to the validity of BKW as well as JCZ3, this virial form has not been used presently. This version of Cheetah, with the default LLNL library, is mainly appropriate for high explosives. It was shown (Bauer 1995) that the use of the code for dense gaseous explosives required a more suitable library of thermophysical data. The recent version was implemented with these data and it will be shown how this update has turned the code into a very appropriate tool for the specific needs of ram accelerator studies. Unlike high explosive detonation products, those involved in the detonation of dense explosives, as it has already been emphasized, include a series of dissociated species, e.g., O, H, OH, which happen to be in non-negligible amounts. Should these species be ignored, the corresponding results would be strongly over-estimated. Moreover, the molecular interaction data were also adjusted on the basis of existing values that were derived from Quatuor. In other words, the default library, namely that of LLNL, was implemented with additional data from ISL. Most of the values were unchanged, but the main addition was that of H, O, and OH. The corresponding thermophysical data of these components are presented in Table 4, where units are cc/gmol, angström, and K, respectively for BKW covolumes, molecular radius, and molecular interaction energy ratio. These values were derived from the numerous set of data that has been acquired over the last decades.

Table 4. Thermophysical data of added species in the Cheetah/ISL library

Specie	B_i (BKW)	r_i	ϵ/kT
OH	413.0	3.00	100.0
H	76.0	3.01	100.0
O	120.0	3.02	100.0

3.3 Results The calculation was conducted for stoichiometric methane, oxygen, and nitrogen mixtures, namely, $CH_4+2O_2+YN_2$. Actually, ram accelerator propellants are fuel rich but, according to a lack of experimental data related to a stable detonation of this type of mixture, (Bauer et al. 1993,1996) the calculation was not considered here. As a reference, calculations performed with another thermochemical code, namely Quartet, were provided as well. These values together with experimental data are presented in Table 5.

It shows that the calculations made with Cheetah (Ch) and, more especially those based on the ISL library, using either BKW or JCZ3 e.o.s., are in extremely good agreement with experimental data over a wide range of dilutions and initial pressures. As it shows in this table, this includes the domain of ram accelerator propellants as well as that of mixtures at a slightly higher pressure.

The data provided by Quartet (Q), on the basis of the JCZ3 e.o.s., are also in very good agreement with experimental values. One can observe that the P-Y e.o.s. is no longer appropriate beyond p_0 =10 MPa. Neither is the IG e.o.s. for any of these high pressure systems. Nonetheless, the corresponding calculated values are given as a reference.

Cheetah offers a wider variety of possible calculations and, to some extent, it should be preferably employed for our present concern.

4. Summary and conclusion

This paper was aimed at highlighting the key features of a thermochemical code required for the needs of ram accelerator studies. Among the various codes that have now become available, a more specific focus was given to the latest version of Cheetah. It includes several real gas equations of state which are appropriate for most calculations involving ram accelerator propellant mixtures at any initial pressure. Although it has not been described in details in this paper, other very useful calculations can be conducted with this code, such as those yielding specific values of Hugoniot curve or adiabat, mechanical energy, JWL coefficients, in addition to the current CJ characteristics. It is a very flexible code and its latest version has been implemented with a library, named ISL, that includes few more species, i.e., a series of radicals and dissociated products, together with their thermophysical properties. It turns the code into a very flexible and reliable tool for the use of ram accelerator studies.

Table 5. Calculation of detonation velocities (m/s) for $CH_4+2O_2+YN_2$

Y	0	2.11	3.12	4.69	5.18	5.27	3.92	3.93	5.50	5.51
p_0 (MPa)	3.5	3.5	3.5	3.5	3.5	3.5	10	40	10	30
Exp. (average)	2646	2322	2226	2093	2073	2056	2346	3070	2220	2740
IG (Ch/LLNL)	2655	2250	2146	2011	1983	1981	-	-	-	-
P-Y (Q)	-	-	-	-	-	-	2365	3445	2269	2922
Rel. Dev. (%)	-	-	-	-	-	-	-0.81	-12.21	-2.21	-6.64
JCZ3 (Q)	2664	2303	2208	2082	2056	2055	2367	3345	2277	2823
Rel. Dev. (%)	-0.70	0.82	0.81	0.50	0.80	0.02	-0.90	-8.96	-2.57	-3.03
JCZ3 (Ch/LLNL)	2766	2354	2249	2111	2083	2081	2391	3265	2299	2807
Rel. Dev. (%)	4.55	1.38	1.03	0.88	0.51	1.24	-1.92	-6.35	-3.56	-2.45
JCZ3 (Ch/ISL)	2673	2317	2222	2094	2068	2067	2343	3176	2255	2750
Rel. Dev. (%)	1.04	-0.22	-0.18	0.07	-0.22	0.56	0.13	-3.45	-1.58	-0.36
BKW (Ch/LLNL)	2786	2372	2266	2126	2097	2096	-	-	-	-
Rel. Dev. (%)	5.31	2.15	1.80	1.60	1.18	1.97	-	-	-	-
BKW (Ch/ISL)	2708	2339	2242	2111	2084	2083	-	-	-	-
Rel. Dev. (%)	2.36	0.73	0.72	0.88	0.55	1.34	-	-	-	-

References

Austing JL, Selman JR (1988) Computation of Rankine-Hugoniot curve from real gas equation of state. In: Proc 13th Int Pyro Seminar, Chicago, Il, USA

Bauer P, Brochet C (1983) Properties of detonation waves in hydrocarbon-oxygen-nitrogen mixtures at high initial pressures. AIAA Progress in Aero Astro 87:231-243

Bauer P (1985) Contribution à l'étude des caractéristiques de détonation de mélanges explosifs gazeux à haute pression initiale. Doctorat-ès-Sciences dissertation. Univ Poitiers, France

Bauer P, Presles HN, Heuzé O, Brochet C (1985) Equation of state for dense gases. Archivum Combustionis 5:315-320

Bauer P, Presles HN, Heuzé O, Fearnley PJ, Boden JC (1986) Influence of hydrocarbon additives on the detonation velocity of methane-air mixtures at elevated initial pressures. AIAA Progress in Aero Astro 106:321-328

Bauer P, Brochet C, Presles HN (1988) Detonation study of gaseous mixtures at initial pressures reaching 10 MPa. Archivum Combustionis 4:191-196

Bauer P, Dunand M, Presles HN (1991) Detonation characteristics of methane oxygen and nitrogen mixtures at extremely high initial pressures. AIAA Progress in Aero Astro 133:56-62

Bauer P, Legendre JF, (1993) Detonability limits of propellant mixtures used in the Ramac. In: Proc 14th Int Symp on Ballistics, Quebec, Canada, pp 389-398, also ISL Rep # PU361/93

Bauer P, Presles HN, Heuzé O, Legendre JF (1994a) Prediction of detonation characteristics of dense gaseous explosives on the basis of virial equations of state. In: Proc 20th Int Pyro Seminar, Colorado Springs, CO USA, pp 93-106, also ISL Rep # PU308/95

Bauer P, Giraud M, Legendre JF, Catoire L (1994b) Detonability limits of methane-oxygen mixtures at elevated initial pressures. Propellants, Explosives, Pyrotechnics 19:311-314, also ISL Rep # CO222/92

Bauer P (1995) Cheetah. Report LCD-395. Univ Poitiers, France

Bauer P, Legendre JF, Knowlen C, Higgins AJ (1996) Detonation of insensitive dense gaseous mixtures in tubes. In: Proc 32nd Joint Propulsion Conference, AIAA paper 96-2682, also ISL Rep # PU304/97

Bauer P, Knowlen C, Bruckner A (1998) Real gas effects on the prediction of ram accelerator performance. Shock Waves, to be published

Braithwaite M, Byers Brown W, Minchinton A (1996a) The use of ideal detonation computer codes in blast modeling. In: Mohanty (ed) Rock fragmentation by blasting, Balkema, Rotterdam, Netherland, pp 37-44

Braithwaite M, Byers Brown W (1996b) Description of condensed phase detonation based on chemical thermodynamics. In: Explosives, the service of man. The Nobel Heritage, Glasgow, UK, pp 109-120

Buckwalter DL, Knowlen C, Bruckner AP (1998) Real gas effects on thermally choked ram accelerator performance. In: Takayama K, Sasoh A (eds) Ram Accelerators, Springer-Verlag, Heidelberg, pp 125-134

Byers Brown W (1987) Analytical representation of the excess thermodynamic equation of state. J Chem Phys 87:566-570

Byers Brown W, Amaee A (1992) Review of equations of state of fluids valid to high densities. Report # 39/1992, Dept Chemistry, Univ Manchester, UK

Carnahan NF, Starling KE (1969) Equation of state for non attracting rigid spheres. J Chem Phys 31:635-638

Cowperthwaite M, Zwisler WH (1976) The JCZ equations of state for detonation products and their incorporation in the TIGER Code. In: Proc 6th (Int) Symp on Detonation. Office of Naval Research, Arlington, VA USA

Edwards JC, Chaiken RF (1974) Detonation calculations with a Percus-Yevick equation of state. Comb and Flame 22:269-272

Fried LA (1996) Cheetah user's manual. Rep # UCRL-MA-117541

Giraud M, Legendre JF, Simon G, Catoire L (1992) Ram accelerator in 90-mm-caliber : First results concerning the scale effect in the thermally choked propulsion mode. ISL Rep # CO210/92

Giraud M, Legendre JF, Henner M (1998) RAMAC in subdetonative propulsion mode - State of the ISL studies. In: Takayama K, Sasoh A (eds) Ram Accelerators, Springer-Verlag, Heidelberg, pp 65-77

Gordon S, Mac Bride BI (1971) Computer program for calculation of complex chemical equilibrium conditions, rocket performance, incident and reflected shocks and Chapman-Jouguet detonations. NASA Rep # SP-273

Greenlee CL, Butler PB (1993) Influence of product species on thermochemical equilibrium calculation for energetic materials. MS degree dissertation, Univ Iowa, USA

Henner M, Giraud M, Legendre JF, Berner C (1998) CFD computations of steady and non reactive flow around fin-guided RAM projectiles. In: Takayama K, Sasoh A (eds) Ram Accelerators, Springer-Verlag, Heidelberg, pp 325-332

Hertzberg A, Bruckner AP, Knowlen C (1991) Experimental investigation of ram accelerator propulsion modes. In: Takayama K (ed) Shock Waves, Vol I. Springer-Verlag, Heidelberg, pp 17-25

Heuzé O (1985) Contribution au calcul des caractéristiques de détonation de substances explosives gazeuses ou condensées. PhD dissertation. Univ Poitiers, France

Heuzé O, Presles HN, Bauer P (1985) Computation of chemical equilibria. J Chem Phys 38:4734-4737

Heuzé O (1986) Equations of state of detonation products. Phys Rev 34:428-432

Heuzé O, Bauer P, Presles HN, Brochet C (1986) Equations of state for detonation products and their incorporation into the Quatuor code. In: Proc 8th (int) Symp on Detonation. Naval Surface Weapon Center, Silver Spring, MD USA, pp 762-769

Heuzé O, Presles HN, Bauer P (1987a) Quatuor: thermochemical code for computing thermodynamic properties of detonation and combustion products. In: Proc 12th Int Pyrotechnics Seminar. Ed. SERIEP, Paris, France, pp 91-96

Heuzé O, Bauer P, Presles HN (1987b) Compressibility and thermal properties of gaseous mixtures at a high temperature and a high pressure. High Temperatures, High Pressures 19:611-620

Heuzé O, Bauer P (1994) A Simple method for the calculation of detonation characteristics of dense gaseous explosives. In: Proc 20th Int Pyro Seminar, Colorado Springs, CO USA, pp 81-92

Jacobs SJ (1969) On the equation of state for the detonation products at high density. In: Proc 12th Symp (Int) on Comb, p 501

Kee RJ, Rupley FM, Miller JA (1990) The Chemkin thermodynamic database. Rep # SAND87-8215B

Kemp MK, Thompson RF, Zigrang DJ (1975) Equation of state with two constants. J Chem Education, p 802

Kihara T, Hikita (1953) Equation of state for hot dense gases and molecular theory of detonation. In: Proc 4th Symp (Int) on Comb, p 458

Kistiakowski GB, Wilson EB (1941) Report on the prediction of detonation velocity of solid explosives. Office of Sci Res and Dev Rep # OSRD-69

Knowlen C (1991) Theoretical and experimental investigation of the thermodynamics of the thermally choked ram accelerator. PhD dissertation. Univ Washington, Seattle, WA USA

Knowlen C, Bruckner AP (1991) A Hugoniot analysis of the ram accelerator. In: Takayama K (Ed) Shock Waves Vol I. Springer-Verlag, Heidelberg, pp 617-622

Knowlen C, Bruckner A, Bauer P (1996) Ram accelerator operation in the superdetonative velocity regime. AIAA paper 96-0098

Knowlen C, Bruckner AP (1997) Facility upgrade for high pressure ram accelerator experiments. In: Proc 3d Int Workshop on Ram Accelerators, No 9, Sendai, Jpn, July 16-18

Kruczynski D (1993) Analysis of firings in a 120-mm-ram-accelerator at high pressures. In: Proc 1st Int Workshop on Ram Accelerators, Saint-Louis, France

Lebovitz JL (1964) Exact solution of generalized Percus-Yevick equation for a mixture of hard spheres. Phys Rev 133:895-899

Legendre JF, Giraud M, Bauer P (1993) Effect of inert additives on the detonation properties of dense gaseous explosives. In: Proc 1st Int Workshop on Ram Accelerators, Saint-Louis, France. ISL Rep # PU359/93 .

Legendre JF, Giraud M, Bauer P, Voisin D (1995) 90L35 Detonation tube experiments: Influence of diluent nature on the detonation characteristics of dense methane-based gaseous explosive mixtures. In: Proc 2nd Int Workshop on Ram Accelerators. Seattle, WA USA, also ISL Rep # PU348/95

Legendre JF (1996) Contribution à l'étude de la sensibilité et des caractéristiques de détonation de mélanges explosifs gazeux denses à base de methane utilisés pour la propulsion dans les accélérateurs à effet stato. Ph.D dissertation. Univ Poitiers, France

Legendre JF, Giraud M, Bauer P, (1998) Detonation properties of dense methane, oxygen, and diluent gaseous mixtures: Application to ram accelerators. Shock Waves, to be published

Mader CL (1979) Numerical modeling of detonations. Univ California Press, Berkeley, CA USA

Percus JK, Yevick GJ (1958) Analysis of classical statistical mechanics by means of collective coordinates. Phys Rev 110:1-13

Redlich O, Kwong JNS (1949) On the thermodynamics of solutions. An equation of state. Fugacities of gaseous solutions. Chem Rev 44:233-237

Stull DR, Prophet H (1971) Thermochemical Tables. Second Edition. NRDS-NBS37, National Bureau of Standards, Washington DC, USA

Schmitt RG, Butler PB, French N (1993) Chemkin real gas: a FORTRAN package for the analysis of thermodynamics and chemical kinetics in high pressure systems. Rep # UIME-PBB 93-006

Schmitt RG, Butler PB (1995) Detonation properties of gases at elevated initial pressure. Comb Sci and Tech 106:167-191

Facilities and experiments

High acceleration experiments using a multi-stage ram accelerator

J.E. Elvander, C. Knowlen, A.P. Bruckner
Aerospace and Energetics Research Program, University of Washington, Seattle WA 98195, U.S.A.

Abstract. Methods of maximizing projectile acceleration to obtain high velocities with the thermally choked ram accelerator have been investigated. A low mass (50 gm) projectile geometry was developed that can be reliably accelerated up to the Chapman-Jouguet detonation speed of the propellant in a 38 mm bore ram accelerator. Experiments were performed in $CH_4/O_2/He$ propellants wherein Mach number and heat release of combustion were varied to obtain very high projectile acceleration. Using 5.0 MPa fill pressure in a 2-m long first stage, average accelerations of over 38,000 g were routinely achieved, allowing the projectile to be accelerated from 1320 to 1800 m/s. Two and three-stage ram accelerator investigations culminated in a staging configuration that produced an average acceleration of $\approx 35,000$ g over three 2-m long stages, resulting in a peak velocity of 2404 m/s in a distance of 6 m.

Key words: Velocity, Mach number, Acceleration, Mixture map, Propellant

1. Introduction

Since its inception in 1983, the ram accelerator has been presented as a device with the potential to accelerate projectiles to velocities of 8 km/s or higher (Hertzberg, et al. 1988). Based on this potential, a variety of applications have been proposed, including direct space launch (Bogdanoff 1992), aeroballistic ranges, hypervelocity impact studies, and scramjet research (Bruckner, et al. 1992, Naumann and Bruckner 1994). The highest velocity observed in the 16-m long ram accelerator at the University of Washington (UW) to date, however, is 2.7 km/s. The goal of the research presented here was to develop an experimental methodology which enables efficient exploration of ram accelerator propellants and staging configurations, in order to increase the velocity capability of the fixed-length UW facility.

This investigation was focused on the thermally choked propulsive mode, wherein the flow is decelerated to subsonic conditions on the aft section of the projectile and combustion occurs just behind its base, leading to thermal choking at full tube area. Theory indicates that the maximum velocity obtainable in the thermally choked propulsive mode is the Chapman-Jouguet (CJ) detonation speed (Bruckner et al. 1991). Since the maximum CJ speed of combustible gases is ≈ 4 km/s, this limits the ultimate velocity potential of the thermally choked ram accelerator. For superdetonative velocities, it is necessary to utilize supersonic combustion processes on the projectile body (Seiler et al. 1995, Knowlen et al. 1996, Sasoh et al. 1995).

2. Theoretical considerations

The approach to attain high velocity in a ram accelerator described here is based on a straightforward analysis of one-dimensional flow theory applied to the thermally choked propulsive mode. The conservation equations of mass, momentum and energy with steady, inviscid and calorically perfect flow are applied to a control volume surrounding a projectile. Looking at the upstream and downstream flow conditions with a fixed quantity of heat release added to the control volume, yields the thermally choked thrust equation in its simplest form:

$$\tau = \frac{F}{p_1 A} = M_1[2(\gamma+1)(1+\frac{(\gamma-1)}{2}M_1^2+Q)]^{\frac{1}{2}} - (1+\gamma M_1^2) \quad (1)$$

where F is the thrust acting on the control volume, p_1 is the fill pressure, A is the tube cross-sectional area, and τ is the nondimensional thrust. The upstream Mach number is M_1, the ratio of specific heats γ is assumed constant, and the nondimensional heat release is expressed as $Q = \Delta q/c_{p1}T_1$, where Δq is the heat release of combustion, c_{p1} and T_1 are the specific heat capacity at constant pressure and the upstream temperature of the propellant, respectively.

The variation of τ with M_1 and Q is shown in Fig. 1, for a propellant having constant $\gamma = 1.45$. It can be seen that as M_1 increases, τ goes through a maximum and then decreases. In general, τ and the maximum thrust Mach number both increase as Q is increased. For the flow area ratio at the projectile throat (0.42) used in this investigation, ram accelerator operation below $M_1 = 2.6$ is not possible because the flow cannot pass through the throat without choking, thus unstarting the projectile and driving a shock wave ahead of it. Consequently, the ram accelerator generally operates with M_1 greater than the maximum thrust Mach number.

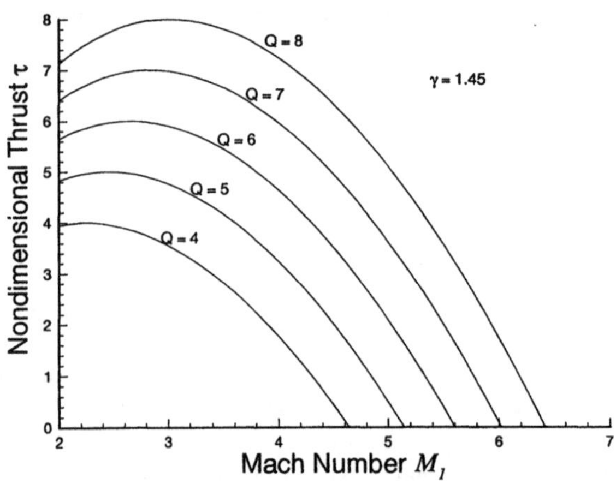

Fig. 1. Nondimensional thrust for various values of Q

Applying Newton's second law yields the following for acceleration: $a = F/m = \tau p_1 A/m$. It is evident that the acceleration acting on the projectile is inversely proportional to projectile mass m and proportional to p_1, A, and τ. The fill pressure is constrained by hardware considerations; i.e., the UW ram accelerator is currently limited to $p_1 = 5.0$ MPa. Similarly, A is fixed as well. The remaining variables, m and τ, are controllable and can be used to optimize the performance.

3. Experimental facility

The experiments described here were performed with the 38 mm bore ram accelerator facility at the UW (Elvander et al. 1996). This facility consists of a helium gas gun, 16-m long test section, and a decelerator section. The ram accelerator test section consists of eight 2-m long tube segments joined with threaded collars. Diaphragms are placed between the tube joints to contain the pressurized propellant, and to separate propellants loaded in adjacent stages. Located every 40 cm along the test section are instrument ports, in which electromagnetic sensors and pressure transducers are placed to determine time-of-passage of the projectile and the flow field pressure at the tube wall. The propellant composition control is based on sonic orifice metering and is supported by gas chromatography. Absolute mixture accuracy is within 5%, and the relative precision when varying mixtures is within 1%.

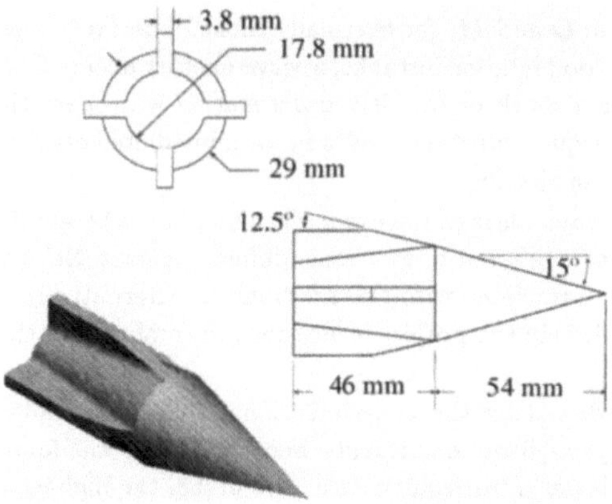

Fig. 2. Experimental low mass projectile

The projectile configuration used in these experiments is shown in Fig. 2. It is fabricated in two parts: a hollow nose and body which thread together. An annular groove, located at the throat where the two pieces join, contains a magnet that enables the projectile to be tracked with the electromagnetic sensors as it moves down the tube. The projectiles were manufactured from aluminum alloy (7075-T6) and had a mass of 50 gm.

4. High velocity methodology

The approach to attain high velocity described here is based on the use of a low mass projectile geometry and short (2-m long) stages of various $CH_4/O_2/He$ propellants at $P_1 = 5.0$ MPa. In addition, the projectile is launched using the maximum breech pressure of the gas gun to provide the highest possible entrance velocity to the test section. The propellant is systematically varied between experiments to determine the thermochemical properties (M_{in} and Q) which enable the highest reliable acceleration. Experimental considerations are discussed in this section, as well as a methodology that minimizes the number of experiments needed to determine the optimal staging configuration.

4.1 Projectile mass reduction The projectile shown in Fig. 2 is the result of experimental efforts to reduce its mass (Imrich 1995). Prior to this development, "standard" UW ram accelerator projectiles had 10° half-angle noses, 71-mm long bodies, and four or five fins (Hinkey et al. 1992). A series of experiments were performed wherein the nose angle, body length, fin number and thickness, nose wall thickness and material were varied to determine how these geometric parameters affect the capability of the projectile to successfully operate in the thermally choked propulsive mode. The peak velocity and average acceleration were found to be very sensitive to variations in the nose cone angle and body length. The projectile geometry chosen for this study was that which had demonstrated the highest acceleration in the subdetonative velocity regime while still being able to reach the CJ speed of the propellant without unstarting. This projectile has a 15° nose cone half-angle and a 46 mm body length, its throat and base cross-sectional areas are the same as the standard UW projectile. These modifications reduced the mass by 34%.

4.2 Propellant variation Although the one-dimensional theory of the thermally choked propulsive mode accurately predicts the performance of a driving ram accelerator projectile, it does not predict if ram acceleration will occur. A projectile will not always operate in any combination of fuel, oxidizer, and diluent. The limitations on Q and M_1 for thermally choked operation have been explored by Higgins et al. (1993). If Q is too high, for instance, a wave unstart occurs from the pressure rise behind the projectile driving a shock or overdriven detonation wave past the projectile throat (Shultz et al. 1998). The subsequent unstart results in an immediate cessation in thrust and sometimes the destruction of the projectile.

Unstarts can also occur if the flow over the projectile is at too low a Mach number, wherein the flow reaches the sonic condition before the throat; referred to as a sonic diffuser unstart (Schultz et al. 1998). Detonation waves are not typically associated with this mechanism. Alternatively, it is possible for the heat release to be too low, wherein the projectile outruns the combustion; this is called a wave fall-off.

Thus, limitations exist which are not predicted by the one-dimensional thermally choked theory. Such limitations are dependent on the propellant constituents. For instance, it was found that in a 2.5-MPa propellant having $CH_4/O_2/N_2$ at a fuel equivalence ratio of 2.8, the highest Q which can propel a projectile for over 2 m was $\sim$ 5.5 (Higgins et al. 1993), whereas for 2.5-MPa propellant of stoichiometric CH_4 and O_2 diluted with CO_2 (Hertzberg et al. 1988), the highest Q was $\sim$ 8.5.

4.3 Mixture maps As discussed above, the nondimensional thrust, τ, is a function of upstream Mach number, M_1, and Q. If the entrance velocity, V_{in}, into the ram accelerator stage is known, then τ is primarily a function of the propellant chemistry, which determines Q and the acoustic speed. These two parameters are closely coupled as a result of this chemistry dependence. By utilizing a mixture map, it is possible to visualize several variables' dependence on propellant chemistry (Elvander et al. 1996, Elvander 1997).

Figure 3 shows such a map, for a 5.0 MPa propellant of $CH_4/O_2/He$ having V_{in} = 1320 m/s, in which lines of constant Q (iso-Q), M_{in} (iso-M_{in}), and acceleration (iso-g) are plotted as a function of mixture composition. A point in the plane corresponds to a specific molar ratio of CH_4 and He with respect to $2O_2$. It can be seen from this mixture map that as the amount of dilution is decreased and the propellant is made less fuel rich (moving to the lower left of the plane), Q increases.

With V_{in} fixed, the entrance Mach number, M_{in}, becomes a function of mixture composition and, hence, location in the CH_4-He plane. As the amount of helium is increased in the propellant, the acoustic speed increases which causes M_{in} to decrease; whereas adding CH_4 has the opposite effect, as evident by the slope of the iso-M_{in} lines in Fig. 3.

Using Eq. (1), it is possible to predict the thrust acting on a projectile of known mass in a given propellant. The ultimate goal, however, is to maximize the velocity gain per length of tube, which requires the maximum average acceleration feasible. As shown in Fig. 1, the thrust decreases as the Mach number increases, which motivated the use of an average acceleration, a_{ave}, over a distance of 2 m to compare the potential performance of different propellant formulations in the mixture map. Given m and p_1, a_{ave} for a 2-m long ram accelerator stage is then a function only of the propellant chemistry, and can be plotted as an iso-g contour on the mixture map, as shown in Fig. 3. The nearly vertical iso-g curves indicate that a_{ave} in this propellant is much more sensitive to variations in CH_4 concentration than He.

A mixture map with plots of iso-g lines can be used to determine how to vary the propellant in order to achieve higher velocity in the thermally choked propulsive mode; i.e., move perpendicular to the iso-g lines in the direction of higher acceleration. Once the Q and M_{in} limitations for a particular propellant are experimentally determined, the mixture map enables one to determine what fuel-diluent ratios will yield higher a_{ave} while maintaining the required values of Q and M_{in}. The acceleration performance can then be maximized by conducting experiments with ever

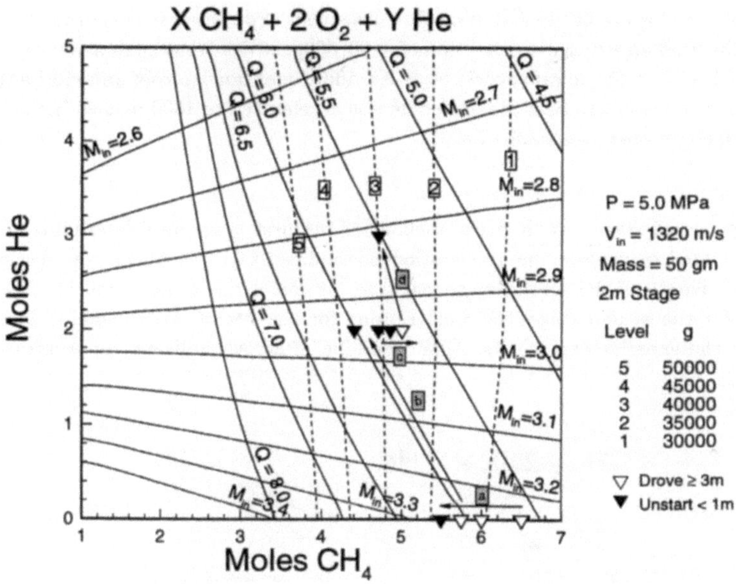

Fig. 3. $CH_4/O_2/He$ mixture map with iso-Q, iso-M_{in}, and iso-g lines. Plotting symbols indicate results for first stage experiments

decreasing changes in chemistry. The application of mixture maps is discussed in greater detail in Elvander et al. (1996) and Elvander (1997).

5 Experimental results

Mixture maps facilitated the development of an experimental configuration that enabled high velocity ram accelerator operation in the UW facility. In lieu of an ideal propellant gradient, stage lengths of 2 m were used to keep the M_1 variation in each stage as small as possible, in order to maximize a_{ave}. This investigation explored the operating characteristics of three different stages, the results of which are presented here.

5.1 High velocity first stage The first set of experiments was conducted to find a $CH_4/O_2/He$ propellant in which the ram accelerator could be started with projectiles having an entrance velocity $V_{in} = 1320$ m/s. The propellant for each experiment was loaded in the first 8 m of the test section so that its full operating Mach number range could be examined. Experimental results are plotted on the mixture map discussed in previous section (Fig. 3). Open triangles represent experiments that successfully accelerated for over 2 m, whereas the solid triangles indicate that an unstart occurred before the projectile left the first tube. Propellants using just CH_4 and O_2, without any He diluent, were investigated by reducing the excess CH_4 content until an unstart occurred in the first 2 m (arrow (a)). Helium was then added and CH_4 was removed to decrease M_{in} while keeping Q relatively constant (arrow (b)). Starting of the ram accelerator could not be accomplished with this propellant until Q was lowered 5% by the addition of more CH_4 (arrow (c)). A further attempt to reduce M_{in} while increasing the potential acceleration was unable to successfully start the ram accelerator (arrow (d)).

The highest average acceleration over the first 2 m was the criterion used to choose the propellant for the starting stage (Elvander et al. 1996). Since the mixture map (Fig. 3) indicated that no further substantial gains in a_{ave} seemed possible by varying the chemistry, the $5CH_4$+$2O_2$+$2He$

propellant was selected for use in the first stage for subsequent experiments. This propellant provided the projectile with an average acceleration of $\approx 38,000g$, which resulted in a velocity boost of 480 m/s in the first 2 m of the ram accelerator. An additional bonus, from an aerothermodynamic heating point of view, was that the projectile was accelerated to 1800 m/s without having to operate at a Mach number greater than 4.

5.2 High velocity second stage With the propellant of the first stage now determined, experiments were conducted to explore the acceleration possibilities of a two-stage ram accelerator configuration. The length of the second stage was 6 m for this series of experiments. Based on V_{in} = 1800 m/s for the second stage, the mixture map for this set of experiments is shown in Fig. 4 along with the experimental results. Again the labeled arrows indicate the progression of experiments.

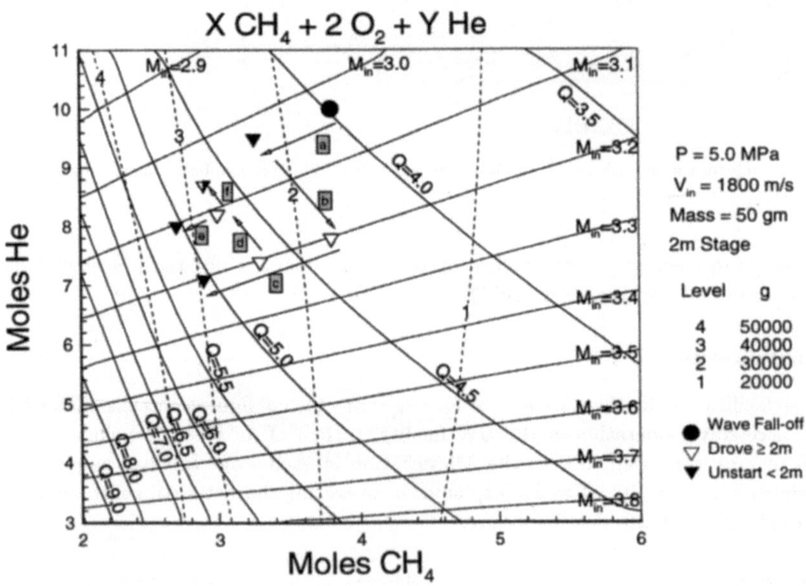

Fig. 4. Mixture map and results of second stage high velocity experiments

Since the operating limits in the second stage were not known a priori, the second stage propellant was initially chosen to have Q and M_{in} properties that were within the operating regime demonstrated by the high velocity first stage propellant. As it turned out, this test propellant was not energetic enough to support ram acceleration when the projectile underwent the transition from the first stage, and a wave fall-off was observed (circular plotting symbol in Fig. 4). Increasing Q while staying on the same iso-M_{in} line (arrow (a)) resulted in an immediate unstart after transition. It was theorized that perhaps the Mach number was too low and a M_{in} limit was being observed, so the M_{in} was increased while keeping Q constant (arrow (b)). This experiment, at M_{in} = 3.2, was successful. Then Q was increased while keeping M_{in} constant (arrow (c)), yielding higher average acceleration levels, until a wave unstart was observed. After which the M_{in} for the second stage was reduced along an iso-Q line (arrow (d)), resulting in an even higher a_{ave} being demonstrated.

In the last attempts at higher acceleration in the second stage, experiments were conducted along the iso-M_{in} and iso-Q lines that intersect at the coordinates of the highest performance propellant found so far (Fig. 4). The experiment along the iso-M_{in} line (arrow (e)) generated a wave unstart within 2 m of the stage entrance, thus confirming that the limiting Q for these conditions was ≈ 4.7. Along the iso-Q line (arrow (f)), however, there were mixed results in that one experiment successfully accelerated the projectile for over 2 m with a_{ave} = 42,000 g, yet a wave unstart occurred when a repeat attempt was made. Thus semi-solid triangle was used to represent these experiments, and it was decided that this propellant was too unreliable for subsequent use.

The propellant used in the successful experiment at the intersection of arrows (d), (e), and (f) in Fig. 4 was chosen for the second stage. This experiment was repeated for validity. This propellant, $3CH_4+2O_2$ +8.2He, produced $a_{ave} \approx$ 36,000 g, which resulted in a velocity gain of 350 m/s in the 2-m long second stage. Note that in just ten experiments, a high-performance second stage propellant was found, and the limiting Q and M_{in} values for thermally choked ram accelerator operation were determined. This experimental series demonstrates the usefulness of mixture maps to guide the development of new propellants and staging configurations.

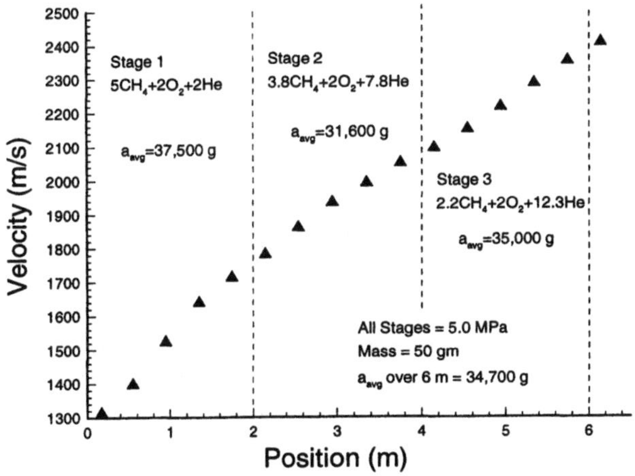

Fig. 5. High velocity third stage experiment

5.3 High velocity third stage Some preliminary work was performed to develop a high velocity third stage. In one experiment, a 4-m long third stage propellant composed of $2.2CH_4+2O_2$ +12.3He, with $Q = 4.7$ and $M_{in} = 3.2$, successfully accelerated a projectile for over 2 m; however, it could not be repeated (Elvander et al. 1997). The velocity-distance data from the successful experiment are shown in Fig. 5. It was found that a less energetic propellant, $3.8CH_4+2O_2$+7.8He, had to be used in the second stage when its length was shortened to 2 m, which reduced a_{ave} and lowered the corresponding velocity gain by 50 m/s. The a_{ave} in the third stage was $\approx$ 35,000 g and the velocity gain in 2 m was 300 m/s, resulting in a peak velocity of over 2400 m/s in this experiment.

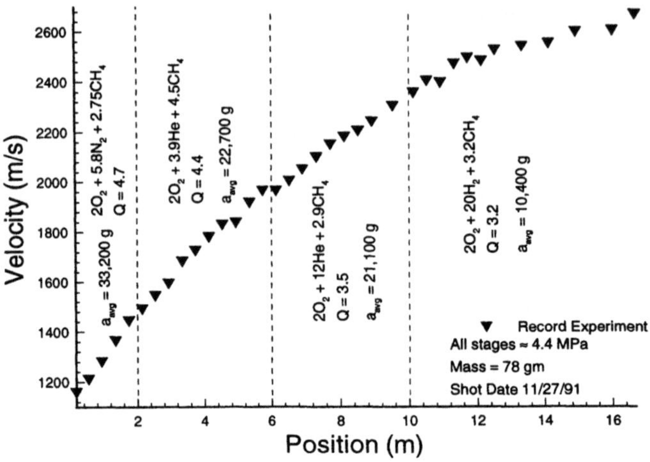

Fig. 6. Record velocity ram accelerator experiment

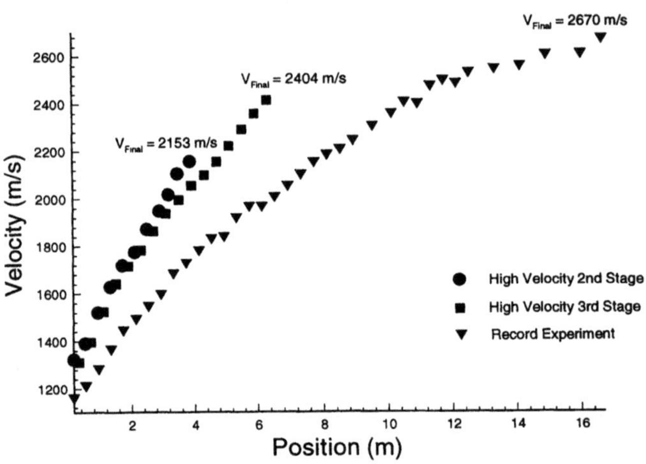

Fig. 7. Comparison of high velocity experiments to record experiment

6 Discussion

It was observed in the two-stage experiments that a_{ave} for the 2-m long first stage was $\approx 5\%$ less than demonstrated when the propellant was loaded into 8 m of tube. The reasons the performance of 2-m long, first and second stages was not observed when the propellant was loaded in longer lengths of test section are not totally understood at this time. The propellant gas handling system has since been modified and future investigations of this issue are planned.

Data from the highest velocity (2670 m/s) experiment to date in the UW ram accelerator facility are shown in Fig. 6. The projectile in that experiment had a mass of 78 gm and was accelerated through four stages containing propellants having increasing acoustic speed and decreasing Q. It was found in these early multi-stage experiments that operation at high velocities was only possible after reducing the propellant Q in the third and fourth stages. Thus, the acceleration average for the 16-m long test section was much less than that generated in the first few meters of ram acceleration. For the experiment of Fig. 6, the average acceleration in the fourth stage was approximately one-third that observed in the first.

The experimental necessity of reducing Q after the projectile has been accelerated by multiple ram accelerator stages is most likely the result of trying to operate with a severely eroded projectile (Hinkey et al. 1992). The thermochemistry of the high acoustic speed propellants may also play a role by enhancing the kinetic rates sufficiently to warrant the addition of more diluent to prevent premature propellant ignition. Another hypothesis is that the aerothermodynamic heatnig experienced by the projectile, prior to its entry to high velocity stages, may have increased its surface temperature to the point where the hot walls play a significant role in the propellant ignition (Seiler et al. (1995), Chew and Bruckner 1994). These are topics in need of more investigation.

A comparison of the best results from the recent two- and three-stage experiments with the record velocity data is shown in Fig. 7. The increased accelerations due to the use of low mass projectiles (50 vs. 78 gm) and higher fill pressure (5.0 vs. 4.4 MPa) are evident in the steeper slopes of the velocity-distance data from the more recent experiments. Reducing projectile mass also enabled the use of higher entrance velocity to the first ram accelerator stage. These factors have resulted in the projectiles reaching a velocity of $\approx$ 2150 m/s in 4 m and $\approx$ 2400 m/s in 6 m of ram acceleration for the two- and three-stage test section configurations, respectively, whereas the record velocity experiment required almost 12 m of test section to achieve 2400 m/s. Since an average acceleration of only 24,000 g is needed to attain 3000 m/s in the length available in UW facility, the potential to reach this milestone is very high as indicated by the results of this investigaiton.

7 Conclusions

A method for attaining very high velocities with the thermally choked, subdetonative ram accelerator has been described, along with prior and current experiments. Using the thermally choked one-dimensional theory, mixture maps have been developed which allow one to readily identify paths to high acceleration operation. The mixture map proved to be a significant tool to aid the experimental development of a first stage propellant ($5CH_4$+$2O_2$+2He) that could accelerate a projectile from 1320 to 1800 m/s. A second stage propellant ($3CH_4$+$2O_2$+8.2He) was also found using this approach, which propelled the projectile to 2150 m/s with an average acceleration of $\approx 37,000$ g over 4 m. A three-stage configuration, using $2.2CH_4$+$2O_2$+12.3He in the third stage, was able to propel the 50 gm projectile to a velocity greater than 2400 m/s in 6 m, thus demonstrating an average acceleration of $\approx$ 35,000 g and halving the length previously required to attain this velocity.

Acknowledgments

This work was supported by the U.S. Army Research Office under AASERT grant DAAL03-02-G-100.

References

Bogdanoff DW (1992) Ram accelerator direct space launch systems: New concepts, J Prop and Power 8:481-490

Bruckner AP, Knowlen C, Hertzberg A (1992) Applications of the ram accelerator to hypervelocity aerothermodynamic testing. AIAA paper 92-3949

Bruckner AP, Knowlen C, Hertzberg A, Bogdanoff DW (1991) Operational characteristics of the thermally choked ram accelerator. J Prop and Power 7:828-836

Chew G, Bruckner AP (1994) A computational study of the nose heating in the ram accelerator. AIAA paper 94-2964

Elvander JE, High velocity operation of the thermally choked ram accelerator. MSAA thesis, Dept Aero Astro, Univ Washington

Elvander JE, Knowlen C, Bruckner AP (1996) High velocity performance of the thermally choked ram accelerator. AIAA paper 96-2675

Elvander JE, Knowlen C, Bruckner AP (1997) Recent results of high acceleration experiments in the University of Washington ram accelerator. AIAA paper 97-2650

Hertzberg A, Bruckner AP, Bogdanoff DW (1988) Ram accelerator: A new chemical method for accelerating projectiles to ultrahigh velocities. J AIAA 26:195-203

Higgins AJ, Knowlen C, Bruckner AP (1993) An investigation of ram accelerator gas dynamic limits. AIAA paper 93-2181

Imrich TS (1995) The impact of projectile geometry on ram accelerator performance. MSAA thesis, Dept Aero Astro, Univ Washington

Knowlen C, Higgins AJ, Bruckner AP, Bauer P (1996) Ram accelerator operation in the superdetonative velocity regime. AIAA paper 96-0098

Naumann KW (1996) Thermomechanical constraints on ram accelerator projectile design. AIAA paper 96-2678

Naumann KW and Bruckner AP (1994) Ram accelerator ballistic range concept for softly accelerating hypersonic free-flying models. J Aircraft 31:1310-1316

Sasoh A, Knowlen C, Bruckner AP (1995) Effect of finite rate chemical reactions on ram accelerator thrust characteristics. AIAA paper 95-2492

Schultz E, Knowlen C, Bruckner AP (1998) Overview of subdetonative ram accelerator process. In: Takayama K, Sasoh A (eds) Ram Accelerator, Springer-Verlag, Heidelberg, pp 189-203

Seiler F, Patz G, Smeets G, Srulijes J (1995) Gasdynamic limits of ignition and combustion of a gas mixture in ISL's RAMAC30 scram accelerator. In: Sturtevant B et al. (eds) Proc 20th Int Symp Shock Waves, World Scientific, Singapore, pp 1057-1062

RAMAC in subdetonative propulsion mode: State of the ISL studies

M. Giraud - J.F. Legendre - M. Henner
ISL, French-German Research Institute of Saint-Louis, BP 34, F-68301, Saint-Louis Cedex, France

Abstract. The RAMAC is an aerothermochemical launcher aimed at accelerating a projectile up to hypersonic velocities under smooth conditions. In most cases, the ram tube is a smooth cylindrical bore, filled with a dense gaseous reactive mixture in which is injected a supersonic under-calibrated projectile. The mixture composition and projectile shape are two key-elements for the success of the ram process. Composed preferably by an axisymmetric biconical body guided by fins, the projectile shape has been designed in order to control the flow expansion once it has been compressed at the nose, to trigger the combustion and control its appropriate location at the rear of the projectile in the thermally choked propulsion mode, also referred to as subdetonative propulsion mode. Two smooth-bore experimental tools, a RAMAC 30 and a RAMAC 90, in 30 and 90 mm caliber respectively, are alternately used with the same gas handling, gas chromatography and data acquisition system in order principally to study the scaling effects. These two ram facilities recently received some important modifications among which the installation of new preaccelerators allowing enhanced RAMAC entrance performances, i.e. the launching of higher masses at the same initial velocity of about 1340 m/s.

The present state highlights (1) the effect of the diffuser geometry on the RAMAC performances, i.e., the importance of the profile, size and number of the guiding fins, and also the effect of the Mach number; (2) the reproducibility of the RAMAC process by using a multistage ram, and the efficiency of the latter to obtain higher velocities with a "quasi-constant" Mach number; (3) the necessity to maintain the Mach number at a relatively low level to avoid too fast an ablation and erosion of the fins due to high thermal and mechanical ram loads and finally to obtain a better impulse.

Key words: RAMAC, Subdetonative, Experiments, Performances, Metrology

1. Introduction

The feasibility of the ram accelerator as a means of achieving high velocities has been extensively demonstrated in 38 mm caliber at the University of Washington (Seattle, WA, USA) since 1986 (Hertzberg et al. 1986). Experiments conducted at ISL with a RAMAC 90 have proved the scalability of a ram accelerator in the thermally choked propulsion mode or subdetonative propulsion mode (Giraud 1992, Giraud et al. 1992a, 1992b, 1993a, 1993b, 1993c, 1995b), In order to study very carefully the scaling effects for the whole ram process, taking the geometrical, physical and chemical conditions into consideration, experiments are carried out with two smooth-bore RAMAC facilities, in 30 mm and 90 mm calibers, or RAMAC 30 and RAMAC 90, respectively. The mixture composition and projectile shape are two key-elements for the success of the ram process.

Concerning the reactive mixture, a thorough knowledge of the sensitivity to detonation of different dense gaseous mixtures as well as a validated method to compute combustion characteristics provide a valuable tool for the choice of the most appropriate gas mixture (Legendre 1996), (Henner et al. 1997a).

To date, CFD computations (3D Navier-Stokes code) have been conducted in steady and non-reacting flow (Henner et al. 1997b) in order to determine the most appropriate projectile shape for the comparative experiments necessary for a better understanding of the scaling.

In this paper, the results of experimental investigations in the subdetonative propulsion mode are given for an appropriate reactive gas mixture, the entrance velocity of the projectile being about 1335 m/s (Mach number 3.8), the central part of the latter and the sabot remaining unchanged. The parameters are only the filling pressure (in the range 3 - 4.5 MPa), the shape (with or without a chamfered leading edge), the size and the number of fins (3, 4 or 5). No on-board or external ignition devices have been used. The ignition of the reactive gas mixture at the rear of the projectile is achieved by a natural process.

Fig. 1. RAMAC 90 installation

2. Experimental Conditions

2.1 The RAMAC facilities They are located in the launch room of the aeroballistic range facility at ISL (Fig. 1). Both are mainly composed of the ram accelerator itself, the gas handling system, the data acquisition system, the gas chromatography system and the X-rays radiography system (Fig. 2).

The RAMAC 30 is a modular launcher, every module being a 3-m-long smooth-bore. The preaccelerator is a conventional 30L100 (i.e. 30 mm caliber, 100 calibers long) smooth-bore powder gun which can accelerate a total mass of 0.3 kg (projectile and sabot) at velocities up to 1400 m/s. The peak acceleration is below 40 kG's. The next element connected to the gun muzzle is a 2.25-m-long interface organized in three parts like the one used in 90 mm caliber. The ram section itself is composed of two modules (for a total length of 200 calibers) connected to the interface (Fig.3).

The RAMAC 90 is also a modular launcher, every module being a 3-m-long smooth-bore. The pre-accelerator is a conventional 90L60 smooth bore powder gun which can accelerate a total mass of 2.5 kg at velocities up to 1800 m/s. The peak acceleration is below 45 kG's. The 4-m-long

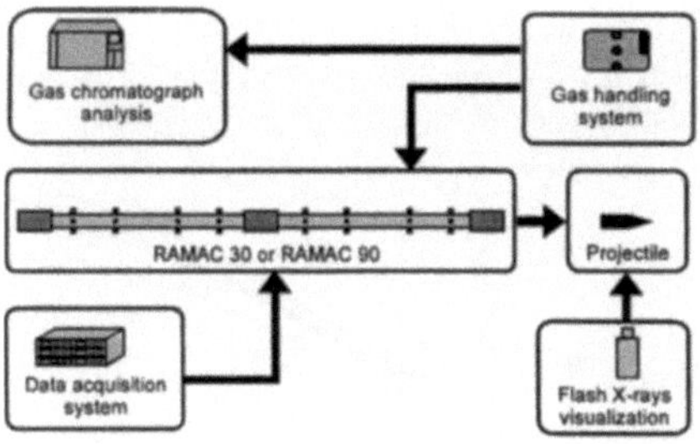

Fig. 2. RAMAC experimental setup

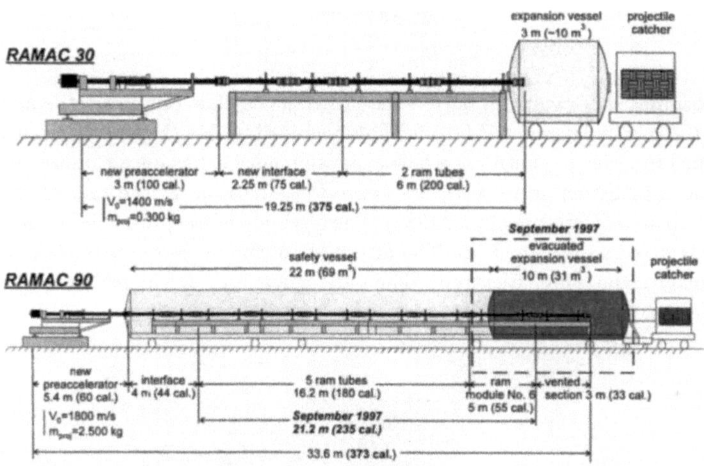

Fig. 3. ISL smooth-bore RAMAC facilities

Fig. 4. Fin-guided ram projectile

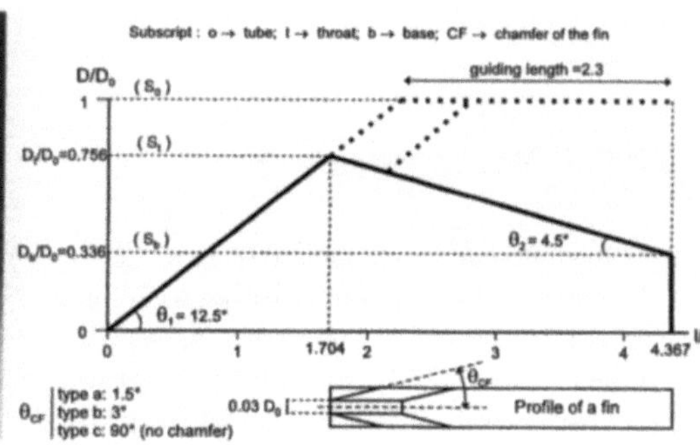

Fig. 5. Basic shape of the projectile

interface organized in three parts is largely described in (Giraud et al. 1992b, 1993c). Finally, the ram section itself is composed of 5 modules, with a total length of 180 calibers which is relatively close to the RAMAC 30 value, condition necessary for the study of the scaling effect. Its length will be increased, by September 1997, to 235 calibers by adding a 6th module (55-caliber-long) (Fig.3).

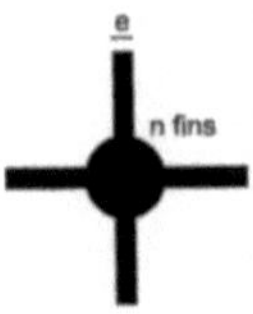

A → RAMAC30 B → RAMAC90

Fin Model No.	Number of fins n	e [mm] A	e [mm] B	e (cal D_0)	θ_{CF} [°]
1	3	4.7	14	(0.156)	a) 1.5, b) 3, c) 90
2	4	3	9	(0.100)	c) 90
3		4.7	14	(0.156)	a) 1.5, b) 3, c) 90
4	5	2.4	7.2	(0.080)	c) 90
5		3	9	(0.100)	a) 1.5, b) 3, c) 90
(6)		4.7	14	(0.156)	c) 90

(Model No. 6 has not been tested)

Fig. 6. Fin models

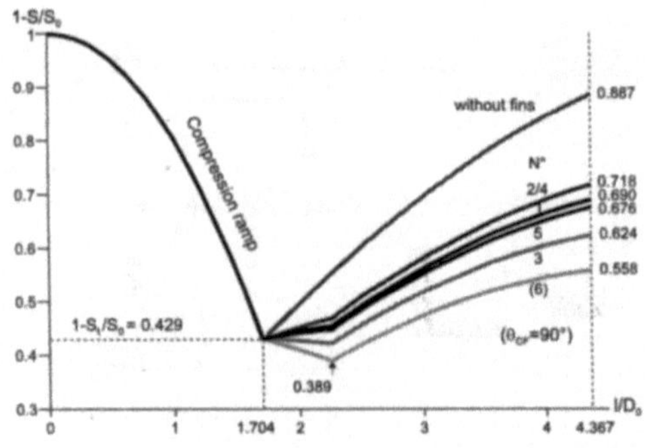

Fig. 7. Diffuser area 1-S/S_0 for $\theta_{CF} = 90°$, S=cross sectional area

2.2 The projectile It is composed of two main parts : the projectile itself or ram projectile and its sabot. The ram projectile is an axisymmetric biconical body guided by fins (Fig. 4). Basically, the shape has been designed in order to control the flow expansion once it has been compressed at the nose, to trigger the combustion and control its appropriate location at the rear of the projectile in the recirculation zone (Hertzberg et al. 1986). The forebody of the projectile is made of aluminium alloy (Dural) fitted with a steel tip. The afterbody of the projectile is made of a low-density material : a magnesium alloy (Elektron). The basic shape and size in caliber of the ram projectile are given in Fig. 5.

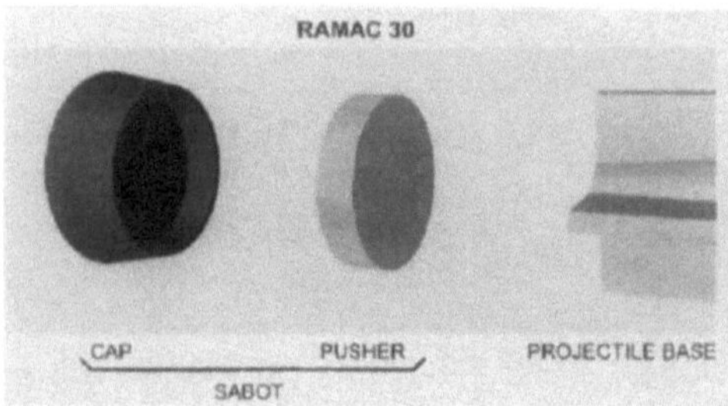

Fig. 8a. RAMAC 30 sabot configuration

Fig. 8b. RAMAC 90 sabot configuration

Each fin can be chamfered ($\theta_{CF} = 1.5°$ or $3°$) or not ($\theta_{CF} = 90°$). For each model considered, the number of fins and their thickness are indicated in Fig. 6. The mass of the projectile equipped with the fin model No. 2c or No. 4c is 60 g in small caliber (RAMAC 30) and 1.340 kg in large caliber (RAMAC 90) respectively. The total length according to the caliber is 131 mm and 393 mm respectively. The afterbody of the projectile being made of magnesium alloy and in order to take into account the material of the fins affected at their rear by the combustion products (length = 2/3 cal.) and also on their leading edges by the ablation (length = 1/3 cal.), the guiding length is 2.3 calibers (Giraud et al. 1995b). The balloting motion of the projectile noted with a short projectile (Giraud et al. 1993b) has been therefore drastically reduced. The residual yaw angle can reach 1° instead of 4° previously. A 2 mm deep step machined at the top of the leading edge of every fin (Fig. 4) of the RAMAC 90 projectile is used as cutting tool for the numerous thick PVC diaphragms (Giraud et al. 1993c). The diffuser area of the basic shape, 1-S/S_0 (without the fins) and its distribution along the afterbody are modified by the presence of the fins. The position of the throat can be situated at different locations according to the fin model (Fig. 7).

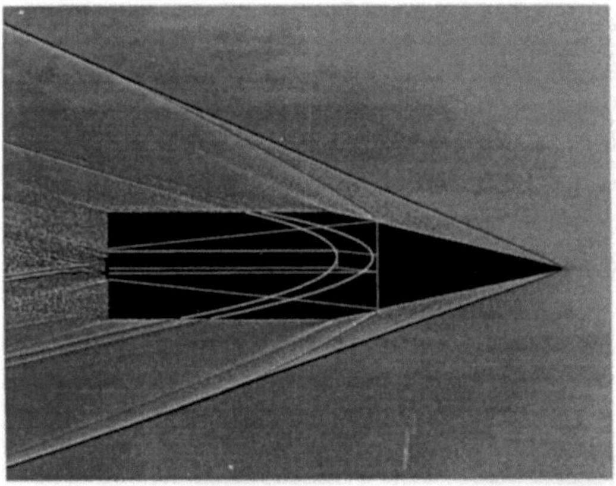

Fig. 9. RAMAC 30 projectile in free flight at Mach=3.4, fin model No 2c : 4-fin, $\theta_{CF} = 90°$, e=0.1D_0

The reduction of the diffuser area can reach about 37 % at the base of the projectile and the corresponding radius of the equivalent circular body is increased by ratio 2. Consequently, the ram performances can be affected by the effects of the fins due to their number and their geometrical characteristics on the flow conditions (pressure, temperature) (Henner et al. 1997b).

Strickly necessary for the launching in the preaccelerator, the central body of the projectile being subcalibrated, the sabot is also used as ignitor of the reactive gas mixture in the first ram tube (Hertzberg et al. 1986), (Giraud et al. 1995a, 1995b). The success of the ram starting by natural process and its good reproducibility depend on the distance between the projectile and the sabot. This distance depends on the residual pressure at the muzzle of the preaccelerator and on both length and design (vents, vacuum, etc...) of the interface. Afterwards, it will also depend on the initial pressure in the first ram tube. Consequently, the design of the sabot is different according to the caliber. The basic and empirical rule is the respect of a distance between the projectile and its sabot at the RAMAC entrance equal to about 2/3 caliber. This distance becomes about 1 caliber at the ignition time. A good reproducibility of the starting process has been obtained with the two following sabot designs by taking the preaccelerator performances and the interface characteristics into consideration. The natural process of the ignition is obtained (Giraud et al. 1995b) by the excellent coupling between the adequate aerodynamic conditions (pressure, temperature) just at the rear of the projectile in the reactive gas mixture and the local conditions (shock induced) : in front of the pusher (full disk) aerodynamically separated from the projectile in 30 mm caliber (Fig. 8a) or in the central part of the ring mechanically separated from the projectile 1.25 m before the RAMAC entrance in 90 mm caliber (Fig. 8b, Giraud et al. 1995a). The masses of projectile and sabot are given in Table 1. We note that the theoretical values of the masses given by the geometrical scale are not respected. The masses of the projectile and of the sabot in large caliber are proportionally smaller than the ones used in small caliber, respectively 1.340 kg instead of 1.620 kg for the projectile and 340 g instead of 864 g for the sabot.

The shadowgraph (Fig. 9.) taken 10 meters after the RAMAC 30 muzzle shows an excellent ballistic attitude of the projectile in free flight and highlights the complex structure of the flow around the afterbody due to the fins. If any balloting motion of the projectile occurs in the ram tubes, it is caused by the torque created by the thrust which is not rigorously centred in its base. The basic composition of the ternary reactive gas mixture chosen for the present experiments is $3CH_4+2O_2+10N_2$. It is not detonable (Giraud et al. 1993a, 1993b, 1995b) according to experiments carried out in detonation tubes with direct condensed explosive initiation and confirmed in RAMAC 90 with normal shock wave generated by a full-caliber piston initiation technique (Legendre 1996, Legendre et al. 1997). The realization of the reactive gas mixture in various

Table 1. RAMAC 30 and RAMAC90 projectiles characteristics, fin models No 2c or No 4c

	RAMAC 30	RAMAC 90	Ratio	Theoretical values
Caliber	30mm	90mm	λ=3	/
Mass of the projectile	60g	1.340kg	22 instead of λ^3=27	1.620kg
Mass of the sabot	32g	0.340kg	10.6 instead of λ^3=27	0.864kg
Sabot/projectile ratio	53%	25%	/	53%
Total mass	92g	1.680kg	18 instead of λ^3=27	2.484kg

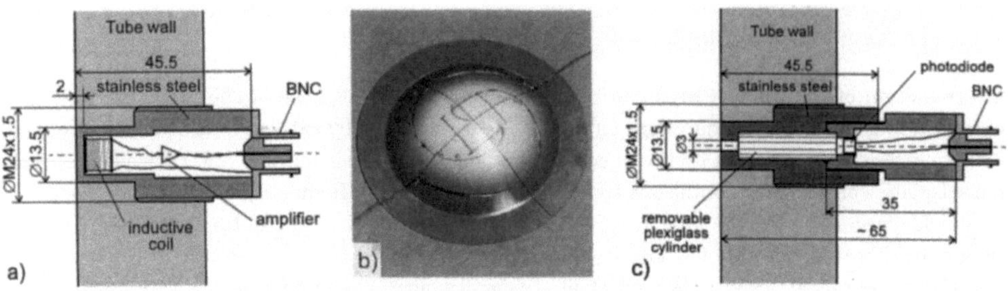

Fig. 10. a, Magnetic transducer; b, instrumented diaphragm; c, optoelectronic gauge

amounts of each component is achieved by means of remote computer-controlled thermal mass-flow regulators. Samples are collected during the filling phase and analyzed afterwards by gas chromatography (Giraud et al. 1993c). The analysis provides the composition of the gas mixture with a relative precision of 2 % on the equivalence ratio and 0.5 % on the inert dilution ratio. To date, the RAMAC 30 has been organized only in single stage and the RAMAC 90 either in single stage or in two stages (two successive gas mixtures separated by a 3 mm thick PVC diaphragm).

2.4 The metrology: instrumentation and diagnostic techniques To date, the number of instrumented cross-sections is respectively 12 in 30 mm caliber and 44 in 90 mm caliber (Giraud 1992, Giraud et al. 1992b, 1993c, Simon et al. 1995). The corresponding number of instrumented ports is 26 and 127 respectively.

The projectile and the pusher of the sabot are both equipped with a magnetic ring, the shape and size being different with respect to the caliber. Accordingly, their trajectories, i.e. distance (x) - time (t) histories, are unambiguously reconstructed through magnetic transducers (Fig. 10a). The distance between the base of the projectile and the front of the sabot is known at each time, particularly at the RAMAC entrance time and at the ignition time. By means of a classical mathematical smoothing method, velocities and accelerations are deduced from these experimental values (x, t) obtained prior to and just after the RAMAC entrance and during the whole process in the RAMAC stages.

2.3 The reactive gas mixture

The RAMAC 90 stages being obturated at both ends with thick diaphragms made of PVC, the latter have been instrumented by a very simple device (a copper thread mounted in a double spiral, Fig.10b) in order to determine through their bursting conditions, either by the nose of the projectile or by an "air" piston in front of the projectile, the type of the flow around the projectile (a started diffuser or an unstarted diffuser, respectively).

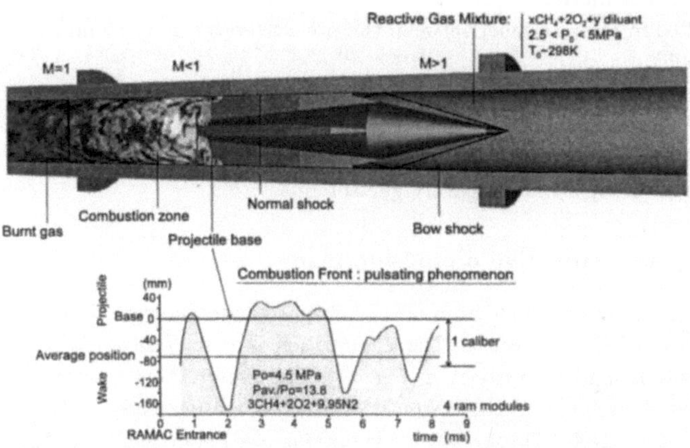

Fig. 11. Subdetonative propulsion mode : combustion front oscillations

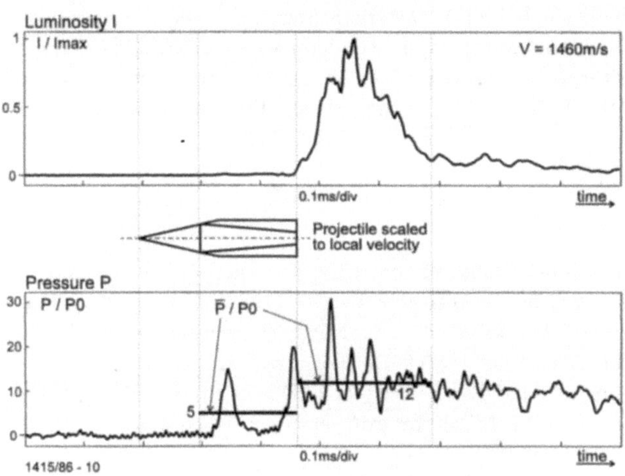

Fig. 12. RAMAC 90 experiment : position of observation x=2.7m after ram section entrance, fin model No 4c : 5-fin, θ_{CF} = 90°, e=0.08D_0, 3.1CH_4+2O_2+ 9.85N_2, P_0=4.0MPa

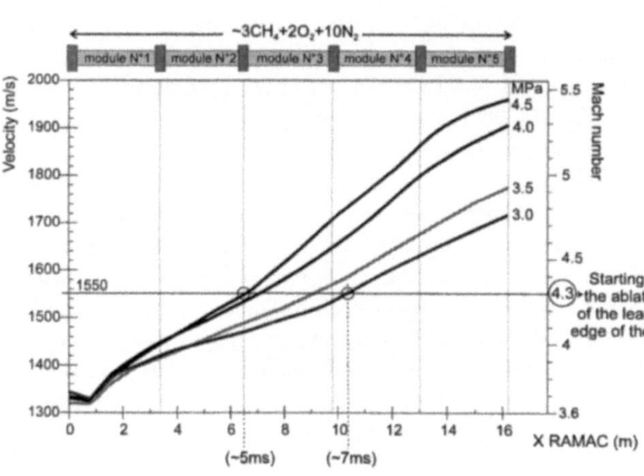

Fig. 13. RAMAC 90 experiments : effect of initial pressure P_0 on velocity performance, 3CH_4+2O_2 +10N_2, fin model No 4c : 5-fin, θ_{CF} = 90°, e=0.08D_0

The pressure is measured by classical piezoelectric gauges such as Kistler or PCB (Legendre 1996). The combustion front is detected by photodetector with a large spectral response : 400 to 1100 nm. The optical gauge is largely described in (Simon et al. 1995, Legendre 1996, Fig.10c).

In order to maintain the ram tubes motionless and to avoid any mechanical shock on them, a recoil-absorbing device is installed in the interface, between the preaccelerator muzzle and ram section entrance (Simon et al. 1995).

In order to control the projectile integrity after the whole ram process, an X-rays photography is taken 0.5 m after the RAMAC 30 muzzle and 3 m after the RAMAC 90 muzzle respectively. The X-Rays source used is a 300 keV under 25 kV high-voltage supply. The synchronization is ensured either by a magnetic trigger or by a microwave trigger antenna.

3. Experimental results in the subdetonative propulsion mode

3.1 General considerations Observations of signals from optoelectronic gauges indicate that the combustion is initiated at the rear of the projectile. The combustion is mainly situated and stabilized in the recirculation region behind the projectile (Fig. 11). The energy there is released in large-scale turbulent structures. The flow and the combustion process are unsteady and are affected by the projectile acceleration itself. The pressure oscillations are important (Fig. 12) and consequently the location versus time of the combustion front can change very rapidly. Its motion is similar to a pulsating phenomenon, but no characteristic frequency could ever have been established. We consider that the propulsion mode is a pure subdetonative one when the average position of the combustion front is situated at the base of the projectile or behind it during the whole ram process. The variation of the distance between the base and the combustion front can reach about 1 caliber, from one experiment to the other.

3.2 Performances

3.2.1 Reproducibility The excellent reproducibility of the RAMAC performances in terms of velocity, acceleration and ballistic efficiency has been largely described (Giraud et al. 1995b, Fig. 13). To date the best performance has been obtained in RAMAC 90 by using the projectile equipped with fin model No 4c, 5-fin, $\theta_{CF} = 90°$, e=0.08D_0. For a ram section length of 16.2 m, the ram projectile of 1.340 kg being injected at a velocity of 1330 m/s into a $2.95CH_4+2O_2$ $+10N_2$ mixture under a filling pressure of 4.5 MPa, the exit velocity reached 1986 m/s. The reproducibility of the velocity gain was about 1%.

3.2.2 Effect of the diffuser area and of the Mach number As noted in 2.2, the influence of the fins by a reduction of the diffuser area can be very important according to the model used (Fig. 5-6-7). For instance, by using the fin models No 3 (4-fin), in the same initial conditions :
(1) the diffuser did not start with type c ($\theta_{CF} = 90°$), the entrance velocity being 1340 m/s ;
(2) the flow was choked at Mach 4.4 for types a ($\theta_{CF} = 1.5°$) and b ($\theta_{CF} = 3°$).

Consequently, the aerothermodynamically blocked diffuser will yield the initiation (either direct or by a deflagration-to-detonation transition process) of a detonation, propagating ahead of the projectile during the so-called phenomenon of "unstart". Nevertheless, no blocking of the diffuser has been observed with fin model No 2c during the whole ram process, the exit Mach number being about 5.5. A large number of experiments have been carried out with the different fin models (Fig. 5-6) with or without a chamfer on their leading edge, in both 30 and 90 mm caliber. The results are similar. The sensitivity of the diffuser choking as a function of its cross-sectional area (Seddon et al. 1985) and also of the leading edge profile of the fins has been clearly highlighted. It appears that the stronger the shocks interactions, the better the ram efficiency but the shorter the useful Mach number field for the RAMAC process (Figs. 14 and 15). By means of two kinds of experiments, conducted respectively in an inert gas (Fig. 16) and in a reactive gas mixture (Fig. 17), through the measurement of the total drag coefficient, two critical Mach numbers have

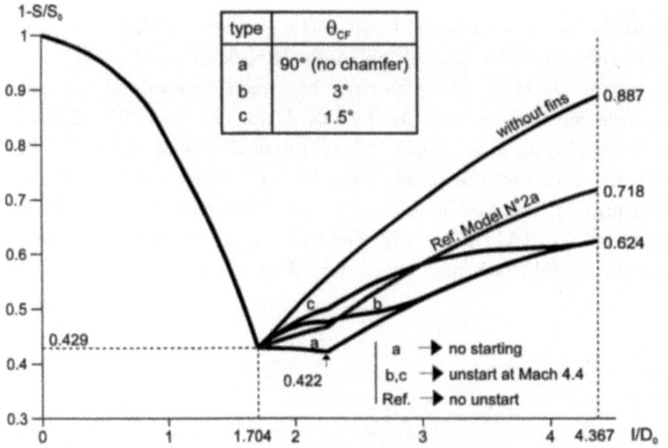

Fig. 14. RAMAC 90 experiments : effect of θ_{CF} on the diffuser area, fin model No 3 : 4-fin, e=0.156D_0

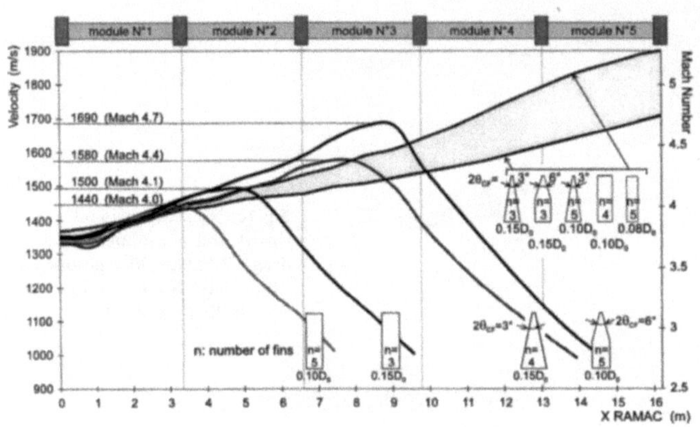

Fig. 15. RAMAC 90 experiments : effect of fins on velocity performance, $3CH_4+2O_2+10N_2$, P_0=4.0MPa

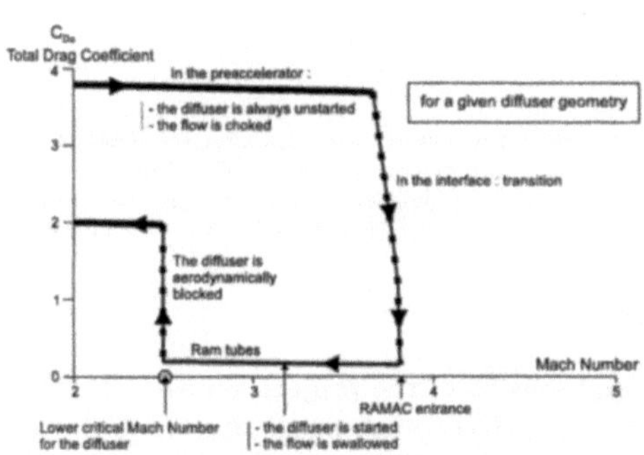

Fig. 16. Total drag coefficient : RAMAC 90 experiments in inert gas

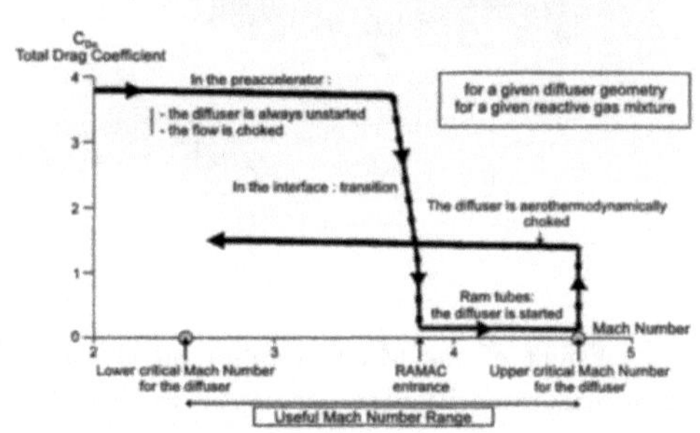

Fig. 17. Total drag coefficient : RAMAC 90 experiments in reactive gas mixture, 2.5≤P_0≤ 4.5MPa

been determined: A lower one when the diffuser is just aerodynamically blocked (obtained by deceleration of the projectile) and an upper one when the diffuser is aerothermodynamically choked (Fig. 15). Finally, for a given diffuser geometry, i.e., for a given projectile geometry and for a given reactive gas mixture, a useful Mach number range has been clearly identified. The success of the whole ram process will depend on the respect of this Mach number field.

3.2.3 Effect of the Mach number on the non dimensional thrust The greater the Mach number, the smaller the total thrust in the subdetonative propulsion mode (Bruckner et al. 1991) and the higher the total drag of the projectile (Fig. 18). The drag has been determined through experiments conducted in non-reactive gas mixtures and also by calculation (Nielsen 1960, Stivers 1971, Mac Donnal 1972, Krasnov 1978, Dupuis et al. 1989). When the total drag of the projectile is subtracted from the total thrust produced by the combustion, the net thrust can be reduced to zero. To avoid such a situation and to maintain a substantial net thrust level, the mixture composition along the entire length of the RAMAC is adjusted in order to maintain the well-adapted Mach number situated in the useful Mach number as described in 2.2.

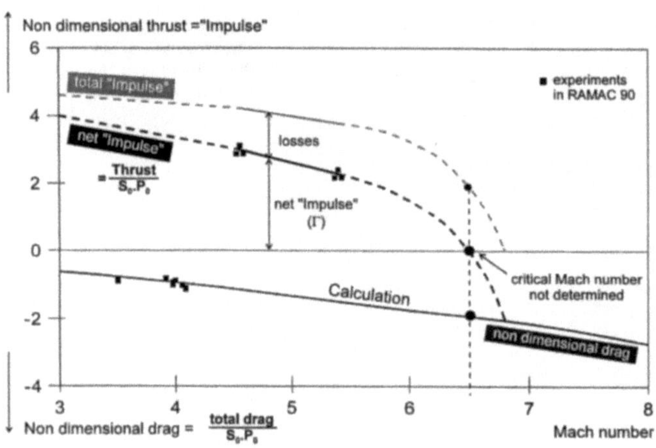

Fig. 18. Non-dimensional thrust and non-dimensional drag : RAMAC 90 experiments : fin model No 4c : 5-fin, $\theta_{CF} = 90°$, e=0.08D$_0$

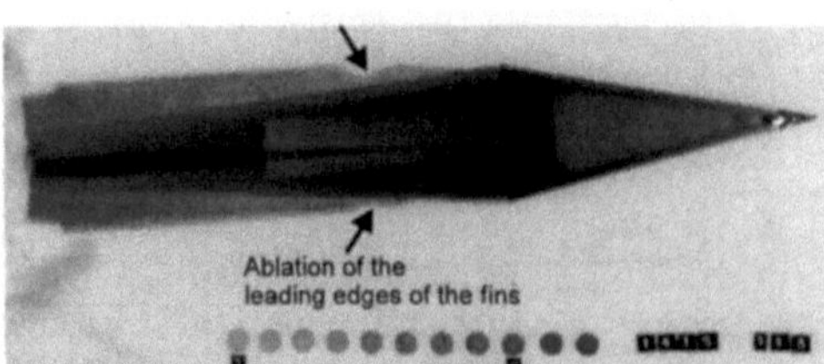

Fig. 19. X-ray picture taken 3m after the RAMAC90 muzzle : fin model No 1a : 3-fin, $\theta_{CF} = 1.5°$, e=0.156D$_0$, guiding length=2.3D$_0$, Mach=5, reactive mixture : $3.25CH_4+2O_2+9.80N_2$, P_0=4.0MPa

3.2.4 Effect of the Mach number on the ablation and erosion of the projectile material The structural integrity of the projectile is controlled by an X-rays picture taken just after the RAMAC muzzle (chapter 2.4 and Fig. 19). The severe aerothermodynamic heating of the projectile material is such that its melting temperature can be reached rapidly according to the Mach number. The present experiments being conducted with projectile afterbody made of magnesium alloy (chapter 2.2), two phenomena appear after a very short flight time in the 5-6 ms range :

(1) the ablation of the leading edge of the fins due to the flight conditions (initial pressure, velocity) ;

(2) the erosion and ablation of the projectile rear part caused by high temperature and pressures along the body and in the combustion zone in contact with the projectile.

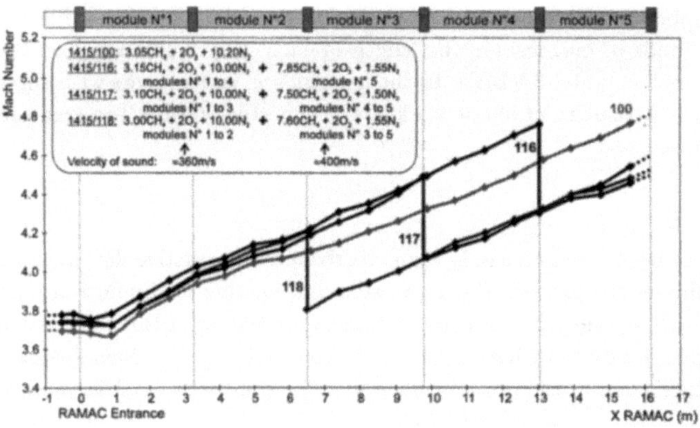

Fig. 20. RAMAC 90 multi-stage experiments : P_0=3.5MPa, fin model No 2c : 4-fin, $\theta_{CF} = 90°$, e=0.10D_0

The total material loss points to about 12 % of the projectile mass after a 10 ms flight time (Giraud et al. 1995b). It is difficult to quantify exactly the lost material and its recession velocity . Indeed, it changes the flow conditions as well as the combustion process (not a rigorously monophasic mixture). The ablation and combustion of magnesium contributed for one part to the projectile acceleration and subsequent unstarts could be provoked by an unsteady phenomenon close to the real projectile throat (Fig. 7). In order to avoid the problems encountered with the high thermal and mechanical ram loads applied on the given projectile material, the Mach number has to be maintained at the well-adapted level according to the useful Mach number field described in 2.2.

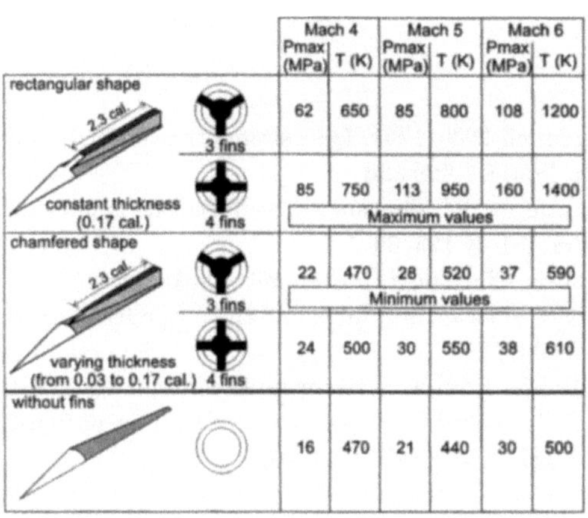

	Mach 4 Pmax (MPa)	Mach 4 T (K)	Mach 5 Pmax (MPa)	Mach 5 T (K)	Mach 6 Pmax (MPa)	Mach 6 T (K)
rectangular shape, constant thickness (0.17 cal.), 3 fins	62	650	85	800	108	1200
rectangular shape, constant thickness (0.17 cal.), 4 fins	85	750	113	950	160	1400
chamfered shape, varying thickness (from 0.03 to 0.17 cal.), 3 fins	22	470	28	520	37	590
chamfered shape, varying thickness (from 0.03 to 0.17 cal.), 4 fins	24	500	30	550	38	610
without fins	16	470	21	440	30	500

Fig. 21. Computed maximum pressures and temperatures at the same location in the longitudinal plane between two fins (initial conditions P_0=4.5MPa, T_0=298K)

3.2.5 Solution to increase the exit velocity The solution is now well known (Hertzberg et al. 1986, Giraud et al. 1995b, Kruczynski 1996). It consists in using a multistage RAMAC in which the mixture composition is adjusted in such a manner that the sound velocity follows the increase of the projectile velocity in order to maintain the Mach number quasi-constant. To date only the RAMAC 90 has been organized in two stages (chapter 2.3) and the reproducibility of the experiments conducted with it is excellent. Fig. 20 shows how to adjust the length of each stage to obtain the most appropriate Mach number, the mixture composition being well-adapted. In order to maintain the Mach number under the value 4.3 (indicated in Fig. 13) to avoid the ablation of the leading edge of the fins made of magnesium, the length of each stage will depend on the initial pressure: for instance, 8 meters under 3.5 MPa. In the near future, experiments with three or four stages will be conducted. A velocity of about 2.5 km/s is expected without too great an ablation/erosion of the projectile.

4. CFD

To date, the numerical study has been conducted only for a steady and non-reactive flow in order to highlight the effect of the fins in the process of compression and heating of the flow around the projectile when entering the first ram tube and just before the initiation of the combustion (Henner et al. 1997b, Fig. 21). Computations have been carried out with the 3D Navier-Stokes code TASCflow. In the near future, computations will be done with the code in its reactive version.

5. Conclusion

The present state of the RAMAC in the subdetonative propulsion mode highlights
(1) the effect of the diffuser geometry on the RAMAC performances, i.e. the importance of the profile, size and number of the guiding fins, and also the effect of the Mach number,
(2)the reproducibility of the RAMAC process by using a multistage ram, and the efficiency of the latter to obtain higher velocities with a "quasi-constant" Mach number,
(3) the necessity to maintain the Mach number at a relatively low level to avoid too fast an ablation and erosion of the fins due to high thermal and mechanical ram loads and finally to obtain a better impulse.

Further experiments will be conducted in order to obtain a better understanding of the scaling effects and also to study the effect of the material used for the projectile afterbody on the combustion and consequently, on the RAMAC performances.

References

Bruckner AP, Knowlen C, Hertzberg A, and Bogdanoff DW (1991) Operational characteristics of the thermally choked ram accelerator. J Prop Power 7:828-836

Dupuis A, Giraud M (1989) Sur les méthodes pratiques d'évaluation du coefficient de trainée de frottement d'un projectile en vol libre. ISL Rep ISL-R 123/89

Giraud M (1992) First results concerning the scale effect on the thermally choked propulsion mode. European Symposium STAR - Properties of Reactive Fluids and their Applications to Propulsion, Poitiers, France, ISL Rep ISL- PU 307/93

Giraud M, Legendre JF, Simon G, Catoire L (1992a) Ram accelerator in 90 mm caliber ; first results concerning the scale effect in the thermally choked propulsion mode. 13th International Symposium on Ballistics, Stockholm, Sweden, ISL Rep ISL-CO 210/92

Giraud M, Legendre JF, Simon G (1992b) Ram accelerator studies in 90 mm caliber. 43rd Meeting of the Aeroballistic Range Association, Columbus, OH, USA, ISL Rep ISL- CO 233/92

Giraud M, Legendre JF, Simon G (1993a) RAMAC 90: Experimental studies and results in 90 mm caliber, length 108 calibers. First International Workshop on ram accelerator - ISL, Saint-Louis, France, ISL Rep ISL-PU 360/93

Giraud M, Legendre JF, Simon G (1993b) RAM accelerator in 90 mm caliber or RAMAC 90. experimental results concerning the transdetonative combustion mode. 14th Int Symp on Ballistics, Quebec, Canada, ISL Rep ISL- PU 363/93

Giraud M, Legendre JF, Simon G, Mangold JP, Simon H, Kauffman H (1993c) RAMAC 90 : facility and diagnostic methods. 44th Meeting of the Aeroballistic Range Association, Munich, FRG, ISL Rep ISL- PU 362/93

Giraud M, Simon H (1995a) Sabot for projectiles of ram accelerators and projectiles equipped with such a sabot. United States Patent Number 5, 394, 805

Giraud M, Legendre JF, Simon G, Henner M, Voisin D (1995b) RAMAC in 90mm caliber or RAMAC 90 : starting process, control of the ignition location and performances in the thermally choked propulsion mode. 2nd International Workshop on ram accelerators, Seattle / WA, USA, ISL Rep ISL- PU 349/95

Henner M, Legendre JF, Giraud M, Bauer P (1997a) Initiation of reactive mixtures in a ram accelerator. AIAA Paper 97-3173, ISL Rep to be published

Henner M, Giraud M, Legendre JF, Berner C (1998) CFD computations of steady and non reactive flow around fin-guided RAM projectiles. In: Takayama K, Sasoh A (eds) Ram Accelerators, Springer-Verlarg, Heidelberg, pp 325-332

Hertzberg A, Bruckner A, Bogdanoff DW (1986) The ram accelerator : a new chemical method of achieving ultrahigh velocities. 37th Meeting of the Aeroballistic Range Association, Quebec, Canada

Krasnov NF(1978) Aerodynamics. Vysshaya Shkola Publishers, Moscow, NASA report NASA TT-F-765

Kruczynski D, Liberatore F, Hewitt J, Tuerk J (1996) Staged ram accelerator experiments with unique projectile geometries. US Army Research Laboratory, ARL report ARL-TR-1219

Legendre JF (1996) Contribution à l'étude de la sensibilité et des caractéristiques de détonation de mélanges explosifs gazeux denses à base de méthane utilisés pour la propulsion dans les accélérateurs à effet stato. Ph.D dissertation, University of Poitiers, Poitiers, France

Legendre JF, Bauer P, Giraud M (1998) RAMAC 90 : detonation initiation of insensitive dense methane-based mixtures by normal shock waves. In: Takayama K, Sasoh A (eds) Ram Accelerators, Springer-Verlarg, Heidelberg, pp 223-231

Mc Donnal (1972) USA Stability and Control DATCOM / Chapter 4.2.3 Body drag. Douglas Corporation - Douglas Aircraft Division - October 1960 - revised February 1972

Nielsen JN (1960) Missile Aerodynamics. McGraw-Hill Inc.

Seddon J, Goldsmith E (1985) Intake Aerodynamics - An account of the mechanics of flow in and around the air intakes of turbine-engined and ramjet aircraft and missiles. Collins, London

Simon G, Mangold JP, Simon H, Kauffmann H (1995) Instrumentation de l'accélérateur par effet stato - RAMAC 90. ISL Rep ISL-RT 509/95

Stivers LS (1971) Calculated Pressure Distributions and Components of Total Drag Coefficients for 18 Constant-Volume, Slender Bodies of Revolution at Zero Incidence for Mach Numbers from 2.0 to 12.0 with Experimental Aerodynamic Characteristics for Three of the Bodies. NASA report NASA TN D-6535

Presentation of the rail tube version II of ISL's RAMAC 30

F. Seiler, G. Patz, G. Smeets, J. Srulijes
French-German Research Institute of Saint-Louis (ISL), F-68301 Saint-Louis, France

Abstract. The new concept of accelerating a projectile flying in a tube at supersonic speed by self-synchronized ignition of a comustible gas mixture, the ram accelerator, has generated considerable interest over the last years in different countries. In France, ISL has performed among other experiments in a 30-mm-caliber ram accelerator, called RAMAC 30, mainly at superdetonative flight speeds. As accelerator tube a ram-tube equipped with rails for guiding smooth cylindrical projectiles was in use. In 1993 and 1994 the rail tube version I with four rails and since beginning of 1997 the rail tube version II with five inner rails is tested. In 1995 and 1996 the smooth bore technique was proved and the results are presented by Patz et al. (1997).

Key words: Ram acceleration, rail tube, cylindrical projectiles

1. Introduction

The ram accelerator concept was developed and tested successfully in a 38-mm-device by Hertzberg et al (1986) at the University of Washington, Seattle, USA, in 1986. In 1988, based on the need of ISL for a hypersonic launching facility, the decision was taken to build two ram accelerators: a 30-mm-tube, called RAMAC 30, and a 90-mm-one, RAMAC 90 (see Giraud et al. 1993). The RAMAC 30 facility is used for basic research, mainly in the superdetonative flight regime, with the objective of investigating the ignition and combustion phenomena with regard to the gas mixtures to be used as well as the projectile design. Scaling and geometry effects are of main interest, wherefore the RAMAC 30 was tested on the one hand with the smooth bore technique in 1995 and 1996 (see Patz et al. 1997) and on the other hand with rail equipped tubes: version I with four rails in 1993 and 1994 (see Seiler et al. 1995) and version II with five rails beginning in 1997.

Smeets (1998) published a new concept for a ram accelerator with guiding tube rails for firing rail stabilized projectiles to replace the fin stabilized projectiles originally used at the University of Washington which are accelerated in a cylindrical bore, Herzberg (1989). The rail tube idea has some advantages, e. g., no sabot necessary as required for fin guided projectiles, simple projectile geometry, and possibility of varying the inner tube geometry. Therefore it was decided to test the rail tube principle in the RAMAC 30 with rail guided projectiles. The first rail experiments proving this technique started in 1993 and continued until the end of 1994. To bypass the gasdynamic problems of subdetonative ignition and transition to superdetonative combustion, the direct firing into the superdetonative combustion mode is investigated in the RAMAC 30 used as a scram accelerator.

2. Principle of scram acceleration

Fig. 1 explains the principle of the scram accelerator process with supersonic combustion. A vehicle consisting of a cylindrical centerbody with conical portions at its front and rear ends propagates through a combustible gas mixture filled in a cylindrical tube having a diameter greater than that of the centerbody. By mean of fins, (Patz et al. 1995), or rails, (Seiler et al. 1995), which are not shown in Fig. 1, the projectile is guided in a centered position inside of the tube. This tube containing the combustible gas mixture is closed at both ends by diaphragms which are destroyed by the moving projectile.

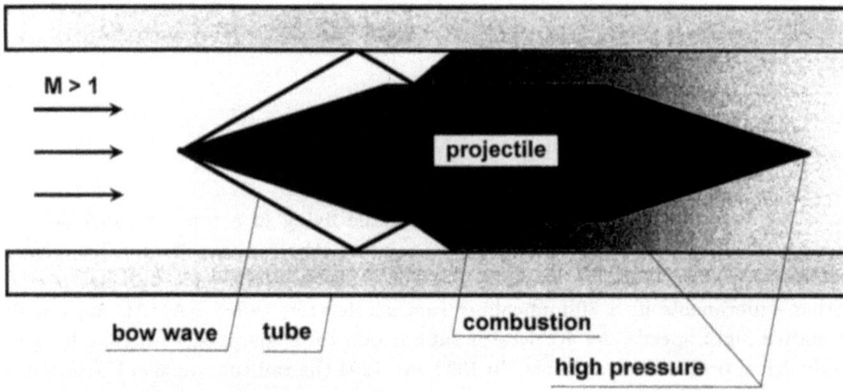

Fig. 1. Principle of the scram accelerator process with superdetonative combustion

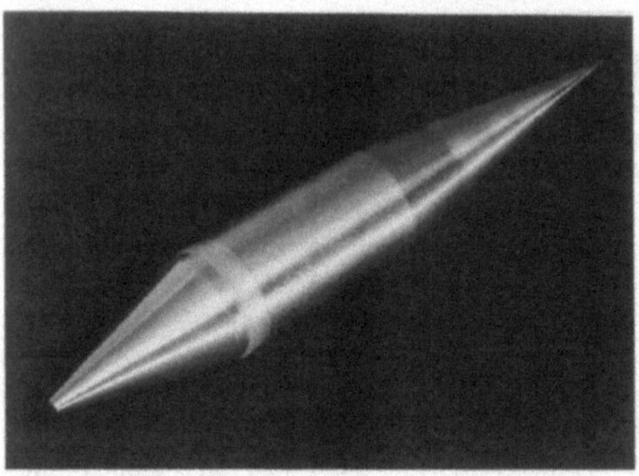

Fig. 2. Cylindrical ram-projectile used in rail tube version I

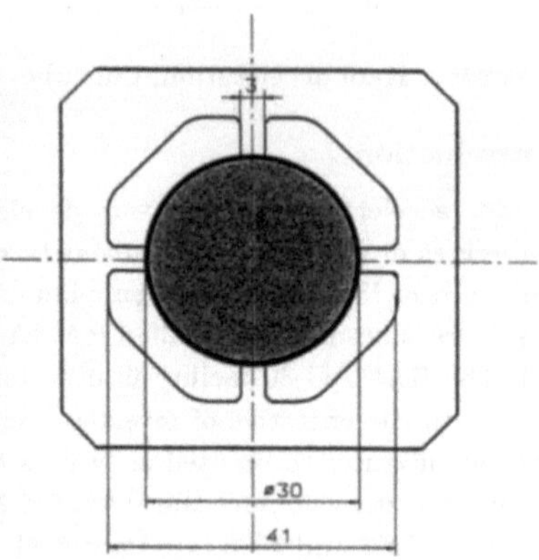

Fig. 3. Cross-section in rail tube version I

3. Rail tube version I of RAMAC 30

3.1 Description of the facility

The RAMAC 30 design follows in principle the prototype ram accelerator of Hertzberg et al (1986) in which fin stabilized bodies are accelerated by combustion in a circular ram tube. The first RAMAC 30 design, however, was chosen for testing the alternative concept of a ram tube with inner rails in combination with cylindrical and finless projectiles. The circular projectile of Fig. 2 has no fins and is guided in a tube with four inner rails, see Seiler et al. (1993). The front cone angle is for the projectiles used 14-16 degrees and the back angle is similar. The combustor zone of constant diameter has a length of 45-60 mm and the projectile mass is 125-135 g. The tube cross-section is given in Fig. 3. The total cross section area of the rail tube is 1241 mm^2. A direct view on the ram-tube is seen in Fig. 4. At the end of the ram accelerator, the projectiles are hitting a set of replaceable steel plates inside of a piston which moves backwards inside of the catcher tube after impact.

In the first experiments we used one ram-section of 3.6 meters length. In the conventional powder gun having 1.8 meters tube length, aluminium (Dural alloy: AlMgCu1) projectiles partially fitted with an inner magnesium core with total masses of about 130 g are accelerated to a muzzle velocity of about 1800 m/s being the initial velocity at the entrance to the ram-section. This

Fig. 4. Rail tube I between two dump tanks

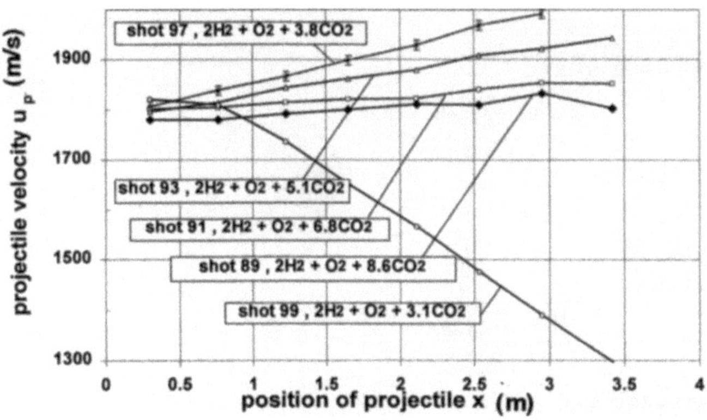

Fig. 5. Flight velocities obtained with hydrogen based mixtures in the rail tube ram accelerator version I

allows to start in the superdetonative ram accelerator mode, with combustion at the cylindrical part of the projectile as it is shown in Fig. 1.

3.2 Experimental results

The first experiments leading to positive results were carried out with the above mentioned projectiles with an inner magnesium body which is covered by an aluminum cowling present in the combustor region as well at the fore- and afterbody. We investigated in the RAMAC 30 three different gas mixtures based on hydrogen (H_2), methane (CH_4) and ethylene (C_2H_4), mainly having a stoichiometric ratio of fuel and oxidizer and carbondioxide CO_2 as diluent: (1) ethylene-oxygen-carbondioxide, (2) methane-oxygen-carbondioxide, (3) hydrogen-oxygen-carbondioxide. Best ignition behaviour as well as avoidance of unstart effects in limiting the heat release according to the theory described in section 3.3 were obtained with all three gas mixtures. The maximum velocity increase in a 2.7 MPa gas mixture was about 200 m/s (see Fig. 5, shot no. 97). More CO_2-contents (shot no. 89, 91, 93) gives less heat release and consequently less projectile acceleration. A diminishing CO_2-part increases the heat production by combustion followed by an unstart as shown for shot no. 99.

The results shown in Fig. 5 for hydrogen based mixtures and in Fig. 6 for hydrogen, methane and ethylene fuels were determined using the signals from the electromagnetic sensors identifying

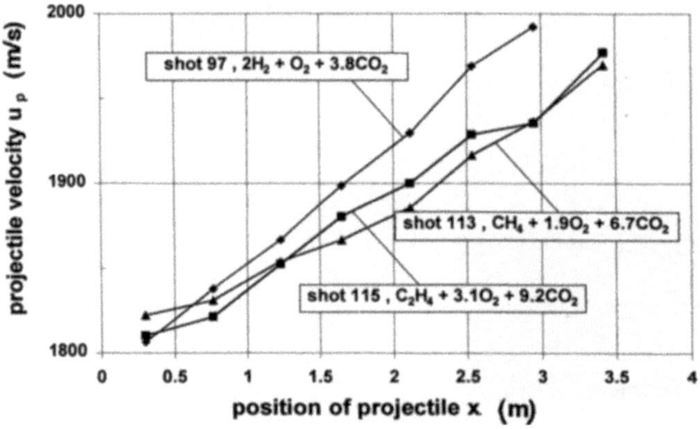

Fig. 6. Flight velocities with mixtures based on hydrogen, methane and ethylene in the rail tube ram accelerator version I

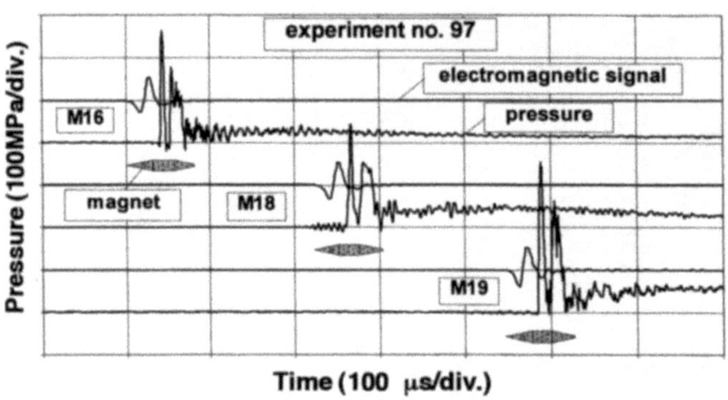

Fig. 7. Electromagnetic and pressure signals received for shot no.97

the passage of the magnet to better than one microsecond, see Fig. 7 for shot no. 97 and Fig. 8 for no. 89.

3.3 Limits of ignition and combustion

The experiments show that in the superdetonative mode there exist stability limits. One limit is set by thermal choking followed by an unstart which means that the combustion wave moves in front of the projectile. Another limit for a failure is found to be the decoupling of the ignition and combustion from the projectile.

Figure 7 shows a set of typical signals of experiment no. 97 recorded with the electromagnetic sensors and the wall pressure gauges for three measuring locations M16, M18 and M19 along the tube. The pressure signals are correlated in time directly to the position of the projectile. They show a very strong pressure increase by a front shock placed at the beginning of the combustor zone with a peak pressure of more than 100 MPa. This first wave is probably produced by the reflection of the incident shock wave at the tube wall forming the first reflected shock. In the expansion generated at the corner of the projectile shoulder the pressure strongly decreases. A second wave follows initiating a high pressure region. This second jump is maybe located where the third (or fourth) reflected shock is formed from the second (or third) one inducing combustion which looks similar to the well-known detonation phenomena, described by Chapman and Jouguet (see Zierep 1990).

Increasing the carbondioxide content makes the combustible gas mixture more and more phlegmatic and it needs a third wave to be ignited, as seen in Fig. 8 for shot no. 89 at measuring station M19. Then the gas reaction occurs behind the projectile and the pressure gain is lost. Decoupling of the combustion or no combustion depends on the mole parts of the inert component of the gas mixture, i.e., the CO_2-content in the $H_2/O_2/CO_2$ mixture used herein. This outcome defines an upper dilution limit of the combustible gas mixture for getting the mixture ignited in the combustor before the expansion in the divergent back region takes place.

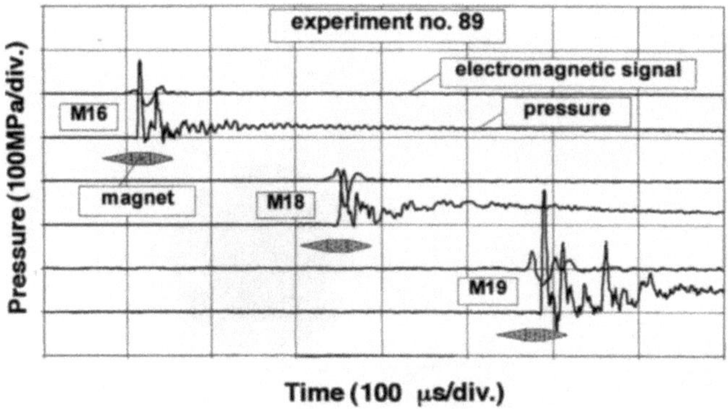

Fig. 8. Electromagnetic and pressure signals received for shot no.89

To avoid thermal choking followed by an unstart, the heat release must be adapted to the flow Mach number M in front of the combustion. For a one-dimensional flow, as present in the channel between projectile and wall, Zierep (1990) gives a relation describing the maximal heat input q as follows:

$$\left(\frac{q}{c_p T}\right)_{max} = \frac{(M^2-1)^2}{2(\chi+1)M^2} \tag{1}$$

In this relation, coupling the normalized heat release with the flow Mach number M, the parameter χ is given as $\chi = c_p/c_v$, with c_p the specific heat at constant pressure and c_v that at constant volume. T is the gas temperature in front of combustion. Fig. 9 shows the distribution of the normalized heat release as a function of the flow Mach number M, calculated with the code developed by Smeets et al. (1992). We see two regions: a zone in which no stable combustion occurs and the flow is thermally choked and a zone with stable combustion by limiting the heat input. Five points are inserted representing the firings no. 89, 91, 93, 97 and 99. Looking at Fig. 5 for shot no. 99 we have a projectile unstart and also predicted by the above given relation. The other four shots are at the border (no. 97) and inside (no. 89, 91, 93) the allowed region, giving a projectile acceleration as shown in Fig. 5.

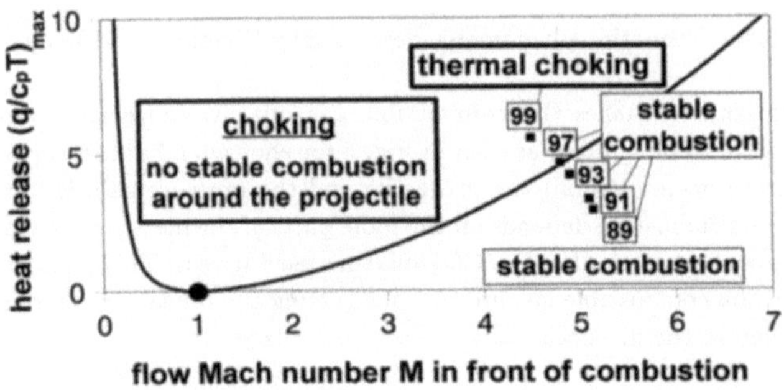

Fig. 9. Maximum heat release inside of the combustor channel for superdetonative combustion

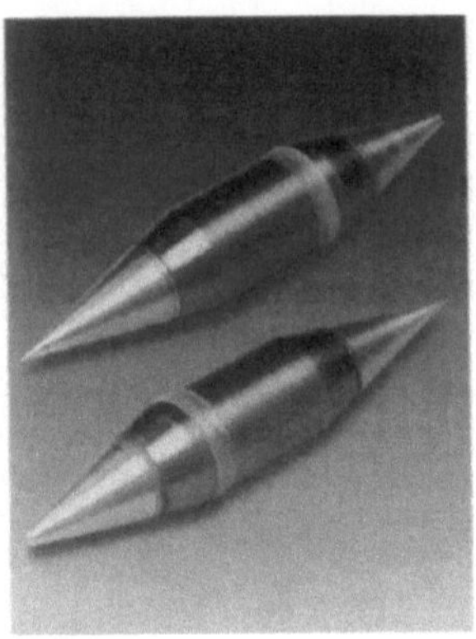

Fig. 10. Rail tube version II projectile

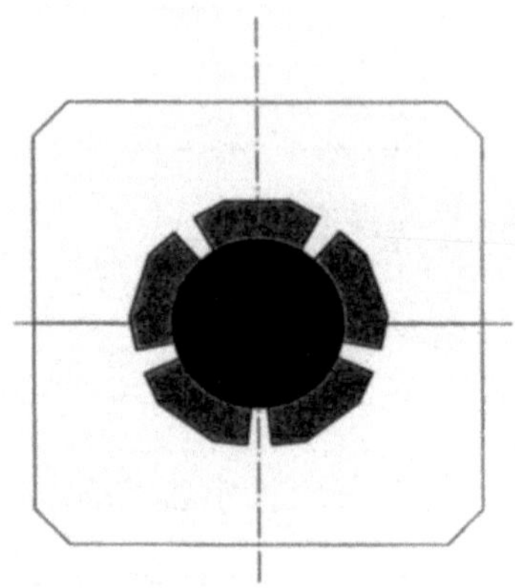

Fig. 11. Cross-section of rail tube version II with five rails

4. Rail tube version II of RAMAC 30

4.1 RAMAC 30 facility

In rail tube version II the tube is equipped with five inner rails, see Fig. 11. The total tube area is is about 1381 mm^2. The ram-tube consists of two tubes with a total tube length of 4.7 m. The rail tube II placed between the two dump tanks is seen in Fig. 12 with all the equipment necessary for RAMAC 30 operation: (1) valves for gas filling, (2) pressure gauges for pressure measurement, and (3) electromagnetic sensors for determination of projectile position.

4.2 Experimental results

Heat transfer calculations done by Seiler et al. (1997) taking into account the boundary layer formation at the fore- and midbody of the ram projectile predicted that, e.g., for firing no. 97 the midbody surface begins to melt because the surface temperature exceeds melting temperature of the aluminum used. Surface melting is followed by surface ablation causing the midbody diameter to decrease and the midbody diameter of 30-mm-caliber is not more guaranted for projectile guidance. Canting of the projectile begins which is a source for the development of an asymmetric flow around the projectile and might cause the combustion or detonation wave to move in front of the projectile.

This theoretical outcome is supported with experimental data of RAMAC 30, see Patz et al. (1997). They report in smooth bore technique that projectiles fabricated of magnesium, aluminum and even titanium begin to melt at the combustor midbody surface at flow conditions similar to those in rail tube version I. Though we gathered positive results in rail tube version I with aluminum projectiles, it seems that in rail tubes melting and ablation is not of that importance as in smooth bores with fin-guided projectiles, mainly because in rail tubes melting of the rails is not present. But nevertheless for rail tube version II to avoid melting processes which are always coupled with surface erosion giving surface ablation, other materials than aluminum have to be used.

Fig. 12. Rail tube II

Probably the best way to prevent melting from projectile material is to use steel, as predicted by Seiler et al (1997). For this reason we designed a new cylindrical projectile consisting in the inner part of a plastic body (Delrin) of diameter 24 mm at which at the front and back part an aluminum forebody, respectively, an aluminum afterbody is screwed. In order to protect the plastic core a steel cowling of 3 mm wall thickness forms the constant diameter combustor, see Fig. 10. The first firings performed with this new designed projectile had the aim to get at the muzzle of the powder gun 1800 m/s as entrance velocity into the ram tube to initiate superdetonative combustion. Unfortunately the steel protected projectiles have a mass of about 200 g and the effort to increase the projectile velocity to the requested initial velocity at the entrance to the ram accelerator tube could not be satisfied with the projectile with a 3 mm thick steel cowling.

To diminish the projectile mass, the steel was replaced by a titanium cowling giving a mass reduction of about 50 g to m $\approx$ 150 g instead of m $\approx$ 200 g in case of steel cowling. By this projectile mass reduction using titanium with the same thickness as in case of steel cowling (3 mm), the initial projectile velocity could be enhanced to 1772 m/s for firing no. 197. Ram acceleration was achieved with this titanium protected projectiles, as given in Fig. 13 for shot no. 196 ($2H_2$+O_2+$7.5CO_2$) and shot no. 197 ($2H_2$+O_2+$6CO_2$). The firing in inert gas, shot no. 198 ($2H_2$+N_2+$5.2CO_2$) is also included in Fig. 13. The reactive firing no. 196 succeeded to projectile acceleration with immediate gas ignition and stable combustion inside of the combustor zone. The velocity increase is nearly 4% of the initial velocity of 1739 m/s with Δu = 64 m/s. This result is comparable to that obtained in RAMAC 30 rail tube version I for firings no. 89 or 91.

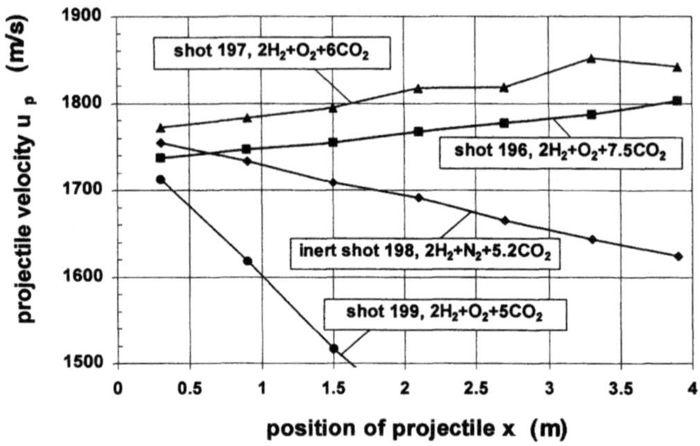

Fig. 13. Velocity distribution in rail tube version II

5. Conclusions

5.1 Rail tube version I

The experiments performed in 1993 and 1994 in the rail tube version I of RAMAC 30 at ISL constitute a significant breakthrough in superdetonative ram combustion. The following points can be highlighted: (1) Cylindrical bodies with conical front and back parts are accelerated in a rail tube. In this case, pre-acceleration in a gun without sabot was possible. (2) There is no problem of an unstart with hydrogen, methane and ethylene based combustible gas mixtures in limitating the heat release. Seiler et al. (1997) have done extensive calculations for firing no. 97 showing that at the cone of the aluminum projectile no melting occurs. Less melting phenomena are present in the midbody region, but a small amount of melting and ablation seem to be tolerable at the midbody. The positive experimental outcomes shown in Fig. 5 support this discussion, though surface melting occuring at the body surface has been calculated with the heating model given by Seiler et al. (1997).

5.2 Rail tube version II

The first projectiles used in rail tube version II of RAMAC 30 consist of an inner plastic core (Delrin) with a fore- and afterbody fabricated of aluminum (Dural alloy: AlMgCu1). The midbody is protected with an outer steel or titanium cowling. The photography of Fig.10 shows the design of the actual projectile in case of steel. With the projectile with a steel cowling it was not possible to get the required initial projectile velocity of 1800 m/s for entering directly into superdetonative combustion. The projectile mass of about 200 g is too large to be accelerated with the existing powder gun to the desired gun muzzle velocity. Using titanium instead of steel succeeded in mass reduction to about 150 g and to a projectile injection with nearly 1800 m/s into the ram accelerator tube. But due to melting and surface erosion present with titanium other materials must be found for carrying out future positive acceleration cycles.

6. References

Giraud M, Legendre J-F, Simon G (1993) Ramac 90: Experimental studies and results in 90 mm caliber, length 108 calibers. Proc 1st Int Workshop on Ram Accelerator, RAMAC I, ISL, France

Hertzberg A, Bruckner AP, Bogdanoff DW (1986) The ram accelerator: A new chemical method of achieving ultra-high velocities. 37th ARA-Meeting, Québec, Canada

Hertzberg A (1989) Thermodynamics of the ram accelerator. In: Kim YW (ed) Proc 17th Int Symp on Shock Waves and Shock Tubes, Bethlehem, USA, pp 2-11

Seiler F, Patz G, Smeets, G, Srulijes J (1993) Status of ISL's RAMAC 30 with rail stabilized projectiles. Proc 1st Int Workshop on Ram Accelerator, RAMAC I, ISL, France

Seiler F, Patz G, Smeets G, Srulijes J (1995) Gasdynamic limits of ignition and combustion of a gas mixture in ISL's RAMAC 30 scram accelerator. In: Stortevant B et al (eds) Proc 20th Int Symp on Shock Waves, World Scientific, Singapore, pp 1057-1062

Seiler F, Patz G, Smeets G, Srulijes J (1995) The rail tube in ram acceleration: Feasibility study with ISL's RAMAC 30. Proc 2nd Int Workshop on Ram Accelerator, RAMAC II, Univ Washington, Seattle, USA

Seiler F, Gatau F, Mathieu G (1998) Prediction of surface heating of a projectile flying in RAMAC 30 of ISL. In: Takayama K, Sasoh A (eds) Ram Accelerators, Springer-Verlag, Heidelberg, pp 151-158

Smeets G (1988) Ram-Nachbeschleuniger für aus konventionellen Kanonen verschossene Vollkaliber-Projektile, ISL-Report N 603/88

Smeets G, Gatau F, Srulijes J (1992) Rechenprogramm für Abschätzungen zur Ram-Rohrbeschleinigung, ISL-Report RT 507/92

Patz G, Seiler F, Smeets G, Srulijes J (1995) Status of ISL's RAMAC 30 with fin guided projectiles accelerated in a smooth bore. Proc 2nd Int Workshop on Ram Accelerator, RAMAC II, Univ Washington, Seattle, USA

Patz G, Seiler F, Smeets G, Srulijes J (1998) The behaviour of fin-guided projectiles superdetonative accelerated in ISL's RAMAC 30. In: Takayama K, Sasoh A (eds) Ram Accelerators, Springer-Verlag, Heidelberg, pp 89-95

Zierep J (1990) Strömungen mit Energiezufuhr, G. Braun Verlag, Karlsruhe, Germany

The behaviour of fin-guided projectiles superdetonative accelerated in ISL's RAMAC 30

G. Patz, F. Seiler, G. Smeets, J. Srulijes
French-German Research Institute of Saint-Louis (ISL), F-68301 Saint-Louis, France

Abstract. The new concept of accelerating a projectile flying in a tube at supersonic speed by self-synchronized ignition of explosive gas mixture filled in a tube closed at its ends by diaphragms, the ram accelerator, has generated considerable interest over the last years in different countries: especially in the United States, Israel, Japan, France and others. In France, the ISL performed, among others, experiments in a 30-mm-caliber ram accelerator, called RAMAC 30, at superdetonative flight speeds. The results and difficulties found in the version with a smooth bore and fin guided projectiles are described herein and can be summarized as follows: Using a conventional gun as a preaccelerator for injecting a projectile into the ram tube, combustion has been achieved in a single stage tube, flying in the superdetonative regime from the beginning on. Combustion was stabilized at the projectile body with hydrogen and methane based gas mixtures. Unfortunately, the thrust has been very small and significant acceleration could not be obtained Damage of projectile was observed on x-ray pictures, especially of the fins, probably by melting and burning when magnesium, aluminum or titanium alloys has been used. Steel projectiles can survive the ram acceleration cycle, but the projectile mass is too high to observe considerable acceleration.

Key words: Superdetonative mode, fin-guided projectile, erosion, projectile materials

1. Introduction

The ram accelerator concept was developed and tested successfully in a 38-mm-device by Hertzberg et al (1986) at the University of Washington, Seattle, USA, in 1986. In 1988, based on the need of ISL for hypersonic launching facilities, the decision was taken to build two ram accelerators: a 30-mm-tube, called RAMAC 30, and a 90-mm-one, RAMAC 90, see Giraud et al (1993). The RAMAC 90 is designed to accelerate masses of several kilograms to velocities of up to 3 km/s. The RAMAC 30 facility is used for basic research, mainly in the superdetonative flight regime, with the objective of investigating the ignition and combustion behaviour of different gas mixtures, the influence of projectile geometry and new tube concepts, as published by Seiler et al (1993).

In the ram accelerator at the University of Washington (1989) the process always starts with subsonic combustion behind the projectile flying initially at speeds lower than the Chapman-Jouguet detonation velocity of the combustible gas mixture. For higher velocities, it passes the transdetonative mode and finally switches into the superdetonative combustion mode where combustion occurs in the supersonic flow in the slit between the projectile and the tube wall. To bypass the gasdynamic problems of subdetonative ignition, the direct firing into the superdetonative combustion mode (scram accelerator) is investigated in the RAMAC 30. Two ram-tube versions have been meanwhile tested: (a) the rail tube ram accelerator and the "classical" smooth bore ram-tube. The results gathered with a tube equipped with inner rails for projectile stabilization are described by Seiler et al (1995). In this paper the smooth bore concept in RAMAC 30 is discussed and the results obtained in 1995 and 1996 are presented.

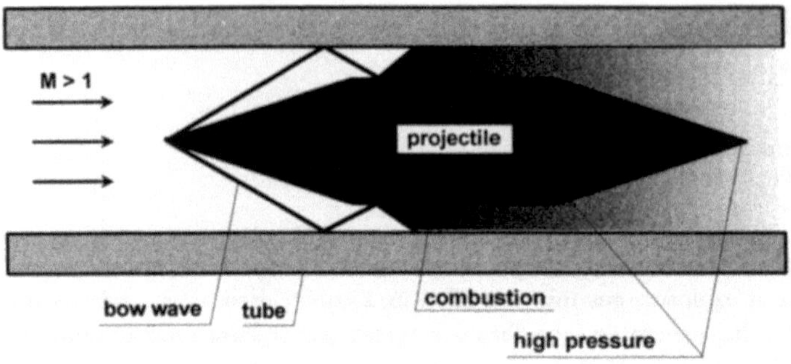

Fig. 1. Principle of superdetonative gas combustion

Fig. 2. Smooth ram tube (left) and sabot stripper tube (right)

2. Principle

Fig. 1 explains the principle of the superdetonative ram accelerator process. A projectile consisting of a cylindrical centerbody with conical portions at its front and rear ends propagates through a combustible gas mixture filled in a cylindrical tube closed at both ends by diaphragms. The inner diameter of the tube is greater than that of the projectile centerbody. By means of fins, see Hertzberg et al (1986), or rails, see Seiler et al (1995), which are not shown in Fig. 1, the projectile is centered inside of the tube.

3. Smooth bore version

3.1 RAMAC 30 facility set-up

After having shown the good application of the rail equipped accelerator tube concept for ram acceleration, we have investigated the conventional smooth bore technique of Hertzberg et al (1986) in 1995 and 1996, for the superdetonative case, with a projectile guided by fins fixed at the body of the projectile.

Concerning the 30-mm-caliber smooth ram-tube, we used two 3 m tubes attached to each other with a total length of 6 m and a powder gun as pre-accelerator. In Fig. 2 the ram-tube is seen at the left hand side. Between ram-tube and pre-accelerator a sabot stripper tube (at right) is needed. It consits of two concentrically arranged tubes: a bigger diameter outer tube and an inner "clarinet tube" with the bore diameter. This stripper tube with 1.5 m length is filled with a gas of high compressibility, e.g., CO_2.

Fig. 3. Ram projectile made of aluminum with five guiding fins

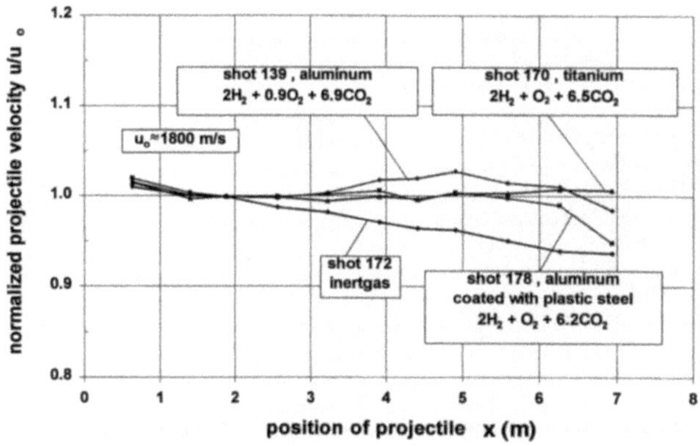

Fig. 4. Velocity distribution in RAMAC 30 with different ram projectiles

The projectile is injected into the smooth ram tube with about 1800 m/s to be superdetonative relative to the combustible gas mixture used. The geometry used for the aluminum and titanium projectiles is depicted in the photography of Fig. 3, showing the fins fixed at the body of the projectile with constant diameter and the conical front and back parts. In this case of fin guided projectile a sabot is necessary for pre-acceleration. The projectiles are made of aluminum and titanium with four or five fins having a thickness of 2 - 2.5 mm. The front cone angle is 14 - 16 degree, the back angle is similar. The combustor zone with constant diameter has a length of about 50 mm with a diameter of 20 mm. The projectile mass is 80 - 85 g with aluminum and about 110 g with titanium. The sabot has a mass of about 32 g.

3.2 Experimental results

The results of shot no. 139 (aluminium), 170 (titanium), 178 (aluminum with plastic steel coating) and 172 (inert firing) are presented in Fig. 4. There the projectile velocity is shown as a function of the position of the projectile inside of the tube. For firings no. 139, 170 and 178 the velocity initially decreases in the sabot stripper tube and then increases compared with the inert

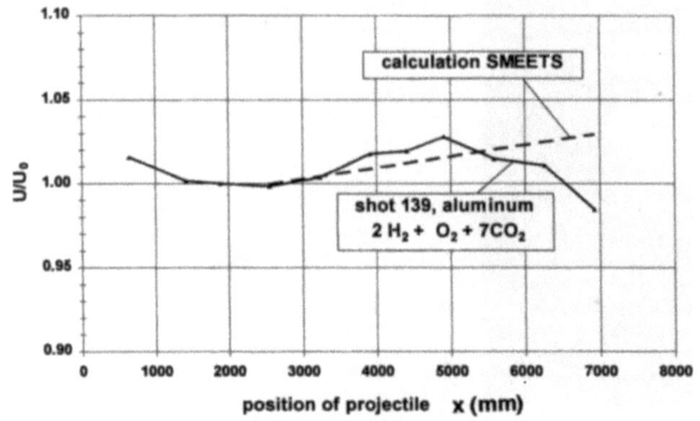

Fig. 5. Aluminum projectile firing no.139 compared with calculation

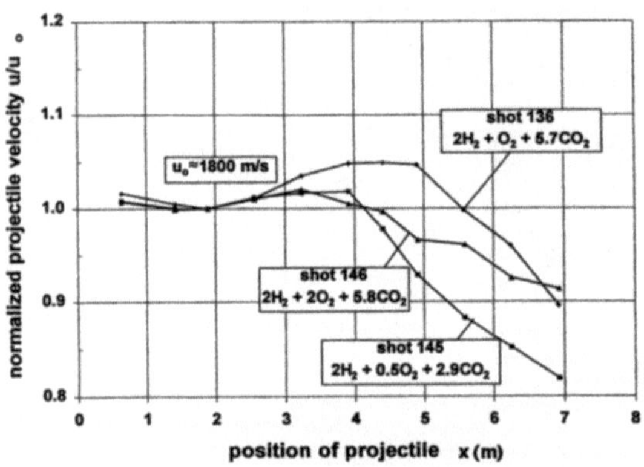

Fig. 6. Aluminum projectiles fired with fuel lean, rich and stoichiometric hydrogen based mixtures

firing no. 172, where the oxygen has been replaced by nitrogen. The combustion starts inside the combustor producing thrust on the back side of the projectile. But, due to the high combustion temperature the heat flux into the ram projectile is enormous, in some cases maybe initiating melting processes and chemical reactions between the projectile material and the oxygen as well as the diluent CO_2 present in the combustible gas mixture. This chemical reaction produces an additional heat release which surpasses the maximum allowable heat input, see Seiler et al (1995), and generates an unstart which begins after a projectile travel of about 4 - 5 m.

A comparison with calculation of Smeets (1992) is shown in Fig. 5. The calculated velocity rise is smaller compared with that given in the experiment no. 139. The higher experimentally obtained acceleration present with pure aluminum is caused by burning of the projectile body and/or fins generating an additional heat source besides the heat release by combustion. The consequence is an undesired velocity gain by burning of projectile material which in most cases is followed by an projectile unstart.

The velocity distribution gathered with aluminum projectiles fired in fuel rich (no. 145), stoichiometric (no. 136) and fuel lean (no. 146) hydrogen based gas mixtures are drawn in the diagram of Fig. 6. There exists no significant difference in the velocity behaviour of these projectiles. In all three cases the supersonic combustion starts with a projectile acceleration. But after about 3 - 4

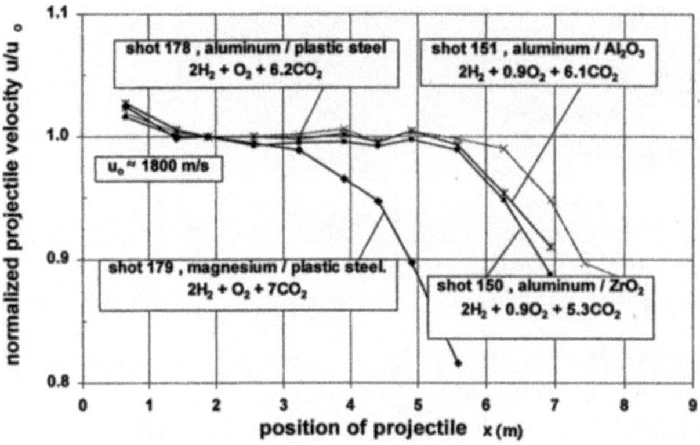

Fig. 7. Velocity distribution with coated projectiles

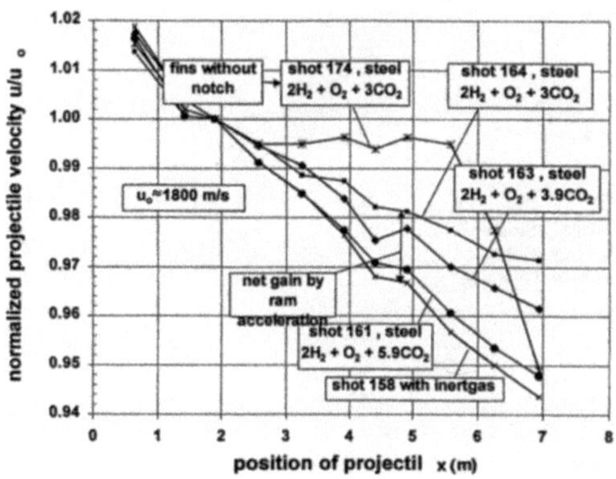

Fig. 8. Firings with steel projectiles

m the acceleration stops and is changed into a strong projectile deceleration. The reason for this failure is that the fins and the body are damaged by melting and burning. In each of this three firings practically the whole projectile is burned because nearly no impact has been obtained in the steel plates of the catcher tube.

In Fig. 7 a comparison is shown between a shot fired with aluminum projectiles, plasma coated with zirconium oxide ZrO_2 (no. 150) or aluminum oxide Al_2O_3 (no. 151). Firing 178 was done with an aluminum projectile coated with a plastic steel layer. The coating thicknesses are in the range of about 200 μm. The protected aluminum projectiles survive longer in time than those of pure aluminum. Due to the protection of its outer surface these projectiles are not burned up totally on 6 m tube length as also could be seen from the impact into the steel plates. The aluminum projectile survives a flight of 3 - 4 m. After 6 m flight the aluminum projectile nearly vanishes. The same result is present with a coated projectile made of magnesium, firing no. 179.

In Fig. 8 we can see the results of some firings with steel projectiles. To save mass (mass of projectile was about 115 g), the 3 fins of the projectiles have been machined with square notches. The recorded velocity distributions show a strong decrease of velocity due to the higher drag at the notches but also a net gain compared to the shot into inert gas for all firings. As we can also

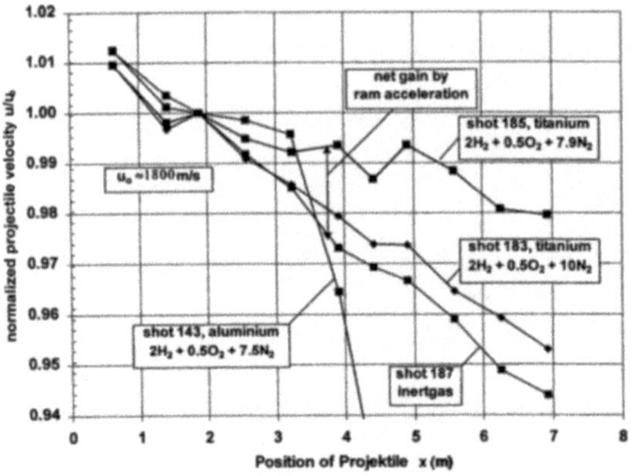

Fig. 9. Firings with nitrogen as diluent

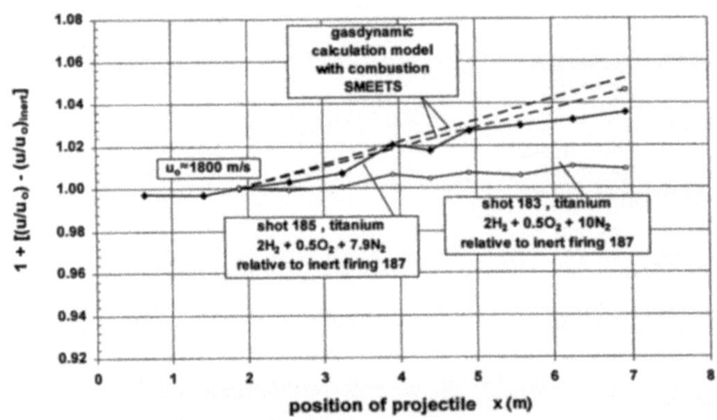

Fig. 10. Firings with nitrogen as diluent compared with calculated velocity distributions

see there is a smaller gain compared to the shot no. 174 without notches. But there we had an unstart after about 5 m.

Fig. 9 shows the results of some firings into gas mixtures with nitrogen as diluent. The titanium projectiles show a rather good behaviour whereas the shot with the aluminium projectile gives an unstart shortly after the beginning of the ram cycle. Fig. 10 again shows the comparison between the experimental results in Fig. 9 with the corresponding calculations. The gain by the ram cycle also increases with decreasing dilution with N_2, but there exists only a narrow range of possible N_2 dilution between no ram acceleration and unstart by detonative reaction of the gas mixture. The projectile velocities are shown compared to shot 158 fired in inert gas.

4. Conclusions

In conclusion, the firings in ISL's RAMAC 30 with a smooth ram tube and fin guided projectiles carried out in 1995 up to 1996 could demonstrate:

(1) Direct firing into the superdetonative combustion mode with fin guided projectiles with ignition and stable combustion at the projectile body without unstart is possible.

(2) Controlled sabot separation in a sabot stripper tube.

(3) Weak acceleration with hydrogen and methane based combustible gas mixtures diluted with carbondioxide was obtained. The velocity gain was small due to necessary dilution and ineffective projectile back shape geometry.
(4) The strong heat flux cause aluminum and titanium projectile damage, especially at the projectile fins by melting and burning due to chemical reactions of projectile material with the oxygen and the diluent (CO_2).
(5) The additional heat input by burning material is an unwelcome heat source and leads to an unstart with a detonation wave moving in front of the projectile. Projectile canting may support the unstart.
(6) Varying the fuel content from fuel rich to fuel lean with hydrogen and methane based mixtures gave no better results in view of the previous points.
(7) Results with coatings on aluminum with zirconium oxide (ZrO_2) and aluminum oxide (Al_2O_3) have shown that coating can improve the life-time but not along the whole ram cycle.
(8) Steel projectiles can survive the whole ram cycle but the ram acceleration is very small due to high projectile mass.

References

Bruckner AP, Burnham EA, Knowlen C, Hertzberg A, Bogdanoff D W (1991) Initiation of combustion in the thermally choked ram accelerator. In: Takayama K (ed) Shock Waves, Proc 18th Int Symp on Shock Waves, Springer-Verlag, Heidelberg, pp 623-630

Giraud M, Legendre J-F, Simon G (1993) Ram accelerator in 90 mm caliber or RAMAC 90: Experimental results concerning the transdetonative combustion mode, 14th Int Symp on Ballistics

Hertzberg A, Bruckner AP, Bogdanoff DW (1986) The ram accelerator: A new chemical method of achieving ultra-high velocities. 37th ARA-meeting, Québec, Canada

Hertzberg A (1989) Thermodynamics of the ram accelerator. In: Kim YW (ed) Proc 17th Int Symp on Shock Waves and Shock Tubes, AIP Conf Proc 208, pp 2-11

Knowlen C, Burnham EA, Kull AE, Bruckner AP, Hertzberg A (1990) Ram accelerator performance in the transdetonative velocity regime. 41st ARA-meeting, San Diego, USA

Seiler F, Patz G, Smeets G, Srulijes J (1993) Status of ISL's RAMAC 30 with rail stabilized projectiles. Proc Int Workshop on Ram Accelerators, RAMAC I, ISL, France

Seiler F, Patz G, Smeets G, Srulijes J (1995) The rail tube in ram acceleration: Feasibility study with ISL's RAMAC 30. Proc 2nd Int Workshop on Ram Accelerators, Univ Washington, Seattle, USA

Smeets G, Gatau F, Srulijes J (1992) Rechenprogramm für Abschätzungen zur Ram-Rohrbeschleunigung, ISL-Report RT 507/92

High performance ram accelerator research

D. L. Kruczynski
UTRON Inc. 8506 Wellington Rd., Manassas, VA 20109, USA

Abstract. Some applications for ram acceleration require high projectile accelerations or energy increases within constrained tube lengths. This may be accomplished through increased combustion pressure, increased propellant chemical energy, and/or optimized projectile geometry. Taken together these items serve to enhance overall ram accelerator performance. In addition, reduction of overall accelerator length can be accomplished by reducing accelerator entrance velocity requirements and by eliminating unnecessary components such as vent sections. This paper summarizes these effects and presents continuing large caliber experimental efforts towards developing "high performance" ram accelerators.

Keywords: Starting process, Obturator, Unstart

1. Introduction

Ram acceleration is initiated by injection of a projectile, similar in shape to the center body of a ramjet engine, into a tube filled with a gaseous fuel/oxidizer/diluent, or simply propellant. As the sub-caliber fin guided projectile enters these gases, at supersonic speeds, shock and viscous heating occurs. Properly "timed", this ignites and sustains combustion on the aft section and behind the projectile. This energy release occurs continuously as the projectile accelerates. Figure 1 shows the process.

While the energy release is continuous, the propulsive efficiency is primarily a function of the projectile's Mach number, the propellant energy content, the propellant pressure, and to some extent the projectile geometry. By segmenting the accelerator tube with diaphragms the propellant mixture may be adjusted to ensure "efficient" operation. This process, typically referred to as "staging", has allowed the 38-mm diameter bore accelerator at the University of Washington (UW) to obtain projectile velocities in excess of 2.6 km/s using a 16-m long accelerator operating at relatively low combustion pressures, on the order of 100 MPa (Knowlen et al. 1994). As the process was successfully scaled from 38-mm to 90-mm it was reported that the energy level of the propellant had to be reduced to ensure operation without unstart (Giraud et al. 1992). For a ram accelerator, an unstart refers to combustion or detonation moving upstream of the projectile's midbody, in which case the projectile decelerates rapidly. It was thought, that this was a fundamental physics issue related to the induction time of the propellant. That is, given the same basic projectile geometry and Mach number, the energy released behind the throat (midbody) of the smaller projectile would be released on the forebody of a longer projectile. Based on this assumption, even less energetic propellants were used in initial experiments at 120-mm bore diameter. While scaling the ram acceleration process to larger bore size was relatively straightforward, there was concern that reductions in propellant energy would adversely affect performance. This concern follows from the simple fact that the mass of a projectile increases approximately by the cube while the area available for propulsion forces increases only by the square. Therefore, if anything, more energetic propellants are desirable at larger scales.

As additional experiments were performed, it became clear that the projectile material may in some cases burn, and in so doing release significant energy into the process (Giraud et al. 1995, Naumann 1995, Liberatore 1995)). This was particularly true of the initial 90-mm experiments which were conducted with magnesium projectiles.

Recently, successful operation of a 120-mm bore ram accelerator was demonstrated using propellants with energy levels similar to those used in smaller scale ram accelerators (Kruczynski

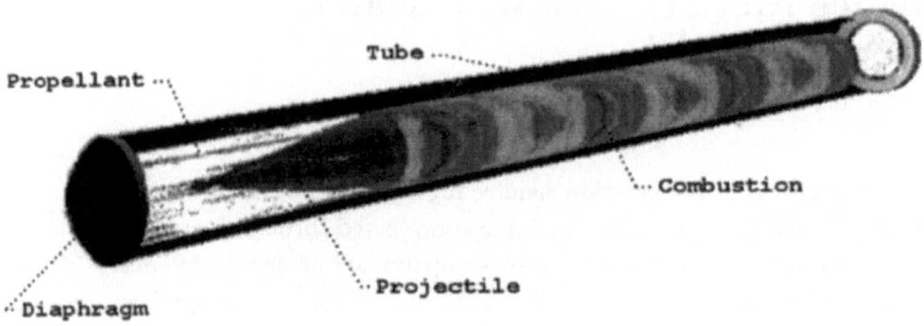

Fig. 1. Ram acceleration process

1996). These experiments, using aluminum projectiles, also used significantly higher propellant pressures.

For some applications it is desirable to maximize the projectile's acceleration and reduce the accelerator length. Achieving this goal will require changes to current ram acceleration design and operation in addition to increased propellant energies.

2. Reducing accelerator length

2.1 Shortening the pre-launcher The fundamental requirement for a ram projectile to be traveling at supersonic speeds before ram operation is initiated requires that a substantial pre-acceleration be provided to the projectile by a separate propulsion source (such as a light gas or solid propellant gun). Some theoretical work has been done examining the feasibility of eliminating the pre-launcher by essentially accelerating the propellant towards the projectile (US patent 1992), however this technique has not, to the authors knowledge, been experimentally demonstrated. However, reducing the length of the pre-accelerator is feasible. The minimal length of the pre-accelerator is controlled by two related factors.

2.2 Minumum ram "start" velocity The pre-accelerator must provide minimum projectile velocity for successful ram process "starting". For current projectile designs and propellants the "starting" velocity is above 1100 m/s. Investigations are ongoing with other propellants and geometries to lower the initial velocity requirement by 20 to 30 percent (Knowlen et al. 1997). With reduced initial velocity requirements, a pre-launcher of only a couple of meters in length will be required to provide the necessary initial projectile velocity.

2.3 Starting dynamics Successful transfer of the projectile from the pre-accelerator into the accelerator proper is a complex process involving multiple variables (Stewart et al. 1997, Nusca 1997). Parameters which effect this process are the back pressure from the pre-accelerator gun, the obturator sealing efficiency and mass, the amount of vacuum in the pre-accelerator, the accelerator diaphragm thickness (strength) and the accelerator propellant pressure. The interplay of these variables controls the conditions during the projectile entrance into the accelerator and thus the ability to initiate proper combustion in the accelerator ("starting").

Of the above parameters, the one directly effected by shortening of the pre-accelerator is the backpressure behind the obturator. The obturator must be separated from the projectile base during entrance into the ram accelerator for proper "starting". Separation of the obturator

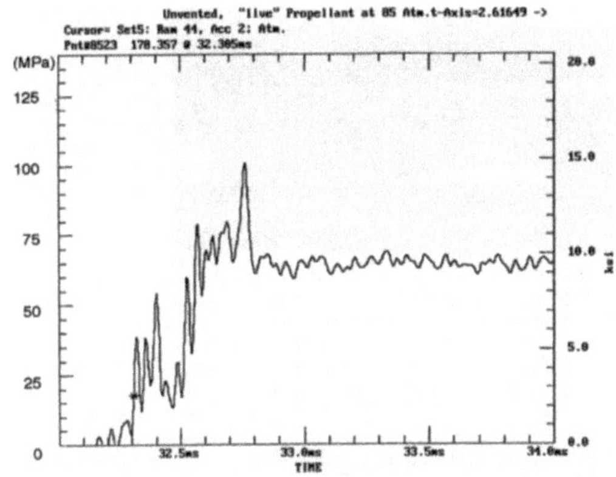

Fig. 2. "Started" pressure profile of Shot 44, 0.6m into the accelerator

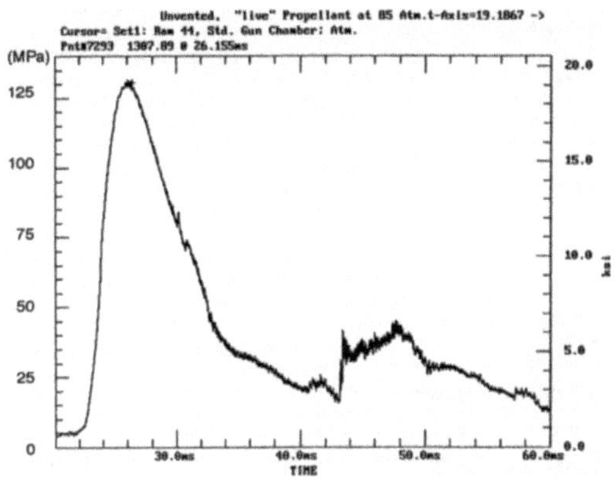

Fig. 3. Shot 44 pre-accelerator chamber pressure with reflected shock wave

is controlled by the pressure in-balances "upstream" and "downstream" of the obturator as it approaches and enters the accelerator proper. Interplay between the accelerator pressures and the pre-accelerator pressures has been experimentally recorded and will be addressed in the next section.

It is not clear at this time if the need for a minimal entrance velocity or the need for obturator separation will be the driving force in determining the absolute minimum pre-accelerator length required. Recent modeling and experimental results indicate that as launch tubes are shortened and venting is eliminated the accelerator combustion process itself becomes an important driver in separation of the obturator and subsequent ram acceleration start up. The process may become controlled by ram combustion dynamics and gas dynamics versus pure gas dynamics of previous ram accelerator experiments. These results are discussed further in the next section.

2.4 Eliminating Venting Historically, ram acceleration facilities have included a section of tube between the pre-accelerator and the ram accelerator, typically referred to as the vent section. This tube was essentially a highly perforated extension of the pre-accelerator varying from one to several meters in length. Prior to firing the perforations were either temporally sealed or the entire vent section was placed in an evacuated chamber so that the pre-accelerator could be evacuated. Some level of evacuation of the pre-accelerator was needed to preclude excessive shock/pressure

Fig. 4. 120-mm high performance ram accelerator

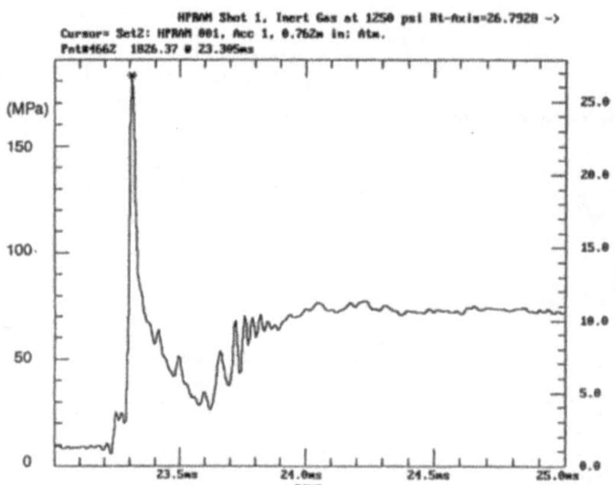

Fig. 5. "Unstarted" pressure profile of Shot 1, 0.762m into the accelerator

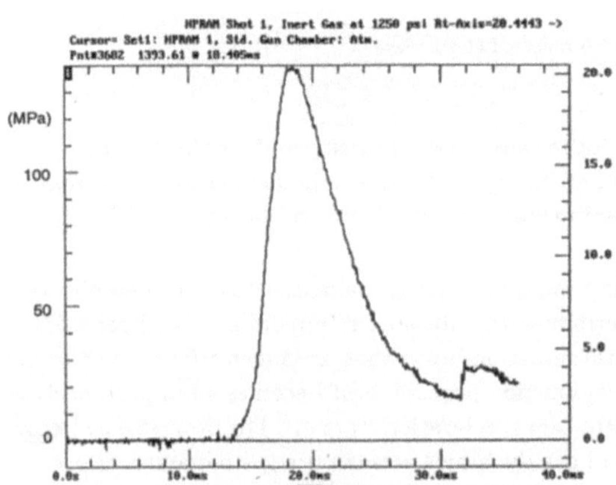

Fig. 6. Shot 1 pre-accelerator chamber pressure with reflected shock wave

buildup in front of the projectile which could burst the accelerator entrance diaphragm prematurely (Stewart et al. 1997).

A vented tube section was thought to promote successful starting by reducing levels of pressure in front of the projectile from incomplete vacuum or launch gun gas "blowby" and by reducing the backpressure behind the projectile obturator prior to entrance into the accelerator. While venting does perform these functions, other variables previously mentioned such as the vacuum level and diaphragm thickness are probably more dominant (Stewart et al. 1997).

Successful ram acceleration operation can be obtained with the vented tube section removed as demonstrated in small caliber experiments (Stewart et al. 1997). To determine if "ventless" ram acceleration was feasible at larger scales, an experiment in the 120-mm ram accelerator was performed. A 2.4m vented section between the pre-launcher and the accelerator proper was replaced by a tube section of simular length but without vent holes. This unvented section essentialy served only as an extension of the pre-accelerator tube. Other conditions for this experiment are show in Table 1.

Table 1. Conditions for no vent test.

Propellant Mixture	Fill Pressure	Heat Capacity	Sound Speed	CJ Speed
(Molar)	(MPa)	(Q/C_pT)	(m/s)	(m/s)
$2.6CH_4+2O_2+10.2N_2$	8.5	3.6	359	1508

Shot44, 1 accelerator tube 4.7m long; projectile 10-deg cone, 0.261m long, 4-deg afterbody, 0.261m long, mass = 4.276kg, material = 7075-T6; obturator mass = 0.511kg.

Proper ram acceleration was achieved with the projectile accelerating from 1244 to 1453 m/s. A pressure measurement from the accelerator (Figure 2) shows a nominal pressure profile for a succesful ram accelerator test. While little difference was seen in the accelerator pressures for this experiment, as compared to a vented test, there was a distinct difference in the pressure profile recorded in the pre-accelerator (standard gun) combustion chamber. As shown in Figure 3 a shock reflection returning from the base of the separating obturator is clearly seen late in the pressure time history starting at about 44 ms. In previous 120-mm experiments using a vented section, no influence on the pre-accelerator chamber pressure from the ram acceleration process was recorded (no second pressure rise).

Clearly, the lack of a vent section tends to more closely associate the two launch processes. As the pre-accelerator is further shortened these processes may begin to interact directly and more dramatically.

3. 120-mm high performance ram accelerator

With support from the Office of Naval Research and the NSWC, UTRON personnel designed and constructed a 120-mm ram accelerator capable of operating at combustion pressures in excess of 680 MPa. The facility was designed to take advantage of the demonstrated ability of ram acceleration to produce nearly continuous pressures at levels which vary nearly linearly with fill pressure (Kruczynski 1993). Continued investigations with increased energy(Q/C_pT) propellants will also be experimented in this facility. To further demonstrate accelerator length reduction, no vent or transition section is used between the pre-accelerator and the high performance ram accelerator tube. For operational simplicity no "staging" (multiple propellant) operation is expected.

The facility is shown in Figure 4. The projectile is brought up to injection speed (near Mach 3) using a solid propellant gun. The solid propellant gun directly connects to the accelerator. Since accelerator propellant pressures may approach 34 MPa the entrance diaphragm design allows single thick diaphragms or multiple thinner diaphragms, with intermediate fill pressures, to reduce the load on the projectile nosetip during accelerator entrance. The constant diameter, 4.57 m long, accelerator tube is instrumented with high pressure Yuma E30FM pressure gages evenly distributed down its length. Doppler radar tracks the projectile inbore and at exit, while high speed movies verify projectile integrity at exit. The propellant gases are pre-mixed, sampled by an online gas chromatograph to ensure proper constituent levels, and then pumped to the accelerator.

3.1 Inert firings in the new facility Historically, inert gas firings have been conducted in ram accelerators to ensure that the obturator can be discarded in time to establish supersonic flow through the projectile throat (i.g. start the projectile). Since there is no ram combustion this "cold" starting is simply a gas dynamic process. Cold starting does not ensure that the ram acceleration process will work with energetic gas since other ram starting failure modes are possible with energetic gases (Schutz and Bruckner 1996). Previously, it was thought that failure to "cold" start the ram process ensured that the process would not work with energetic ("hot") gases, all other conditions the same. Some recent experiments however, have started "hot" where they previously failed to start "cold" (Bruckner and Knowlen 1997).

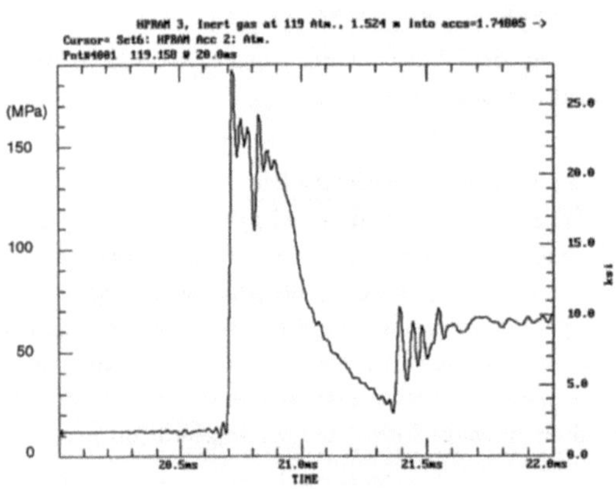

Fig. 7. "Unstarted" pressure profile of Shot 3, 1.524m into the accelerator

Indeed, as the ram accelerator vent sections are removed, pre-accelerators shortened, and entrance velocities reduced, "cold" starting may not be feasible. The combustion pressures at ram acceleration onset may be required to separate the pre-accelerator obturator in time to prevent unstarts. Modeling results show that the pressures on the accelerator side of the obturator for "hot" shots may be an order of magnitude higher than for "cold" shots (Nusca 1997).

As part of initial shakedown testing in the new facility inert gas firings were conducted as shown in Table 2.

Note that test 1 is essentially a repeat of shot 44 (detailed previously) except that the accelerator gas is inert and the 2.4 m transition section is removed. There is no venting in either case. As shown in the accelerator tube pressure of Figure 5, the inert shot 1 unstarted almost immediately. Figure 6 shows the standard gun pressure in which, like shot 44, a reflected shock wave travels back "upstream" into the standard gun. In part, to determine if a higher accelerator fill pressure would allow cold starting in the new facility shot 3 was performed with a 40 percent increase in pressure. It also unstarted almost immediately as shown in the pressure profile of Figures 7. Figure 8 shows the standard gun pressure profile (with reflected wave) from shot 3.

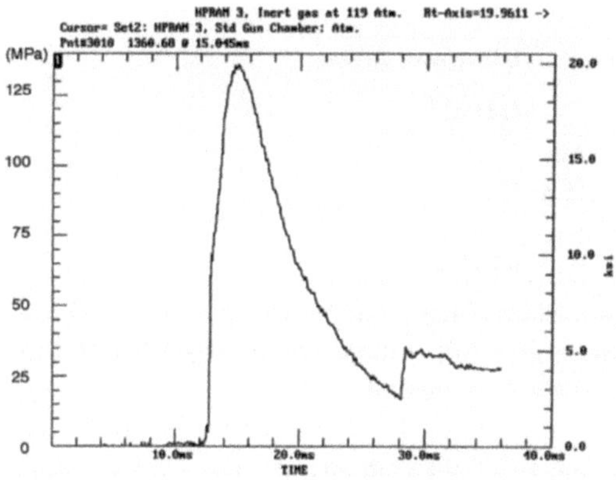

Fig. 8. Shot 3 pre-accelerator chamber pressure with reflected shock wave

Table 2. Inert test series

Test	Accelerator Pressure (MPa)*	Projectile Entrance Velocity (m/s)	Entrance Mach Number
1	8.5	1120	3.4
3	11.9	1125	3.4

* Nitrogen fill; Projectile 10-deg cone, 0.261m long, 4-deg afterbody, 0.261m long, mass = 4.276kg, material = 7075-T6; Obturator mass = 0.511kg.

To better understand the dynamics of the three shots, particularly as related to the wave activity in the standard gun some comparisons are made in Table 3.

Examining Table 3 reveals no clear relationship between the returning time or magnitude of the reflected shock wave into the standard gun chamber which would account for shot 44 "starting" and shots 1 and 3 "unstarting". There is an increase in pressure behind the obturator in shots 1 and 3 which would be expected since the pre-launcher is 2.4 m shorter. Note that for timing comparisons the wave time of shot 44 has been adjusted (shortened) to account for its 2.4 m longer pre-accelerator using modeled gas sound speed properties near exit.

It appears from this limited data that rate of separation of the obturator during the entrance into the accelerator is more important than the time of separation. The timing of the wave data indicates that the obturator in the "unstarted" inert gas shots initially separated from the projectile base at least as quickly as in the "live" propellant "started" case. However, since the "live" gas shot had much higher pressures on the accelerator side of the obturator and slightly lower pressures on the gun side, it most likely separated from the projectile base at a much faster rate, once separated.

Although still not fully understood, this interaction between the pre-accelerator and accelerator as the two are more closely coupled provides another diagnostic tool for analyzing the integrated performance of the complete launcher system.

Table 3. Comparisons of Shot 44, 1 and 3

Round	Fill press. (MPa)	Peak press. (MPa)[b]		Wave time (ms)[c]		Obturator press. (MPa)[d]
		Cham	Wave		Adj	
44	85[a]	130.8	27.0	21.0	18.4	51.1
1	8.5	139.3	24.6	17.2	—	54.4*
2	11.9	136.1	35.7	15.7	—	53.1*

[a], "Hot Propellant"; [b], pressures from initial propellant (Cham) and returning wave; [c], difference in time from Cham pressure rise to Wave pressure rise; Adj, adjustment for longer launch tube; [d], pressure behind obturator at accelerator entrance; *, estimated

4. Conclusions

Significant progress has been in reducing accelerator length by eliminating the need for a vented tube section, reducing the required injection velocity, and increasing propellant energy and pressure. A new 120-mm ram accelerator facility specifically designed for high performance operation is operational and undergoing initial testing.

Acknowledgements

Dave Siegel of the Office of Naval Research and Gill Graff of the NSWC for support of this study.

References

Bruckner AP, Knowlen C (1997) Private communication

Giraud M, Legendre JF, Catoire L (1992) Ram accelerator in 90-mm caliber, first results concerning the scale effect in the thermally choked propulsion mode. 13th Int Symp on Ballistics

Giraud M, Legendre JF, Simon G, Henner M, Voisin D (1995) RAMAC in 90-mm caliber or RAMAC90: Starting process, control of the ignition location and performance in the thermally choked propulsion mode. In: Proc 2nd Int Workshop on Ram Accelerators

Knowlen C, Higgins AJ, Bruckner AP (1994) Investigation of operational limits to the ram accelerator. AIAA paper 94-2967

Knowlen C, Schultz E, Bruckner AP (1997) Investigation of low velocity starting techniques for the ram accelerator. AIAA paper 97-3174.

Kruczynski DL (1993) New experiments in a 120-mm ram accelerator at high pressures. AIAA paper 93-2589

Kruczynski DL (1996) Experimental investigation of high pressure/performance ram accelerator operation. AIAA paper 96-2676

Liberatore F (1995) The effects of real material behavior on ram accelerator performance. Proc 2nd Int Workshop on Ram Accelerators 1995.

Naumann KW (1995) Thermal stress due to aerodynamic heating of projectiles during acceleration in a ram accelerator tube. Proc 2nd Int Workshop on Ram Accelerators

Nusca M (1997) Computational simulation of the ram accelerator using a coupled CFD/interior ballistic code approach. AIAA paper 97-2653

Stewart J, Knowlen C, Bruckner AP (1997) Effects of launch tube gases on starting of the ram accelerator. AIAA paper 97-3175

US patent (1992) Method and apparatus for zero velocity start ram acceleration. Number 5,097,743, March 24

Schutz E, Bruckner AP (1996) Ignition of the ram accelerator: propellant chemistry, Mach number, throat area, and obturator configuration effects. 33d JANNAF Combustion Meeting, CPIA Publ 653, Vol 1, pp 409-433

Ignition study for low pressure combustible mixture in a ram accelerator

X. Chang, S. Matsuoka, T. Watanabe S. Taki

Department of Mechanical Engineering, Hiroshima University, 1-4-1 Kagamiyama Higashi-Hiroshima, 739, Japan

Abstract. An ignition tube as a new section is inserted between the prelauncher and the ram acceleration section of the ram accelerator facility in Hiroshima University to study the ignition process under a sub-detonative condition. Because of the low fill pressure (around 0.5MPa) in ram acceleration section, it's hard to ignite the premixed combustible gas which has the appropriate mixing ratio to keep the combustion behind the projectile in thermal choking propulsion mode. For this reason, a short ignition tube is employed in which hot (energetic) combustible gas is included to help the ignition. In the ram acceleration section, gas mixture colder (less-energetic) than it in the ignition tube is used to accelerate the projectile in sub-detonative mode. A 56 m/s velocity gain has been observed in a 2 m long ram acceleration section with $1.4CH_4+2O_2+4.3CO_2$ mixture at 0.45 MPa.

Key words: Ignition, Low pressure, Hot start, Rectangular projectile

1. Introduction

An experimental research with a rectangular bore ram accelerator has started since 1993 in Hiroshima University (Chang et al. 1994) although such kind of studies performed with cylindrical bore has been carried out in the University of Washington (Hertzberg et al. 1988), Army Research Laboratory, USA (Kruczynski et al. 1991) and French-German Institute of Saite-Louis, France (Giraud et al. 1992). The rectangular bore is considered for the fundamental research of ram acceleration phenomenon by visualizing the flow field around the projectiles flying with ram acceleration. Another purpose is to remove the fins on the projectiles flying in the cylindrical tube which effect the flow field, the ignition process and the combustion in a ram accelerator.

The ram accelerator in Hiroshima University is designed to be operated at the pressure lower than 0.5 MPa. The combustion in the ram accelerator is initiated by the impact of a plate, so-called sabot, behind a projectile. Because of the low operation pressure, the methane-oxygen gas mixture must be hot (energetic) enough to be ignited by the sabot impact. However, the combustion can not be maintained behind the projectile and generates a detonation wave passing through the projectile throat resulting in a "diffuser unstarted" in such a hot mixture. On the other hand, if the gas mixture is diluted by inert gas to keep the detonation occurrence away, it will not be ignited by the sabot impact. Hundreds shots were made to try to reach an operational region, but all failed.

In order to make sure the ignition and the following combustion in an appropriate mixture, an ignition tube was installed upstream the ram acceleration section. This tube contains energetic mixture for the ignition. Therefore the tube must be short enough otherwise the combustion of the energetic mixture will cause "diffuser unstarted" in it, before the projectile is ejected. The gas mixture in ram acceleration section is set colder (non-energetic) by inert dilution for the ram acceleration operation. This colder gas can not be ignited by the impact, but can be ignited by the burning gas ejected from the ignition tube behind the projectile.

2. Experiment facility

Hiroshima University ram accelerator is shown in Fig. 1. The projectile prelauncher is a two stage light gas gun which has the potential to launch a combination of a projectile and obturator weighing about 0.010 kg up to about 1.1 km/s. A driver gas dump tank is connected to the launcher to allow the driver gas behind the obturator running away. An ignition tube is newly added to be connected to the dump tank. This tube is 200 mm long and has the same cross-section as the launch tube and the ram tube shown in Fig. 1. The configuration of the rectangular bore of the launch tube, ignition tube and ram tube is shown in Fig.2. A measurement station consisted of a pressure transducer, a magnetic detector and a optical sensor is located at the tube center to acquire flow and combustion information around the projectile. Figure 3 shows the detail of the station. As the ram acceleration section, a 2 m long ram tube is connected to the ignition tube. Three measurement stations as same as one on the ignition tube are located 270 mm, 1070mm and 1770mm away from the entrance. A series of magnetic detectors is also mounted along the ram tube to verify the projectile position and velocity and hence acceleration. The final dump tank is connected to the ram tube by a deceleration tube to dump the burned gas in the ram tube.

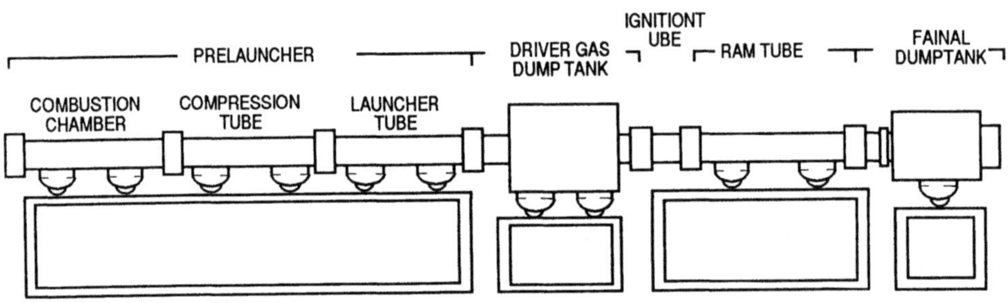

Fig. 1. Schematic of Hiroshima University ram accelerator

A sharp-wedge projectile as shown in Fig. 4 is used to fit the rectangular bore ram accelerator. The grooves on both side of the projectile is designed to ride on the guide rails on the inside walls of the launch, ignition and ram tubes in order to center the projectile flying in these tubes. The obturator consists of two pieces, so-called driver sabot and ignition sabot. The driver sabot is used to accept the driver force of the prelauncher. It is trapped and hence stopped in the driver gas dump tank. The ignition sabot located between the driver sabot and the projectile is launched together with the projectile into the ignition tube to initiate the combustion.

3. Experiment results and discussions

A series of experiments were done in methane-oxygen mixture with carbon-dioxide dilution. Typical experimental conditions are shown in Table 1. The gas mixture in the ignition tube is $1.4CH_4+2O_2+2CO_2$ which can be ignited by the impact of ignition sabot with the entrance velocity around 1 km/s and keeps combustion away from detonation until the projectile and the sabot enter the ram tube. The mixture in the ram tube is $1.4CH_4+2O_2+4.3CO_2$ colder than the

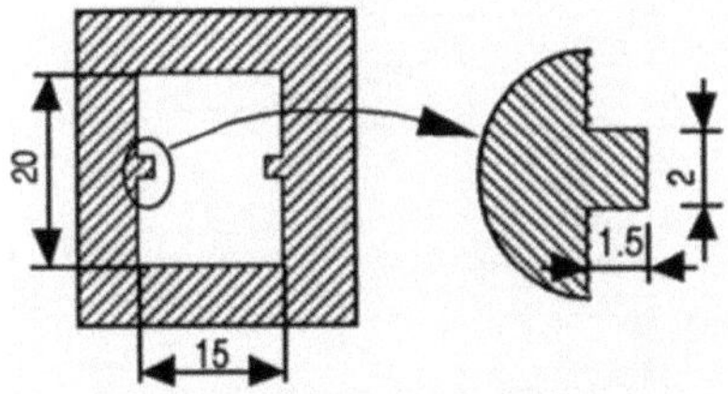

Fig. 2. Configuration of the rectangular bore

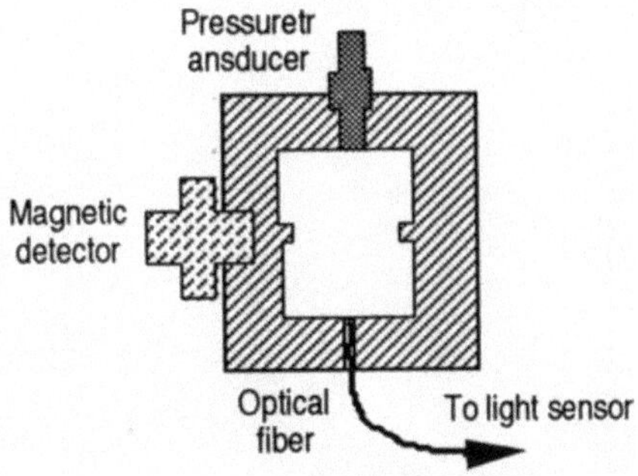

Fig. 3. Detail of the measurement station

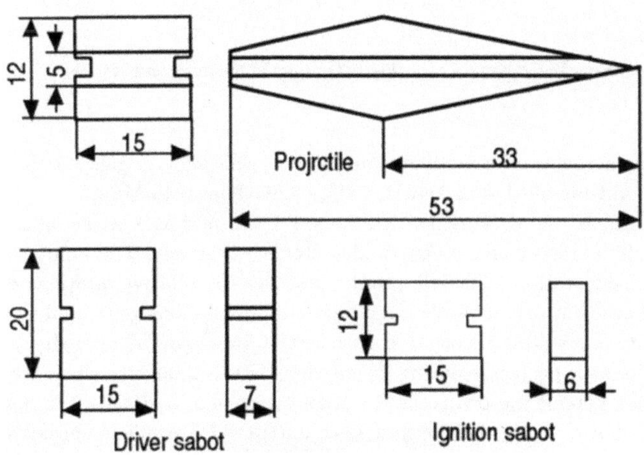

Fig. 4. Details of projectile, driver sabot and ignition sabot

Table 1. Typical experimental conditions

	Composition	Pressure(Pa)
Combustion chamber	CH_4+2O_2	0.39×10^6
Compression tube	He	0.13×10^6
Launch tube	Air	120
He dump tank	Air	120
Ignition tube	$1.4CH_4+2O_2+2O_2$	0.3×10^6
Ram tube	$1.4CH_4+2O_2+4.3O_2$	0.45×10^6
	Mass(kg)	Geometry and dimensions (mm)
Free piston	0.33	$\phi 60 \times L100$
Projectile	6.2×10^{-3}	Sharp-wegde
Driver sabot	2.2×10^{-3}	$H20 \times W15 \times L6$
Ignition sabot	1.4×10^{-3}	$H12 \times W15 \times L6$

gas in the ignition tube. As the diluent gas CO_2 was used to decrease the sound velocity of the mixture and to avoid diffuser unstarted.

The projectile, ignition sabot and driver sabot are made of Lexan. They weighing about 0.010 kg are launched to 1020 m/s at the muzzle of the prelauncher. Figure 5 shows a projectile velocity

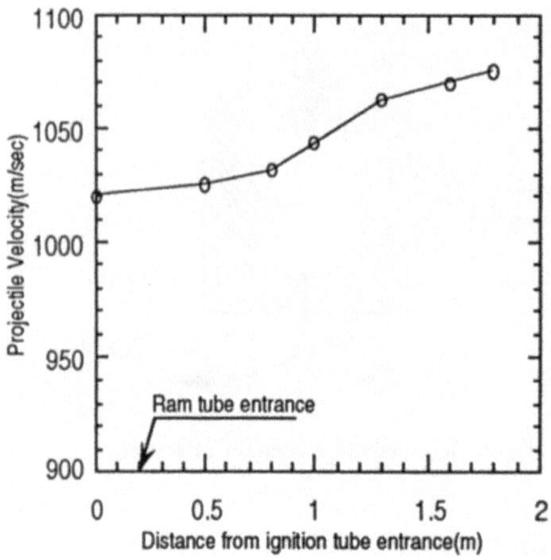

Fig. 5. Projectile velocity distribution along the ignition and ram tubes

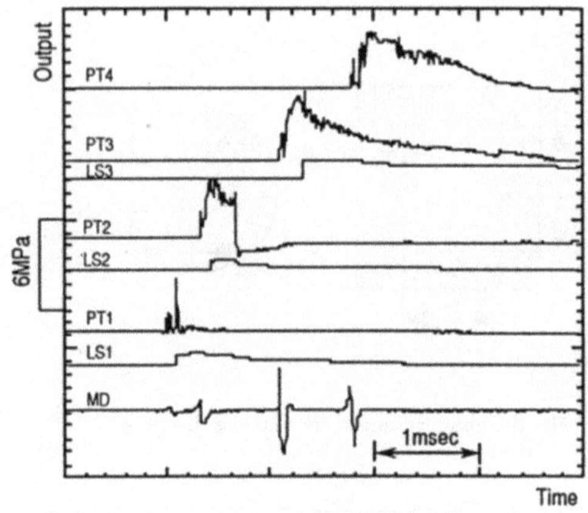

Fig. 6. Outputs of measurement stations

distribution under the experimental conditions shown in Table 1. The velocity gain for this shot is 56 m/s in the 2 m long ram tube filled with $1.4CH_4+2O_2+4.3CO_2$ at 0.45 MPa.

Figure 6 shows the outputs at the measurement stations. PT, LS and MD in the figure represent pressure transducer, light sensor and magnetic detector, respectively. The number after PT and LS corresponds to each station. The MD profile indicates the relative positions of the projectile at each station. These outputs show that the high pressure regions generated by combustion is behind the projectile, resulting thrust to the projectile. However, LS lines shows that the combustion goes away from the projectile during its travelling. Even though the high pressure area is still kept immediately behind the projectile. In order to discuss in details, the outputs of station 2 is magnified as shown in Fig. 7. The first peak on the PT2 profile is considered to be the pressure raise of the oblique shock wave formed at the projectile tip. The second peak will be the normal shock standing on the rear of the projectile. The high pressure region behind the second peak is generated by the combustion. The perturbation on the high pressure region may be due to the instability of the combustion. The rapid pressure decrease must be caused by the expansion wave after the ignition sabot passed. LS2 line shows that the combustion does not occur immediately behind the projectile base. The flame front is observed about 100 μs later than the projectile throat passed. Nevertheless the pressure wave is still driven onto the projectile.

4. Conclusions

A successful shot has been made by having an ignition tube installed between the prelauncher and the ram tube. It is playing an important role in initiating a non-energetic gas mixture in the ram tube which is impossible to be ignited by sabot impact. Experimental results show that even a two-dimensional projectile without fins can also be accelerated in a ram accelerator under a low pressure condition such as 0.45 MPa. A 56 m/s velocity gain has been observed with a Lexan projectile weighing 6.2×10^{-3} kg. The acceleration obtained can be calculated as about 30 km/s^2.

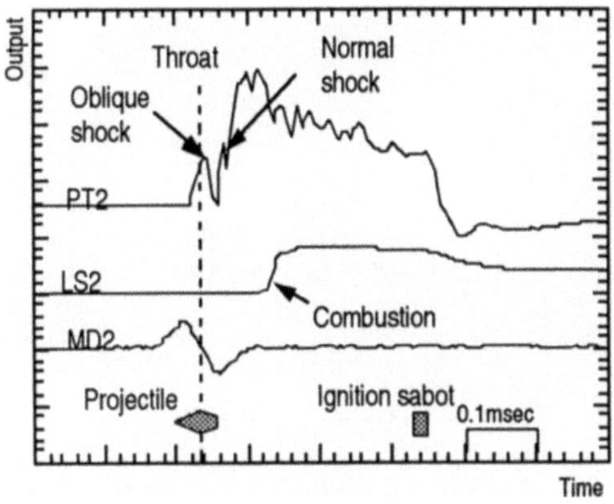

Fig. 7. Detail view of ST2 outputs

Acknowledgement

The authors would like to thank students in our group Y. Hamaguchi, R. Ueno and Y. Maruyama for their helps in performing experiments.

References

Chang X, Kanemoto H, Taki S (1994) A rectangular ram accelerator is being made at Hiroshima University. ISTS 94-a-05

Hertzberg A, Bruckner AP, Bogdanoff DW (1998) Ram accelerator: A new chemical method for accelerating projectiles to ultrahigh velocities. AIAA J 26:195-203

Kruczynski DL (1991) Requirements, design, construction, and testing of a 120-mm in-bore ram accelerator. 28th JANNAF Combustion Meeting, CPIA Publication 573, Vol.1

Giraud M, Legendre JF, Simon G, Catoire L (1992) Ram accelerator in 90 mm Calibe. ISL CO 219/92

Thermally choked operation in a 25-mm-bore ram accelerator

A. Sasoh, S. Hirakata, J. Maemura, Y, Hamate, K. Takayama
Shock Wave Research Center, Institute of Fluid Science, Tohoku University, 2-1-1 Katahira, Aoba-ku, Sendai 980-77, Japan

Abstract. The system of a 25-mm-bore ram accelerator (RAMAC25) was installed at Shock Wave Research Center, Institute of Fluid Science, Tohoku University. The total length of the ram acceleration tube is 6 m. It uses a powder gun as a pre-accelerator. The propellant mixture is supplied through a system composed of mass flow controllers and backpressure regulators. Pressure signals are recorded through multiplexer systems, thereby saving the number of channels of the data storage device. Experimental study was conducted, changing the amount of the smokeless powder in the pre-accelerator, the component and pressure of the mixture, the initial pressure in the launch tube of the pre-accelerator, the design of the sabot and disk, and the thickness of the diaphragm etc. Ram acceleration was realized using a methane-based mixture. Results of the operation experiments are presented here.

Key words: Starting process, Diaphragm, Precursor shock wave

1. Introduction

In the ram accelerator operation, or in the momentum transfer from combustion-driven waves to the projectile, there exist two characteristic length scales in the system; one is related to the device dimension such as the bore diameter; another is related to the chemical reactions. Without any dissipation effects taken into account, gasdynamic similarity exists that the same normalized thrust is obtained with the fill pressure and the heat release of the mixture, the Mach number and geometry of the projectile being unchanged. Among the existing ram accelerators, the bore diameter varies from 20 mm to 120 mm. Basically, they use mixtures of similar components. A majority uses fin-guided, conically nosed projectiles. The fill pressure ranges from 2.5 to 10 MPa. So far, scaling effects on ram accelerator operation have not been clarified. An important thing in varying the dimension of the device is that the ratio of the device dimension to the length scale related to the chemical reaction changes. As is shown by Sasoh and Knowlen (1997a), this ratio can affect the thrust performance of the device. So far it is not clear how a small ram accelerator such as of 25 mm in bore dia. works, nor how this theoretically predicted tendency is realized in experiments.

Based on the square-cubic law, the smaller the length scale of the device the higher the acceleration becomes. This may suggest that in order to improve the current level of the ram accelerator performance, a smaller device should be developed. However, this may contradict against the advantage of the ram accelerator that it is suitable as a heavy-mass launcher because the duration of the acceleration is long while the peak acceleration is low (Hertzberg et al. 1988). Another aspect of the small devices is that it is weaker against the same level of pressure or mechanical load.

As is stated above, the small-bore ram accelerator has a potential of obtaining higher performance although its structural load is more critical. In this study, experimental study was conducted using a 25-mm-bore ram accelerator installed at Shock Wave Research Center, Institute of Fluid Science, Tohoku University.

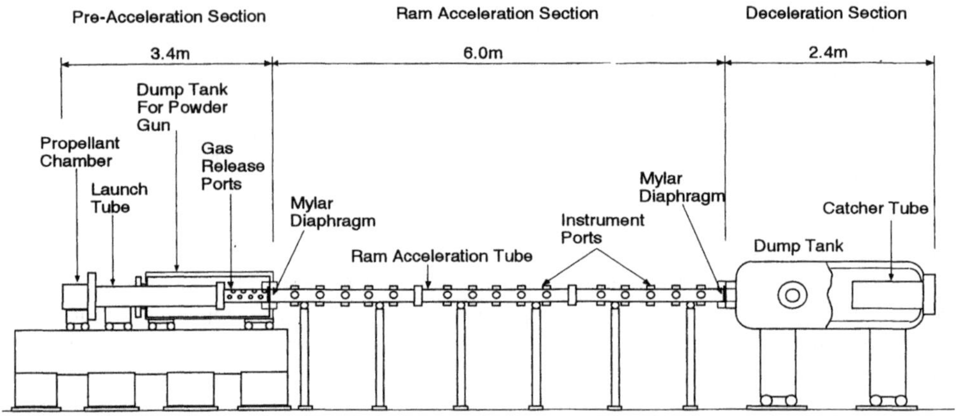

Fig. 1. 25-mm-bore ram accelerator at Shock Wave Research Center

2. RAMAC25 system

2.1 RAMAC25

Figure 1 schematically illustrates the 25-mm-bore ram accelerator used in the experiment (Sasoh et al. 1996a, 1996b, 1997b). The Shock Wave Research Center started ram accelerator operation experiments September 1995. Up to July 1997, 115 hot shots and 52 cold shots have been conducted. The bore diameter and the length of the ram acceleration tube are 25 mm and 6 m, respectively. A projectile is initially accelerated up to 1.25 km/s using a powder gun. In the powder gun, 25 to 30 gram singly-based smokeless powder (NY500, Nippon Fats and Oil Co.) is used. The length of the launch tube of the powder gun is 3 m. At the exit of the launch tube, twenty four 8-mm-dia. ventilation holes are made. The ram acceleration tube is composed of three 2-m-long, stainless steel tubes. Locations where a pressure and the magnetic induction produced by the passage of the projectile are measured are apart from each other by a separation distance of 0.4 m.

2.2 Projectile and sabot

Figure 2 shows the designs of the projectile, sabot and disk used in the experiment. The projectile used in this study has a conical nose of a half apex angle of 10 degrees. Its aft body has four fins which center the projectile. The nose is made of aluminum alloy, A7075-T651. The aft body is made either of the aluminum alloy or magnesium alloy, AZ31(F). The aluminum aft body can withstand the initial impact loaded by the powder gun even if the mass of smokeless powder amounts to 30 grams. However, if the material of the aft body was the magnesium alloy, its base was damaged by the impact with this amount of the smokeless powder.

In this device, a sabot made of polycarbonate which had perforation holes could not be used because it did not withstand a large initial impact made by the powder gun. If the sabot made of polycarbonate did not have the perforations, that is, a solid one, it was not broken by the shot. However, a condition for ram acceleration was not found using this solid sabot. The configuration of the sabot has been one of the most critical design factors in this device. Currently, the magnesium alloy is used as the material of the sabot and disk. Using this material, the sabot withstood against the initial impact of the powder gun. We tried to make the sabot as thin as possible in order to increase the tolerance for the ram acceleration. It was found that replacing the stainless steel diaphragm of the powder gun with one made of Mylar, the diaphragm rupture

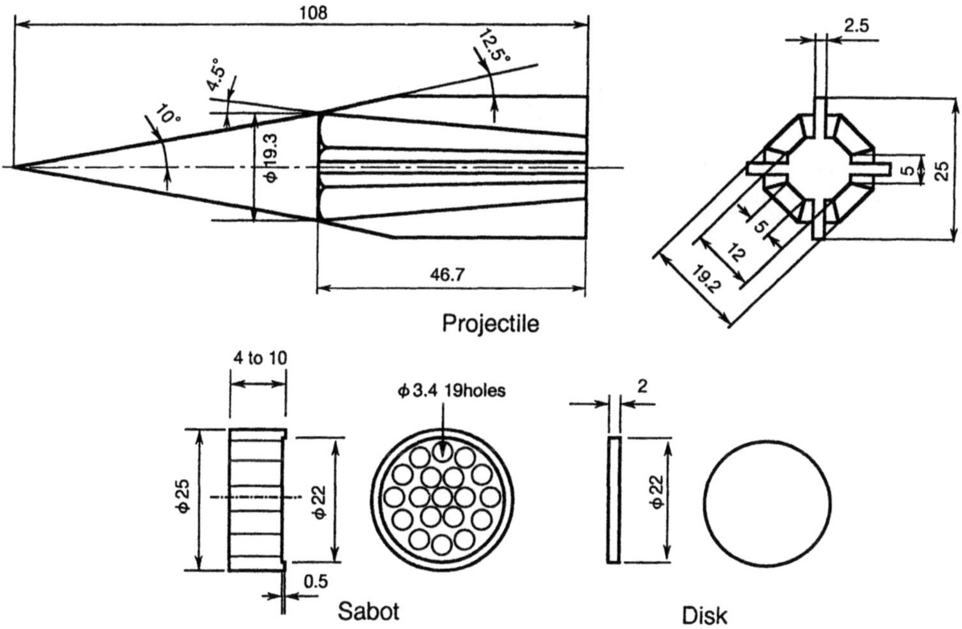

Fig. 2. Designs of projectile, sabot and disk

pressure, which was measured by a static method, was decreased from 35 MPa to 5 MPa. With this Mylar diaphragm, the thickness of the sabot with the perforations could be reduced up to 5 mm without causing its serious damage.

2.3 Gas handling system

The gas handling is conducted at about 10 m horizontally apart from the ram accelerator. The location of the control is shaded by a wall of the laboratory. After the propellant of the powder gun is loaded, the propellant mixture is remotely supplied to the ram accelerator tube.

Figure 3 illustrate the mass flow control system used for the propellant mixture supply. The system employs the combinations of a mass flow controller (Brooks 5850E) and a back pressure regulator (Tescom 26-1725-24) for each species. This system has the following advantages:

(1) It regulates the mass flow rate in an active manner. Therefore, once the calibration is conducted, it becomes easy to handle the system.

(2) The pressure difference through the mass flow controller is only 0.3 MPa at most. The flow regulation is done not by using a sonic orifice but by subsonic flow through the orifice. In this system, the gas in a cylinder can be utilized up to a pressure only 0.5 MPa higher than the fill pressure of the ram acceleration tube.

2.4 System of measurements and data acquisition

The velocity of the projectile was measured by the method of time of flight. The projectile has a rubber ring magnet which was magnetized under a static magnetic field of 0.8 T produced by an electrical magnet for a minute. The magnetic induction produced by the passage of the projectile is sensed by pick-up coils.

Time variations of pressures in the ram acceleration tube are measured using piezoelectric pressure transducers. Using a similar system to that of University Washington, pressures measured

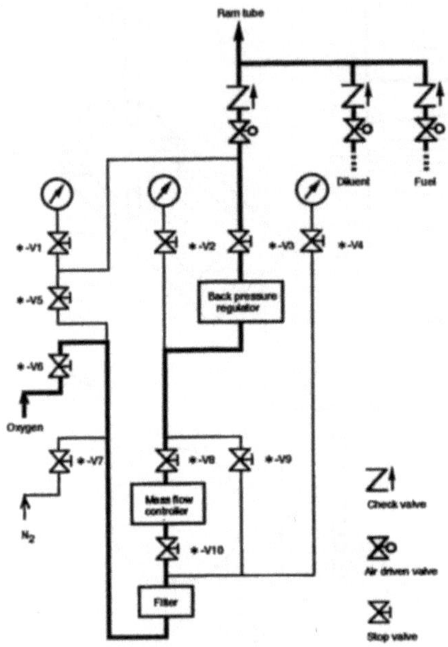

Fig. 3. Mass flow control system

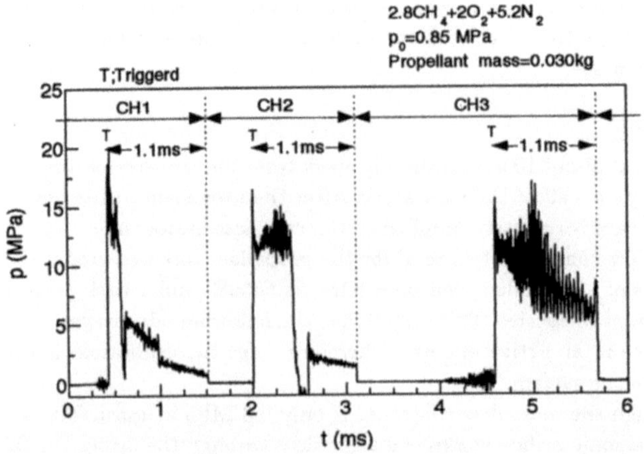

Fig. 4. Examples of recorded pressure data acquired using the multiplexer

at plural positions are multiplexed and recorded into a single channel of the data storage device. The data acquisition system is composed of a data storage device (DM7200, Iwatsu Co.) and a personal computer. Figure 4 shows an example of the recorded time variations of the pressures in the ram acceleration tube. Initially, the multiplexer system reads the signal of CH1. After 1.1 ms (variable value) since the system is triggered by the signal of CH1, it starts reading the signal of CH2. The same procedure continues up to eight channels. This system saves the number of channels of the data storage device and in the present experiment enables us to record eight pressure data into four channels of it.

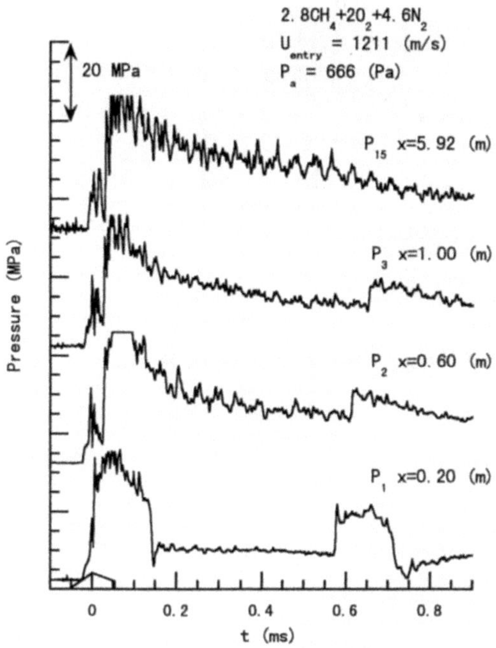

Fig. 5. Pressure histories of ram acceleration operation

3. Operation experiments in RAMAC 25

Figure 5 shows the pressure histories of ram acceleration operation. The mixture was $2.8CH_4+2O_2+4.6N_2$. Its non-dimensional heat release, speed of sound and Chapman-Jouguet (C-J) detonation speed are 4.93, 366 m/s and 1771 m/s, respectively. The fill pressure was 2.5 MPa. The amount of the smokeless powder was 30 grams. The speed of the projectile determined from the passage periods at $x = 0.2$ m and 0.6 m, where x designates a distance form the entrance of the ram acceleration tube, was 1211 m/s. The nose and the aft body of the projectile was made of the aluminum alloy.

At $x = 0.2$ m, the mixture around the projectile is overdriven by the sabot. Around the nose of the projectile, the pressure profile does not change from that observed when the tube is filled with inert gas, implying that no significant heat release does not occur there. The pressure drop at $t = 1.43$ ms is associated with the passage of the sabot. After the passage of the sabot, the pressure reduced to a lower level. It is not clear what caused the second pressure jump, that is the shock wave observed at $t=1.86$ ms. It could be either a shock wave driven by the sabot which is suddenly decelerated by the high pressure on the downstream side or that driven by the disk which is supposed to be separated from the sabot immediately after entering the ram acceleration tube.

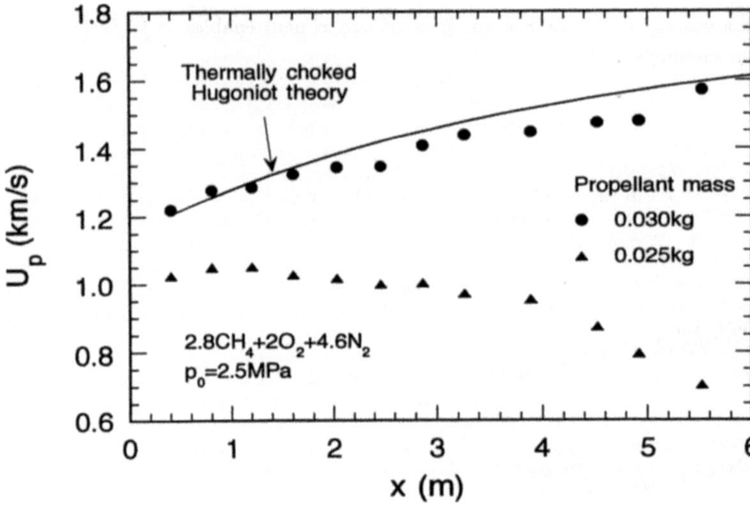

Fig. 6. Projectile velocity profiles: symbols, experiment; solid line, ram accelerator Hugoniot theory

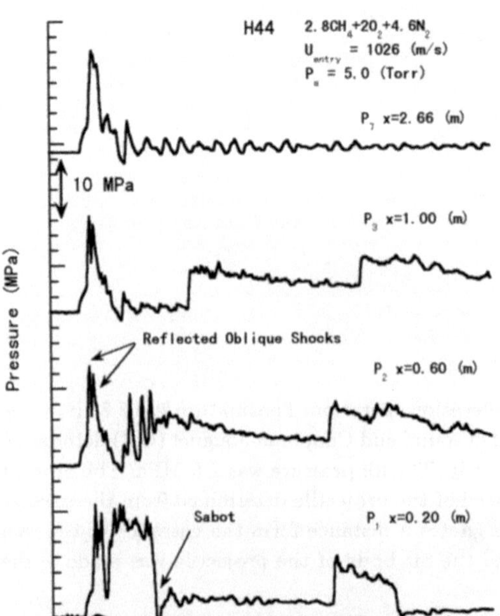

Fig. 7. Pressure histories of operation without ignition

The pressure histories measured downstream definitely exhibit that ram acceleration was realized in this experiment. Those show typical characteristics of the thermally choked operation (Bruckner et al. 1991). Around the nose, the pressure spikes correspond to oblique shock wave reflections on the tube wall. The pressure level there is almost the same as that of cold shots where inert gas is used as the test gas. Behind the throat, the pressure becomes highest and then decreases. The maximum value is, although in the experiment the data storage device overscaled, almost at the same level of that of the C-J detonation, 36.0MPa. The decrease in pressure behind the wave is the typical profile of subsonic heat release. It follows from these results that a thermally choked ram accelerator operation was realized in this experiment.

Figure 6 shows by closed circles the velocity profile measured in the same experiment. That calculated using the thermally choked ram accelerator Hugoniot analysis (Knowlen 1991, Sasoh et al. 1996b) is also plotted by the solid line. The experimentally measured velocity qualitatively agreed with but slightly lower than that was calculated.

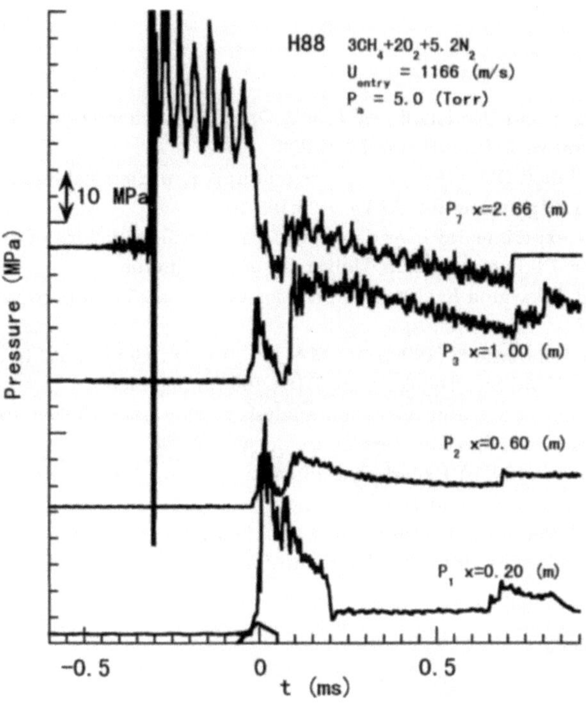

Fig. 8. Pressure histories of operation of combustion-driven unstart

The velocity profile in the case of ignition failure is shown also in this figure by the closed triangles. The corresponding pressure histories are shown Fig. 7 In this case, the amount of the smokeless powder was 0.025 kg and the projectile entry speed near the entrance was about 1030m/s. Because of this low entry speed, the mixture was not ignited. The projectile coasted supersonically in the tube and then at around x=4m, the deceleration of the projectile suddenly increased because choking occurred at the throat, thereby resulting in unstart as a supersonic diffuser.

Figure 8 shows pressure histories of an operation of immediate unstart leading to an overdriven detonation propagating ahead of the projectile.

4. Summary

The system of the 25-mm-bore ram accelerator was installed and its operation experiments were conducted. Ram acceleration was experimentally realized. Further study is now being conducted to obtain wide ranges of the ram acceleration operation. In particular, an intensified study on the starting processes is being conducted by visually diagnosing the diaphragm rupturing process of the projectile, see Sasoh et al. 1998.

5. Acknowledgement

The authors would like to thank Dr. O. Onodera, Messrs H. Ojima and T. Ogawa of the Shock Wave Research Center, Institute of Fluid Science, Tohoku University, for their assistance in conducting the present experiments.

6. References

Bruckner AP, Knowlen C, Hertzberg A and Bogdanoff DW (1991) Operational characteristics of the thermally choked ram accelerator. J. Prop Power 7:828-836

Hertzberg A, Bruckner AP and Bogdanoff DW (1988) Ram accelerator: A new chemical method for accelerating projectiles to ultrahigh velocities. AIAA J 26:195-203

Knowlen C (1991) Theoretical and experimental investigation of the thermodynamics of the thermally choked ram accelerator. Ph.D dissertation, University of Washington

Sasoh A, Hirakata S, Ujigawa Y and Takayama K (1996a) Operation tests of a 25-mm-bore ram accelerator. AIAA paper 96-2677

Sasoh A, Hirakata S, Ujigawa Y and Takayama K (1996b) RAMAC 25 - ram accelerator at Shock Wave Research Center. Mem Inst Fluid Sic, Tohoku Univ 7:161-190 (in Japanese)

Sasoh A and Knowlen C (1997a) Ram accelerator operation analysis in thermally choked and transdetonetive propulsive modes. Trans Jpn Soc Aero Space Sci 40:130-148

Sasoh A, Hirakata S and Takayama K (1997b) Experimental investigation on ram accelerator starting processes. J Jpn Soc Aero Space Sci 46:37-45

Sasoh A, Hirakata S, Maemura J and Takayama K (1998) Fluid dynamic processes of diaphragm rupturing by a conically-nosed projectile. Proc 21st Int Symp Shock Waves, Great Keppel, July 1997

37-mm bore ram accelerator of CARDC

Liu Sen, Z.Y. Bai, H.X. Jian, X.H. Ping, S.Q. Bu
Hypervelocity Aerodynamics Institute (HAI), China Aerodynamics R&D Center (CARDC), Mian Yang, Sichuan 621000, P.R. China

Abstract. A 37-mm bore ram accelerator has been built at Hypervelocity Aerodynamics Institute (HAI) of CARDC. Subdetonative acceleration mode (i.e. the thermally-choked mode) has been demonstrated, and projectiles are accelerated from 1,100 m/s to 1,450 m/s through a 4.8m-long accelerator, with averaged acceleration ranging from 7,000 to 11,000 g. Description of this 37-mm ram accelerator facility is given in this paper, together with typical results of cold/hot shot tests.

Key words: Thermally-choked mode, Experiment

1. Introduction

Ram accelerator is a ramjet-in-tube concept of hypervelocity projectile launcher, proposed by Prof. A. Hertzberg et al. in 1980s (Hertzberg et al. 1988). Two ram accelerators have been built in U.S.A., namely the 38-mm bore ram accelerator at University of Washington (UW, Hertzberg et al. 1988), and 120-mm bore ram accelerator at Army Research Laboratory (ARL, Kruczynski et al. 1996). Two other accelerators (90-mm and 30-mm bore) are working at the French-German Institute of Saint Luise (ISL, Naumann et al. 1988). All of these four ram accelerators have demonstrated positive acceleration, among which the highest projectile velocity is about 2.7 km/s at UW, while the heaviest projectile is the nearly 4 kg one at ARL. In Japan, cold shot tests have been conducted with two ram accelerators at University of Hiroshima(15×20 mm rectangular bore) and Tohoku University(25-mm bore, A Sasoh et al. 1996).*

Here at Hypervelocity Aerodynamics Institute (HAI) of China Aerodynamics R.& D. Center (CARDC), research in ram accelerator was started in early 1995, and a 37-mm bore ram accelerator has been built and subdetonative acceleration has been realized, with averaged acceleration ranging from 7,000 to 11,000g. Projectiles are accelerated from 1,100 m/s to 1,450 m/s.

2. Description of 37-mm ram accelerator

Structural designation and fabrication of this 37-mm bore ram accelerator was started in May 1995, and the whole facility consists of four sub-systems(as shown in Fig 1) are following:

(1) Subsystem of initial launcher/projectile. A 37-mm bore anti-aircraft gun is adopted as initial launcher (Figure 2), providing the projectile (similar to that at UW and weighed 110 gram) with an entrance velocity of about 1,100 m/s.

(2) Subsystem of accelerator/venting tubes. With an O.D. of 100 mm, the length of accelerator and venting tubes are 2 × 2.4 m and 2.5 m, respectively.

(3) Subsystem of gas handling. Based on partial pressure law, a relatively primitive gas handling system is used to provide the needed propellant gas mixture(mixture of methane, oxygen and nitrogen for present study) for the ram accelerator. Propellant gases are first mixed in pressure bottles, then pumped into accelerator tubes. Proper mixture concentration is insured by operating a gas chromatographer.

(4) Subsystem of measurement and data acquisition. Ten measurement sections distributes along the accelerator and venting tubes; On each section, there are one pressure transducer and one

* Recently, ram acceleration has been realized in these Japanese ram accelerators.

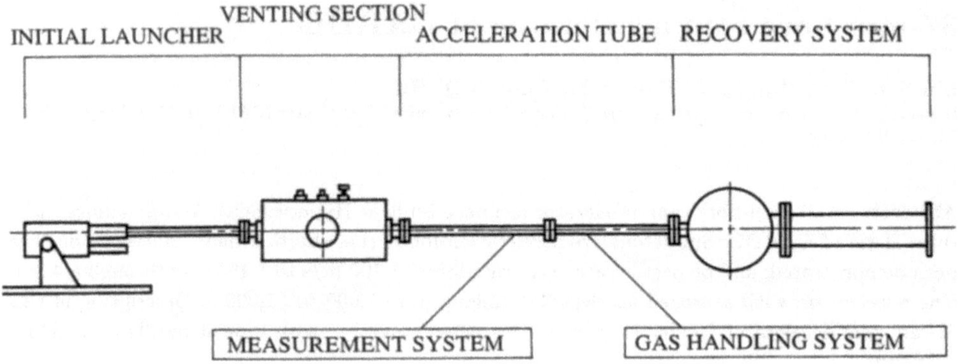

Fig. 1. 37-mm bore ram accelerator system

Fig. 2. 37-mm bore initial launcher

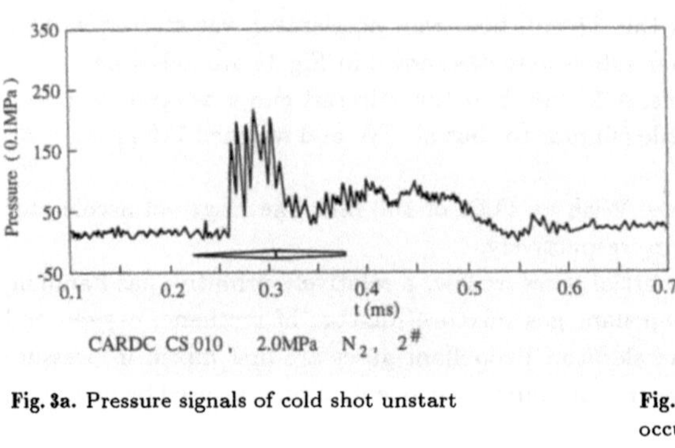

Fig. 3a. Pressure signals of cold shot unstart

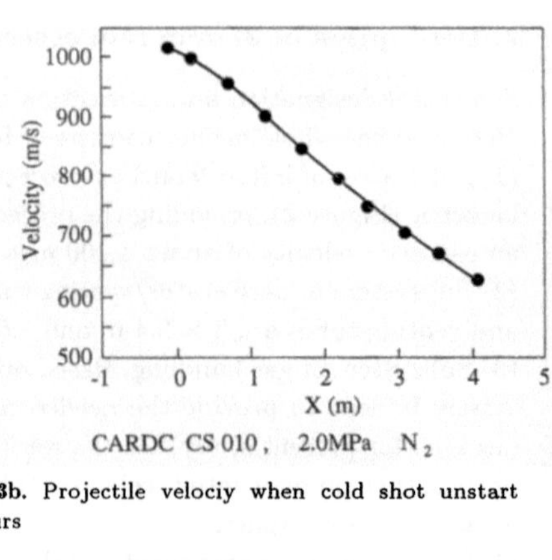

Fig. 3b. Projectile velociy when cold shot unstart occurs

electro-magnetic detector. Signal output of transducers and detectors are magnified if necessary, before recorded by a 12-channel data acquisition system.

3. Results of cold/hot shot tests

After the facility was installed in early 1996, cold and hot shot tests were conducted. So far, 15 cold shots and 30 hot shots have been tested.

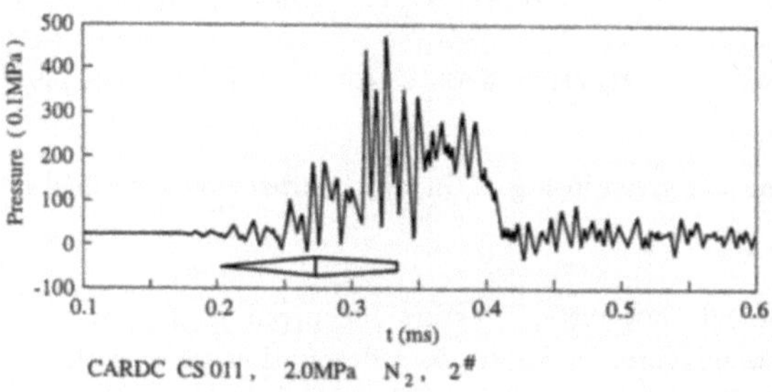

Fig. 4a. Pressure signals of successful cold shot

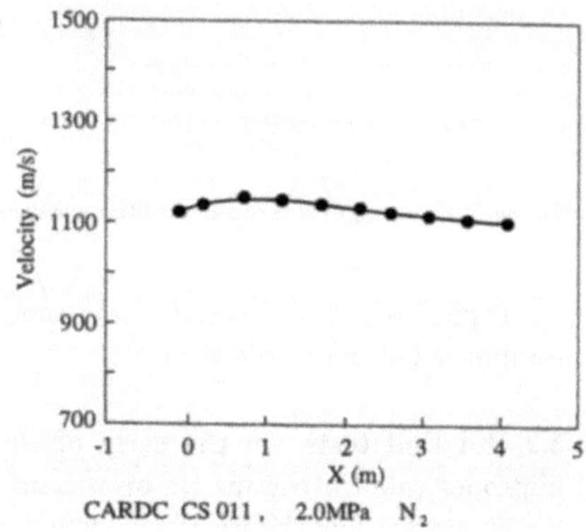

Fig. 4b. Projectile velocity of a successful cold shot

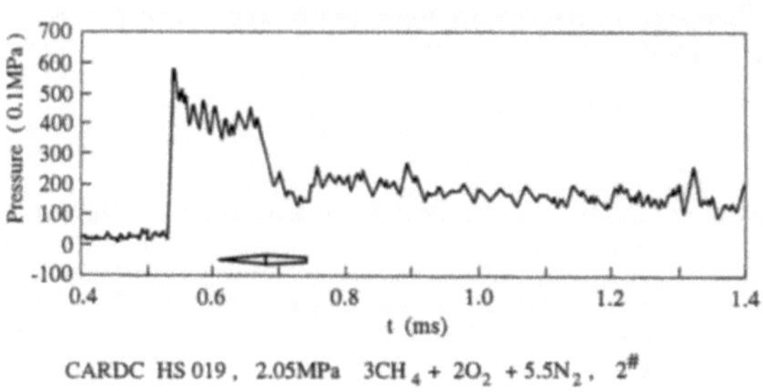

Fig. 5a. Pressure signals when hot shot unstart occurs

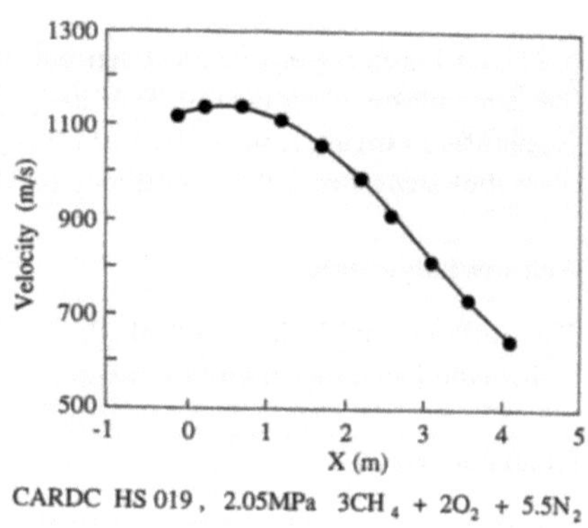

Fig. 5b. Projectile velocity when hot shot unstart occurs

3.1 Cold shot tests

The aim of cold shot tests is to examine the reasonability of this 37-mm ram accelerator design. Usually, about 2.0 MPa nitrogen is pumped into accelerator tubes in these shots. Because of the failure of projectile or sabot, cold shot "unstart" occurred at first. As show in Fig. 3, a shock wave exists up-stream to projectile throat, resulting in projectile's rapid deceleration.

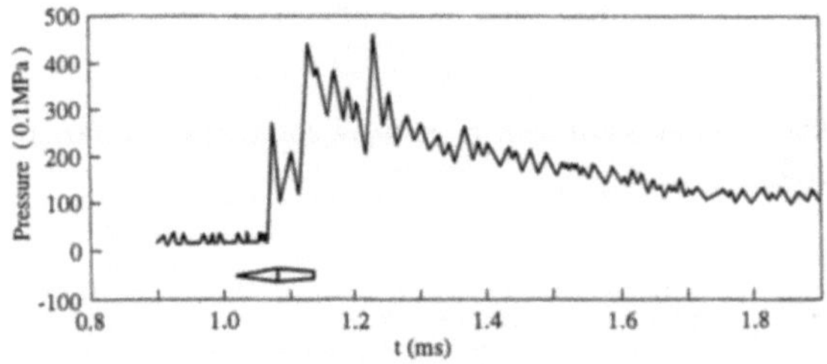

CARDC HS 031, 2.2MPa $3CH_4 + 2O_2 + 5.8N_2$, 3#

Fig. 6a. Pressure signals of sub-detonative acceleration

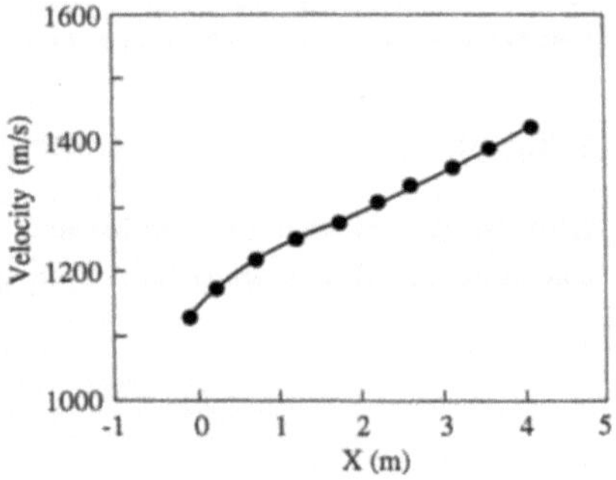

CARDC HS 031, 2.2MPa $3CH_4 + 2O_2 + 5.8N_2$

Fig. 6b. Projectile accelerates along accelerator tube

Typical result of successful cold shot test are given in Fig. 4, in which supersonic flow field is established at projectile throat.

3.2 Hot shot tests In the early phase of hot shot tests, "unstart" occurred as the result of improper sabot structure or propellant gas mixture. In Figure 5, a detonation wave is shown clearly proceeding the projectile, which decelerates projectile abruptly.

As shown in Figure 6, positive ram acceleration is realized finally, in which the projectile enters accelerator tube at 1,100 m/s, and exits at about 1,450 m/s; corresponding to averaged acceleration of about 9000 g.

4. Concluding remarks

A 37-mm bore accelerator has been built at HAI of CARDC. Subdetonative acceleration mode (i.e. the subsonic combustion thermally-choked mode) has been demonstrated, with averaged acceleration ranging from 7,000 to 11,000 g. Research in best propellant gas mixture and projectile/sabot structure will be continued on this facility.

Acknowledgement

The authors would like to thank Prof. A.Hertzberg, Prof. A. P. Bruckner and their group for their helpful suggestion and discussion.

References

Hertzberg A et al (1988) Ram accelerator: A new chemical method for accelerating projectile to ultra-high velocities. AIAA J 26:195-203

Kruczynski D et al. Flow visualization of steady and transient combustion in a 120-mm ram accelerator, AIAA paper 94-3344

Liu sen et al(1997) Ram accelerator cold shot test of CARDC-RAMAC37. J Fluid Exp, Meas and Contl (in Chinese) Vol. 11(4)

Naumann K (1996) Thermomechanical constraints on ram-accelerator projectile design limitations, possibilities, solutions. AIAA paper 96-2678

Sasoh A et al (1996) Operation tests of a 25-mm bore ram accelerator. AIAA paper 96-2677

Performance prediction

Real gas effects on thermally choked ram accelerator performance

D. L. Buckwalter, C. Knowlen, A. P. Bruckner
Aerospace and Energetics Research Program, University of Washington, Seattle, WA, 98195, USA

Abstract. A one-dimensional performance analysis code has been modified to account for real gas effects in the combustion zone of the thermally choked ram accelerator. Previous computational studies utilized the ideal gas equation of state and underpredicted experimentally measured thrust values as a function of Mach number. The post-combustion pressure in the ram accelerator is on the order of a few ten MPa and in this regime it is no longer appropriate to ignore intermolecular interactions. Including a compressibility factor to account for these real gas effects provides a much better match with experimental data. The current research code incorporates the following equations of state: ideal gas, Boltzmann, Percus-Yevick, and a direct solution to the virial expansion in which the virial coefficients are computed using the Lennard-Jones and Stockmayer potentials. The ideal gas equation of state performs well for initial fill pressures up to 1.0 MPa, while the Boltzmann and direct virial model perform adequately for fill pressures up to 5.0 MPa. The Percus-Yevick model also performs well for ram accelerator fill pressures up to 5.0 MPa and is expected to accurately model fill pressures up to a few ten MPa.

Key words: Ram accelerator, Performance analysis, Real gas effects, Thermally choked

1. Introduction

The ram accelerator is a chemically driven mass launcher that operates on principles similar to a ramjet (Hertzberg et al. 1988). It produces thrust in one of three primary propulsion modes and has accelerated projectiles to speeds up to 2.7 km/s in the University of Washington 38 mm-bore facility. In the thermally choked propulsive mode, shown in Fig. 1, the flow behind the projectile is thermally choked by the heat release of combustion (Bruckner et al. 1991, Hertzberg et al. 1991).

Original performance modeling of the thermally choked propulsive mode assumed the gases act ideally. This assumption neglects intermolecular interactions and is typically valid for low pressure and/or high temperature systems. The thermally choked ram accelerator produces a pressure ratio of 10-20 and for initial pressures of 1.0 MPa or higher the pressure in the combustion zone is on the order of a few ten MPa. At these elevated pressures it is no longer acceptable to neglect the intermolecular interactions. Consequently, ideal gas analyses substantially underpredict experimental thrust and velocity profiles at fill pressures greater than 1.0 MPa.

Initial attempts to account for real gas effects in the numerical modeling of the thermally choked propulsive mode were performed by Bauer and Knowlen (1995). That scheme computed an initial solution with an ideal gas assumption. The results were used to determine a compressibility factor which was then used to compute a modified non-dimensional thrust profile. (The non-dimensional thrust is defined as the thrust acting on the projectile normalized by the product of the initial fill pressure and the tube cross sectional area.) The resultant modified thrust profiles agreed much more closely with experimental data than the original ideal gas solutions.

The current research directly couples the real gas effects to the chemical equilbrium calculations on which all performance critieria are based. In particular, a compressibility factor is included in the equation of state and also corrects the equilibrium combustion routines through the fugacity (Denbigh 1992). Current experimental efforts intend to operate the ram accelerator at fill pressures up to 20 MPa, which will result in post-combustion pressures on the order of 300-400 MPa. Predicting performance criteria under these conditions requires an appropriate equation of state that accounts for real gas effects in the combustion zone.

2. Theoretical method

The mass, momentum, and energy conservation equations for quasi-steady, one-dimensional flow are applied to a control volume (Bruckner et al. 1991, Knowlen et al. 1991) as illustrated in Fig. 2. As shown, F is the thrust on the projectile and Q is a non-dimensional heat release parameter defined as:

$$Q = \frac{h_{f_1} - h_{f_2}}{c_{p_1} T_1} \tag{1}$$

where h_f is the enthalpy of formation at 0 K, c_p is the specific heat at constant pressure, and T is the temperature of the fill gas. The subscripts "1" and "2" refer to the properties at the entrance and exit locations of the control volume. The flowfield is assumed to be adiabatic and steady, with the combustion products exiting the control volume at the local speed of sound relative to the projectile, i.e., thermally choked.

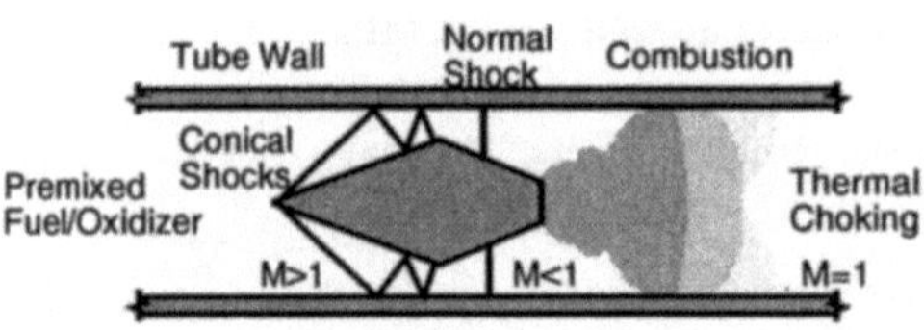

Fig. 1. Ram accelerator operating in the thermally choked propulsive mode

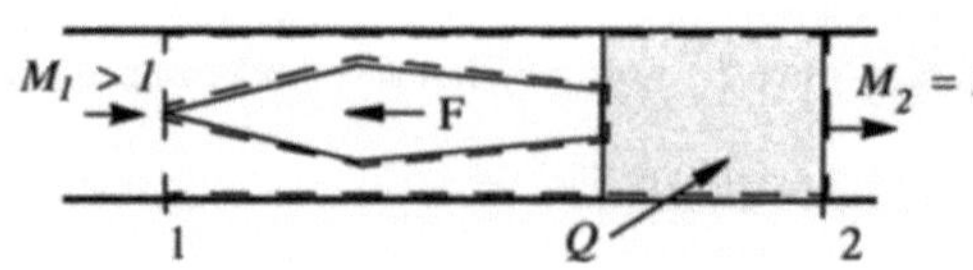

Fig. 2. Control volume model used for the numerical analysis

Real gas effects are included in the conservation equations via a compressibility factor, σ, which modifies the equation of state, speed of sound, pressure ratio, energy equation, mole fractions, and the equilibrium constants of the combustion reaction (Heuze et al. 1987, Buckwalter et al. 1996). With this corrective term, the equation of state can be written in the form:

$$Pv = \sigma \hat{R} T \tag{2}$$

where P is the pressure, T is the temperature, v is the specific volume, and $\hat{R}$ is the gas constant. The sound speed, a, of a reactive, real gas mixture can be written as

$$a^2 = \Gamma R T \tag{3}$$

where the "adiabatic gamma," Γ, is defined as (Heuze et al. 1987)

$$\Gamma = \frac{dh}{de}\Big|_s. \tag{4}$$

Combining the real gas equation of state with the equations for mass continuity and the speed of sound yields a relation for the ratio of the final pressure, P_2, to the initial fill pressure, P_1, of

$$\frac{P_2}{P_1} = \frac{\sigma_2}{\sigma_1} \frac{M_1}{M_2} \sqrt{\frac{\Gamma_1 R_2 T_2}{\Gamma_2 R_1 T_1}} \tag{5}$$

where M is the Mach number at the desired location. The equilibrium routine iterates on the total enthalpy of the system to ensure conservation of energy. Real gas effects are incorporated by adding corrective terms to the ideal gas total enthalpy and internal energy, resulting in an energy equation of

$$h_1 + h_{f_1} + \frac{1}{2}u_2^2 + \tilde{h}_1 = h_2 + h_{f_2} + \frac{1}{2}\Gamma_2 R_2 T_2 + \tilde{h}_2 \tag{6}$$

where

$$\tilde{h} = \left(v - T\frac{\partial v}{\partial T}\Big|_p\right) dp \tag{7}$$

Next, the equilibrium calculations are modified to account for the change in the equation of state. Corrections are made to the computations of the mole fractions and the equilibrium constants, K_p. The corrected equilibrium constants for non-ideal gas mixtures, K_f, are determined from the fugacity, f. The K_f values reduce to the familiar K_p for an ideal gas. The value of K_f can be determined from

$$K_f = K_p \prod \chi_i^{\nu_j} \tag{8}$$

where

$$\chi_i = \frac{f_i}{P_i} \tag{9}$$

f_i is the fugacity, P_i is the partial pressure of the given species, and ν_j is the stoichiometric coefficient for the production of one mole of species. The stoichiometric coefficients are defined as positive for products and negative for reactants. The fugacity of the given species can then be evaluated at the desired pressure via (Denbigh 1992).

$$\ln(f_i) = \int_0^P \frac{\sigma_i}{P} dp \tag{10}$$

The present analysis assumes that the initial species exhibit ideal behavior at the entrance to the control volume, i.e. $\sigma_1 = 1.0$. The actual compressibility factor for the initial conditions of the ram accelerator is typically slightly less than unity, but its contribution appears minimal for mixtures under current investigation.

3. Equation of state

Computing the compressibility factor for a given equation of state is the basis for incorporating real gas corrections. Numerous equations of state have been developed based on generalized, empirical, and theoretical considerations. Generalized models often have two constants which are determined by the generalized behavior of the substance (Black et al. 1985). Empirical equations of state typically have a larger number of empirical constants which are based on experimental data rather than the generalized behavior of the gas. Neither generalized nor empirical models were implemented due to their complexity or limited range of applicability.

The virial equation of state is derived from principles of kinetic theory. Virial expansions have proven to be reliable for very large pressure ranges and often lead to analytic solutions which can be readily included in numerical algorithms. The virial expansion for the compressibility factor is

$$\sigma = 1 + \frac{B(T)}{v} + \frac{C(T)}{v^2} + \frac{D(T)}{v^3} + \ldots \tag{11}$$

where $B(T)$, $C(T)$, and $D(T)$ are the 2nd, 3rd, and 4th virial coefficients, respectively, and are temperature dependent. The 2nd virial coefficient can be viewed as the correction to account for 2-body interactions and has the largest effect on σ. Similarly, the 3rd coefficient accounts for 3-body interactions, the 4th for 4-body interactions, and so on. As the order of the coefficient increases, it has a smaller effect on σ. An advantage of the virial formula is that it is accurate for a larger range of pressures and densities than the generalized or empirical equations.

4. Compressibility factor

The current analysis code is designed such that all real gas effects are resolved from the compressibility factor. An equation of state is implemented by changing the way σ is computed and keeping all other calculations consistent (Buckwalter et al. 1996). Only equations of state of a virial form are currently incorporated into the program. The current code is capable of using either an ideal gas, Boltzmann, or Percus-Yevick equation of state. In addition, a direct computation for the second and third virial coefficients from the Lennard-Jones and Stockmayer potentials was incorporated and is referred to as the Virial equation for the remainder of this paper.

4.1. Boltzmann The Boltzmann virial expansion adequately predicts Chapman-Jouguet properties when the pressure of the combustion products does not exceed 200 MPa (Bauer et al. 1985). This equation of state treats the individual molecules as hard spheres and the mixing rule only accounts for interactions of similar species. The Boltzmann expansion for the compressibility factor is computed by the formula

$$\sigma = 1 + x + 0.625x^2 + 0.287x^3 + 0.193x^4 \tag{12}$$

where x is defined as

$$x = \sum_i \frac{X_i B_i}{v_i} \tag{13}$$

B_i is the covolume, X_i is the mole fraction, and v_i is the specific volume of speicies i (Heuze et al. 1985).

4.2. Percus-Yevick The Percus-Yevick equation of state can be considered a summation of a virial development in which the compressibility factor is computed from

$$\sigma = \frac{(1 + z + z^2)}{(1 - z)^3} \tag{13}$$

The non-dimensional factor, z, is given by

$$z = \frac{\pi}{6} r^{*3} \frac{\hat{N} n}{V} \tag{14}$$

where $\hat{N}$ is Avogadro's number, n is the number of moles, and V is the volume (Heuze et al. 1985). The characteristic distance term, r^*, is derived from the interaction law

$$r^* = \lambda \left(\sum_i \sum_j X_i X_j r_{ij}^3 \right)^{1/3} \tag{15}$$

where

$$r_{ij} = \frac{r_i + r_j}{2} \tag{16}$$

In these equations, r_i and r_j are the molecular diameters of the two interacting species and λ is an adjustable constant. All of the computations in this report were obtained using $\lambda = 0.75$.

4.3. Virial equation The virial expansion for the compressibility factor to third order is given by

$$\sigma = 1 + \frac{B(T)}{v} + \frac{C(T)}{v^2} \tag{17}$$

The Virial equation of state directly computes the virial coefficients from a potential function rather than modeling them. If the gas is assumed to behave ideally, the internal energy of the gas is fully determined by the kinetic energy of the particles. When accounting for real gas effects, the internal energy must also include a potential energy, which is determined by the proximity of the particles to one another. The second virial coefficient of a gas can be determined from

$$B(T) = -2\pi\hat{N}\int_0^\infty \left[e^{-\phi/kT} - 1\right] r^2 dr \tag{18}$$

where k is the Boltzmann constant and ϕ is the molecular potential function (Hirschfelder 1954). The integration variable, r, is the distance between the centers of the particles. These coefficients can be computed by assuming a specific potential function, ϕ, that accounts for the attraction/repulsion of the molecules. The present analysis utilizes the Lennard-Jones potential function for non-polar species and the Stockmayer potential for polar species.

The Lennard-Jones potential function is given as

$$\phi(r) = 4\varepsilon\left[\left(\frac{r_0}{r}\right)^{12} - \left(\frac{r_0}{r}\right)^{6}\right] \tag{19}$$

where r_0 and ε are constants that depend on the specie under consideration. Given this function for ϕ, the integration can be performed analytically by expanding the terms in an infinite series (Hirschfelder 1954). The Stockmayer potential is required to compute the virial cōefficients for polar species. It is similar to the Lennard-Jones potential but includes an extra term to account for polarity and is given as

$$\phi(r) = 4\varepsilon\left[\left(\frac{r_0}{r}\right)^{12} - \left(\frac{r_0}{r}\right)^{6}\right] - \frac{\mu^2}{r^3}g \tag{20}$$

where g is a function of the mutual orientation of two polar molecules and μ accounts for the strength of the dipole. The integration of the Stockmayer potential with respect to r results in an infinite series of gamma functions. The third virial coefficients for both the Lennard-Jones and Stockmayer potentials are extremely difficult to calculate. Instead they can be reduced to the form

$$C(T) = \left(\frac{2}{3}\pi\hat{N}r_0^3\right)^2 C^* \tag{21}$$

where the values for C^* can be determined from tables for the given potential functions (Hirschfelder 1954).

5. Results and discussion

The present performance modeling is intended to support ram accelerator experiments by providing theoretical thrust and velocity data for initial fill pressures up to 20 MPa. Once the equations of state were implemented, they were validated with experimental Chapman-Jouguet detonation speeds (CJ speeds) and ram accelerator thrust and velocity profiles.

The first validation compared computed CJ speeds for various mixtures with experimental values. The experiments performed at the University of Washington initiated detonations by firing pistons into ram accelerator tubes filled with mixtures of CH_4, O_2, and N_2 in which both the relative number of moles of reactants and the initial fill pressure were varied. The results, listed in Table 1, show the increased CJ speeds predicted by the inclusion of the real gas effects at high pressures and also show that all the models matched the ideal gas solutions at low pressures.

There were no experimental data available for the CH_4-Air mixture at 0.30 and 0.10 MPa. These cases were run to ensure that all real gas models produce results similar to the ideal gas

Table 1. Computed and experimental Chapman-Jouguet detonation speed

Press (MPa)	Mixture	Ideal (m/s)	Boltz (m/s)	P-Y (m/s)	Virial (m/s)	Exp (m/s)
20	$0.99CH_4+2O_2+3.9N_2$	2113	2705	2813	*	2625†
10	$0.99CH_4+2O_2+3.9N_2$	2100	2357	2396	2447	2346†
5.0	$2.95CH_4+2O_2+5.7N_2$	1672	1779	1822	1836	1850‡
2.4	$2.8CH_4+2O_2+5.7N_2$	1685	1737	1753	1759	1770‡
3.42	$CH_4+2O_2+7.52N_2$	1861	1985	1958	1961	1942‡
2.03	$CH_4+2O_2+7.52N_2$	1856	1925	1911	1911	1905‡
0.95	$CH_4+2O_2+7.52N_2$	1846	1875	1872	1872	1852‡
0.61	$CH_4+2O_2+7.52N_2$	1841	1861	1857	1857	1835‡
0.30	$CH_4+2O_2+7.52N_2$	1831	1840	1838	1838	x
0.10	$CH_4+2O_2+7.52N_2$	1813	1816	1816	1816	x

† Bauer et al. 1991
‡ Bauer et al. 1996
x No data available
* Numerical solution unstable

values at low pressures where the intermolecular interactions should be negligble. As expected, the calculations are nearly identical at the lower pressures, with only a 0.5% difference among the values for the 0.30 MPa case and a 0.2% difference for the 0.10 MPa case.

The ideal gas equation of state performs adequately for initial fill pressures below 1.0 MPa. In these cases it has less than 1% error in computing the CJ speed. As the initial pressure is raised, the ideal gas CJ speed drops significantly below the experimental values, having a 2.6% error at 2.0 MPa and a 9.7% error at 5.0 MPa. These errors become even larger at higher fill pressures with errors of 10.5% and 19.5% at 10 and 20 MPa, respectively.

The Boltzmann model performed well over the entire range of pressures and mixtures. It had an error less than 2% for pressures up to 2.4 MPa and a maximum error of 3.8% for the 5.0 MPa case. It also matched experimental data better than the Percus-Yevick and Virial models for the stoichiometric mixtures at 10 and 20 MPa. The Percus-Yevick and Virial equations of state performed similarly to one another up to 5.0 MPa. Both produced slightly high values of CJ speed at lower pressures, as was the case with the Boltzmann model; however, each performed slightly better than the Boltzmann model for the fuel-rich 5.0 MPa test case, with the Percus-Yevick model having an error of 1.5% and the Virial model deviating by less than 1%. For the near stoichiometric mixtures at higher pressures, both tended to have more problems matching experiments. The Percus-Yevick CJ speeds are high and have errors of 2.1% and 7.2% at 10 and 20 MPa, respectively. These solutions could be refined by more careful selection of the constant, λ, but the intent is to produce results without prior knowledge of the solution, so λ was fixed. The Virial model had problems for the approximately stoichiometric mixtures. It had a 4.3% error at 10 MPa and the numerical scheme became unstable at 20 MPa. Thus a more rigorous algorithm must be included to solve for the virial coefficients under these conditions.

The next validation step computed non-dimensional thrust and velocity profiles and compared them with experiments. The selected experiments were performed at the University of Washington at initial pressures of 2.5 and 5.0 MPa and provided data for which the projectile ran the length of the ram accelerator tube. The 2.5 MPa experiment was conducted with a 63 g projectile and a propellant mixture of $2.8CH_4+2O_2+5.7N_2$, and the 5.0 MPa experiment used a 109 g projectile and a mixture of $3CH_4+2O_2+5.7N_2$. The entrance velocity for each of these experiments was approximately 1100 m/s. The experimental thrust and velocity profiles were determined from time-distance data. The theoretical velocity curves were obtained by integrating the thrust vs. Mach number profile for the known projectile mass.

The non-dimensional thrust vs. Mach number and velocity vs. distance profiles for the 2.5 MPa test are shown in Figs. 3 and 4, respectively. Even at this moderate fill pressure, the real gas equations show noticeable improvement. As anticipated, the ideal gas model underpredicts both thrust and velocity values. All the real gas models matched the experimental data well, with the Percus-Yevick and the Virial agreeing slightly better than the Boltzmann. In Fig. 4, the experimental velocity deviates significantly from the theoretical values after about 6 m of travel. This region of enhanced velocity, and, consequently, thrust seen in Fig. 3, is due to the projectile making a transition from the thermally choked propulsive mode to the transdetonative mode (Hertzberg et al. 1991). This transition typically occurs when the projectile reaches approximately 90% of the CJ speed. The current simulations only attempt to predict performance in the thermally choked propulsion regime and are not applicable in the transdetonative mode.

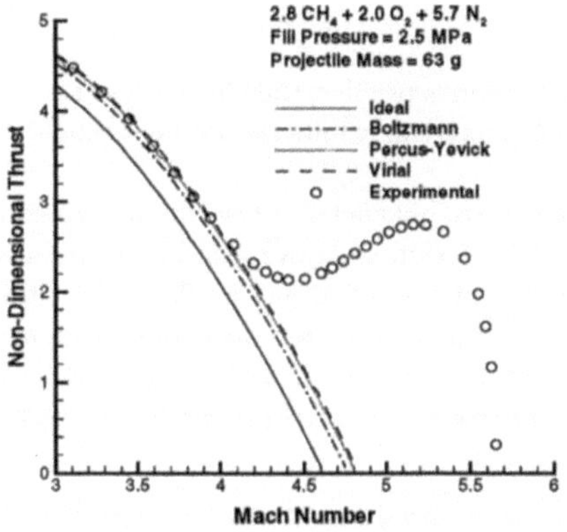

Fig. 3. Non-dimensional thrust vs. Mach number for a fill pressure of 2.5 MPa

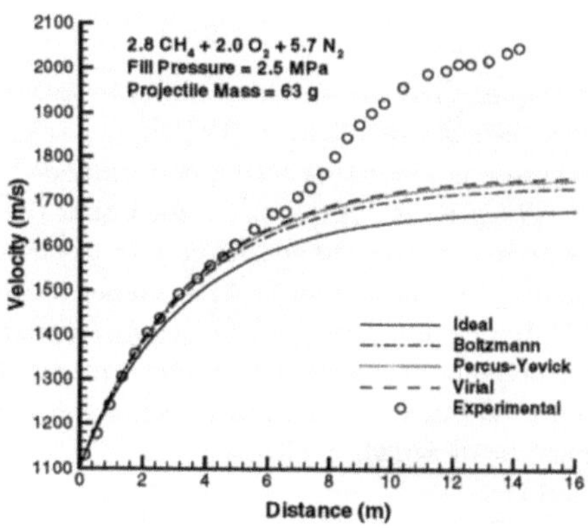

Fig. 4. Velocity vs. distance for a fill pressure of 2.5 MPa

The thrust and velocity profiles for the 5.0 MPa test case are shown in Figs. 5 and 6, respectively. As in the 2.5 MPa case, the real gas equations of state agree better with experimental data than the ideal gas assumption. The Boltzmann model again provides lower values of thrust and velocity than the Percus-Yevick model, which in turn slightly underpredicts the Virial model. This 5.0 MPa case indicates that these equations of state can provide reasonably accurate modeling at higher pressures.

These results show that the real gas equations of state predict an enhanced performance compared to the ideal gas calculations. This enhanced performance is the result of the increase in compressibility and the shift in the chemical equilibria of the mixture. The differences in thrust increase at higher pressures, primarily due to an increasingly larger value of σ, which increases the pressure in the combustion region and enhances thrust. There is also a secondary thrust enhancement due to a shift in chemical equilibria which results in more heat release (Buckwalter et al. 1996).

There are plans to operate the ram accelerator at fill pressures up to 20 MPa. These tests will use a 110 g titanium projectile with an entrance velocity of 1150 m/s. To provide preliminary performance data for these shots, each equation of state was run for a mixture of $2.8CH_4+2O_2+5.7N_2$ at fill pressures of 10 and 20 MPa. The corresponding velocity profiles are shown in Figs 7 and 8.

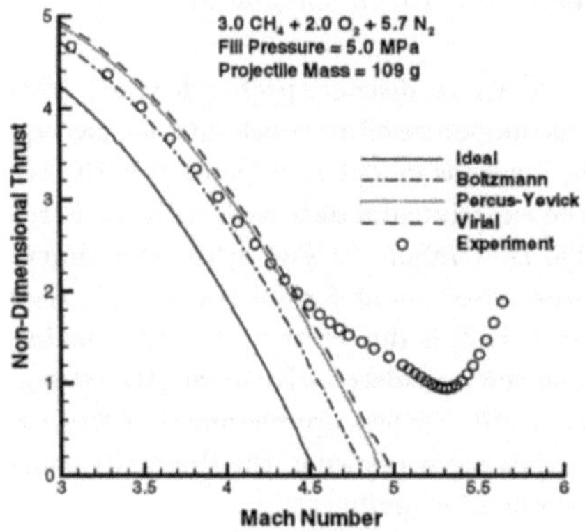

Fig. 5. Non-dimensional thrust vs. Mach number for a fill pressure of 5.0 MPa

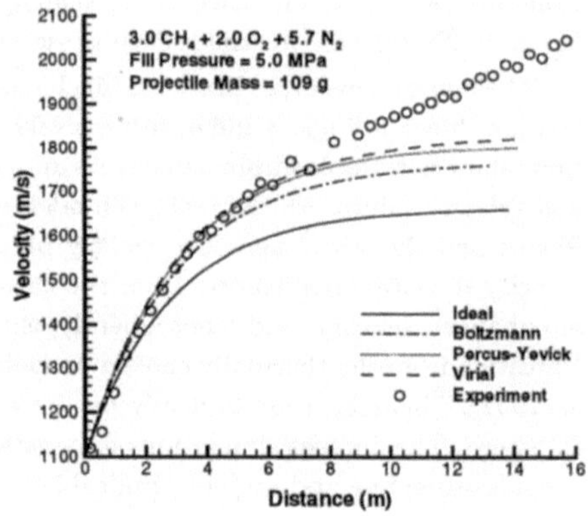

Fig. 6. Velocity vs. distance for a fill pressure of 5.0 MPa

Although there are no current data to validate these performance profiles, the current lower pressure tests indicate that the Boltzmann and Percus-Yevick numerical results should be reasonably accurate in predicting actual performance.

The plots in Figs. 7 and 8 show that the real gas effects are beneficial to the ram accelerator performance parameters. After 4 m of travel in the 10 MPa example above, the Percus-Yevick model predicts a velocity 17% higher than that predicted by the ideal gas model and in the 20 MPa case, it predicts a velocity 47% higher. Although it was known that operation at higher fill pressure translates into higher thrust, the real gas computations give an indication of how much additional thrust can be expected and the pressure levels which the projectile and tubes must be designed to withstand.

6. Conclusions

It is necessary to account for real gas effects to accurately predict performance characteristics of the thermally choked ram accelerator due to the high pressure combustion products. The current one-dimensional performance analysis code used at the University of Washington to model the thermally choked ram accelerator propulsive mode can choose any of three real gas equations of state or the ideal gas option. Analyses show that the ideal gas equation of state works well for fill pressures below 1.0 MPa and the Boltzmann model works well for fill pressures up to 20 MPa for stoichiometric mixtures. For fuel-rich mixtures, the Percus-Yevick and Virial equations both perform well up to 5.0 MPa. For the stoichiometric mixtures at very high pressures, however, these two models overpredict experimental values for Chapman-Jouguet detonation speeds.

From current data, the Boltzmann or Percus-Yevick models are expected to provide the best performance calculations for ram accelerator experiments with initial fill pressures above 5.0 MPa. The Percus-Yevick model in particular matches experimental data very well for these mixtures up to 5.0 MPa. It has a variable parameter that can be used to tune it for a desired pressure range. The Virial model currently implemented did well for the test cases shown here but may be somewhat limited at higher pressures. Although it has a more physical foundation than the other schemes, it is increasingly difficult to compute the 4th, 5th and higher order virial coefficients. It will be left up to the upcoming high pressure ram accelerator experiments to provide data that will determine which equation of state is most applicable in the 20 MPa regime.

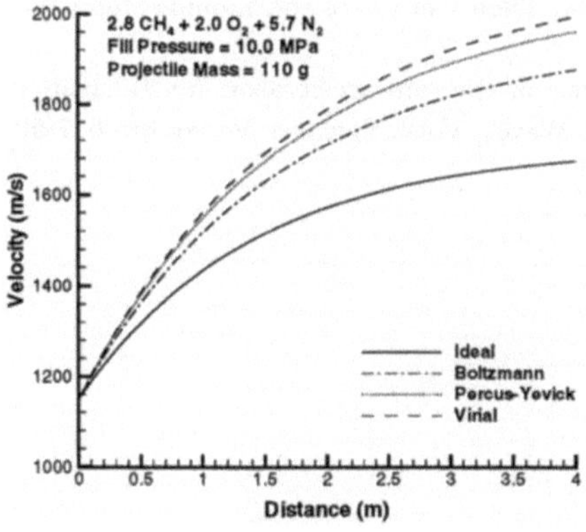

Fig. 7. Velocity vs. distance for a fill pressure of 10 MPa

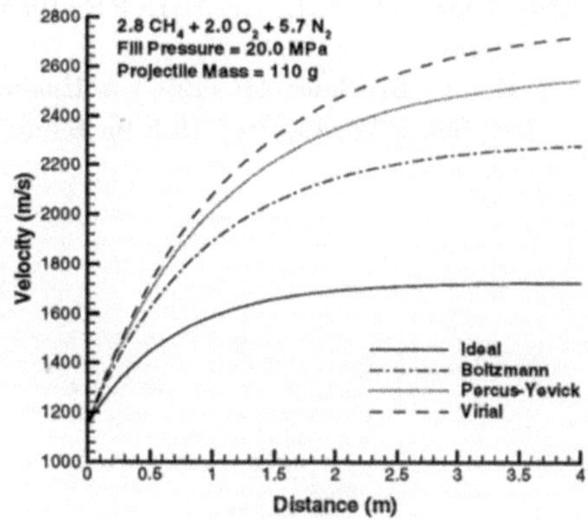

Fig. 8. Velocity vs. distance for a fill pressure of 20 MPa

Acknowledgements

This work was supported in part by the U.S. Army Research Office under Grant #DAAL03-92-G-0100. The authors are deeply indebted to Pascal Bauer for his assistance with the implementation of the real gas corrections, and Jesse Stewart for his help with experimental data and various other aspects of this research. Finally, the first author would like to thank the U.S. Air Force Palace Knight Office and the Advanced Propulsion Division at Wright Patterson AFB for their support.

References

Bauer P, Brochet C, Heuze O, Presles HN (1985) Equation of state for dense gases. Archivum Combustionis 5:35-45

Bauer P, Knowlen C (1995) Real gas effects on the operating envelope of the ram accelerator. Proc 2nd Int Workshop on Ram Accelerators

Bauer P, Legendre JF, Knowlen C, Higgins A (1996) Transition of detonation on insensitive dense gaseous mixtures in tubes. AIAA paper 96-2682

Bauer P, Presles N, Dunand M (1991) Detonation characteristics of gaseous methane-oxygen-nitrogen mixtures at extremely elevated initial pressures. 12th Int Colloquium on Dynamics of Explosions and Reactive Systems (ICDERS), AIAA Progress in Aero Astro 13:56-62

Black WZ, Hartley JG (1985) Thermodynamics, Harpers & Row

Bruckner AP, Knowlen C, Hertzberg A, Bogdanoff DW (1991) Operational characteristics of the thermally choked ram accelerator, J Prop Power 7:828-836

Buckwalter DL, Knowlen C, Bruckner AP (1996) Ram accelerator performance analysis code incorporating real gas effects. AIAA paper 96-2945

Denbigh K (1992) The Principles of Chemical Equilibrium, Cambridge University Press

Hertzberg A, Bruckner AP, Bogdanof DW (1988) Ram accelerator: A new chemical method for accelerating projectiles to ultrahigh velocities. AIAA J 26:195-203

Hertzberg A, Bruckner AP, Knowlen C (1991) Experimental investigation of ram accelerator propulsive modes. Shock Waves 1:17-25

Heuze O, Bauer P, Presles HN, Brochet C (1985) The equations of state and their incorporation into the Quator Code. Int Symp Detonations, p 762

Heuze O, Bauer P, Presles HN (1987) Compressibility and thermal properties of gaseous mixtures at high temperatures and high pressures. High Temperatures - High Pressures 19:611-620

Hirschfelder JO, Curtis CF, Byrd RB (1954) Molecular Theory of Gases and Liquids, John Wiley & Sons

Knowlen C, Bruckner AP (1991) A Hugoniot analysis of the ram accelerator. In: Takayama K (ed) Shock Waves, Proc 18th Int Symp on Shock Waves, Vol I, Springer-Verlag, pp 617-622

Numerical investigation on subdetonative mode ramjet-in-tube

M.M. Morales, M.A.S. Minucci, J.B. Channes-Jr., A.G. Ramos, D. Bastos-Netto
Combustion and Propulsion Laboratory, National Institute for Space Research, Rodovia Presidente Dutra, km 40, Cachoeira Paulista, SP, Brazil CEP 12630-000

Abstract. Numerical simulations on a subdetonative mode ramjet-in-tube were made according to the proposed specifications on the flight model geometry and test section for RAMAC III Benchmark Test. They were made also according to others configurations using different model geometries plus small variations in combustible mixture composition and different initial pressures. For the numerical simulation it is supposed that a pre-launcher section pushed the obturator plus projectile combination to a 1.4 km/s, that is the entrance velocity in the ramjet-in-tube combustion section. The combustion was then initiated by the stagnation flow on the obturator before it was discarded by gasdynamic drag. A simplified model based on the one dimensional inviscid flow and equilibrium chemical reaction computation was tested. The various theoretical and experimental results encountered in the available technical literature were confronted with the numerical modelling developed in the present paper and the results correlated well.

Key words: Subdetonative mode, One-dimensional analysis, Non-equilibrium effects

1. Introduction

The present paper was inspired by the work developed at University of Washington (Hertzberg et al. 1988) and motivated by the ongoing supersonic combustion research at LCP (Guimarães et al. 1997).

The problem of obtaining hypersonic velocities in the process of projectile acceleration has been studied over the last decades and, particularly, the recently released technique called "ram accelerator" or "ramjet-in-tube" has been studied since mid 1983.

Ultimately being a ramjet, where the central body (projectile) is free to move and its cowl (tube wall) is fixed, the ram accelerator enables the study of the supersonic combustion problem in ground facilities. The problem of getting the fuel mixed and burned with the incoming air is bypassed due to the fact the projectile (central body) travels in a premixed homogeneous medium. As a consequence such a concept would complete the already operational LCP facility for testing supersonic combustors for scramjets.

The work developed here is encouraged by the demonstrated capability and simplicity of one-dimensional flow analysis applied to ramjet and rocket motor performance calculations (Williams 1988, Heiser 1994).

In this paper, as a first study in this fascinating area, only the thermally choked mode is considered because of its relative simplicity and very good agreement between theory and experiments (Hertzberg 1989).

2. Theoretical modelling

The following assumptions were made:
1. The ramjet-in-tube is operating in the subdetonative mode, i.e., the velocities achieved by the flight model are below the Chapman-Jouguet detonation velocity of the combustible mixture, calculated for the initial conditions of the undisturbed flow.
2. A pre-launcher section pushes the projectile plus obturator combination to a velocity of 1.4 km/s, that is the required velocity of injection in the combustible mixture, and the combustion

was then initiated by the stagnation flow on the obturator before it was discarded by gasdynamic drag.

3. Referential frame is fixed in the flight model body.
4. The flow is one-dimensional, steady, adiabatic and parallel to the freestream (x-direction).
5. All the properties of the flow upstream of the projectile nose tip are known.
6. The undisturbed combustible mixture behaves as an ideal gas.

With these assumptions the conservation equations for the steady one-dimensional inviscid reactive flow over the projectile are:

$$\frac{\partial(\rho u)}{\partial x} = 0 \tag{1}$$

$$\rho u \frac{\partial u}{\partial x} + \frac{\partial p}{\partial x} = 0 \tag{2}$$

$$u\frac{\partial p}{\partial x} - \gamma p\rho\left(u\frac{\partial \rho}{\partial x}\right) = q\rho\frac{(\gamma - 1)}{M_o^2}(1-\lambda)\, e^{\frac{\Theta\rho}{pM_o^2}} \tag{3}$$

$$u\frac{\partial \lambda}{\partial x} = (1-\lambda)\, e^{\frac{\Theta\rho}{pM_o^2}} \tag{4}$$

where

$$q = \frac{\rho_o q_o}{p_o}, \quad \Theta = \frac{E_o \rho_o}{p_o}, \quad M_o^2 = \frac{\rho_o u_o^2}{p_o \gamma}, \quad 0 \le \lambda \le 1 \tag{5}$$

Equation (1) describes the conservation of mass, where ρ and u are the density and the velocity in the x-direction, respectively. The conservation of momentum in the x-direction is given by equation (2), where p is the pressure. Equation (3) describes the energy conservation, with assumption of an ideal gas and one-step Arrhenius kinetics, where the γ is the ratio of specific heats, q and Θ are the dimensionless heat of reaction and dimensionless activation energy, respectively; q_o and E_o are the dimensional heat of reaction and the dimensional activation energy, respectively, and λ is the mass fraction of the reaction product gas, with $\lambda = 0$ and $\lambda = 1$ corresponding to no reaction and complete reaction, respectively. The o subscript indicates a undisturbed initial flow condition. From (5) it is clear that M_o is the flight Mach number. In this work, because only the thermally choked mode ramjet-in-tube is modelled, a great simplification was considered in the governing equations: the exit flow Mach number, i.e., the flow velocity to local sound velocity ratio behind the projectile is equal to unity. Equation (4) is a rate law for Arrhenius kinetics with simple depletion.

In order to describe the acceleration process of the projectile it was used the following equations:

$$F = \int p(x)\, dA_x \tag{6}$$

$$\frac{dU}{dx} = \frac{F}{m} \tag{7}$$

Where F is the thrust, A_x is the variable cross section of the projectile along the x-direction, U and m are the velocity and mass of the projectile, respectively. The integral appearing in the equation (6) is evaluated over the control volume around the entire projectile and the reaction zone until the point of thermal choking.

In order to take in to consideration the non-equilibrium gas flow effects, a balance between the calculation made with $\lambda = 1$, i.e., considering infinitely fast chemical kinetics (equilibrium chemistry flow), and with $\lambda = 0$, i.e., considering infinitely slow chemical kinetics (frozen chemistry flow) was made. The non-dimensional thrust is the ratio between the dimensional thrust F and the product of the undisturbed upstream pressure p_o and the tube cross section.

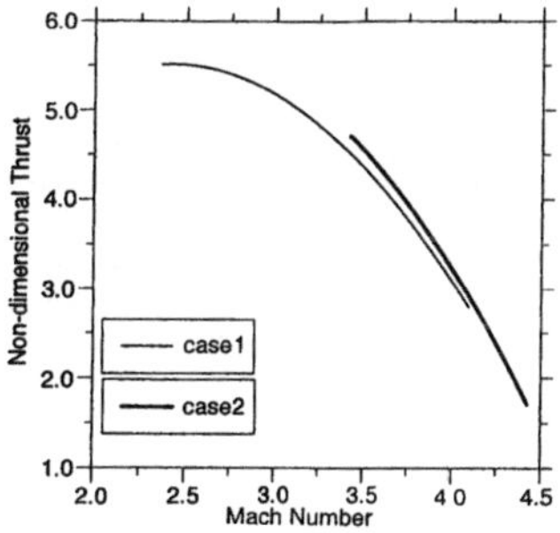

Fig. 1. Non-dimensional thrust versus Mach number for the combustible mixture $CH_4 + 2O_2 + 6CO_2$, at 1.2 MPa, with projectile mass = 0.075 kg, considering equilibrium gas flow and input velocities of 690 and 1000 m/s, using the LCP code.

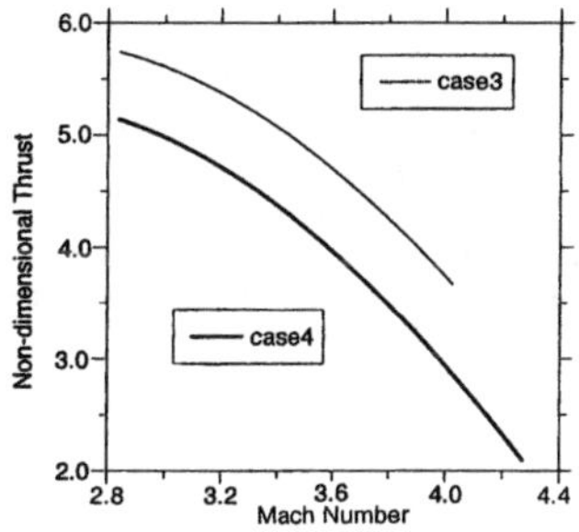

Fig. 2. Non-dimensional thrust versus Mach number for the Case 3 combustible mixture $2.5CH_4 + 2O_2 + 5N_2$, at 1.2 MPa, and Case 4 combustible mixture $2.5CH_4 + 2O_2 + 6N_2$, at 2.0 MPa, with projectile mass = 0.075 kg, considering equilibrium gas flow and input velocity of 1020 m/s, using the LCP code.

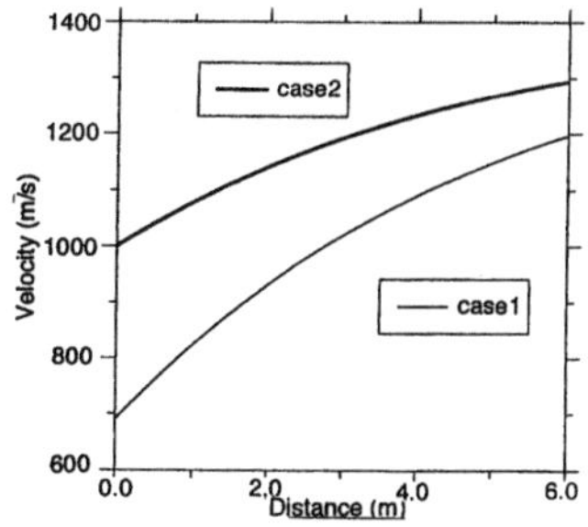

Fig. 3. Velocity versus distance for the combustible mixture $CH_4 + 2O_2 + 6CO_2$, at 1.2 MPa, with projectile mass = 0.075 kg, considering equilibrium gas flow and input velocities of 690 and 1000 m/s, using the LCP code.

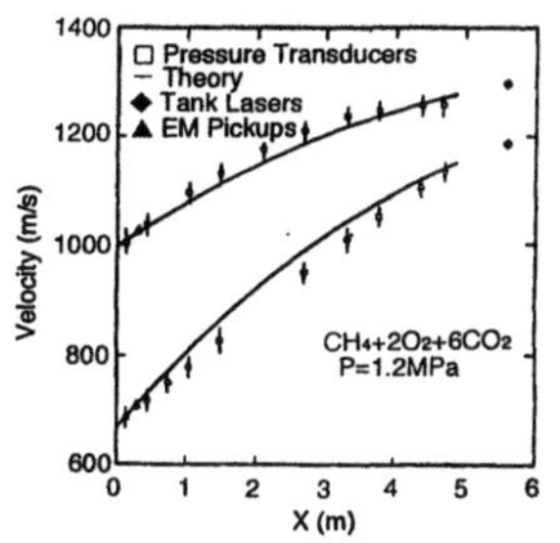

Fig. 4. Comparison of experimental and theoretical velocity profiles in ram accelerator for $CH_4 + 2O_2 + 6CO_2$ propellant mixture and initial velocities of 690 and 1000 m/s. Projectile mass = 0.075 kg (extracted from Hertzberg et al. 1988).

3. Numerical Results

The governing equations, in one-dimensional inviscid form, are solved with the aid of a LCP-modified version of the NASA-SP-273 code (Gordon and McBride 1971), in order to consider non-equilibrium gas flow for the ramjet-in-tube performance calculations; this modified version was linked to a main Fortran routine that solved the governing equations presented in previous section, in a straightforward manner using explicit integration.

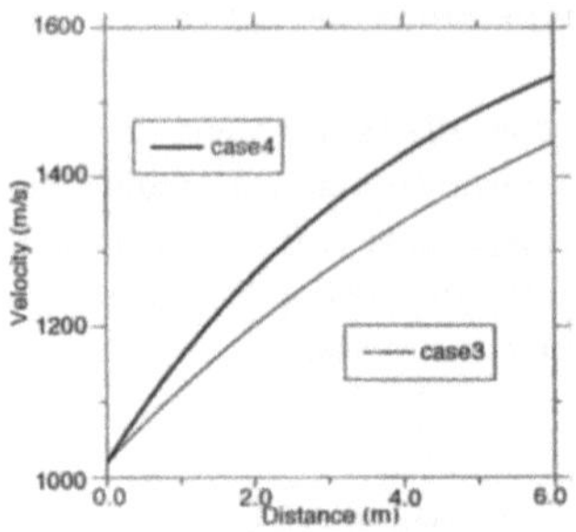

Fig. 5. Velocity versus distance for the Case 3 combustible mixture $2.5CH_4 + 2O_2 + 5N_2$, at 1.2 MPa, and Case 4 combustible mixture $2.5CH_4 + 2O_2 + 6N_2$, at 2.0 MPa, with projectile mass = 0.075 kg, considering equilibrium gas flow and input velocity of 1020 m/s, using the LCP code.

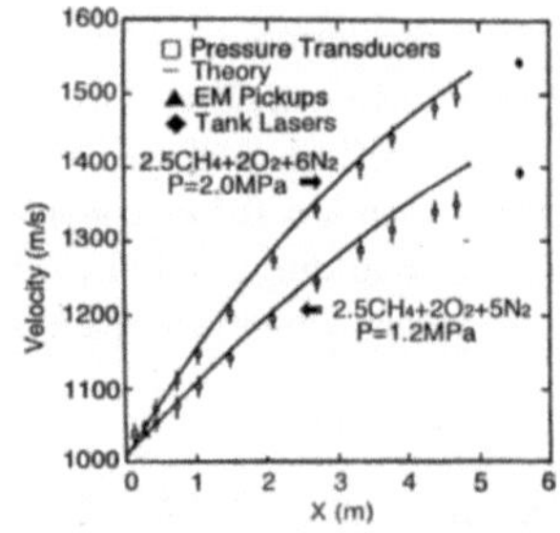

Fig. 6. Comparison of experimental and theoretical velocity profiles in ram accelerator for $2.5CH_4 + 2O_2 + 5N_2$ propellant mixture at 1.2 MPa and $2.5CH_4 + 2O_2 + 6N_2$ propellant mixture at 2.0 MPa. Initial velocities = 1020 m/s; projectile mass = 0.075 kg (extracted from Hertzberg et al. 1988).

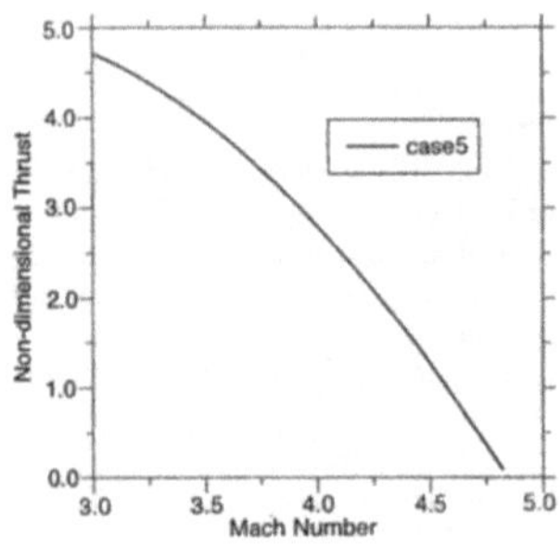

Fig. 7. Non-dimensional thrust versus Mach number for the Case 5 combustible mixture $2.8CH_4 + 2O_2 + 5.7N_2$, at 2.5 MPa, with projectile mass = 0.063 kg, considering non-equilibrium flow and input velocity of 1100 m/s, using the LCP code.

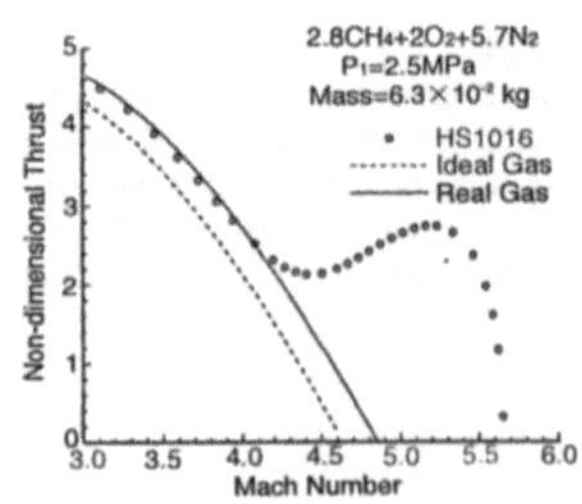

Fig. 8. Non-dimensional thrust coefficient for $2.8CH_4 + 2O_2 + 5.7N_2$ propellant mixture at a fill pressure of 2.5 MPa (extracted from Buckwalter et al. 1996).

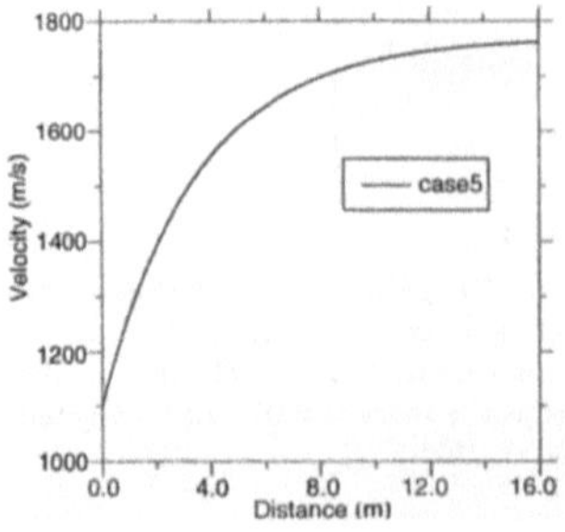

Fig. 9. Velocity versus distance for the Case 5 combustible mixture $2.8CH_4 + 2O_2 + 5.7N_2$, at 2.5 MPa, with projectile mass = 0.063 kg, considering non-equilibrium gas flow and input velocity of 1100 m/s, using the LCP code.

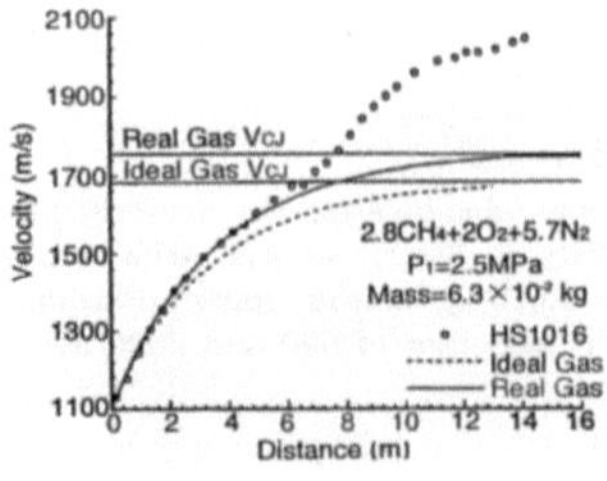

Fig. 10. Velocity profile for $2.8CH_4 + 2O_2 + 5.7N_2$ propellant mixture at a fill pressure of 2.5 MPa (extracted from Buckwalter et al. 1996).

Numerical and experimental results presented in previous studies of the ramjet-in-tube concept using one-dimensional analysis including real gas effects (Buckwalter et al. 1996) and not including real gas effects (Hertzberg et al. 1988) were used to validate the LCP code. After these validation, the code was used to calculate the performance of a hypothetical flight model in a ramjet-in-tube following the combustible mixture and projectile geometry specified for the RAMAC III Benchmark Test.

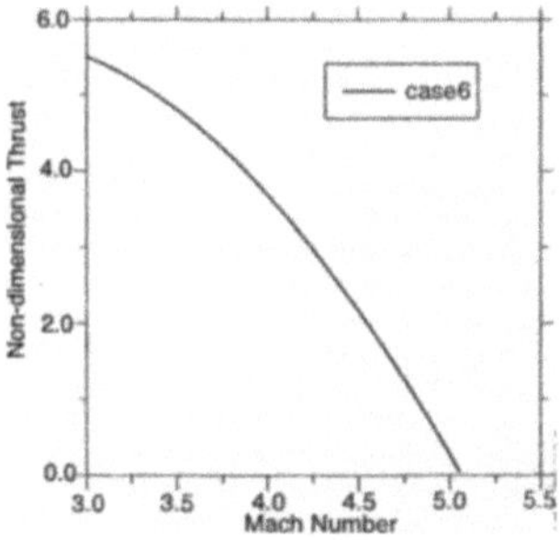

Fig. 11. Non-dimensional thrust versus Mach number for the Case 6 combustible mixture $3CH_4 + 2O_2 + 5.7N_2$, at 5.0 MPa, with projectile mass = 0.109 kg, considering equilibrium gas flow and input velocity of 1100 m/s, using the LCP code.

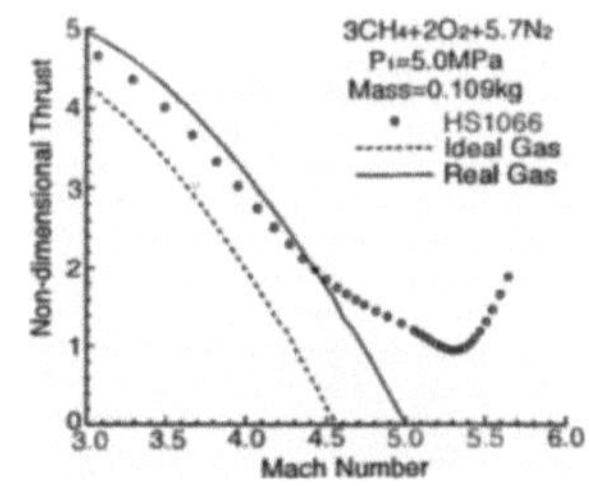

Fig. 12. Non-dimensional thrust coefficient for $3CH_4+$ $2O_2+$ $5.7N_2$ propellant mixture at a fill pressure of 5.0 MPa (extracted from Buckwalter et al. 1996).

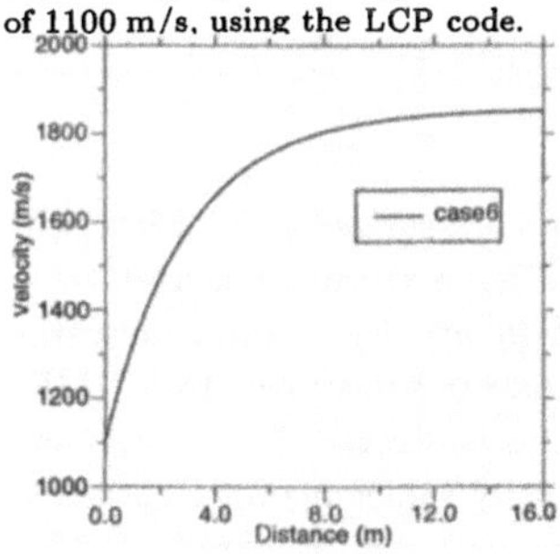

Fig. 13. Velocity versus distance for the Case 6 combustible mixture $3CH_4+$ $2O_2$ + $5.7N_2$, at 5.0 MPa, with projectile mass = 0.109 kg, considering non-equilibrium gas flow and input velocity of 1100 m/s, using the LCP code.

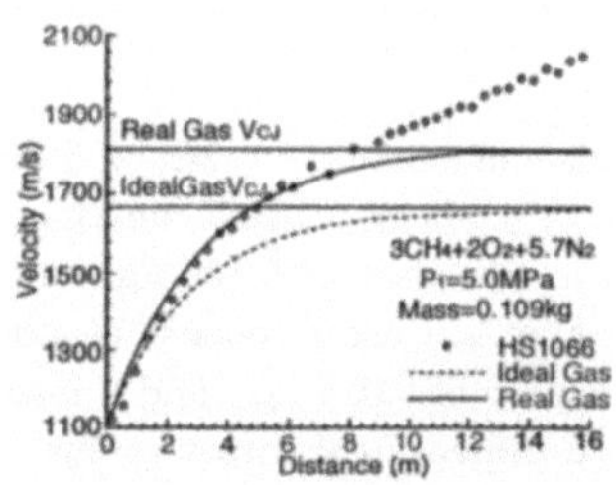

Fig. 14. Velocity profile for $3CH_4+$ $2O_2+$ $5.7N_2$ propellant mixture at a fill pressure of 5.0 MPa (extracted from Buckwalter et al. 1996).

Cases 1, 2, 3 and 4 are the LCP code solutions for the ramjet-in-tube configuration used by Hertzberg et all., at University of Washington (Hertzberg et al. 1988). Case 1 consists of a CH_4 + $2O_2$ + $6CO_2$ combustible mixture pressurized to 1.2 MPa, and the projectile entrance velocity in the ram section was 690 m/s. Case 2 consists on the same conditions except that the projectile entrance velocity was 1000 m/s. Case 3 consists of a $2.5CH_4$ + $2O_2$ + $5N_2$ mixture combustible pressurized to 1.2 MPa, and for case 4 the mixture was $2.5CH_4$ + $2O_2$ + $6N_2$ pressurized to 2.0 MPa. In both last cases the projectile entrance velocity was 1020 m/s. The projectile mass in the these four cases was 0.075 kg. In all cases considered in the present paper, the initial temperature of the pressurized combustible mixture was 300 K.

Figures 1 and 3 show, respectively, the non-dimensional thrust versus Mach number and velocity versus distance curves travelled by the projectile inside the combustion section for cases 1 and 2.

Figures 2 and 5 show, respectively, the non-dimensional thrust versus Mach number and velocity versus distance curves for cases 3 and 4.

In the above cases, λ was made equal to unity, equilibrium chemistry flow, and no balance was made. Comparing these curves to the curves presented by Hertzberg, here as figures 4 and 6 (extracted from Hertzberg et al. 1988), we verify that the results are very closed related, as it was expected.

To evaluate the non-equilibrium effects, i.e., considering a balance between frozen and equilibrium flow, the test conditions presented by Buckwalter et all. (1996) were simulated in cases 5, 6, and 7.

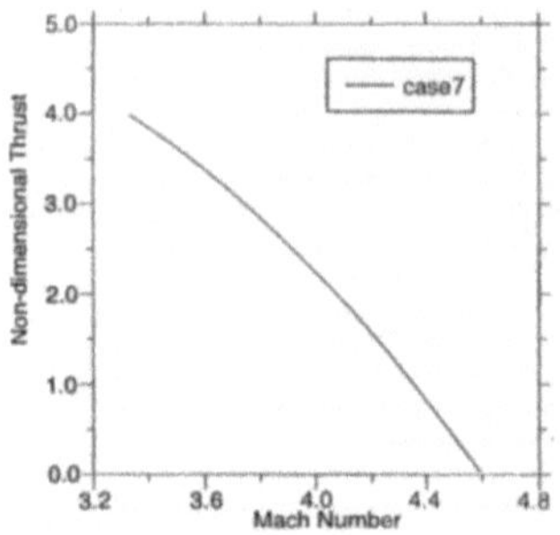

Fig. 15. Non-dimensional thrust versus Mach number for the Case 7 combustible mixture $6CH_4 + 2O_2 + 2H_2$, at 5.1 MPa, with projectile mass = 0.084 kg, considering non-equilibrium gas flow and input velocity of 1500 m/s, using the LCP code.

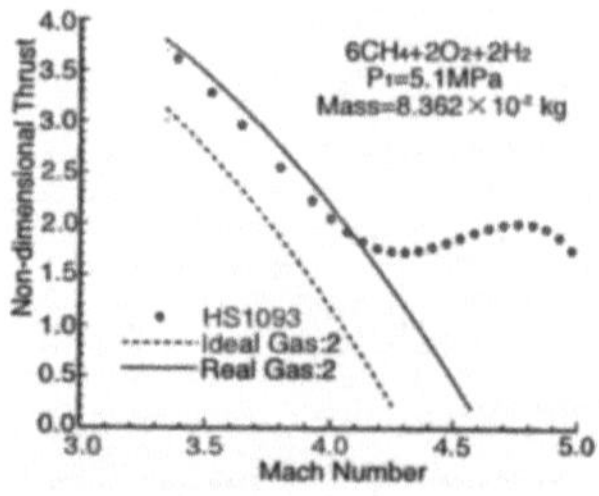

Fig. 16. Non-dimensional thrust coefficient for $6CH_4 + 2O_2 + 2H_2$ propellant mixture at a fill pressure of 5.1 MPa (extracted from Buckwalter et al. 1996).

Case 5 consists of a $2.8CH_4 + 2O_2 + 5.7N_2$ combustible mixture pressurized to 2.5 MPa, with projectile mass of 0.063 kg. Case 6 consists of a $3CH_4 + 2O_2 + 5.7N_2$ mixture pressurized to 5.0 MPa, with projectile mass of 0.109 kg. In both cases, the insertion velocity in the combustion section was 1100 m/s. Case 7 consists of a $6CH_4 + 2O_2 + 2H_2$ mixture pressurized to 5.1 MPa, with projectile mass of 0.084 kg , and a insertion velocity of 1500 m/s. Figures 7,9,11,13,15 and 17 show the non-dimensional thrust versus Mach number and velocity versus distance curves for the cases 5, 6 and 7. Comparing these results with the curves presented by Buckwalter et al., here as figures 8, 10, 12, 14, 16 and 18 (extracted from Buckwalter et al. 1996) we see that the case of mixture pressurized to 2.5 MPa (case 5) correlated well, but for the cases of mixture pressurized to 5.0 and 5.1 MPa (cases 6 and 7), there is a significative difference in initial portion of non-dimensional thrust and velocity curves; this difference is likely because the compressibility factor correction used by Buckwalter et al. and not present in the LCP code.

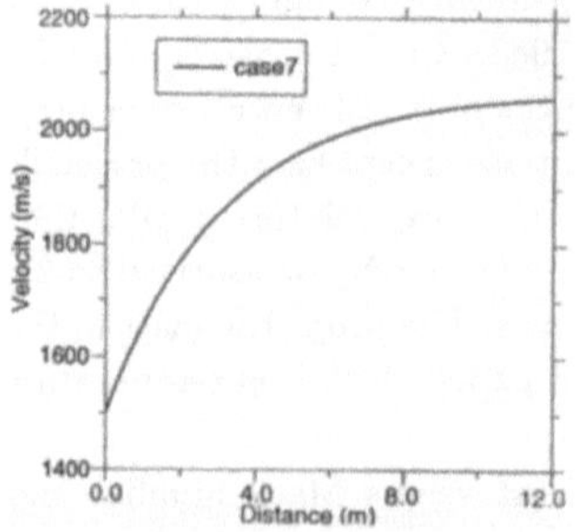

Fig. 17. Velocity versus distance for the Case 7 combustible mixture $6CH_4 + 2O_2 + 2H_2$, at 5.1 MPa, with projectile mass = 0.084 kg, considering non-equilibrium gas flow and input velocity of 1500 m/s, using the LCP code.

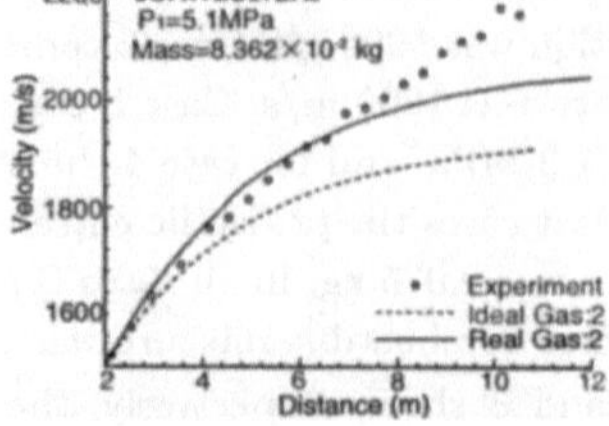

Fig. 18. Velocity profile for $6CH_4 + 2O_2 + 2H_2$ propellant mixture at fill pressure of 5.1 MPa (extracted from Buckwalter et al. 1996).

The LCP code simulation results for the RAMAC III Benchmark Test are presented in the figures 19-22 (cases 8 and 9). Case 8 consists on a combustible mixture $3CH_4 + 2O_2 + 7.5N_2$ pressurized at 2.5 MPa and case 9 on mixture $3CH_4 + 2O_2 + 7.5N_2$ pressurized at 5.0 MPa. The projectile entrance velocity is in accordance with the proposed specification (1400 m/s). Figures 23-25 illustrates the flight model inside the tube. The projectile estimated mass is 0.133 kg. The internal diameter of tube is 47.7 mm and the model length is 192.0 mm.

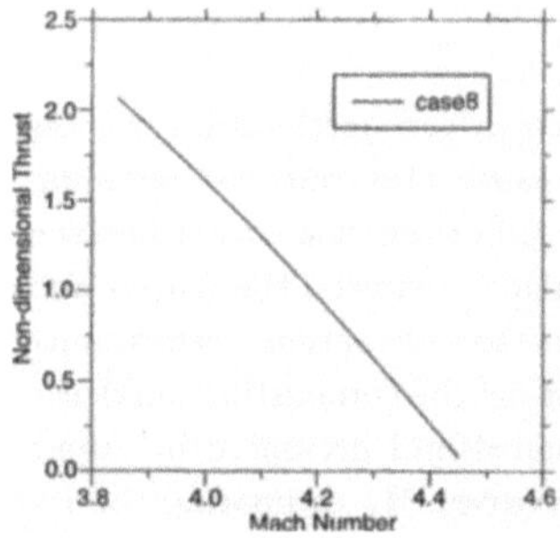

Fig. 19. Non-dimensional thrust versus Mach number for the Case 8 combustible mixture $3CH_4 + 2O_2 + 7.5N_2$, at 2.5 MPa, with projectile mass = 0.133 kg, considering non-equilibrium gas flow and input velocity of 1400 m/s, using the LCP code.

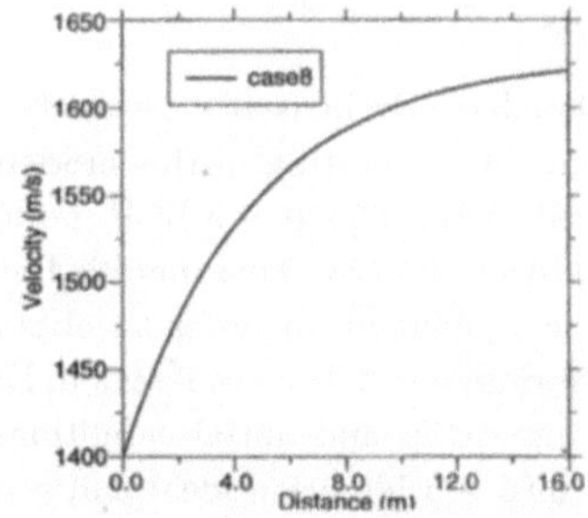

Fig. 20. Velocity versus distance for the Case 8 combustible mixture $3CH_4 + 2O_2 + 7.5N_2$, at 2.5 MPa, with projectile mass = 0.133 kg, considering non-equilibrium gas flow and input velocity of 1400 m/s, using the LCP code.

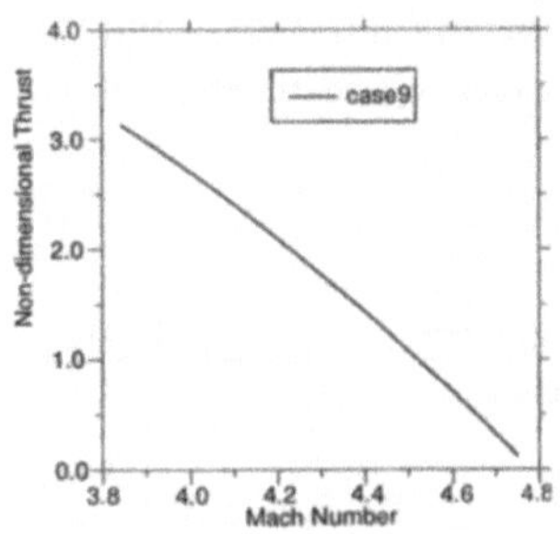

Fig. 21. Non-dimensional thrust versus Mach number for the Case 9 combustible mixture $3CH_4 + 2O_2 + 7.5N_2$, at 5.0 MPa, with projectile mass = 0.133 kg, considering non-equilibrium gas effects and input velocity of 1400 m/s, using the LCP code.

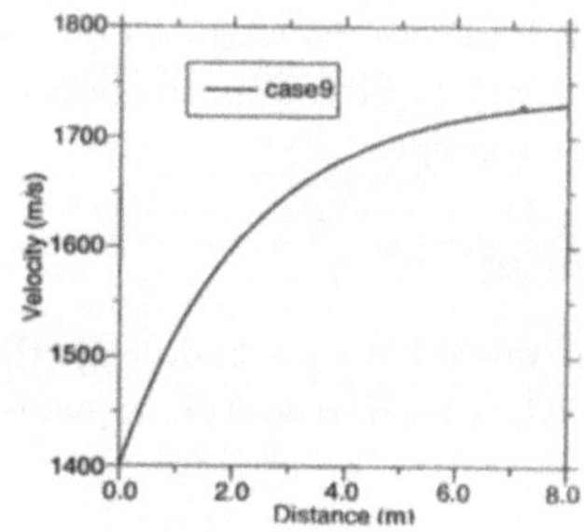

Fig. 22. Velocity versus distance for the Case 9 combustible mixture $3CH_4 + 2O_2 + 7.5N_2$, at 5.0 MPa, with projectile mass = 0.133 kg, considering non-equilibrium gas flow and input velocity of 1400 m/s, using the LCP code.

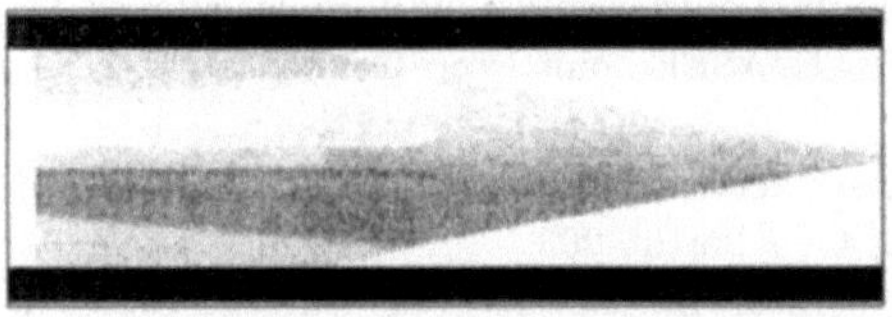

Fig. 23. Schematic of the hypotetical projectile used in cases 8 and 9.

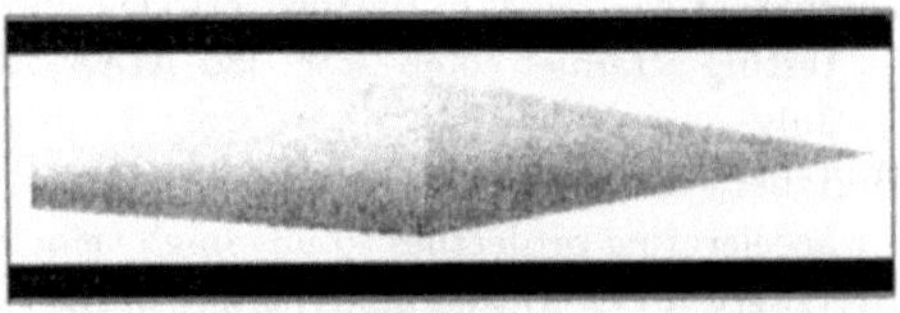

Fig. 24. View of the projectile without fins.

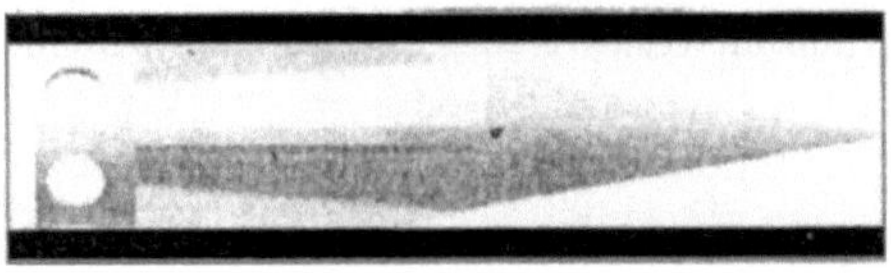

Fig. 25. View of the projectile coupled with obturator.

4. Conclusions

A one-dimensional steady inviscid flow analysis on the combustion driven acceleration process in the ramjet-in-tube operating in the subdetonative mode was made. The conservation equations were solved with the help of a LCP-modified version of the equilibrium calculation Fortran package code NASA SP-273. This modified version includes a balance between the frozen and equilibrium flow properties, in order to obtain a result equivalent to calculation performance with non-equilibrium gas flow consideration. Numerical simulations for the combustible mixture, projectile characteristics and initial conditions that had been studied and presented in others articles, taken into consideration ideal and real gas flow were performed. By comparing the results obtained for the present paper with the various results encountered in the cited articles, it was verified that they correlated well. Then, a performance calculation was made for a hypothetical ramjet-in-tube configuration according to the specification proposed for the RAMAC III Benchmark Test.

At the moment, the authors are developing a bidimensional axisymmetric code for the simulation of inviscid reactive flow over the ramjet-in-tube model to help them in future theoretical investigations. It is also in the works an improvement of the one-dimensional model used in the present paper, as way to extend the analysis for the transdetonative and superdetonative modes of ramjet-in-tube concept.

Acknowledgement

The first author would like to acknowledge the support for his M.Sc. degree program at INPE provided by the Coordenação de Aperfeiçoamento de Pessoal - CAPES/BRAZIL.

References

Buckwalter DL, Knowlen C, Bruckner AP (1996) Ram accelerator performance analysis code incorporating real gas effects. AIAA paper 96-2945

Gordon S, McBride BJ (1971) Computer program for calculations of complex chemical equilibrium compositions, rocket performance, incident and reflected shocks, and Chapman-Jouguet detonations. NASA SP-273

Guimaraes AL, Sinay L, Bastos-Netto D (1997) Liquid fuel burner with oxygen replenishment for testing scramjet combustor. 33d AIAA/ASME/SAE/ASEE Joint Prop Conf, Seattle, USA, July

Hertzberg A, Bruckner AP, Bogdanoff DW (1988) Ram accelerator: A new chemical method for accelerating projectiles to ultrahigh velocities. AIAA J 26:195-203

Hertzberg A (1989) Thermodynamics of the ram accelerator. In: Kim YW (ed) Current Topics in Shock Waves, Proc 17th Int Symposium on Shock Tubes and Shock Waves, AIP Conf Proc, American Inst Phys, pp 2-11

Heiser WH, Pratt DT (1994) Hypersonic airbreathing propulsion, 1st ed, AIAA Education Series, Washington DC

Williams FA (1988) Combustion Theory, 2nd ed, Addison-Wesley Publishing Company, Menlo Park, CA Chapter 11.

On the optimization of thermally-choked ram accelerator systems

X.J. Wang, E. Spiegler, Y. Timnat
Faculty of Aerospace Engineering, Technion-Israel Institute of Technology, Haifa 32000, Israel

Abstract The performance of thermally-choked ram accelerator systems can be optimized by varying the chemical composition of the reacting mixtures and their distribution along the multi-stage launcher. Two different approaches to the problem are presented and demonstrated: a) the practical approach, in which the solution is based on a finite number of available mixtures whose thermo-ballistic properties are known, and b) the mathematical approach, in which numerical optimization methods are used to define the chemical composition of reacting mixtures which will give the highest performance. The importance of including existing practical, system-connected limitations in the optimization procedure is demonstrated using illustrative examples. It is concluded that significant performance gains can be achieved by applying optimization techniques at the preliminary design stage of such systems.

Key words: Optimization, Ram accelerator design

Nomenclature

a	projectile acceleration	A	launcher internal cross section
c	velocity of sound	C_f	thrust coefficient
C_p	specific heat at constant pressure	D	bore diameter
$f(X)$	objective function	F	thrust
$g_j(X)$	constraint function	L	length of the launcher
L_i	length of stages	m	projectile mass
M	Mach number	n	number of objective function variables
n_d	number of diluent moles	n_f	number of fuel moles
n_g	number of inequality constraints	n_o	number of oxidizer moles
P	pressure	$\bar{q}$	non-dimensional heat of reaction
Q_r	heat of reaction	R	gas constant
t	time	T	temperature
TCRA	thermally choked ram accelerator	U	projectile velocity
x	objective function variables	X	vector of objective function variables
X_L	vector of variables lower boundaries	X_U	vector of variables upper boundaries
Subscripts			
1	at control section 1	2	at control section 2
a,b,c or d	mixtures A,B,C,or D	CJ	Chapman-Jouguet
f	final	i	initial
max	maximum	si	self ignition

1. Introduction

The ram accelerator (RA) concept, which is sometimes described as a ramjet-in-a-tube, was first proposed by Prof. Hertzberg's team at the University of Washington (Hertzberg et al. 1988), where most of the pioneering theoretical and experimental research was done (Hertzberg et al. 1988, 1991). Several modes of operation have been suggested with great potential for applications in surface-to-orbit launching of inert payloads (Bogdanoff et al. 1992) and in ground-based testing of hypersonic propulsive cycles (Bruckner et al. 1992). The present paper deals with the optimization of thermally-choked ram accelerators (TCRA) design, by varying the chemical composition of the reacting mixture along the barrel.

2. The TCRA concept

A schematic of a TCRA system is given in Fig. 1. The projectile is injected into the barrel, at $M_1 = 2 - 3$, by a conventional powder or gas gun. The combustion occurs downstream of the projectile base, where the Mach number increases, from an initial subsonic value, to $M_2 = 1$. A normal shock system located on the afterbody, changes its location with the flight Mach number. A one-dimensional stream-tube analysis (Hertzberg, 1988), leads to the following expression for the thrust coefficient:

$$C_f = F/(P_1 A) = M_1 \Gamma (1 + \bar{q} + 0.5(\gamma_1 - 1)M_1^2)^{1/2} - (1 + \gamma_1 M_1^2) \tag{1}$$

where $\Gamma = (\gamma_1/\gamma_2)(2(\gamma_2^2 - 1)/(\gamma_1 - 1))^{1/2}$, $\bar{q} = Q_r/C_{p1}T_1$; $\gamma_1, \gamma_2, C_{p1}$ and Q_r are calculated using the Chemkin-II code (Kee et al. 1994). The instantaneous acceleration is: $a = F/m = C_f P_1 A/m$ and the rate of change of the velocity with distance is given by:

$$dU_1/dx = (dU_1/dt)(dt/dx) = a/U_1 = P_1 A C_f/(mU_1) \tag{2}$$

The length of the barrel, L, needed to accelerate a projectile of a mass m from an initial velocity U_{1i} to a final velocity U_{1f} is given by :

$$L = \int_0^L dx = m/(P_1 A) \int_{U_{1i}}^{U_{1f}} U_1 dU_1/C_f \tag{3}$$

A typical example of the behavior of $C_f, U_1/C_f$ and L as functions of U_1 is given in Fig. 2, for the following set of assumed values: γ_1 =1.4, γ_2=1.3, $\bar{q}$=10, c_1 = 350 m/s, U_{1i} = 600 m/s, m = 1 kg, P_1 = 4.0 MPa and D = 10 cm. The highest velocity that can be reached in theory, is equal to the Chapman-Jouguet detonation velocity of the mixture, i.e. U_1 = 2080 m/sec (M_1 = 5.94), in the present case. The barrel will have to be about 51 m long.

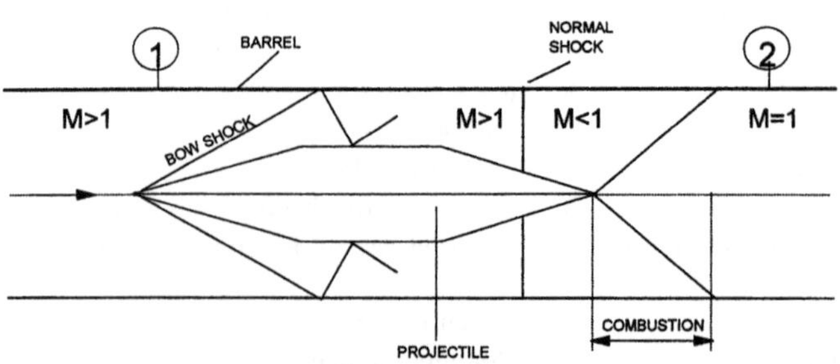

Fig. 1. Schematic of the thermally-choked ram accelerator concept

3. Practical limitations It is worth noting that the barrel length, L, is represented by the hatched area in Fig. 2b and is directly proportional to m and inversely proportional to P_1 and A. Thus, increasing the barrel internal diameter and/or the initial pressure, is an obvious way of decreasing L. However, the peak pressure (behind the normal shock) increases as well and might reach values high enough to burst the barrel open. Practical considerations of this kind limit the operational envelope of TCRAs and must be taken into account in the system optimization process. The most important are: (a) The initial pressure, P_1, must be low enough to keep the peak pressure below a limiting value, P_{max}.(b) The projectile Mach number must be low enough to avoid mixture self-ignition behind the conical shock. In the discussion that follows it will be assumed that self-ignition occurs when the static temperature of the mixture reaches a certain threshold value, T_{si}. (c) The projectile Mach number should be high enough and $\bar{q}$ low enough to avoid "unstart" situations (normal shock in front of the projectile).

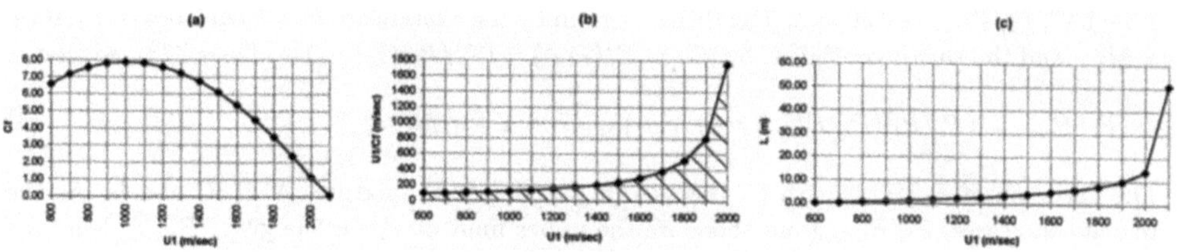

Fig. 2. Typical behavior of $C_f, U_1/C_f$ and L as function of U_1

4. Design optimization

The designer's challenge is to minimize the length and the weight of the barrel needed to accelerate a projectile from U_{1i} to U_{1f}, by varying the chemical composition of the reacting mixture and its initial pressure, along the barrel. In the general case the geometry of the projectile and its weight, the bore diameter, the initial and final velocities, are pre-determined by system and mission considerations. The reacting gas is a mixture of fuel (CH_4, H_2, etc.), oxidizer (O_2 or air) and diluents (N_2, He, CO_2). Usually, a number of mixtures, whose ballistic properties are known, are available from previous projects and their compatibility to the job at hand can be assessed in a comparative way. We shall refer to such a procedure as "practical optimization". A different approach to the problem is to define, by optimization methods, the chemical compositions that maximize the final velocity, U_{1f}, in a launcher of given length. We shall refer to this approach as "mathematical optimization". Both procedures are described and compared below.

5. Practical optimization approach

5.1 Unconstrained Optimization The procedure will be presented for the following mission: An optimized TCRA is to be designed to accelerate a projectile from an initial velocity $U_{1i} = 600$ m/s to a final velocity $U_{1f} = 1800$ m/s. The weight of the projectile is $m = 1$ kg, the nose half-angle $\delta = 20°$, and the bore diameter is $D = 0.1$ m. No constrains are imposed at this stage, and P_1 is assumed to be 10 MPa. Four different reacting mixtures, A , B , C and D, are available; their relevant properties are summarized in Table 1.

The solution is illustrated in Fig. 3. To minimize L (i.e. the area between the curves in Fig. 3a and the horizontal axis), mixture A must be used in the velocity range $600 < U_1$ (m/s)< 1120 ($L_a = 1.37$ m), mixture B for $1120 < U_1 < 1230$ ($L_b = 0.45$), mixture C for $1230 < U_1 < 1400$ ($L_c = 0.83$) and mixture D for $1400 < U_1 < 1800$ ($L_d = 2.58$). This results is in a total barrel length of 5.23 m. In this particular case, the gain is only about 0.60 m; indeed, if only mixture D is used, the total length of the barrel must be 5.84 m.

5.2 Pressure constraints The peak pressure, P_{max}, is obtained just before the section where ignition occurs. Assuming a Rayleigh-type flow in the combustion section we estimate, for $\gamma =$

Table 1. The properties of the 4 reacting mixtures; Tsi values have been assumed for illustrative purpose.

Mixture	Chemical composition	γ_1	γ_2	$\bar{q}$	c_1(m/s)	T_{si}
A	$2O_2$+2.5CH_4+5.6N_2	1.372	1.307	5.26	363	1000
B	$2O_2$+4.5CH_4+2 He	1.375	1.297	4.86	450	900
C	$2O_2$+3.5CH_4+6.5H	1.460	1.384	4.26	547	1100
D	$2O_2$+2.8CH_4+11He	1.516	1.449	3.92	625	700

$1.3-1.45$, $P_2/P_{max} \simeq 0.50 = \lambda$. The thrust is given by the expression: $F = P_2A(1+\gamma_2) - P_1A(1+\gamma_1M_1^2)$, and the thrust coefficient by: $C_f = F/(P_1A) = (P_2/P_1)(1+\gamma_2) - (1+\gamma_1M_1^2)$, so that :

$$P_1/P_{max} = (P_1/P_2)(P_2/P_{max}) = (1+\gamma_2)\lambda/(C_f + 1 + \gamma_1M_1^2) \tag{4}$$

The variation of P_1/P_{max} with U_1 is given in Fig. 4 for the mixtures A, B, C and D. As the velocity increases, P_1/P_{max} (and therefore the upper limit of P_1 for the given P_{max}) decreases monotonically. If the reacting mixture is to be stored at the highest allowable pressure, this should be the value corresponding to U_{1f}. The barrel length, L (see Eq. (3)) can be expressed as:

$$L(P_{max}A/m) = (P_{max}/P_1)_{U_{1f}} \int_{U_{1i}}^{U_{1f}} U_1 dU_1/C_f \tag{5}$$

The quantity $(P_{max}A/m)$ is independent of the reacting mixtures ballistic properties; the right-side of Eq. (5) can therefore be used to compare the barrel lengths needed to accelerate a projectile from U_{1i} to U_{1f}, using different mixtures. The results of such computations are presented in Fig. 5. It is clear that mixture D exhibits the best performance at any velocity. In the specific case when U_{1i} = 600 m/sec, U_{1f} = 1800 m/sec, D = 10 cm, m = 1 kg and P_{max}=100 MPa, we shall find (from Fig. 5b) L = 7.7 m and (from Fig. 4)P_1 = 7.4 MPa.

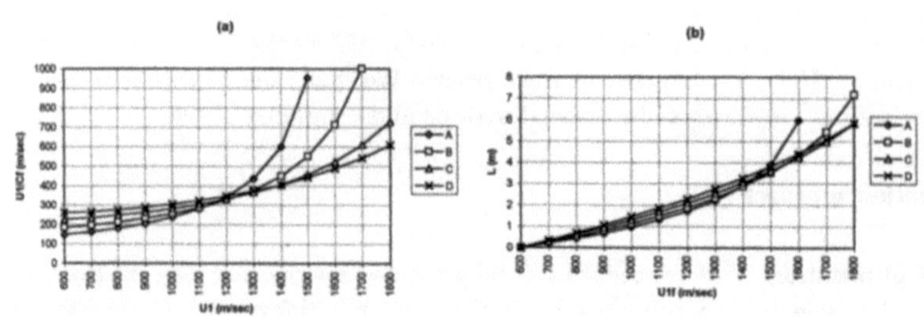

Fig. 3. U_1/C_f and L as function of U_1 for mixture A, B, C and D

5.3 Avoiding premature ignition If the reacting mixture ignites in the fore region of the projectile, drag will be obtained instead of thrust. Premature ignition might happen as a result of increasing the static temperature behind the conical shock beyond the T_{si} value. To illustrate this point, the static temperatures behind the conical shock and the total temperatures, have been calculated as a function of the velocity, for mixtures A, B, C and D; an initial temperature of 300 K and a cone half-angle of 20° were assumed. The results, presented in Fig. 6, show that the static temperatures are well below the respective T_{si} assumed values. However, if ignition is triggered by the high temperatures in the stagnation point region, mixture C must be used, instead of D, at velocities above 1400 m/sec. Using Fig. 5b, we find that the minimum barrel length is 8.95 m.

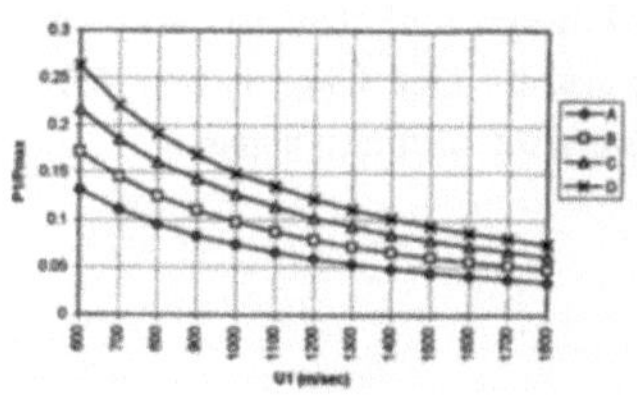

Fig. 4. $P_1/P_{max} = f(U_1)$ for mixtures A, B, C and D

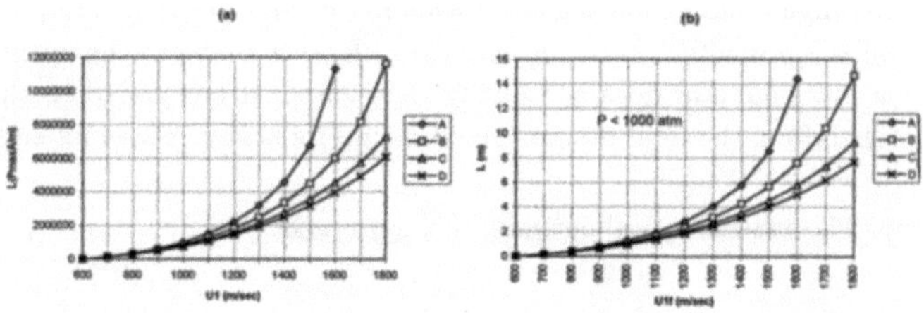

Fig. 5. L=$f(U_1)$ for mixture A, B, C and D (U_{1i} = 600 m/s, P_{max} = 100 MPa)

5.4 Avoiding unstart Unstart (normal shock ejection) occurs when $M = 1$ at the diffuser throat. Unstart must be avoided, because the increase of the pressure on the forebody decelerates the projectile. To do that, the projectile Mach number at which the diffuser throat chokes must be estimated. However, our simple one-dimensional model does not include the diffuser geometry. This problem is best solved by CFD methods. Bruckner et al. (1991) have performed a more detailed (but still one-dimensional) analysis. They have found that, for practical designs, the starting Mach number was in the range 2.3 - 2.6. We shall assume that the starting Mach number for all the mixtures is 2.4. The velocity ranges imposed by the last two restrictions are summarized in Table 2. The two main conclusions are: (a) mixture D cannot be used for the present mission; and (b) U_{1i} must be at least 870 m/sec. Mixtures A, B, and C are to be used in the velocity ranges 870 - 1080, 1080-1310 and 1310 -1800 respectively. Using Fig.5b, the launcher length is found to be 9.4 m.

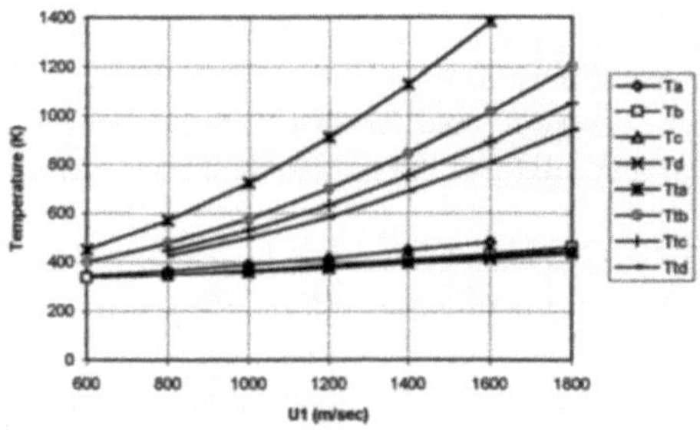

Fig. 6. The static and total temperatures behind the conical shock for mixture A, B, C and D (T_1 = 300 K, $\delta = 20^o$)

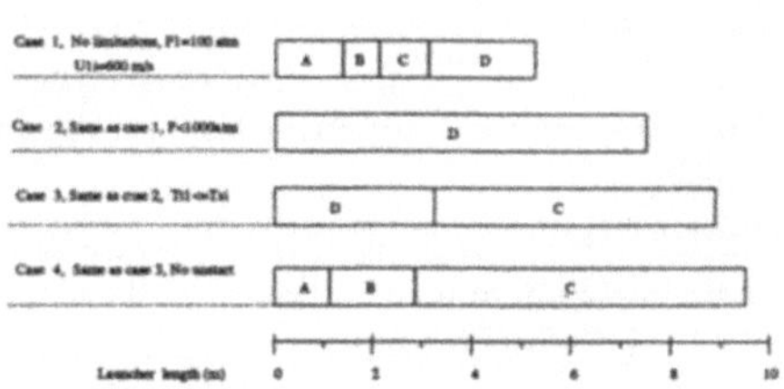

Fig. 7. Comparing optimized solutions with different operational constraints

5.5 The impact of system limitations on the optimization solution The solutions obtained for the four cases considered in sections 5.1 to 5.4 are compared in Fig. 7. As more operational restrictions are imposed on the system, the optimal length of the barrel usually increases. Ignoring the system limitations will lead to a "better" solution (shorter launcher), but the mission will fail. We shall conclude this section with two remarks: 1) The three limitations used to illustrate the problem are not the only ones, nor are they the most restrictive in the general case. Other examples are : (a) making sure that the projectile velocity is such that the flow is indeed choked at the control section 2; (b) avoiding the appearance of "detached" normal shocks in front of the projectile because of too high a nose-cone angle. (c) avoiding normal shock location upstream of the diffuser throat or off the projectile. 2) Sometimes the difference between the optimal design and "the second best" is marginal and does not justify the complications added to the system (see Case 1, Sec. 5.1). Therefore, the final design choice should be always based on engineering common sense.

6. The mathematical optimization approach

6.1 Numerical optimization techniques

A more general solution of optimization problems is to use numerical algorithms (Rao 1984,Haftka 1993) for locating the extreme values of a multi-dimensional objective function, $f(X)$, within defined constraints of its variables, Mathematically, the general constrained optimization problem can be stated as follows:

$$Maximize : f(X); \quad X = (x_1, x_2, ..., x_n); \quad X_L \leq X \leq X_U; \quad g_j(X) \leq 0; \quad j = 1, 2, ...n_g \tag{6}$$

In our case, the function to be optimized (maximized) is the final velocity: $U_{1f} = (U_{1i}^2 + 2\sum_{j=1}^{N} a_j \Delta L_j)^{1/2}; \sum_{j=1}^{N} \Delta L_j = L$. We have assumed that a_j is constant in the interval ΔL_j. We can write:

$$U_{1f} = f(n_f, n_o, n_d, U_{1i}, P_1, A, m, T_1, L_i) \tag{7}$$

Depending on the definition of the problem, some parameters (usually U_{1i}, A, m, and T_1) are constants. In this work an optimization algorithm based on the Variable Metric Method, has been adopted. The penalty functions technique was used for constrained objective functions. Quadratic Interpolation Method or Golden Section method (Rao 1984, Haftka 1993) were used alternatively for one dimensional searching.

6.2 Examples To demonstrate the use of mathematical optimization techniques in the preliminary design of TCRAs, comparative calculations were performed for two tests made at the University of Washington, using non-optimized reacting mixtures. The tests conditions were as follows:

Test 1 (Hertzberg et al. 1988): The 38 mm bore, single-stage launcher was 5 m long, the projectile mass 0.075 kg. and the initial velocity 690 m/sec. The mixture used was $CH_4 + 2O_2 + 6CO_2$ and the initial (storage) pressure was 1.2 MPa.

Test 2 (Bruckner et al. 1991): The 38 mm bore, 12 m long launcher was made of three 4 m long sections separated by membranes. The projectile mass was 0.057 kg. and the initial velocity 1150 m/sec. Three different mixtures were used (see details in Table 3). The initial pressure of all mixtures was 2.1 MPa. The purpose of the computations was to prove (theoretically) that the performance of the system can be increased using optimization algorithms and to estimate the possible gains.

6.2.1 Test 1 Three calculated velocity profiles for Test 1 are presented in Fig. 8, together with the reported experimental values. The lower curve is for the actual test conditions; it predicts a final velocity of 1080 m/sec, as compared with the experimental value of 1120 m/sec. The upper curve is the result of unconstrained optimization. The mixture selected by the optimization model is $1.23CH_4 + 2O_2$ and the predicted muzzle velocity 1820 m/sec, a gain of 740 m/sec over the non-optimized value. The middle curve is the result of a constrained optimization. The limitations imposed on the solution were (a) $2.2 \leq M_1 \leq 4.6$ and (b) $n_{CH4}/n_{O2} \leq 0.5$. The mixture selected is $CH_4 + 2O_2 + 2CO_2$ and the predicted final velocity 1330 m/sec, lower than in the previous case but still 250 m/sec higher than for the non-optimized mixture.

Table 2. Minimum and maximum velocities for the four mixtures

Mixture	No unstart— No pre-ignition
A	$870 < U_a < 1280$
B	$1080 < U_b < 1460$
C	$1310 < U_c < 1880$
D	$1500 < U_d < 1400$

Table 3. Data and computation results for Test 2

Case	Variables	Constraints	Mixtures	L_i (m)	P_1 (MPa)	U_{1f} (m/s)
Experimental simulation	None	None	$2.50CH_4+2O_2+5.60N_2$	4.0	2.1	1566
			$4.50CH_4+2O_2+2.00He$	4.0	2.1	1850
			$3.50CH_4+2O_2+6.50He$	4.0	2.1	2115
No.1	n_f	$L_1 = L_2 = L_3$	$1.06CH_4+2O_2+2.30He$	4.0	2.1	2137
	n_o		$1.05CH_4+2O_2+5.75He$	4.0	2.1	2614
	n_d		$1.02CH_4+2O_2+8.78He$	4.0	2.1	2904
No.2	No.1	$2.4 \leq M_1$	$1.14CH_4+2O_2+2.27He$	3.90	2.1	2130
	L_i	$M_1 \leq 4.7$	$1.07CH_4+2O_2+5.78He$	3.92	2.1	2605
		$L = 12$	$1.02CH_4+2O_2+8.71He$	4.18	2.1	2910
No.3	No.2	No.2	$0.72CH_4+2O_2+2.51He$	3.69	2.9	2182
	P_{1i}	$P_2 \leq 50$ MPa	$0.78CH_4+2O_2+5.99He$	3.92	3.0	2689
			$1.02CH_4+2O_2+9.11He$	4.39	2.9	3059
No.4	No.3	No.3	$5.20CH_4+2O_2+1.10He$	3.70	4.7	1696
		$\bar{q} \leq 6.0$	$3.18CH_4+2O_2+3.89He$	3.92	3.9	2140
			$2.70CH_4+2O_2+4.55He$	4.38	3.5	2354

6.2.2 Test 2 The calculated velocity profiles for Test 2 and the respective experimental profile, are plotted in Fig. 9. Again, the lower curve is for the actual test conditions. The theoretical curves were initiated at the experimentally-determined velocities at the beginning of each propellant mixture; this explains the discontinuities at the transition points. The overall agreement between the calculated and the measured velocities is good.

The curves marked No. 1, 2, 3 and 4 are results of optimization using different constraints and number of variables. The assumptions made and the results obtained are summarized in Table 3. Velocities between 2354 and 3059 m/sec (compared to the non-optimized value of 2115 m/sec) can be achieved. In case No.4, the constraint $\bar{q} \leq 6.0$ is introduced (Knowlen et al. 1995). Results show that this limitation has a strong effect on the final velocity. In No. 4, the optimization results of Mach number is between 2.68 and 4.68. In all four cases, the maximum value of U_{1f}/U_{CJ} is lower than 0.88.

7. Conclusion

Both the engineering and the mathematical approach to the design optimization of TCRAs can lead to a significant increase in the system performance. In order to obtain realistic results, the existing operational constraints must be included in the optimization procedure. In this connection it should be emphasized that the mixtures selected by mathematical optimization methods must be investigated experimentally, to verify that their ignition and combustion properties are adequate. The main parameters in the optimization process are the chemistry of the mixtures and their initial pressures. The strongest constraints have not been identified in this work, because the effect of projectile and barrel geometries is not included in the simple, one-dimensional, stream-tube analysis upon which the expression for the calculation of C_f is based. Avoiding diffuser

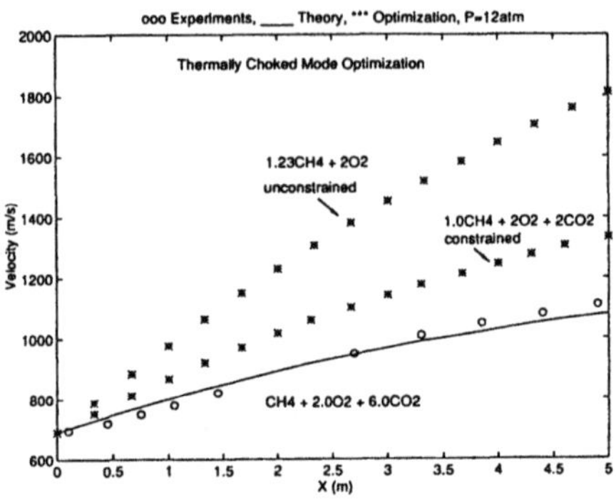

Fig. 8. TCRA optimization test 1

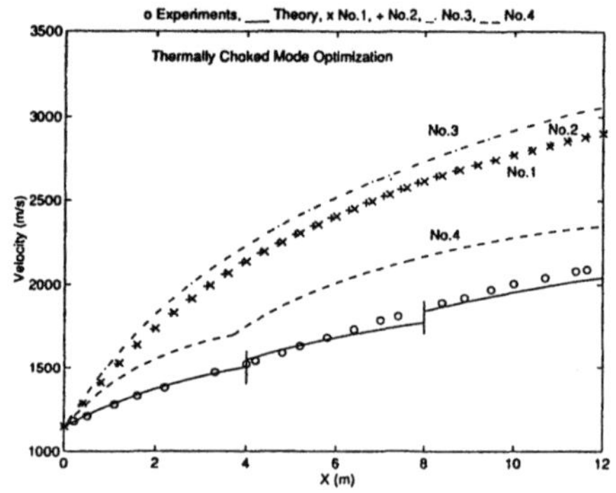

Fig. 9. TCRA optimization test 2

unstart and avoiding normal shock location off the projectile, seem to restrict strongly the range of mixture operational velocities (the first from below the second from above). This may, however, change from case to case. Therefore, the fulfillment of all operational constraints by the selected optimized design should be carefully investigated, using CFD and experimental techniques.

Acknowledgements

The authors would like to thank Professor Josef Rom for his valuable time and helpful discussions.

References

Bogdanoff DW (1992) Ram accelerator direct space launch system: New concepts, J Prop Power 8:481-490

Bruckner AP, Knowlen C, Hertzberg A and Bogdanoff DW (1991) Operational characteristics of the thermally choked ram accelerator, J Prop Power 7:828-834

Bruckner AP, Knowlen C and Hertzberg A (1992) Applications of the ram accelerator to hypervelocity aerothermodynamic testing, AIAA paper 92-3949

Haftka RT and Gurdal Z (1993) Elements of structural optimization, Kluwer Academic Publishers

Hertzberg A, Bruckner AP and Bogdanoff DW (1988) Ram accelerator: A new chemical method for accelerating projectiles to ultrahigh velocities. AIAA J 26:195-203

Hertzberg A, Bruckner AP and Knowlen C (1991) Experimental investigation of ram accelerator propulsion modes. Shock Waves 1:17-25

Kee RJ, Rupley FM and Miller JA (1994) Chemkin-II: A fortran chemical kinetics package for the analysis of gas phase chemical kinetics, Sandia Rep 89-8009B, UC-706

Knowlen C, Higgins AJ and Bruckner AP (1995) Aerothermodynamics of the Ram Accelerator. AIAA paper 95-0289

Rao SS(1984) Optimization Theory and Applications, 2nd edition, Wiler Estern Limited

Prediction of surface heating of a projectile flying in RAMAC 30 of ISL

F. Seiler, F. Gatau, G. Mathieu
French-German Research Institute of Saint-Louis (ISL), F-68301 Saint-Louis, France

Abstract. In a ram accelerator a sharp-nosed-body flies at supersonic velocity through a tube initially filled with a highly compressed combustible gas mixture. By shock compression, i.e., by the bow wave and its reflections at the tube wall, the gas mixture is heated progressively so that it becomes ignited in subdetonative combustion mode at the body's back and in superdetonative mode, investigated mainly in the RAMAC 30 of ISL, in the slit between the projectile midbody surface and the tube wall. The gas combustion causes a temperature rise followed by a gas pressure increase giving a forward thrust to the body of the ram-projectile. Due to the transfer of heat from gas to projectile, the latter's surface temperature increases, on the one hand in the nose region and on the other hand at the body contour of the mid-part of the projectile. This temperature rise can lead to melting processes at the forebody "nose" cone, the mid-part "body" region including fins and at the afterbody "expansion zone". Then ablation of surface material begins which is naturally undesirable not only at the sharp nose but also at the fins and the body of the ram projectile, especially when combustion is localized at the body of the projectile in superdetonative mode. The control of the heating at the nose and the body is necessary for successful ram accelerator operation. Therefore a prediction of the heat flux from gas to projectile surface becomes needful. For this reason a boundary layer and ablation model has been developed by which the heating of the projectile and its melting ablation at nose, fins and body can be estimated for an optimal choice of projectile material at a desired velocity and gas pressure range.

Key words: Ram acceleration, projectile heating, boundary layer and heat conduction equation

1. Introduction

At the high gas pressures present in the flow around a ram projectile and the high projectile velocities, surface heating during the in-tube flight of the projectile will be excessively high. The surface temperature during in-bore movement has to be kept lower than the melting temperature so that melting processes at the projectile surface can be suppressed or limited in order to avoid, e.g., preignition of the combustible gas mixture at the nose followed by a ram unstart. Furthermore, chemical reactions between projectile material and the oxygen as well as the diluent may occur, especially in case the surface is heated and melted by a high heat flux into the projectile. But chemical reactions, i.e., burning of the projectile material, are not treated herein and are not included in the ablation model.

2. Modeling of surface heating

2.1 Assumptions

For calculating the temperature distribution at the surface and inside of the ram-projectile a flow model was developed in the reference frame of the projectile as shown in Fig. 1 for the fin stabilized projectile, see Patz et al (1995). It is assumed that compressible and turbulent boundary layers develop at the following projectile surface regions: (I) at the cylindrical nose cone, (II) in the combustion zone between projectile body and tube wall, (IIf) at the guiding fins in case of the smooth bore, and (III) in the expansion zone.

The flow between conical bow wave and projectile in region (I) is assumed parallel to its surface as well as in regions (II), and (III). It is also assumed that at these surfaces a compressible and

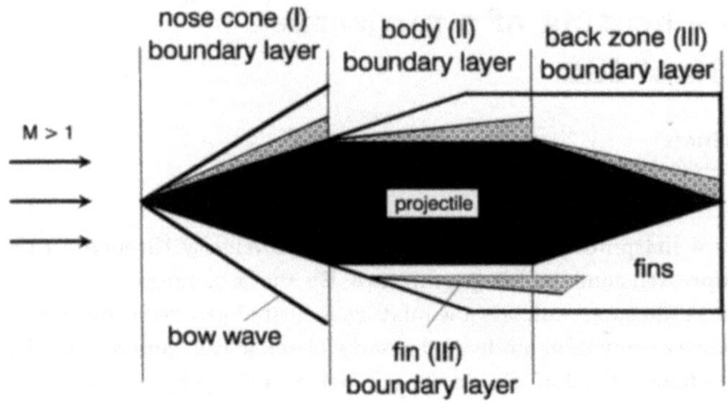

Fig. 1. Boundary layer formation at the fin stabilized projectile

turbulent boundary layer develops similar to that at a flat plate that will be simulated in two dimensions in the flow model. This assumption is justified as long as the boundary layer thickness at the surface is much smaller than the radius of the projectile. Although this assumption fails near the cone tip, this small error is tolerated for obtaining an analytical solution for the description of the whole ram projectile heating.

2.2 Solution of boundary layer equations

Beginning with Prandtl's boundary layer equations, the same differential equation was analytically found for regions (I, II, III) which was solved with the core flow and surface boundary conditions given in these regions (I, II, III). As an analytical solution, the heat flux from gas side to projectile along the projectile's surface is given by Seiler and Mathieu (1995).

2.3 Solution of heat conduction equation

The one-dimensional heat-conduction equation was applied for calculating the temperature increase. Gatau (1997) obtains by integration of the heat conduction equation with the given boundary conditions analytical solutions for the temperature change ΔT inside the projectile material as a function of the heat flux, for both, non-coated walls and coated walls up to three coating layers.

3. Modeling of melting ablation

It is assumed that ablation occurs only by melting erosion with no evaporation. Melting erosion often takes place when hot gas flows with a high stagnation temperature are in contact with colder walls. This process is extensively treated by many authors, see Adams (1959). In this paper here an analytical ablation model for the ablation of a ram-projectile is used, see Seiler (1992), Seiler and Mathieu (1995) and Seiler and Naumann (1993). A similar theory using a numerical calculation model was developed by Naumann (1993).

In the analytical model of Seiler (1992) it is assumed that the sharp-cone geometry (I), the body and fin contours (II, IIf) and the back contour (III) remain approximately unchanged by heating and ablation, i.e., a small amount of ablation occurs. Therefore, the boundary layer formation is considered to be uninfluenced. Due to strong shear stress it is assumed that the melting is whipped away from the cone surface, as soon as it is produced by heat input when the wall temperature exceeds the melting temperature. This means that no liquid layer develops on the solid surface. Heat addition from melt to gas flow is not considered.

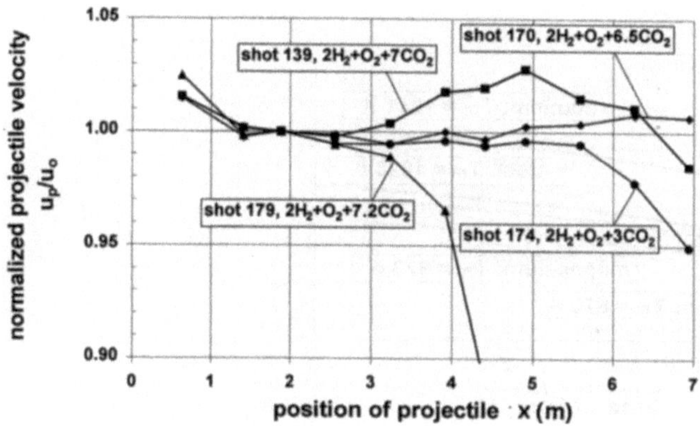

Fig. 2. Velocity distribution gathered experimentally for the smooth bore version

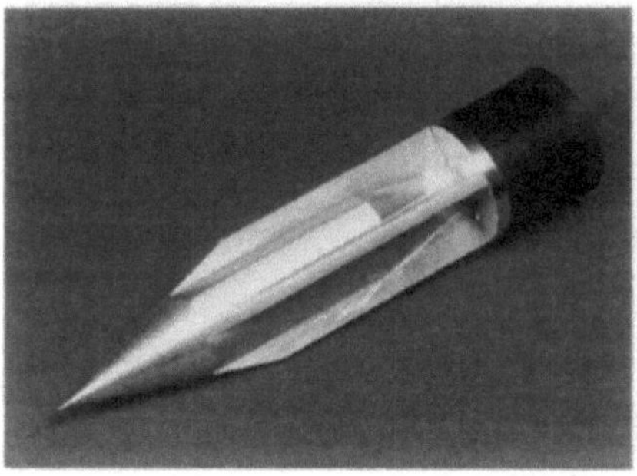

Fig. 3. Aluminum projectile used for firing no. 139

Fig. 4. Steel projectile used for firing no. 174

4. Smooth bore results calculated with the heating and ablation model

4.1 Surface heating

Calculation results for the heating of the ram projectile nose are given for the acceleration cycles in Fig. 2 obtained with the ISL 30-mm-caliber ram accelerator RAMAC 30. The experiments are carried out with fin guided projectiles in the smooth bore version, see Patz et al (1997), of RAMAC 30.

Typical results obtained with the smooth bore version of RAMAC 30 are represented by the following firings: no. 139 (aluminum), no. 170 (titanium), no. 179 (magnesium) and no. 174 (steel). Initially for the four firings the fill pressure was 2.0 MPa. The projectiles used for firings no. 139, 170, and 179 had a half one angle of 16 degrees and a cone length along its surface of 37 mm. The constant diameter combustor region was 50 mm long and the divergent back part surface 26 mm. A photography of the aluminum projectile of shot no. 139 is given in Fig. 3. The steel projectile of shot no. 174 had only three fins and a combustor length of 20 mm, due to spare mass (Fig. 4). In these firings the initial velocity at the beginning of the ram-tube was u_p = 1863 m/s (shot no. 139), u_p = 1740 m/s (shot no. 170), u_p = 1857m/s (shot no. 179), and u_p = 1726 m/s (shot no. 174).

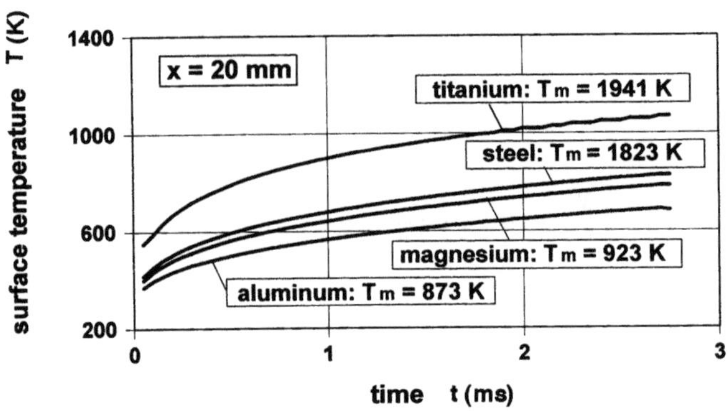

Fig. 5. Forebody (I) surface temperature increase for $x = 20$ mm along the acceleration cycle

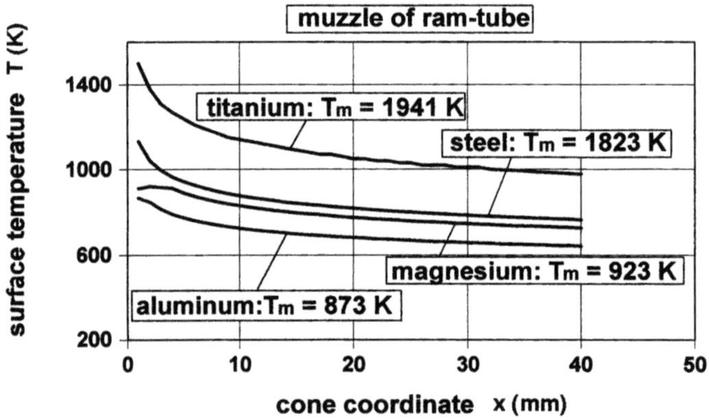

Fig. 6. Surface temperature along the x-coordinate of the cone forebody (I)

In firing no. 139 with the aluminum projectile, acceleration was achieved along the first 3 m inside of the ram-tube, see Fig. 2, of the total tube length of 5.7 m. Then the acceleration turns to deceleration. Looking at shot no. 170 with a titanium projectile the velocity increases, but very smoothly. In both firings the impact into the steel plates, arranged inside of a piston which is placed in the decelerator tube, is weak. These weak impacts indicate that the projectiles have been damaged inside of the ram-tube with an extraordinary mass loss during the ram accelerator cycle. With the projectile made of magnesium (no. 179) the projectile velocity drops immediately after entering the ram-tube, i.e., combustion starting. Herefore no impact into the steel plates is recognized. From this observation we can deduce that the magnesium projectile is totally burned up as an effect of melting and burning in contact with the hot gases present in the combustion region (II). Good results towards erosion resistance have been gathered with projectiles of steel (35 NCD 16). The projectile fired with shot no. 174 survived the ram cycle in good condition and also the impact into the steel plates looks very well.

X-ray pictures taken at the muzzle of the ram-tube will support these outcomes, see Patz et al (1997). The projectile fired for shot no. 139 (aluminum) has lost its fins as well as erosion at the midbody can be observed. A similar result is present for no. 170 (titanium): the fins are burned up and are not more recognized on the x-ray image. The midbody erosion is much smaller compared with that in shot no. 139 using aluminum as projectile material. With the magnesium burning in shot no. 179 a strong unstart phenomenon is coupled and no x-ray picture is available.

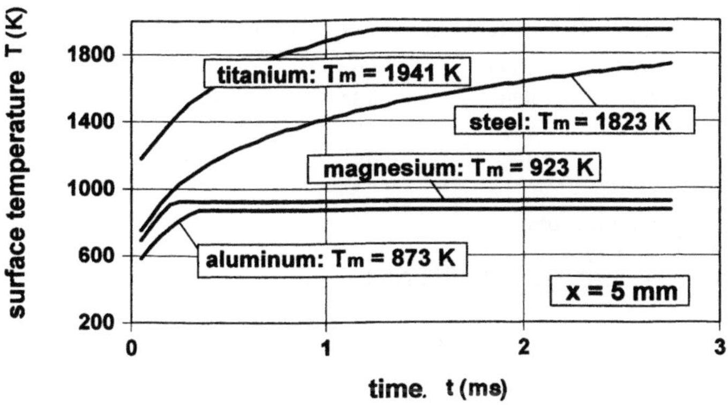

Fig. 7. Midbody (II) surface temperature increase for x = 5 mm along the acceleration cycle

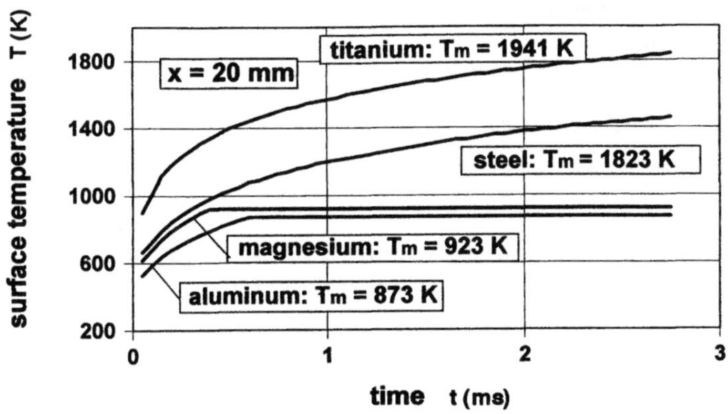

Fig. 8. Midbody (II) surface temperature increase for x = 20 mm along the acceleration cycle

For shot no. 174 (steel) the projectile is seen in the x-ray photography in good shape, i.e., no melting and burning occurred.

In order to get an insight into the heating behaviour of the four materials used: magnesium, aluminum, titanium and steel, a ram cycle was modeled with a constant projectile velocity of 1800 m/s along the whole ram-tube. This model firing reproduces well the real velocity distribution of most firing in smooth bore RAMAC 30.

Heat transfer calculations have been done with input data having constant projectile velocity along the ram-tube with u_p = 1800 m/s, 2.0 MPa fill pressure of the combustible gas mixture: $2H_2+O_2+7CO_2$. The data for the projectile materials used are.

An impression of the distribution of the surface temperature calculated as a function of flight time is given in Fig. 5 for the cone region (I). The cone surface temperature stays for x = 20 mm below melting temperature during the whole modeled ram cycle beginning at t = 0 and ending at t = 2.75 ms.

The surface temperature along the x-coordinate of the cone surface (I) as determined is shown in Fig. 6 for t = 2.75 ms, i.e., the projectile just passes the end cross-section of the ram-tube. No melting is present with titanium and steel as projectile material. For the magnesium and aluminum projectile melting begins near the tip of the nose of the projectile in the region were $x \to 0$.

Table 1. Material properties

Material	ρ (kg/m3)	c_p (J/kg K)	λ(J/m s K)	T_m (K)
magnesium	1800	1046	96	923
aluminum alloy (AlMgCu1)	2790	920	134	873
titanium	4540	471	16	1941
steel (35 NCD 16)	7830	460	38	1823

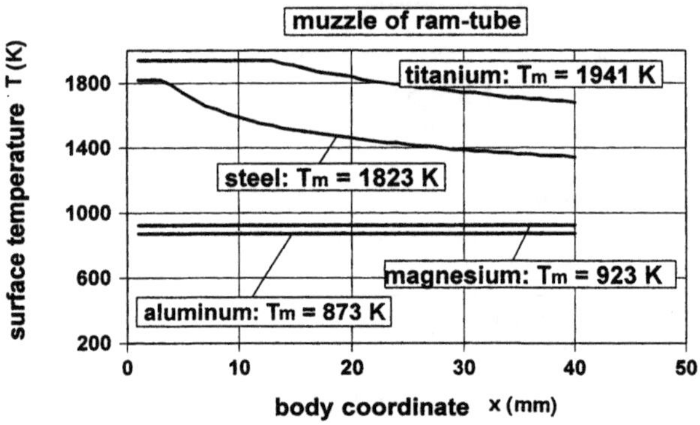

Fig. 9. Surface temperature along the x-coordinate of the midbody(II)

In the midbody region (II) for the gas mixture $2H_2+O_2+7CO_2 \rightarrow 2H_2O+7CO_2$, the following gas conditions have been calculated using the computer code of Smeets et al. (1992): $u_p = 1574$ m/s: $p = 25.2$ MPa, $T_g = 1672$ K, $\rho = 69.2$ kg/m^3, $u = 1574$ m/s. These flow quantities have been taken as boundary conditions in region (II) at the border to the core flow outside of the boundary layer. The calculated temperature increase at the midbody surface is shown as a function of time for $x = 5$ (Fig. 7) and 20 mm (Fig. 8) downstream of the inlet to the combustor region (II). Onset of melting is found for $x = 5$ mm for titanium, aluminum and magnesium. For steel projectiles melting temperature is not reached. More downstream ($x = 20$ mm) the surface temperature is lower than at $x = 5$ mm. Therefore, besides steel, titanium do also not melt during the whole ram firing.

The time integrated heat history is shown in Fig. 9 for that time point where the projectile leaves the muzzle of the ram-tube. For magnesium and aluminum the surface temperature as a function of the body's x-coordinate is higher than melting temperture. The whole midbody melts. Using titanium projectiles, melting is resticted to about one third of the midbody surface.

4.2 Surface ablation

At the cone surface (I) no erosion is calculated because for the four materials investigated, magnesium, aluminum, titanium and steel the surface temperature stays below melting temperature during the whole shooting cycle.

The total erosion e predicted is given for the midbody surface (II) in Fig. 10 and and Fig. 11 as a function of time for the firing modelled with constant $u_p = 1800$ m/s along the entire flight inside of the ram-tube. In Fig. 10 the calculated body erosion is seen for $x = 5$ mm distance from the entrance to the combustor region (II). Magnesium and aluminum show maximum erosion rates of 0.4, respectively, 0.2 mm for $t = 2.75$ ms, i.e., the projectile just exits the ram-tube. The calculations with titanium give less ablation. With steel as projectile material no surface erosion is present. A similar result is shown for $x = 20$ mm (Fig. 11).

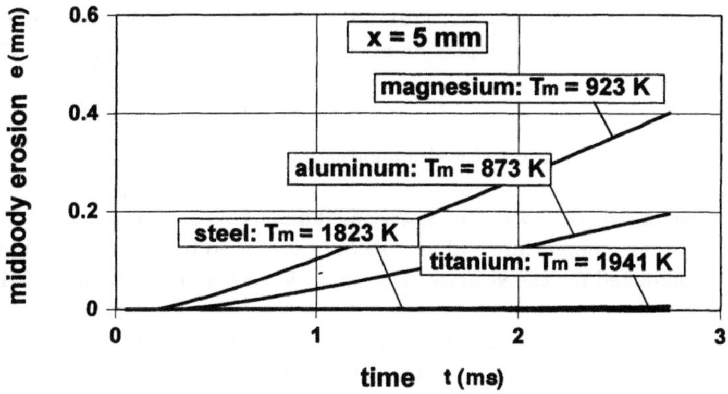

Fig. 10. Erosion along the midbody (II) for $x = 5$ mm along the acceleration cycle

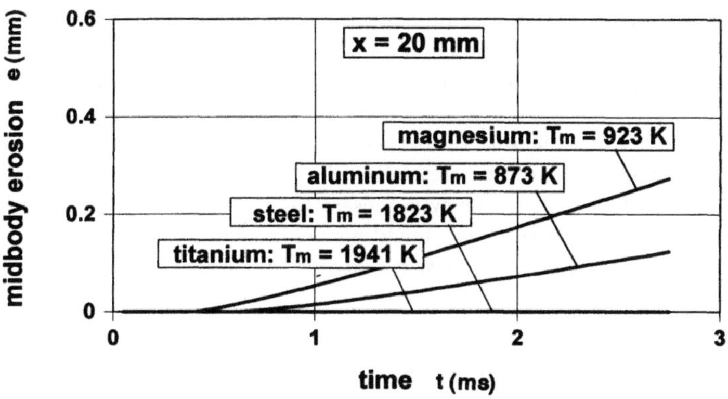

Fig. 11. Erosion along the midbody (II) for $x = 20$ mm along the acceleration cycle

It can be supposed that besides melting erosion chemical reactions, i.e., burning with the oxygen are additionally present, giving as a result more mass loss as predicted with the ablation model. Especially magnesium, aluminum and titanium have a high tendency to react with the oxygen present in the combustible gas mixture. Furthermore, the energy production by projectile material reaction produces an additional undesired heat release inside of the combustor channel which contribute to projectile acceleration, see in Fig. 2 the velocity increase for firing no. 139 (aluminum) and the discussion by Patz et al (1997).

At the projectile fins (IIf) the same ablation as for the body (II) is expected and calculated. Therefore, the results presented in Fig. 10 and Fig. 11 for the body region (II) can be extended to the erosion e along the x-coordinate at the fin's surface (IIf). For $u_p = 1800$ m/s at the end of the ram-tube the erosion e is for the aluminum material used as mentioned at $x = 5$ downstream of the beginning edge of the fins, see Fig. 10: $e = 0.2$ mm. For titanium material the erosion e is nearly zero at the same position. Because melting acts at both sides of the fins, the erosion data must be doubled to define the minimal thickness of the guiding fins required for surviving the shot cycle when melting erosion is present. It came out that the fin thickness of 2.5 mm, e.g., present in experiment no. 139 (aluminum) is enough to survive the ram-tube acceleration. Nevertheless, this firing fails and the fins vanish. Melting of surface material by heating seems therefore not to be the only process ablating the fins and also the body of the ram projectile. As discussed above, burning of aluminum and maybe also titanium (no. 170), i.e., reactions between surface material and oxygen, supported by surface melting, may play an important role in the failure of shot 139.

5. Conclusions

Cone (I): No significant deformation is expected with magnesium, aluminum, titanium and steel projectiles, see Fig. 6.

Midbody (II): Ablation by melting in case of magnesium, aluminum and titanium projectiles is calculated due to the high gas temperature and gas pressure produced by combustion. With steel nearly no ablation is predicted for the velocity range considered (1800 m/s) and 2.0 MPa fill pressure.

Fin (IIf): Ablation at the fins is of the same order as at the body (II). The calculated erosion e is smaller than the fin width in the velocity range of 1800 m/s. Nevertheless, the projectiles have lost their fins during ram firing. That means other erosion mechanisms must be additionally present. It is supposed that chemical reactions between fin material and the oxygen as well as the diluent (here CO_2) are considerd for metals as magnesium, aluminum and titanium, which easy form magnesium oxide, aluminum oxide as well as titanium dioxide, especially at high temperatures as present in the combustion region. Such chemical reactions, if possible, are also present at the midbody surface (II).

Chemical reactions of projectile material with the oxygen and the diluent CO_2 have been found in previous RAMAC 30 experiments, done with projectiles of magnesium. Magnesium strongly reacts with oxygen by burning. The same tendency is present for aluminum and for titanium. Therefore pure magnesium, aluminum or titanium should not be brought into contact with the hot combusted gas mixture, especially in case of melting processes and in case of CO_2 as diluent.

6. References

Adams Mac C (1959) Recent advances in ablation. ARS J

Smeets G, Gatau F, Srulijes J (1992) Rechenprogramm für Abschätzungen zur Ram-Rohrbeschleunigung, ISL-Report RT 507/92

Seiler F (1992) Heating and ablation of a sharp-nosed body flying at hypersonic velocity through a tube filled with highly compressed gas mixture. IUTAM Symp on Aerothermochemistry and Associated Hypersonic Flows, Marseille

Seiler F, Naumann KW (1993) Bow shock wave heating and ablation of a sharp-nosed projectile flying at supersonic velocity inside of a ram accelerator. In: Brun R and Dumitrescu LZ (eds) Shock Waves @ Marseille I:183-188

Naumann K W (1993) Heating and ablation of projectiles during acceleration in a ram accelerator tube. AIAA/SAE/ASME/ASEE 29th Joint Prop Conf and Exhibit, Monterey, Ca, USA

Patz G, Seiler F, Smeets G, Srulijes J (1995) Status of ISL's RAMAC 30 with fin guided projectiles accelerated in a smooth bore. Proc 2nd Int Workshop on Ram Accelerators, Univ Washington, Seattle, USA

Seiler F, Mathieu G (1995) Boundary layer model for calculating the heat transfer into a ram projectile fired in a ram accelerator. Proc 2nd Int Workshop on Ram Accelerators, Univ Washington, Seattle, USA

Gatau F (1997) Conduction thermique dans un solide composé de plusieurs couches homogènes, ISL Rep

Patz G, Seiler F, Smeets G, Srulijes J (1998) The behaviour of fin-guided projectiles superdetonative accelerated in ISL's RAMAC 30. In: Takayama K, Sasoh A (eds) Ram Accelerators, Springer-Verlag, Heidelberg, pp 89-95

Ram accelerator optimization and use of hydrogen core to increase projectile velocity

D. W. Bogdanoff
Thermosciences Institute, NASA Ames Research Center, Moffett Field, CA, 94035-1000, USA

Abstract. The maximum ram accelerator velocities achieved to date are slightly under 3 km/s. This present paper analyzes projectile design and the use of the hydrogen core technique for ram accelerator operation between 3 and 7 km/s. Projectile structural design is studied; hollow, solid and two-material projectiles made of Mg, Al, Ti, steel, Ta and Re are analyzed to determine the maximum accelerations which can be reached. These projectiles are then analyzed for gasdynamic heating and the distances along the launch tube until the projectile surfaces reach the ignition and melting points are determined. The projectile surface is found to reach the ignition and melting temperatures after traversing small fractions of the necessary tube length. The use of a hydrogen core in the ram accelerator tube to reduce the heat flux to the projectile is studied. A new vortex technique is presented. With a projectile with a Ta or Re shell in a hydrogen core, one should be able to make the projectile survive up to launch velocities of 7 to 10 km/s.

Key words: Optimization, Hydrogen core, Increase velocity

1. Introduction

The maximum ram accelerator velocities achieved to date are about 2.7-2.8 km/s (Elvander et al. 1996). Such velocities can also be achieved by other launch techniques, such as combustion-driven guns (Lord 1960, Waldrom et al. 1960), rail guns (Yamori et al. 1996), electrothermal guns (Tidman et al. 1993) and two-stage light gas guns (Canning et al. 1970). The other launch techniques, in particular, two-stage light gas guns, maintain their dominance because (a) they are proven and (b) they possess velocity capabilities much exceeding those demonstrated to date with the ram accelerator. It is therefore very important for higher velocities to be achieved by the ram accelerator, to enable it to become more competitive with other launchers.

Both theoretical and experimental studies (Hertzberg et al. 1986, Bogdanoff 1992, Seiler et al. 1995, Giraud et al. 1995) have shown that gasdynamic heating and melting and burning of projectiles in the launch tube are serious obstacles to the attainment of higher velocities. We address this problem in two ways. First, we determine the projectile materials and construction techniques which yield projectiles capable mechanically of supporting the most rapid acceleration. Then, a gasdynamic heating analysis is performed on the projectiles as they are accelerated from 3 to 7 km/s. The projectile surface was found to reach the ignition and melting temperatures after traversing small fractions of the necessary tube length. The use of a hydrogen core in the ram accelerator tube to reduce the heat flux to the projectile is studied. A new vortex technique is presented. With a projectile with a Ta or Re shell in a hydrogen core, one should be able to make the projectile survive up to launch velocities of 7 to 10 km/s.

2. Mechanical design of projectile

We will consider six different structural materials, magnesium alloy ZK60A-T5, aluminum alloy 7075-T6, titanium alloy Ti-6Al-6V-2Sn, steel alloy 4340, tantalum alloy Ta-10W and rhenium. We will study a hollow cylindrical projectile, a solid one-material projectile and a projectile with an outer shell of one material filled with a solid core of a second material. These projectiles can be analyzed using standard strength of materials techniques.

Two points are worth noting here. First, a solid projectile, on a per unit mass basis, has about 3 times the strength of a hollow projectile, since, upon yielding, the material has no place to go except to flow axially. Second, for the two-material (shell plus core) projectiles studied here, failure also takes place when the core begins to flow axially.

Table 1. Maximum pressures, normalized projectile masses and accelerations for projectiles of various constructions

Case	Projectile construction	Max pressure (MPa)	Normalized proj. mass, M(norm) (g/cm^3)	Normalized proj. accel., A(norm) (m^2/s^2)	Ratio of A(norm)
1	Mg/85/–	23.3	0.60	2,660	
2	Al/85/–	55.9	0.92	4,160	
3	Ti/85/–	117	1.49	5,340	
4	St/85/–	150	2.56	3,980	
5	Ta/85/–	121	5.51	1,490	
6	Re/85/–	207	6.88	2,040	
7	Al/0/–	604	3.30	12,470	
8	Ti/85/–	117	1.49	5,340	1.000
9	Ti/85/Mg	282	3.04	6,320	1.181
10	Ti/95/Mg	263	2.47	7,250	1.357
11	St/85/–	150	2.56	3,980	1.000
12	St/85/Mg	332	4.12	5,490	1.379
13	St/95/Mg	281	2.84	6,710	1.685
14	Ta/85/–	121	5.51	1,490	1.000
15	Ta/85/Mg	321	7.07	3,090	2.071
16	Ta/95/Mg	277	3.88	4,850	3.248
17	Re/85/–	207	6.88	2,040	1.000
18	Re/85/Mg	433	8.44	3,490	1.706
19	Re/95/Mg	316	4.36	4,920	2.407

We estimate the mass of the projectile by considering it to be a rod (or tube) of length L_{eff} and outside diameter d_o. The effective drive pressure of the particular ram accelerator mode being considered is p_{dr}, calculated as acting on the projectile maximum area. With the preceeding quantities known, the normalized projectile acceleration [= (true acceleration) D, where D is the ram tube diameter] can be calculated. The ratio between the maximum pressure applied to the projectile, p_{max}, and p_{dr} will be required and is denoted by R_p. We consider ram accelerator operation in the "Oblique I" superdetonative mode. We take representative values (Knowlen et al. 1987, Brackett et al. 1989) of R_p for this mode as 0.15. For the Oblique I projectile analyzed in Brackett et al. (1989), d_o/D = 0.86 and we can estimate a L_{eff}/D value of 2.6. We will use this L_{eff}/D value, but will decrease d_o/D to 0.76 to allow for boundary layer effects. With the parameters d_o/D, L_{eff}/D and R_p known and using a safety factor of 1.25, we can calculate the maximum pressure capability of a projectile of a given construction as well as its normalized mass [= (true mass)/D^3] and acceleration.

In Table 1, column 2, the projectile construction is described by three entries separated by slashes. The first entry is the shell material. ("St" denotes steel.) The second entry is $R/100$, where R is the ratio of the inside diameter of the shell to the outside diameter; if this entry is zero, the projectile is solid, made of one material. The third entry denotes the core material; if there is no core, this entry is a dash. Cases 1 to 6 are for hollow projectiles. The A(norm) values for these cases reflect the various strength-to-weight ratios of the materials. Ti has the highest A(norm), ~5,300 m^2/s^2. Aluminum and steel follow, with A(norm) values of ~4,200 and ~4,000 m^2/s^2, respectively. Steel is thus fairly competitive. A(norm) for Mg is still lower. The refractory metals, Ta and Re have very much lower A(norm) values, ~1,500 and ~2,000 m^2/s^2, respectively, due to their much higher densities.

In cases 2 and 7 of Table 1, for an Al projectile without a core, we examine the effect of changing from a hollow projectile to a solid projectile. A(norm) is three times higher for a solid projectile. Cases 8 - 19 in Table 1 examine the effect of filling a relatively refractory shell with a core of Mg. For each of the four shell materials, we see, first, the effect of introducing the core with R kept at 0.85 and then the effect of reducing the shell thickness from $R = 0.85$ to $R = 0.95$. For each shell material for cases 8 - 19, in the last column of Table 1, we give the ratios of A(norm) for the various cases with a Mg core to that with the shell alone. This last column gives, then, a measure of the improvement in A(norm) obtained by adding the Mg core and increasing R.

For Ti and steel, the shell materials with higher A(norm) values in cases 1 - 6, the improvement in A(norm) obtained by adding the Mg core is relatively modest, being 36-69% even for the cases with $R = 0.95$. Thus, the advantage gained by filling Ti and steel projectiles with Mg is relatively limited, though somewhat greater for the steel projectile. For the most refractory shell materials, Ta and Re, the improvement in A(norm) obtained by adding the Mg core is much larger, being 140 - 225% for $R = 0.95$. Even for $R = 0.85$, the improvement in A(norm) upon adding the Mg core is 70 - 110%. For the projectiles with Ta and Re shells and Mg cores with R = 0.95, the A(norm) values are about 4,900 m^2/s^2, being thus very competitive with the hollow Ti projectile. Such projectile designs, with Ta or Re shells and Mg cores, may thus be important options to demonstrate higher ram accelerator velocities in reasonably short tubes with projectiles with good heat resisting properties.

3. Heat transfer to projectile

We use Reynolds' analogy (Shapiro 1954), relating the skin friction coefficient and the heat transfer coefficient in boundary layer flows. We use also the simplifying assumption that the Prandtl number is unity. The thermal potential driving the heat transfer at the projectile throat is taken to be the total stream enthalpy at the throat after the combustion energy is released. We consider ram accelerator operation from 3 to 7 km/s in the Oblique I drive mode with the ram tube fill gas gradating from $2H_2$+O_2+$5N_2$ to $2H_2$+O_2+$5H_2$ such as to keep the Mach numbers ahead of the projectile and at the throat after combustion roughly constant at ~8 and ~3.5, respectively. From Bogdanoff and Higgins (1996) and McBride and Gordon (1996), we estimate the average (static) temperature after combustion as ~2300 K. We obtain the value of the skin friction coefficient using the method of van Driest (1951), which gives this coefficient mainly as a function of Mach number, Reynolds number and wall temperature. Coupling the gasdynamic heat flux calculations with a one-dimensional model of heat conduction into the projectile yields the time history of the projectile maximum surface temperature which can readily converted into a wall temperature-distance relation. The projectile is assumed to be injected into the ram tube at an initial temperature of 300 K.

Table 2 shows the values of L/D and $x_{ig}/L = x/L$(ignition) and $x_{me}/L = x/L$(melting) for a number of the projectile constructions shown in Table 1. x is the distance along the ram tube and L is the total length of the ram tube needed to accelerate the projectile from 3 to 7 km/s. The ignition temperatures of steel and Ta are given in van der Bliek (1962) as 1500 K. We have assumed that the ignition temperatures of Ti and Re are also 1500 K. We note that, no matter

Table 2. Normalized distances along ram tube at which surface reaches ignition and melting temperatures

Projectile constr.	Max. pressure (MPa)	L/D at tube exit	x/L(ignition)	x/L(melting)
Ti/85/–	117	3800	0.00046	0.00100
Ti/85/Mg	282	3200	0.00011	0.00025
Ti/95/Mg	263	2700	0.00015	0.00033
St/85/–	150	5000	0.00193	0.00312
St/85/Mg	332	3700	0.00065	0.00107
St/95/Mg	281	3000	0.00108	0.00177
Ta/85/–	121	13,400	0.00076	0.02097
Ta/85/Mg	321	6500	0.00027	0.00852
Ta/95/Mg	277	4100	0.00057	0.01650
Re/85/–	207	9800	0.00069	0.02427
Re/85/Mg	433	5700	0.00030	0.01203
Re/95/Mg	316	4000	0.00078	0.02712
Mg/85/–	23	7500	0.00541	0.00541
Mg/0/–	252	2500	0.00025	0.00025
Al/85/–	56	4800	0.00272	0.00272
Al/0/–	604	1600	0.00011	0.00011

what projectile construction is chosen, the projectile cannot traverse more than 0.5% and 2.7% of the tube length before the projectile wall at the throat reaches the ignition or melting point, respectively.

It might appear, at first sight, that there is therefore no value in comparing the various x/L values in Table 2 among themselves. This is not completely true, however, since if we were to repeat the x/L calculations for thermally choked (TC) projectiles, we would obtain very much larger values of x/L, due to the lower gas kinetic energy, smaller velocity differences and higher R_p for the TC mode. However, the relationship between the results for various projectile constructions would remain, in many ways, as shown in Table 2.

We first consider the x_{ig}/L values for Ti, steel, Ta and Re. We see that steel is by far the best, followed by Ta and Re, with Ti being the worst, due to its poor thermal conductivity. For the same shell material, the hollow projectiles with $R = 0.85$ generally have better x/L values than the projectiles with R = 0.95 and the Mg core, despite their considerably higher L/D values. This is due to the much lower p_{max} for the hollow projectiles, which results in a much lower projectile heat transfer. The projectiles with $R = 0.85$ and the Mg core have the worst values of x/L. The x_{me}/L values for Ti, steel, Ta and Re, in general, follow the same patterns of variation as those of the x_{ig}/L values discussed previously, except that, due to the much higher melting points of Ta and Re, the x/L values for these materials are roughly 10 times the corresponding values for steel.

For the Mg and Al projectiles, the hollow projectiles have much better x/L values, despite their much larger L/D values. Again, this is due to the much lower p_{max} and (hence) q values for the hollow projectiles. Comparing the x_{ig}/L values for all the hollow projectiles, we see that the Al and Mg projectiles have the largest values, on account of their very low p_{max} and q values. Comparing the x_{me}/L values for all the hollow projectiles, we see that the Ta and Re projectiles have the largest values, on account of their very high melting points.

We emphasize that that the discussions of the above two paragraphs, particularly those of the second part of the paragraph just above, do not apply strictly to TC operation of the ram accelerator and the calculations need to be repeated at the appropriate conditions if one wishes to study heat transfer problems in this regime.

We now examine the effect of diameter and pressure scaling on the values of x_{ig}/L (and x_{me}/L.) In van der Bliek (1962), it is shown that the time necessary for a semi-infinite heated wall to reach a given temperature is proportional to q^{-2}, where q is the heat flux. For turbulent flow, q is proportional to $p_{max}^{0.8}D^{-0.2}$, when Reynolds number effects are taken into account. Since, herein, we are always discussing the velocity range of 3 - 7 km/s, distances along the tube will be proportional to time. Thus, we can write $x_{ig} \propto p_{max}^{-1.6}D^{0.4}$. Introducing x_{ig}/L and L/D, we get $x_{ig}/L \propto p_{max}^{-1.6}(L/D)^{-1.0}D^{-0.6}$.

We first consider size (D) scaling of projectiles being accelerated in the same gases at the same pressures through the same velocity range. For such cases, L/D and p_{max} are the same for all projectiles considered. x_{ig}/L and x_{me}/L are seen to decrease as $D^{-0.6}$ as the projectile size increases. Thus, projectile designs which might be satisfactory for small projectiles (e.g., in the TC mode of operation) could well fail when the projectile sizes were increased.

We now consider projectiles of the same size (D) and construction, but we increase L/D and trade this off for a proportional decrease in p_{max}. In this case, $(L/D)p_{max}$ is held constant and the projectile would reach the same final velocity if the same gas composition was always used. In this case, $x_{ig}/L \propto (L/D)^{0.6}D^{-0.6}$. x_{ig}/L would increase as $(L/D)^{0.6}$ as L/D increases. Hence, using longer, lower pressure ram tubes may allow survival of projectiles which could not survive launches in shorter, higher pressure ram tubes.

We now examine the maximum projectile velocities which can be obtained if one can limit the projectile surface temperature. The limiting temperature of the surface of a shell is taken to be 0.90 $\times T_{me}$; ignition and burning of the projectile material will not occur in a hydrogen core. We set the limiting temperature of the projectile surface equal to the stagnation temperature of the hydrogen core (see Bogdanoff and Higgins, 1996); in this way, we can calculate the limiting velocities of projectiles in a hydrogen core. These are 6.3, 6.1, 9.4 , 9.6 and 10.0 km/s for projectiles with Ti, steel, Ta , Re and W shells, respectively. Hence, velocities exceeding orbital velocity can be obtained with projectiles with Ta, Re and W shells in hydrogen cores.

A projectile with a relatively refractory shell material will have some maximum allowable temperature at the inner surface of the shell. This could be due to a limiting temperature of a core material, e.g., Mg, Kevlar or carbon fiber. Also, for a hollow projectile, the inner part of the shell must remain sufficiently cool that the shell maintains adequate strength. Using standard heat conduction theory for a semi-infinite slab, it can easily be shown that the depth in the slab, y, at which at certain temperature, T_y, is reached at certain time, t, after the surface temperature is suddently raised to T_w is proportional to $(\alpha t)^{0.5} \times F(T_w, T_o, T_y)$, where α is the thermal diffusivity, T_o is the initial temperature of the slab and F is some function. In general, for the analyses herein, for different ram tube diameters, D, we can write $t \propto D$. Hence, we can write $y/D \propto \alpha^{0.5}D^{-0.5} \times F(T_w, T_o, T_y)$. Thus, shell materials with lower thermal diffusivities, such as Ti, will be superior (requiring thinner shells). We also note that larger size ram tubes are superior, in that y/D decreases as $D^{-0.5}$ as D increases. Finally, if the maximum allowable temperature at the inner surface of the shell can be increased, the required shell thickness can be decreased.

5. Hydrogen core - implementation

Several earlier techniques proposed for the implementation of a core of hydrogen gas in the ram tube were reviewed in Bogdanoff and Higgins (1996). Two techniques involved balloons to separate the hydrogen core and the surrounding region of fuel-oxidized-diluent (FOD) gas. In two additional techniques, it was proposed to line the ram tube with an annular solid high explosive (HE) layer or an annular layer of deflagrating propellant or oxidizer, perhaps applied as a gel,

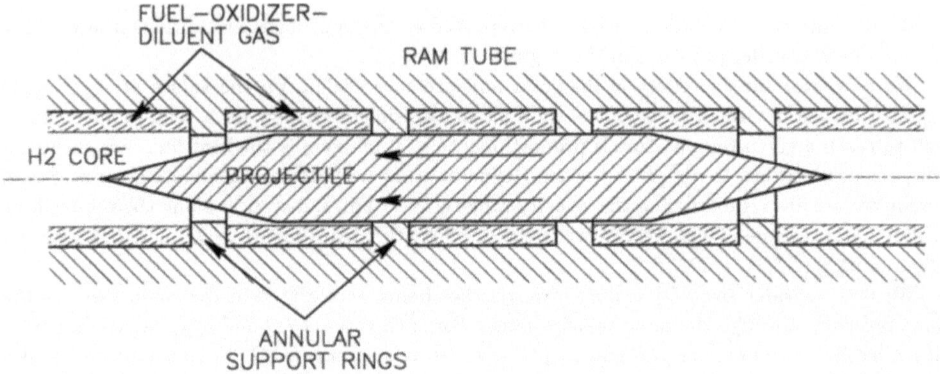

Fig. 1. Proposed new configuration for vortex hydrogen core

and to fill the tube with pure hydrogen. There are many difficulties in the implementation of these techniques, especially if one desires rapid turn-around of the ram accelerator between shots. Many advantages could be gained if the hydrogen core could be set up in the center of the ram tube without introducing any solid (or gel) materials into the tube. This can be done if one uses a vortex to set up the hydrogen core in the ram tube (Hertzberg et al. 1986, Bogdanoff and Higgins 1996). The hydrogen is introduced first, using tangential injection ports, thus setting up the vortex. Then, immediately afterward, the FOD mixture is introduced, taking care to match the interface tangential velocities as much as possible, to minimize mixing between the pure hydrogen and the FOD gas.

It is likely that the projectile will be unstable in the ram tube and must be stabilized by fins on the projectile or rails on the tube. If fins are used on the projectile, they would be outside the hydrogen core and subject to rapid burning and/or melting. If an axisymmetric projectile is used in a tube with longitudinal rails, the required vortex could not be set up. A new vortex core concept, which avoids these difficulties, is shown in Fig. 1. In this concept, the projectile is axisymmetric and is supported away from the ram tube wall by annular rings. This projectile supporting arrangement allows the vortices to be set up between the annular rings. The projectile forward and rear cone angles are 14° and the ratio of the projectile diameter to the tube inside diameter is 0.65. The diameter of the boundary between the hydrogen and the FOD gas is 0.70 times the tube inside diameter. The projectile is 5.33 calibers long. The spacing between the rings is chosen so that there are always at least two rings supporting the projectile.

Figure 2 shows some mechanical details of one possible way of implementing the basic tube geometry shown in Fig. 1. An outer tube serves as a pressure vessel. Inside this outer tube, one slides an alternation of projectile support rings and spacers as shown in Fig. 2. Lengths of the outer tube ("ram tube sections") are joined using the "tube joint blocks" and nuts with right and left hand threads. Gas inlets are located in the tube joint blocks and the gas is distributed along the ram tube through two grooves which extend along the outside walls of the annular support rings and the spacers. The gas is injected tangentially into the ram tube through ports in the spacer sections. The annular support rings and the spacers are kept aligned by inserting a key into a keyway which extends along the exterior of the rings and spacers. The spacer section inside the tube joint block is of a larger diameter than the remaining support rings in order to lock it in position axially.

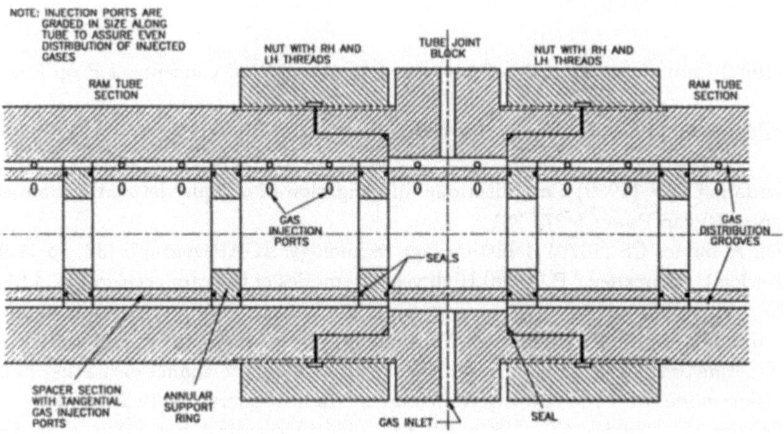

Fig. 2. Details of possible construction of ram tube for implementation of vortex hydrogen core

6. Conclusions

We have analysed some mechanical and thermal aspects of ram accelerator projectile design and the use of the hydrogen core technique for operation between 3 and 7 km/s. In the study of projectile mechanical design, Mg, Al, Ti, steel, Ta and Re were considered as candidate projectile materials. Hollow, solid and two-material projectiles were analyzed - the latter projectiles consisted of Ti, steel, Ta and Re shells and Mg cores. For hollow projectiles, the greatest accelerations could be achieved using Ti projectiles, followed by Al and steel projectiles; Ta and Re projectiles produced much lower accelerations. Solid projectiles yielded ideally three times the accelerations of hollow projectiles. The insertion of a Mg core within Ti, steel, Ta and Re projectiles produced a significant increases in the accelerations for the projectiles, particularly for Ta and Re projectiles.

A model was developed for the gas phase heat transfer from the ram tube gas to the projectile. This was coupled to heat conduction analysis within the projectile to allow estimates to be made of the times and distances along the ram tube at which the projectile surface would reach the ignition and melting temperatures. It was determined that the projectile surface will reach the ignition and melting temperatures after traversing, at most, 0.5% and 2.7%, respectively, of the tube length required to accelerate from 3 to 7 km/sec. For reaching the ignition temperature, steel was found to be the best material, followed by Ta and Re. Ti was found to be very poor, due to its low thermal conductivity. Hollow projectiles were found to traverse the greatest fraction of the required launch tube length before ignition because the required working pressures and heat transfer rates were the lowest. For reaching the melting temperature, the relative performance of the projectile designs followed generally the same trends, except that Ta and Re become very much the best materials, or account of their very high melting temperatures. Larger projectiles were found to have poorer performance, while increasing the tube length and proportionately reducing the ram tube fill pressure was found to improve projectile performance.

Heat soak into the projectile from a fixed surface temperature below the melting point was then considered. Shell materials with lower thermal conductivities and larger projectiles were found to be superior, requiring thinner shell thicknesses.

Implementation of the hydrogen core technique was discussed. A new geometry in which to implement to vortex hydrogen core technique was proposed and described. By using the hydrogen core technique, projectiles with Ta or Re shells should be able to survive ram tube launches at velocities up to 9 - 10 km/s.

References

Bogdanoff DW (1992) Ram accelerator direct space launch system: New concepts. J Prop Power 8:481-490

Bogdanoff DW, Higgins AJ (1996) Hydrogen core techniques for the ram accelerator. AIAA paper 96-0668

Brackett DC, Bogdanoff DW (1989) Computational investigation of oblique detonation ramjet-in-tube concepts. J Prop Power 5:276-281

Canning T N, Seiff A, James CS (1970) Ballistic range technology. AGARDograph 138, pp 11-95

Elvander JE, Knowlen C, Bruckner AP (1996) High velocity modes of the ram accelerator. AIAA paper 96-2675

Giraud M, Legendre JF, Simon G, Henner M, Voisin D (1995) RAMAC in 90 mm caliber or RAMAC 90. Starting process, control of the ignition location and performance in the thermally choked propulsion mode. In: Proc 2nd Int Workshop on Ram Accelerators, Seattle, WA

Hertzberg A, Bruckner AP, Mattick AT, Bogdanoff DW (1986) A chemical method for achieving acceleration of macroparticles to ultrahigh velocities. DOE Rep DOE/ER/13382/1, pp 14-34

Knowlen C, Bruckner AP, Bogdanoff DW, Hertzberg A (1987) Performance capabilities of the ram accelerator. AIAA paper 87-2152

Lampman SR, Zorc TB, Henry SD, Daquila JL, Ronke AW (eds) (1990) Metals handbook, 10th ed, Vol 2, American Soc Metals, pp 481, 482, 493

Lord ME (1960) Performance of a 40-mm combustion heated light gas gun launcher. Arnold Engineering Development Center Rep AEDC-TN-60-176

McBride BJ, Gordon S (1996) Computer program for calculation of complex chemical equilibrium compositions and applications: II. Users manual and program description. NASA Reference Publication 1331

Seiler F, Mathieu G (1995) Boundary layer model for calculating the heat transfer into a ram projectile fired in a ram accelerator. In: Proc 2nd Int Workshop on Ram Accelerators, Seattle, WA

Shapiro AH (1954) The dynamics and thermodynamics of compressible fluid flow. Ronald, New York, p 1100

Tidman DA, Massey DW (1993) Electrothermal light gas gun. IEEE Trans Magnetics, 29:621ff

van der Bliek JA (1962) Further development of capacitance- and inductance-driven hotshot tunnels. Proc 2nd Symp on Hypervelocity Tech, Univ Denver, Denver, CO, p 77

van Driest ER (1951) Turbulent boundary layer in compressible fluids. J Aero Sci 18:145-160, 216

Waldrom HF, McMahan HM, Letarte M (1960) Free flight facilities and aerodynamic studies at Canadian Armament Research and Development Establishment. In: IAS Nat Summer Meeting, Los Angeles, CA, Paper 60-90

Yamori A, Kawashima N, Kohno M, Minami S, Teii S (1996) High quality railgun HYPAC for hypervelocity impact experiments. In: Hypervelocity Impact Symp 1996, Freiburg, Germany

Analysis and experimental results of a fin stabilized subcaliber projectile with a blunt step in the external propulsion accelerator

J. Rom[1], D. Kruczynski[2], M. Nusca[3]
[1]Professor, Lady Davis Chair, Department of Aerospace Engineering, Technion - Israel Institute of Technology, Haifa, ISRAEL
[2]Utron Inc, Aberdeen, MD 21001, USA
[3]Research Engineer, Propulsion and Flight Division, Army Research Laboratory, Aberdeen Proving Ground, MD 21005, USA

Abstract. The paper presents the results of the first proof of concept test of the External Propulsion Accelerator using a subcaliber fin guided projectile. The fin guided subcaliber projectile is fired in the 120 mm Ram Accelerator facility of ARL. The projectile body shape is similar to the free flight projectile body except that the fins do not span the 120 mm tube diameter. The projectile body diameter is 52 mm and the body length is selected so that the flow over the projectile will not be affected by reflected shock waves from the tube wall. A forward-facing step of a 1 mm height is used. The projectile weight is 1.017 kg. The 120 mm RA facility of ARL was modified to accommodate the subcaliber EPA projectile test. The first tube section was filled with nitrogen at initial pressure of 2.0 MPa. The solid 6 m long accelerator tube section is followed by a clear test section. This test section is a 1.8 m long plastic tube filled with the combustible mixture of $CH_4+2O_2+10CO_2$ also at 2.0 MPa. The projectile velocity history is measured by a Doppler radar apparatus. The results of these measurements indicate that after the projectile is clearly separated from the orburator the projectile decelerates in the nitrogen by about 4500 ± 2500 g's. It is found that in the clear tube that is filled with the combustible mixture, the projectile picks up thrust to cancel the drag after a flight time of about 0.4 ms. The projectile then reaches an equilibrium velocity of 1905 m/s that is 1.59 times the detonation velocity of the mixture. The projectile stayed at about this velocity, from 1907-1905 m/s, for 0.7 ms until it exited the tube. High speed photography is used to visualize the combustion on the projectile. The photographs clearly indicate combustion on the projectile, probably initiated by the step and by the leading edges of the root chord of the fins. Calculations using a CFD code are used to verify the position of the combustion on the projectile. The results of this first test indicate that combustion was initiated and stabilized on the rear part of the projectile. It was also shown in this first test that thrust, sufficient to balance the drag, was generated by the external combustion.

Key words: External propulsion accelerator, Blunt step, Subcaliber projectile

1. Introduction

The development of in-tube chemical accelerators is motivated by the fact that the chemical propellants are about three orders of magnitude more compact in weight and size than electromagnetic energy production and storage systems. Therefore, there is a great advantage to develop accelerators of projectiles using chemical propellants. There are two methods for in-tube chemical accelerators utilizing premixed gaseous detonative mixtures; the Ram Accelerator (RA) and the External Propulsion Accelerator (EPA). These in-tube chemical launchers for accelerating projectiles to hypervelocity utilize, in the superdetonative mode of operation, the possibilities of generating continuous thrust by initiating combustion-detonation in the premixed fuel/oxidizer mixture by shock wave interactions. Developments in the research on the External Propulsion Accelerator (EPA) were presented in Rom et al. (1995) and Rom (1996)

The first method proposed for an in-tube chemical launcher was the ram accelerator (RA), originated and developed by A. Hertzberg and his colleagues at the University of Washington (Hertzberg et al. 1988). The concept of RA, operating in the superdetonative mode, is based on utilization of the scramjet cycle, where the projectile acts as a free centerbody and the tube as an extended cowling. The sharp nosed projectile diameter is slightly less than the tube diameter (typically 70% to 80%). Therefore, the nose shock wave is reflected from the tube wall into the projectile centerbody. Under proper conditions this reflected shock wave initiates a combustion-detonation process so that when the products of the chemical reactions are expanded on the rear part of the projectile, thrust is generated. The projectile is centered in the launching tube by 4, 5 or 6 fins attached to the projectile. These fins span the short distance from the projectile body to the tube wall.

Another method for operating the chemical in-tube accelerator is based on the utilization of the external propulsion cycle, proposed by Rom (1990). In EPA the projectile is fired into the launcher tube that is filled with the premixed fuel/oxidizer mixture. However, here the projectile diameter is only about 25% to 40% of the tube diameter. For this subcaliber projectile there is no interaction with the tube wall over the complete length of the projectile. The combustion-detonation is initiated and is confined to the rear part of the projectile only by aerodynamic means. A forward facing step or ramp on the projectile shoulder or a blunt leading edge of a ring wing positioned on the center/rear part of the projectile are used for initiating combustion-detonation. By the interactions of the detonation wave with the nose shock wave an "external combustion chamber" is produced. This external combustion region is confined by the contact layer generated at the intersection point between the nose shock wave and the detached detonation wave ahead of the step. The contact layer separates the high temperature reaction products from the outer layer of non reacting flow. The pressure imposed on the contact region is determined by the pressure jump behind the transmitted oblique shock wave. This pressure is higher than the free stream pressure and is increased as the flow Mach number is increased. Thus the aerodynamically confined combustion region is filled with the hot chemical reaction products compressed to higher pressure by the transmitted shock wave. The hot compressed combustion products are then expanded on the rear part and into the base region of the projectile, producing thrust on the projectile.

The subcaliber projectile can be launched into the large accelerator tube to fly freely in the combustible gas mixture. There, the projectile is stabilized aerodynamically by the use of fins. The fins can be sized to insure a large enough margin of stability. These fins can also be set to induce some rolling rotation motion in order to maintain the projectile on a trajectory close to the tube centerline.

The aerodynamic stabilization of the small projectile in a free flight trajectory along the centerline of the accelerator tube imposes special requirements for the design of the projectile. This design will require relatively large fins and a forward center of gravity position. There are also requirements for a smooth sabot separation in order to minimize the initial disturbances to the flight trajectories. It seemed advantageous to start the tests for proving the concepts of the EPA using a projectile that is guided in its flight in the accelerator tube. Therefore large span fins are attached on the subcaliber projectile in order to center the subcaliber projectile in the accelerator tube. The projectile is centered by wide span fins attached to the subcaliber projectile body extending to the tube wall (fin guided). The body diameter of the subcaliber projectile, for this application, can be in the range of 35% to 45% of the launcher tube diameter. It should be noted that the fin guided EPA projectile is centered in the launching tube in a similar manner to the fin guiding in RA. However, the use of the small diameter body, in the case of EPA, eliminates the strong interaction with the tube wall that can cause high drag forces on the projectile. The present paper presents the results of the first proof of concept test using this subcaliber fin guided projectile. This first test firing was performed at the ARL at the Aberdeen Proving Ground, MD in August 1996. For this purpose, a fin guided subcaliber projectile suitable for firing in the 120 mm RA facility of ARL was used.

The combustion structure on the projectile is photographed by high speed camera through the clear tube wall. Some CFD calculations are performed to evaluate the initiation and establishment of the combustion on the projectile. These CFD calculations are used to verify and to help in the interpretation of the experimental data.

The performance of the projectile flight in the EPA is estimated by the use of the energy balance analysis (Rom 1996). It is indicated in this analysis that, in the EPA, the projectile velocities of several times the detonation velocity of the mixture can be achieved. Although in the present test the thrust generated was just sufficient to overcome the drag, but considering that this is the first firing, it is expected that positive net thrust can be generated in future firings. Thus the EPA emerges as one of the most promising hypervelocity accelerators for space missions etc.

Fig. 1. The 120 mm gun pre-launcher and the transition section

2. Results of the first firing of the proof of concept test program

The objective of the test is to study the establishment of combustion-detonation ahead of the step on the flying projectile. For this purpose the accelerator tube includes a clear tube section filled with the combustible mixture, so that the combustion on the projectile is photographed in this clear tube section. The thrust due to the external combustion is also obtained from the measurements of the velocity of the projectile. The first firing of the proof of concept test program using a subcaliber fin guided projectile was performed at the ARL at the Aberdeen Proving Ground, MD in August 1996.

3. The experimental apparatus

3.1 Accelerator The 120 mm RA facility of ARL was modified to accommodate the EPA subcaliber projectile test. The 120 mm gun, shown in Fig. 1, is used as the pre-launcher for the initial acceleration of the projectile. The pre-launcher gun barrel is followed by the 3 m long transition section. Both the gun barrel and the transition section are made of high strength steel (seen in Fig. 1) and are evacuated to about 330 Pa (2.5 Torr) before the gun firing. The transition section is followed by the first accelerator section, 4.75 m long tube made of steel, that is filled with nitrogen at initial pressure of 2.0 MPa. This solid accelerator tube section is followed by the second accelerator section made of clear acrylic material, shown in Fig. 2. This section is a 1.8 m

Fig. 2. The 120 mm RAM accelerator facility modified to the external propulsion test with the clear tube

Fig. 3. The clear acrylic tube for visualization of combustion

long clear tube filled with the combustible mixture of $CH_4+2O_2+10CO_2$ at 2.0 MPa, shown in Fig. 3.

3.2 The projectile A fin guided subcaliber projectile suitable for firing in the 120 mm RA facility of ARL is shown in Fig. 4. The projectile body shape is similar to the free flight projectile body except that the fins do not span the 120 mm tube diameter. The nose is a 100 half angle and the projectile body diameter is 52 mm and the total length is 286 mm. This length is selected so that the flow over the projectile will not be affected by reflected shock waves from the tube wall. A forward-facing step of a 1 mm height is used. The projectile is made out of aluminum alloy 7075-T6 except for the cone tip that is made of stainless steel. The projectile weight is 1.017 kg. The projectile is positioned on an obturator made of a polycarbonate cylinder with aluminum face plate, weight 0.515 kg. The projectile and the obturator are shown in Fig. 4.

3.3 Instrumentation The projectile instantaneous velocity is measured by a Doppler radar apparatus. The radar beam is reflected into the accelerator tube so that the projectile flight is recorded. Very clear radar track data is obtained in the accelerator sections after the separation of the obturator from the projectile in its flight through the section filled with the nitrogen.

The combustion is photographed using two cameras. The first is a 16 mm high speed framing camera positioned to look at the projectile entrance into the clear tube. The second camera is a 35 mm smear camera focused near the end of the clear tube recording the combustion at the exit of the projectile from the clear tube.

Fig. 4. The fin guided subcaliber projectile with the 1mm step

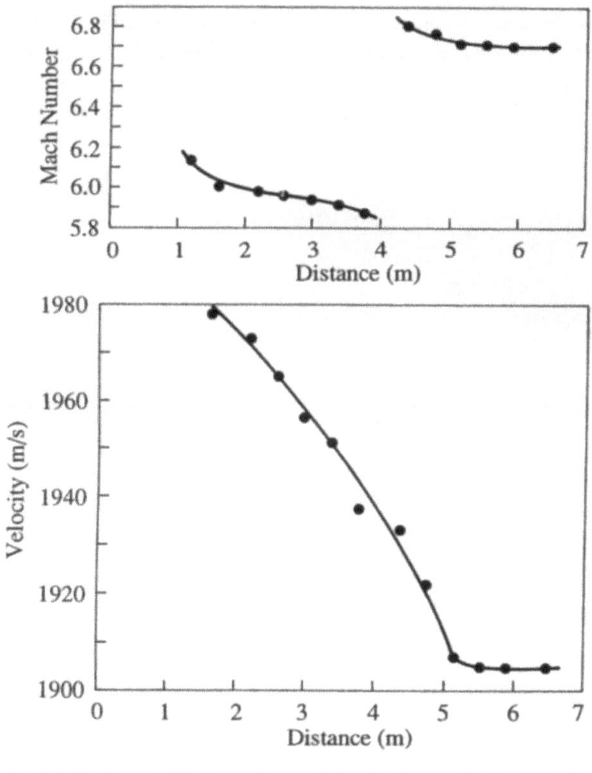

Fig. 5. Velocity and Mach number variation in the accelerator tube for run 45

4. Test results

The measurements of the instantaneous velocity are shown in Fig. 5. This measurement indicates that after the projectile is clearly separated from the obturator the projectile decelerates from an initial velocity of 2024 m/s to 1933 m/s. This deceleration occurs during the flight of the projectile in the solid accelerator tube filled with the inert nitrogen. The measured deceleration level in the nitrogen is about 2500 - 4500 g's. The net drag force acting on the projectile is equal to about 450 kg, as Drag = Weight $\times(a/g)$. It is found that in the clear tube with the combustible mixture, the projectile picks up thrust to cancel the drag after a flight time of about 0.4 ms. The projectile then reaches an equilibrium velocity of 1905 m/s that is 1.59 times the detonation velocity of the mixture. The projectile stayed at about this velocity, from 1907-1905 m/s, for 0.7 ms until it exited the tube. This indicates that thrust was generated on the projectile in the clear tube.

In addition to the velocity measurements, high speed photography is used to visualize the combustion on the projectile. The photographs clearly indicate combustion on the projectile, probably initiated by the step and by the leading edges of the root chord of the fins, shown in Fig. 6. The projectile nose appears obscured by the combination of combustion light and the reflective properties of the acrylic tube material. This effect is common for this type of test. It is the practice to paint the projectile black and to adjust other lighting conditions to reveal details in the non combustion part of the flow. The combustion initiated by the step does not seem to be complete. Apparently the step height is not sufficiently high. It is possible that parts of the step are eroded during the flight through the nitrogen. In any case it seems that the 1 mm step was not sufficiently high to insure full combustion on the step for the case of this methane-oxygen-carbon dioxide mixture. The likely location of the combustion on the projectile is shown in Fig 7. This picture is obtained by superimposing the drawing of the outline of the scaled projectile on the combustion photograph. This photograph indicates that the combustion is initiated at the root of the fins that is positioned above the step. This combustion pattern is compared with results of flow field calculations by the CFD code.

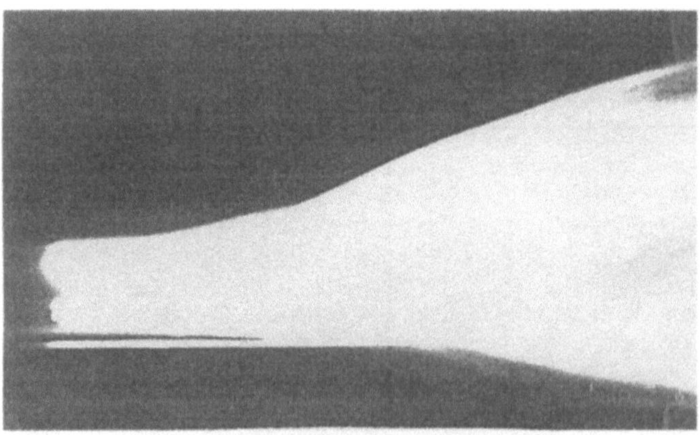

Fig. 6. Photographs of the combustion on the projectile

5. CFD

Computational fluid dynamics (CFD) flow simulations for the RA projectile were performed at the U.S. ARL using the Rockwell Science Center USA (Unified Solution Algorithm) code (Chakravarthy et al. 1985). This code has been used successfully at the ARL for simulation of a full-bore 120 mm Ram Accelerator projectile by Nusca (1994). This CFD code solves the full, 3D, unsteady Reynolds-Averaged Navier- Stokes (RANS) equations. These partial differential equations are cast in conservation form and converted to algebraic equations using an upwind finite-volume formulation. Solution takes place on a mesh of nodes distributed in a zonal fashion around the projectile and throughout the flow field such that sharp geometric corners and other details are accurately represented. The conservation law form of the equations assures that the end states of regions of discontinuity (shocks, detonations, deflagrations) are physically correct even when smeared over few computational cells. The Total Variation Diminishing (TVD) technique

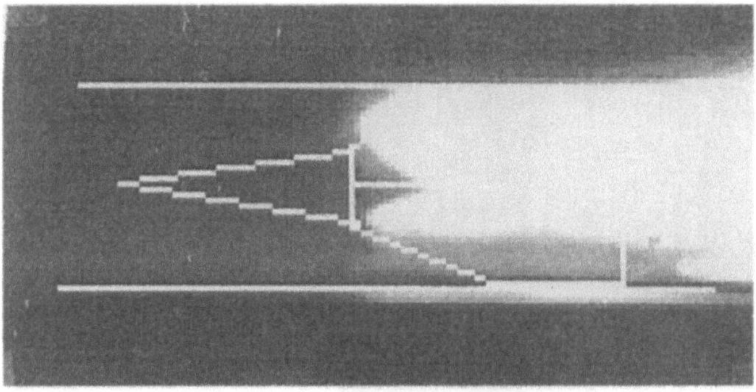

Fig. 7. The position of the combustion outlined by the superposition of the drawing of the projectile

is employed to discretize inertia terms of the conservation equations, while the viscous terms are evaluated using an anbiased stencile. Flux computations across cell boundaries are based on Roe's scheme for hyperbolic equations (Roe 1981). Spacial accuracy of third-order can be maintained in regions of flow field with continuous variation while slope limiting, used near large flow gradients, reduces the accuracy locally to avoid spurious oscillations.

Figure 8 shows the computational mesh (for simplicity, only surfaces are shown) for the projectile which consists of a pointed conical forebody and an afterbody (the step located, at the forebody-afterbody junction has been omitted in this simulation). The dimensions correspond to the subcaliber fin guided EPA test projectile fired at ARL. The grid was generated in zones so that sharp body junctions are accurately represented. Grid clustering was used at the projectile nosetip and other sections where the body slope shanges suddenly. The four fins were also gridded. The grid region downstream of the projectile are not shown in Fig. 8. Approximatly 630000 grid points were used for the 3D simulation; 226 axial (projectile nosetip to downstream boundary which is one-half projectile length from the projectile base), 21 radial (projectile surface to tube wall), and 132 azimuthal (full 360 degrees).

A gas mixture of CH_4+$2O_2$+$10CO_2$ was used for the CFD simulations, however, in the present calculation the flow was not permitted to react (reacting flow simulations for this projectile will be reported in future papers). The initial pressure and temperature of the mixture (before the projectile is injected) were 5.0 MPa. and 300 K, respectively. The projectile velocity was assumed to correspond to a Mach number of 6.3.

Figure 9 shows the computed, frozen flow pressure contours in grey-scale (bright white represents a nondimensional value of about 7 while dark black represents a value of about 0.4) for the projectile geometry displayed in Fig. 8. Important gas dynamic features are labeled. A bow shock extends from the projectile nosetip (producing elevated pressure on the forebody) and intersects the tube wall near the point of maximum fin span. Flow expansion from the forebody to the afterbody causes a localized drop in pressure. Flow through a small gap between the fins, at maximum span, and the tube wall produces a large local pressure. Shocks produced by the leading edges of the fins intersect downstream of the fins; combined with the bow shock reflecting from the tube wall onto the projectile, produce large pressure as well. Therefore, on the last one-half of the projectile afterbody, elevated pressures are observed (except for lower local pressure near the maimum fin span caused by expansion). Figures 10 and 11 show a detailed profile of pressure along the projectile surface and along the tube wall (profiles between the fins as well as near the

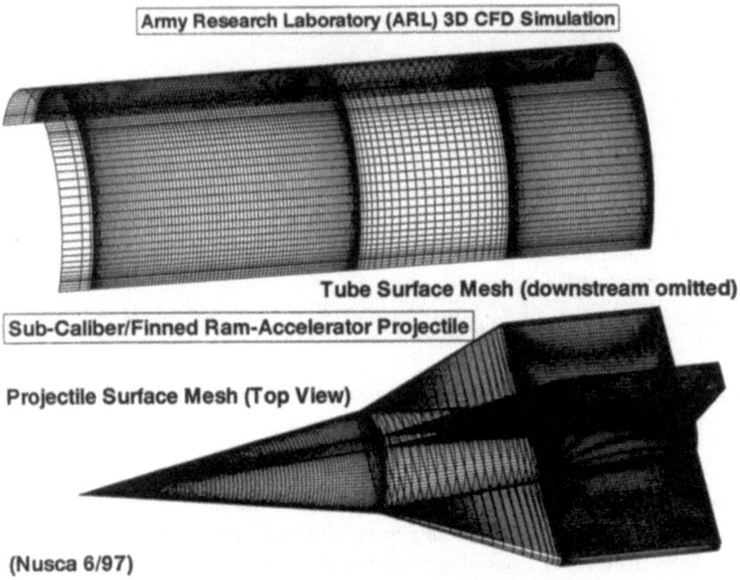

Fig. 8. The grid structure on the finned projectile and the tube surface

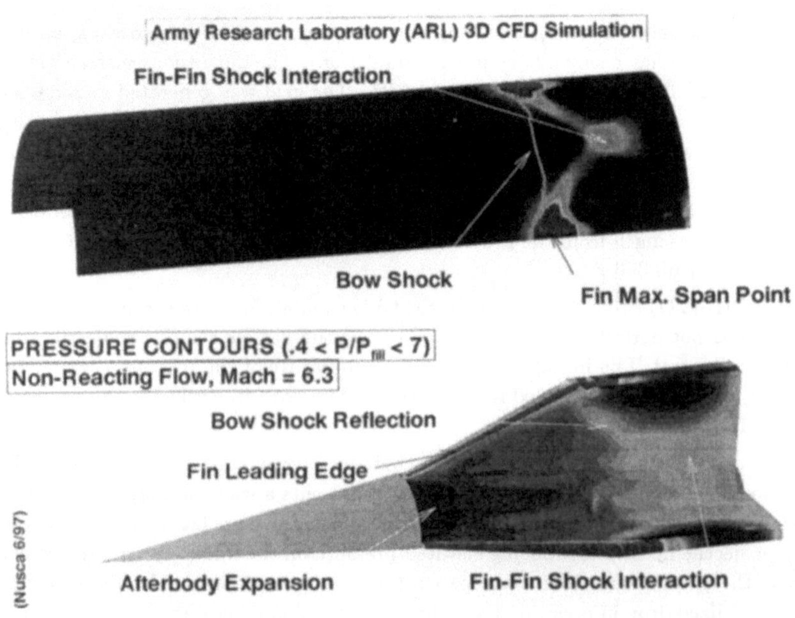

Fig. 9. Pressure contours on the projectile surfaces and on the tube wall due to fin-fin interaction for non reacting flow

fin root chord are also shown). Several gasdynamic phenomena highlighted in Fig. 9 are shown here.

Figure 7 displays a photograph of the combustion on the projectile along with a superimposed sketch of the assumed position of the projectile. Comparing Fig. 9 and Fig. 7 reveals that combustion occures (1) on the high pressure areas caused by projectile leading edges (as well as the projectile forebody step), and (2) along the fin shocks (the angle of illumination in Fig. 7 is larger than the fin shock angle shown in Fig. 9, as expected). The correspondence between Figures 7 and 9 lend credence to the assumed location of the projectile in the photograph.

6. Evaluation of the performance of a fin guided projectile in EPA

The energy balance analysis (Rom 1996, 1997) is applied for the evaluation of the acceleration of the fin guided projectile in EPA. Using the energy balance method it was shown that the acceleration of the projectile can be evaluated from the performance equation for level flight:

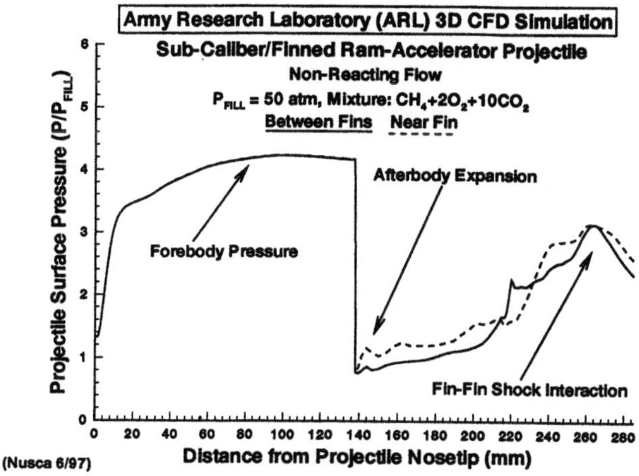

Fig. 10. Surface pressure distribution on the projectile for non-reacting flow

$$\frac{T-D}{W}=\frac{a}{g} \tag{1}$$

In terms of the energy balance relations, Eq. (1) is,

$$\rho_0\left(\frac{\pi(\beta d)^2}{4}\right)Q\eta_p-0.5\rho_0V_\infty^2\left(\frac{\pi d^2}{4}\right)C_D=T-D=\frac{a}{g}W \tag{2}$$

where T is the thrust generated by the combustion, D is the total drag, W is the weight of the projectile and a/g is the acceleration in g's.

It was shown that the maximum projectile velocity (occuring when $T=D$), is

$$\frac{V_{\infty_{\max}}^2}{D_{CJ}^2}=\frac{\beta^2\eta_p}{C_D\left(\gamma^2-1\right)} \tag{3}$$

Using Eqs. 2 and 3 it can be shown that the acceleration is given by the following relations:

$$\frac{a}{g}=\frac{\gamma M_\infty^2C_D}{2}\left[\left(\frac{V_{\max}}{V_\infty}\right)^2-1\right]\frac{p_\infty S_{projectile}}{W} \tag{4}$$

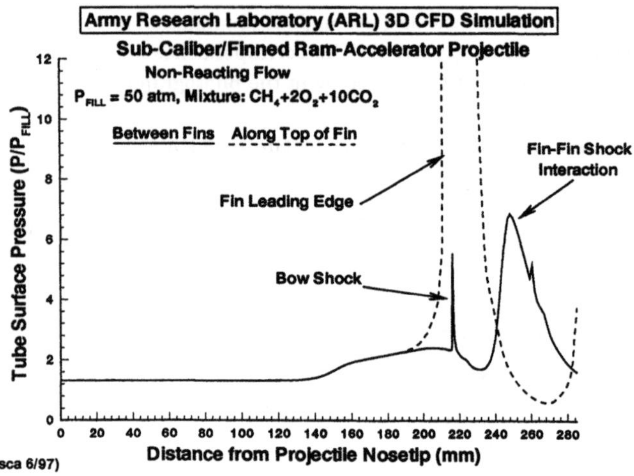

Fig. 11. Surface pressure distribution on the tube wall for non-reacting flow

or

$$\frac{a}{g} = \frac{M_\infty^2 C_D}{2}\left[\left[\frac{\eta_p \beta^2}{(\gamma^2-1)C_D}\right]\left(\frac{M_{CJ}}{M_\infty}\right)^2 - 1\right]\frac{p_\infty S_{projectile}}{W} \tag{5}$$

It is also assumed in this case that the combustion zone thickness increases almost linearly with the increase of the Mach number. As the fin stabilized projectile is considerably larger and heavier than the free flight projectile more thrust is needed for its acceleration. Therefore a 1.5 mm step is considered, in addition to the 1 mm step used in the free flight projectile.

The drag coefficient for the projectile with the step and large fins is estimated to be 0.22 for the 1 mm step and 0.24 for the 1.5 mm step. The mixture composition is the same as the one used in the free flight case: $CH_4+2O_2+10CO_2$. In this case, the orbutrator used for the projectile initial acceleration in the RA operation is also used for the fined projectile. Using conventional gun propulsion it is possible to achieve velocity of about 1800 m/s, which corresponds to initial projectile Mach number of 6.3 in this mixture.

The projectile acceleration evaluated for the 1.0 mm step for propulsive efficiencies of 0.67, 0.5 and 0.3 are shown as Series 1, 2 and 3 in Fig. 12, respectively. The maximum acceleration for the 1 mm step projectile, presented in Fig. 12, is about 14,500 g's. It is obtained at flight Mach numbers in the range of 12 to 16 (Velocity of 3.4 km/s to 4.5 km/s). For the projectile with the 1.5 mm step the maximum acceleration is above 20,000 g's at Mach number 12 decreasing to about 18,000 g's at Mach number 16 for $\eta_p = 0.67$. The maximum velocity for the 52 mm projectile in the 120 mm accelerator tube is 5.7 km/s (Mach number 20) where the acceleration is reduced to zero. The length of the accelerator needed to reach Mach 18 for the projectile with the 1 mm step and for $\eta_p = 0.67$ is 138 m. For the projectile with the 1.5 mm step and $\eta_p = 0.67$, the length of the accelerator needed to reach Mach 18 is reduced to 107 m. The flight duration needed to reach Mach 18 for the projectile with the 1 mm step is about 36 msec. For the projectile with the 1.5 mm step, the flight duration needed to reach Mach 18 is about 28 msec. It can be seen that velocity of 3 km/s and 4.5 km/s can be obtained with a 35 m and 107 m long accelerators for projectiles with the 1.5 mm step. Higher accelerations and shorter accelerator lengths can be obtained by increasing the initial pressure of the mixture. Thus assuming that initial pressures of 20 MPa are feasible, projectile velocities of 3 km/s and 4.5 km/s can be achieved , assuming $\eta_p =$

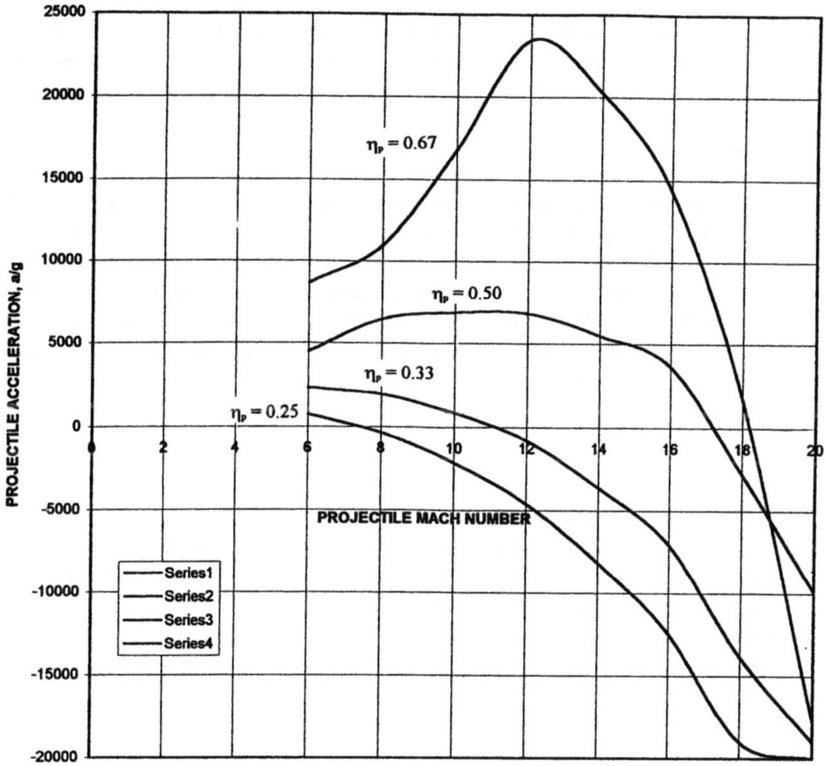

Fig. 12. The calculated variation of the acceleration of the fined projectile with 1mm step as a function of the projectile flight Mach number in the EPA

0.67, with a 9 m and 27 m long accelerators, respectively. Longer accelerator tubes are required for lower propulsive efficiencies.

9. Summary of test results

The results of this first test indicate that combustion was initiated and stabilized on the rear part of the projectile. It was also shown in this test that thrust, sufficient to balance the drag, was generated by the external combustion. This is a first experimental demonstration of the external propulsion concept in the accelerator tube, although the test did not yet produce sufficient thrust for positive acceleration.

The results of these measurements indicate that in the inert nitrogen the deceleration level is about 2500 - 4500 g's. The net drag force acting on the 1.07 kg projectile is equal to about 450 kg. The velocity measurements in the clear tube show that the projectile stayed at a steady velocity, from 1905 - 1907 m/s, for 0.7 ms until it exited the clear tube. This indicates that thrust was generated on the projectile in the clear tube. It is found that in the test gas, in the clear tube with the combustible mixture, the projectile picks up thrust that cancels the drag. The projectile reaches an equilibrium velocity of 1905 m/s that is 1.59 times the detonation velocity of the mixture. There are also photographs that clearly indicate combustion on the rear part of the projectile.

This is a first experimental demonstration of the external propulsion concept in the accelerator tube, although the test did not yet produce sufficient thrust for positive acceleration. It is expected that by few modifications to the projectile shape and using higher pressures in the regular high pressure accelerating tube higher positive thrust and positive acceleration can be obtained. It is therefore proposed to continue the test program for validation of the External Propulsion method.

Acknowledgement

Part of this research program was carried out during the sabbatical of the first author as a visiting professor in Department of Aerospace Engineering at the University of Maryland in College Park, MD. Support for the research was given in part by the Hypersonic Research Center funded by NASA contract NAGW3715, with Dr. I. Blankson as its monitor. Prof. A. P. Bruckner and Dr. C. Knowlen from the University of Washington in Seattle, WA were very helpful in supplying the data from their RA tests. Their help is gratefully acknowledged. It is my pleasure to acknowledge the useful discussions on the EPA technology with Prof. M. Lewis and Prof. A. K. Gupta of the University of Maryland.

References

Chakravarthy SR, Szema KY, Goldberg UC, Gorski JJ, Osher S (1985) Application of a new class of high accuracy TVD schemes to the Navier-Stokes equations. AIAA paper 85-0165

Hertzberg A, Bruckner AP, Bogdanoff DW (1988) Ram accelerator: A new chemical method for accelerating projectiles to ultrahigh velocities. AIAA J 26:195-203

Nusca MJ (1994) Reacting flow simulation for large scale ram accelerator. AIAA paper 94-2963

Roe PL (1981) Approximate reiman solvers, parameter vecors, and difference schemes. J Comp Phys 43:357-372

Rom J (1990) Method and apparatus for launching a projectile at hypersonic velocity, U.S. Patent 4,932,306, June 12

Rom J, Nusca MJ, Kruczynski D, Lewis M, Gupta AK, Sabean J (1995) Recent results with the external propulsion accelerator, AIAA paper 95-2491

Rom J (1996) On the acceleration of projectiles in the ram and external propulsion accelerators by the energy balance analysis, AIAA paper 96-2951

Rom J (1997) Performance limits for projectile flight in the ram and external propulsion accelerators, J Prop and Power 13:583-591

Starting processes

Effects of launch tube shock dynamics on initiation of ram accelerator operation

J.F. Stewart, A.P. Bruckner, C. Knowlen
Aerospace and Energetics Research Program, University of Washington, Seattle, WA, U.S.A.

Abstract. Normal operation of the University of Washington's ram accelerator involves the use of a perforated tube to vent gases from the light gas gun into an evacuated tank before the projectile enters the ram accelerator tubes. Because there is residual air in the launch tube and gases from the gun that blow by the obturator, a shock is generated and reflects between the moving projectile/obturator and the entrance diaphragm. The vent tube perforations relieve the resulting pressure rise, such that it does not become significant until after the projectile has passed the vent section. Experiments were performed to determine the feasibility of eliminating the venting process. It was found, as expected, that the pressure increase from the shock reflections was much higher for ventless operation than for vented operation and in some cases prevented initiation of the ram acceleration process. The effect of increasing residual launch tube pressure on the starting process of the ram accelerator was also investigated. It was found that for initial gun launch tube pressures of 21 kPa or less, successful starting of the ram accelerator was possible at a fill pressure of 5.0 MPa.

Key words: Starting process, Launch tube shock, Launch tube pressure, Venting

1. Introduction

1.1 Starting Process The sequence of events that occurs just prior to the projectile entering the ram accelerator test section is known as the "starting" process and is illustrated in Fig. 1 (Bruckner et al. 1992, Burnham 1993). Ahead of the projectile in the launch tube, there is residual air (a result of incomplete evacuation) and prelauncher driver gas that blows by the "obturator," a full caliber piston placed behind the projectile. In these gases, the obturator and projectile generate a normal shock wave which moves down the launch tube (Fig. 1a), reflects off the diaphragm (Fig. 1b), and travels back toward the projectile and obturator where it is again reflected (Fig. 1c), repeating the process (Fig. 1d,e). The pressure and temperature of the gas in front of the obturator are increased by the shock reflections, eventually causing the backplate to separate from the main body of the obturator (Fig. 1e).

When the projectile enters the combustible gas in the test section, a normal shock moves onto the body of the projectile (Fig. 1f). The perforations in the main body of the obturator vent the shocked gas, weakening the shock and stabilizing it behind the "throat" (the maximum diameter of the projectile is referred to as the throat, because it represents the minimum flow area). It should be noted that the flow is not choked at the throat; the flow is initially supersonic there. In a successful "start," the normal shock remains behind the throat of the projectile. High pressure in front of the obturator causes it to decelerate rapidly as the projectile accelerates, decoupling the obturator from the propulsion process. It was proposed in the study of Burnham (1993) that the propellant mixture is ignited by the large temperature increase in the launch tube resulting from the shock reflections. This hypothesis is supported by observations of gun launch tube behavior using flow visualization techniques, (Sasoh et al. 1996) which clearly show the shock reflections and a region of hot, radiating gas next to the entrance diaphragm. Another possible outcome of the starting process is an "unstart," which occurs when the normal shock is driven in front of the projectile throat. The high pressure, subsonic flow behind the shock chokes at the throat, causing

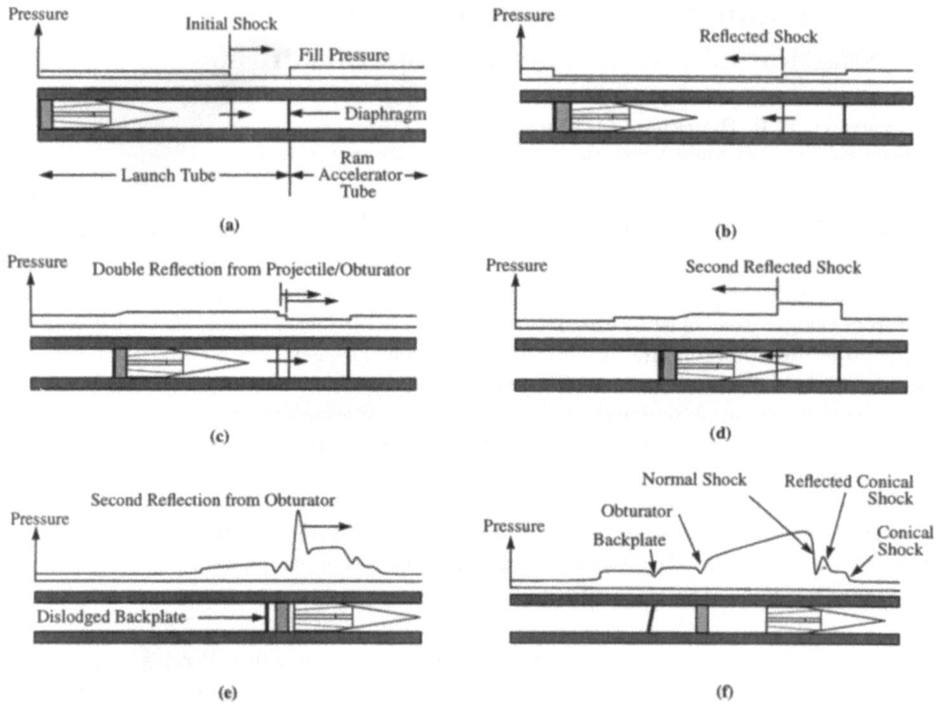

Fig. 1. Schematic illustration of ram accelerator starting process

the projectile to decelerate. An unstart may or may not be combustion driven, depending on the initial conditions of the experiment (Schultz 1997).

1.2 Experimental Setup The launch tube experiments were performed in the University of Washington's (UW) 38-mm-bore ram accelerator, which is described in more detail by Bruckner et al. (1992). A light gas gun is used to accelerate the projectile to the supersonic velocity required to initiate the ram acceleration process. Before entering the 16-m-long ram accelerator test section, the projectile/obturator passes through a perforated vent tube. Two different gun launch tube configurations were used in the experiments described here. One was the nominal launching configuration in which venting is accomplished by 36 perforations of 6.4-mm-diameter. In the second configuration all the vent section perforations were sealed with threaded steel plugs except for one, which was left open to allow the gun launch tube to be evacuated. It is assumed that blocking 35 of 36 perforations (a 97% reduction in venting), makes the system effectively ventless.

In addition to the instrument stations in the ram accelerator and vent sections, small instrumented tube inserts were used to obtain data near the entrance diaphragm. Three different tube insert installation configurations were used. One configuration, shown in Fig. 2a, had the entrance diaphragm located between two single-instrument-station inserts. This allowed measurements 139 and 12 mm ahead of the diaphragm and 12 and 227 mm beyond the diaphragm. A second arrangement is shown in Fig. 2b. In this case both single-station inserts were placed before the ram accelerator test section, allowing data to be taken 171, 44 and 20 mm ahead of the entrance diaphragm and 195 mm beyond the diaphragm. The third configuration, shown in Fig. 2c, used a two-station insert after the entrance diaphragm and the single-station inserts ahead of the di-

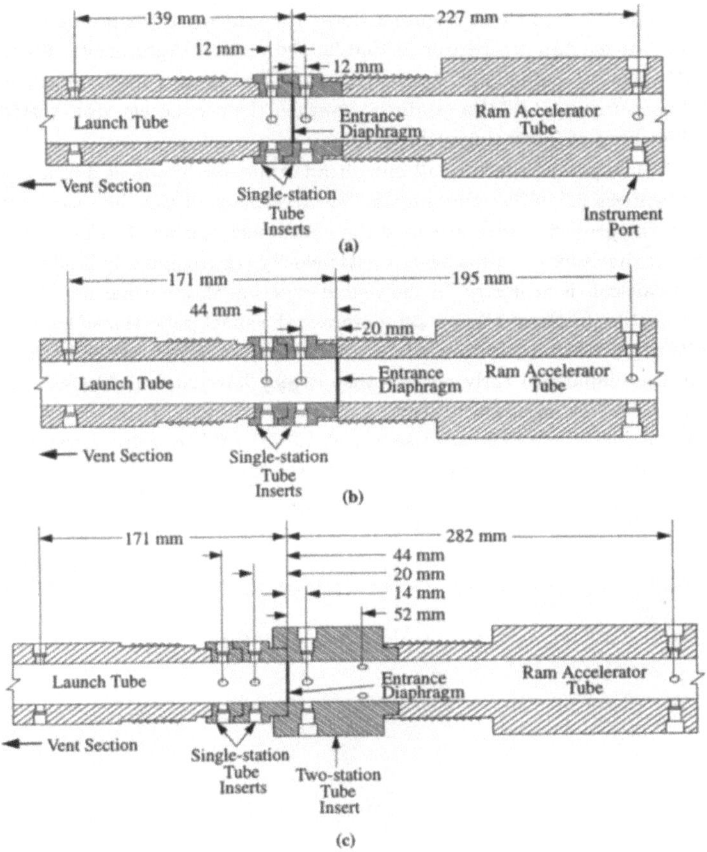

Fig. 2. Launch tube instrumentations

aphragm. In this configuration, instrument stations were located 171, 44 and 20 mm ahead of the entrance diaphragm and 14, 52, and 282 mm beyond the diaphragm.

2. Results and Discussion

For some of the proposed applications of the ram accelerator, the large initial dump tank and the equipment needed to evacuate it are impractical. Therefore, experiments were performed with the perforations in the vent section sealed. The elimination of venting would in turn eliminate the need for the dump tank. Experiments were also performed with increasing amounts of air in the launch tube to evaluate the feasibility of eliminating the evacuation process entirely. In the course of these experiments, it was found that the shock reflections discussed above were having a significant effect on whether the ram acceleration process was successfully initiated. Accordingly, an investigation was conducted to determine the effects of the launch tube gases on the ram accelerator starting process with a ventless launch tube and with increased residual air pressure. All of the experiments were performed using five-finned aluminum projectiles with a mass of 0.076 kg in a mixture of $2.8CH_4+2O_2+5.7N_2$, which has been well characterized.

2.1 Ventless Operation In the ventless experiments, the launch tube was evacuated to approximately 530 Pa (4 Torr) and the single-station tube inserts were installed as shown in Fig. 2a with a 0.7-mm-thick Mylar diaphragm between them. Figure 3 shows the velocity of the projectile as a function of distance of travel in the ram accelerator section for the ventless experiments. When the fill pressure was 2.5 MPa, the projectile failed to start, as indicated by the decelerating velocity profile. When the fill pressure was 3.0 MPa or above, however, the ram accelerator started successfully, as shown by the curves of increasing velocity.

Figure 4 shows the launch tube pressure history 12 mm ahead of the diaphragm as a function of time for the vented and ventless 2.5 MPa experiments. The zero point of the time axis was defined as the time at which the projectile throat passed the instrument station. The projectile profile is drawn for reference. Its horizontal scale is based on its velocity (approximately 1150 m/s) and dimensions, and its vertical scale is arbitrary. In the vented experiment, a normal shock was behind the throat of the projectile. In the ventless case, however, the shock reflections caused a normal shock to be ahead of the throat. In addition, the pressure ahead of the projectile reached the static burst pressure of the diaphragm early enough that it may have ruptured before the projectile's nose tip pierced it, causing an unstart. When the entrance diaphragm thickness was doubled to 1.4 mm, the ram accelerator started successfully at 2.5 MPa fill pressure in ventless configuration. The results of this series of experiments are discussed in more detail in the study of Stewart (1997).

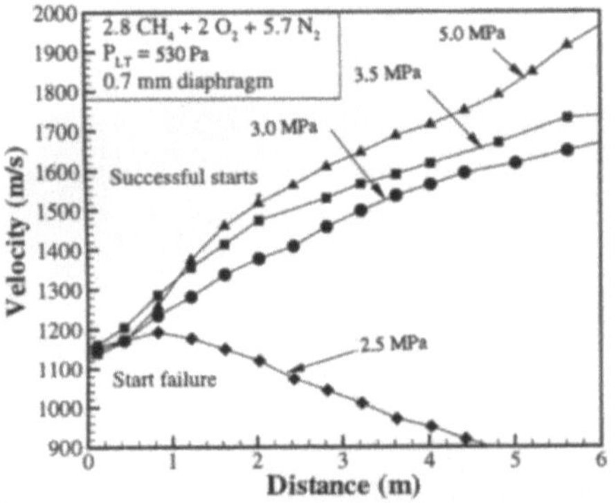

Fig. 3. Velocity versus distance for ventless experiments at various ram accelerator fill pressures

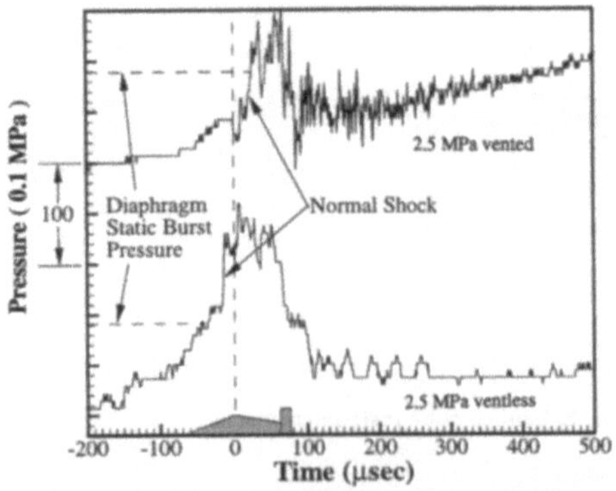

Fig. 4. Pressure versus time 12 mm ahead of diaphragm for 2.5 MPa ventless and vented experiments

2.2 Increased Launch Tube Pressure The ventless experiments indicated that the shock reflections in the launch tube play a significant role in the starting process. It is expected that with increased residual air pressure in the launch tube, the role of these shocks would be even more pronounced. Because the pressure rise from the shock reflections was quite severe in ventless configuration, the experiments with increased launch tube pressure were performed with the vent perforations open (Stewart 1997). In addition, the Mylar entrance diaphragm was thickened to 3.5 mm to protect against premature rupture. Greater resolution in the launch tube pressure measurements was attained by placing both of the single-station instrumented tube inserts ahead of the entrance diaphragm, as shown in Fig. 2b. The launch tube pressure was varied from 7 to 28 kPa. For these experiments, 4 m of the ram accelerator section were used at a fill pressure of 5.0 MPa.

The velocity profiles for the increased launch tube pressure experiments are shown in Fig. 5. For initial launch tube pressures of 21 kPa or less, the ram accelerator started successfully and the projectile accelerated over the entire 4 m test section. For initial launch tube pressures of 24 and 28 kPa, however, the ram accelerator failed to start: the projectile unstarted and promptly decelerated. Comparison of the launch tube pressure traces of these experiments provides some insight on these start failures. Figure 6 presents the launch tube pressure history for the experiment with 7 kPa of residual air in the launch tube. In this figure, the zero of the time axis was defined as the point at which the nose tip of the projectile pierced the entrance diaphragm, as calculated from the projectile's velocity and dimensions and the known distance from the instrument stations to the diaphragm. This experiment exhibited shock reflections which had a much greater magnitude than in the ventless experiments with an evacuated, *i.e.*, 530 Pa (4 Torr), launch tube; however, the 3.5-mm-thick diaphragm withstood the pressure increase and the projectile was able to start successfully. It should be noted that there was little shock activity at the station 171 mm ahead of the entrance diaphragm as a result of the weakening of the shocks as they passed over the vent tube perforations.

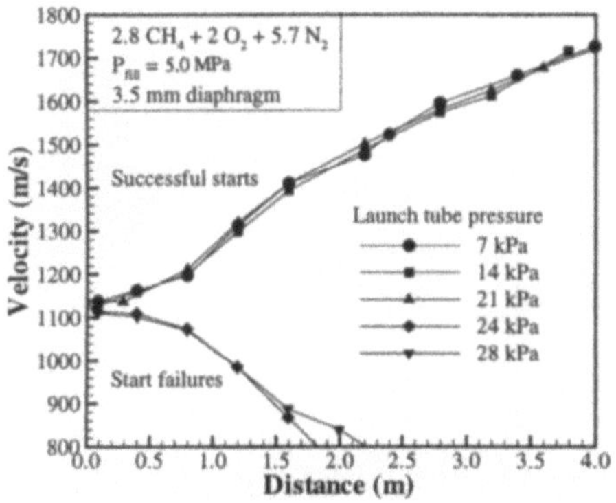

Fig. 5. Velocity versus distance for experiments with increased initial launch tube pressure

It is helpful to examine an x-t diagram of the shock reflections. These data are shown in Fig. 7 and illustrate the travel time of the shocks as a function of distance along the tube. The zero of the distance axis was defined as the location of the diaphragm, and the zero on the time axis was referenced to the time at which the projectile's nose pierced the diaphragm. The motion of the projectile and obturator were determined from the magnetic sensors. The shock motion was determined from the time of arrival of the shock at each transducer station (Fig. 6). For clarity the shocks in Fig. 6 are numbered according to the path they follow in Fig. 7. The first discernible shock generated by the projectile and obturator, path 1 in Fig. 7, was reflected from the diaphragm and traveled back toward the projectile and obturator along path 2. The shock was partially reflected from the projectile throat and traveled along path 3. It should be noted that by the time this reflection had occurred, the projectile nose had pierced the entrance diaphragm. The original reflected shock continued along path 2 until it struck the obturator face, where it was reflected and traveled along path 4, entering the ram accelerator section after the projectile throat. The ram acceleration process was successfully initiated, as shown in Fig. 5. Similar phenomena were observed at 7 kPa initial launch tube pressure in the study of Burnham (1993). The experiment with 14 kPa of residual air in the launch tube also started successfully and showed a shock reflection history similar to the 7 kPa experiment described above, but with a corresponding increase in the magnitude of compression by the shocks.

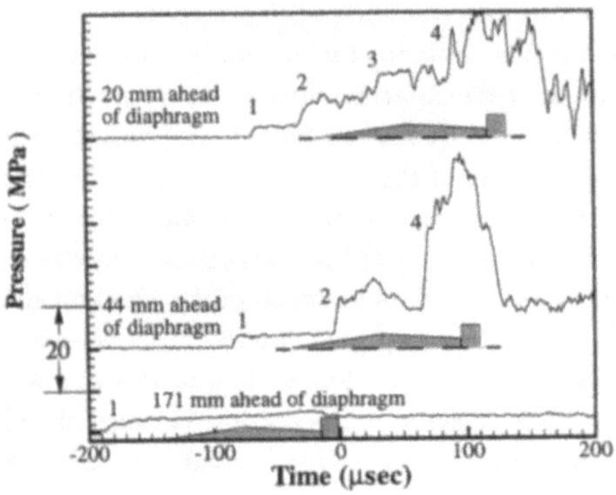

Fig. 6. Launch tube pressure versus time for experiment with 7 kPa residual air in launch tube

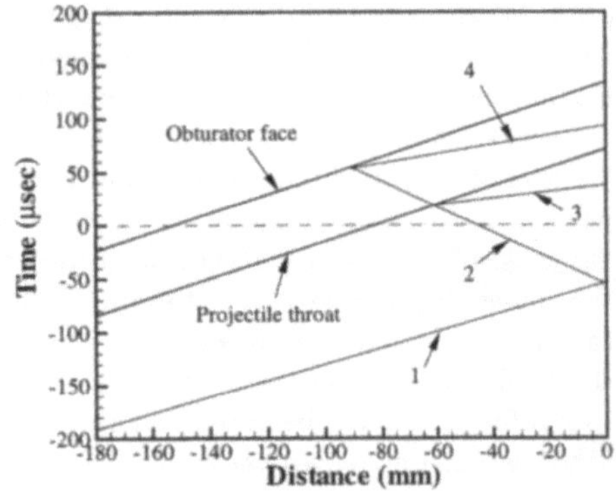

Fig. 7. Time versus distance diagram for experiment with 7 kPa residual air in launch tube

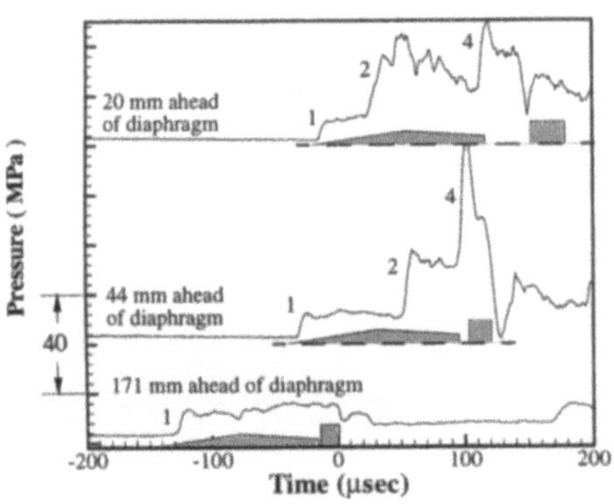

Fig. 8. Launch tube pressure versus time for experiment with 28 kPa residual air in launch tube

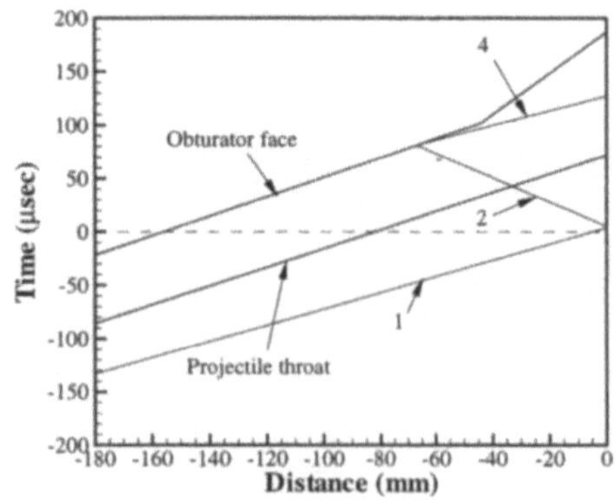

Fig. 9. Time versus distance diagram for experiment with 28 kPa residual air in launch tube

In the experiments with 24 and 28 kPa residual air in the launch tube, a start failure was observed (Fig. 5). Figure 8 shows the launch tube pressure history for the 28 kPa experiment and Fig. 9 shows the x-t history of the launch tube shocks. In this experiment, the shock following path 1 remained near the nose of the projectile, and therefore did not strike the diaphragm until approximately the same time as the nose tip pierced the diaphragm. Because the diaphragm does not instantly open when the projectile nose strikes it, the shock was reflected along path 2 and then reflected from the obturator along path 4. In this experiment there was no observed reflection from the projectile throat. As shown by the projectile and obturator profiles in Fig. 8, the obturator separated from the base of the projectile and decelerated significantly as a result of the large pressure increase on its face. This can also be seen as the change in slope of the obturator path in the Fig. 9. The pressure data from the 24 and 21 kPa experiment were similar to that of the 28 kPa experiment except for the magnitude of the pressure rises. The 24 kPa case also experienced a start failure; however, the case with 21 kPa of residual air in the launch tube started successfully.

The increased obturator separation in the 28 kPa case (also observed at 21 and 24 kPa) explains why the earliest visible shock remains near the projectile nose, as evident in Fig. 8. As discussed above in the starting process, a shock is generated by the projectile/obturator in the residual air and is reflected from the entrance diaphragm (possibly several times). A reflection

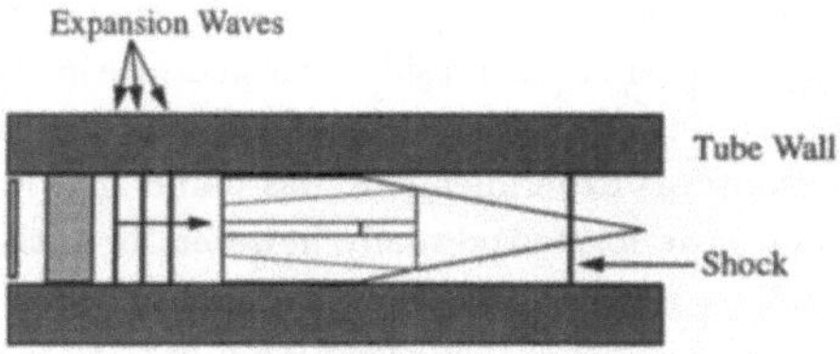

Fig. 10. Effect of obturator separation on shock

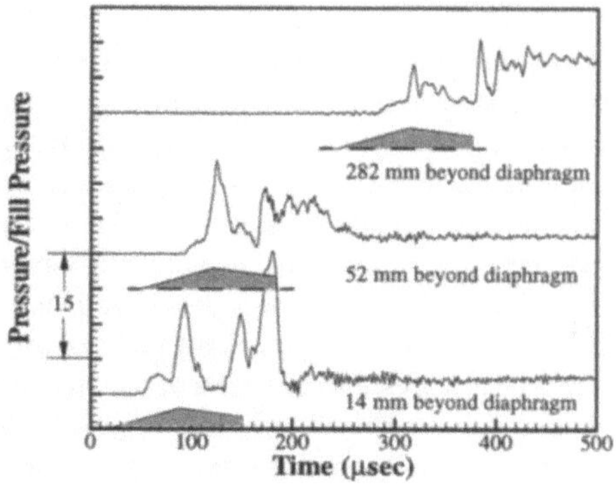

Fig. 11. Pressure versus time beyond diaphragm for experiment with 21 kPa residual air in launch tube

from the obturator moves onto the nose of the projectile as shown in Fig. 10. The pressure increase across the shock causes the obturator and backplate to fall away from the base of the projectile. Because there is now a larger volume between the shock and the obturator, a pressure decrease must occur, and therefore, expansion waves originating at the obturator face move over the projectile. The interaction of these expansion waves with the shock wave serves to weaken it and keep it on the nose of the projectile.

As shown in Fig. 5, the 24 and 28 kPa experiments experienced start failures, while the 21 kPa experiment started successfully. Therefore, the start failure is not solely a result of the shock position relative to the projectile. One possible explanation for the successful start at 21 kPa is that upon entrance to the ram accelerator section, the projectile was able to "swallow" the shock on its nose. This was investigated by repeating the 21 kPa experiment with a two-station tube insert placed after the entrance diaphragm, as shown in Fig. 2c.

Figure 11 shows the data from the pressure transducers beyond the entrance diaphragm. Again, the zero point on the time axis was defined as the time at which the projectile pierced the entrance diaphragm. At the stations 14 and 52 mm beyond the diaphragm, there was a shock just ahead of the throat, that would seem to indicate the projectile was unstarted. Presumably, this shock was transmitted into the ram accelerator section when the launch tube shock struck the entrance diaphragm. By the time the projectile passed the station 282 mm beyond the diaphragm, however, the pressure profile was typical of a successful start. To further investigate the role that combustion plays in the start failures at increased launch tube pressure, a second experiment was performed at 21 kPa with 5.0 MPa of an inert mixture, $2.8CH_4+7.7N_2$, in the ram accelerator section. This non-reacting mixture has properties virtually identical to those of the combustible mixture. It was obtained by simply replacing the O_2 with N_2. The data 14 and 52 mm beyond the diaphragm showed pressure profiles similar to those of the experiment shown in Fig. 11, however, data from the station 282 mm beyond the diaphragm clearly indicated an unstart. These results provide evidence that with increased launch tube pressure the initiation of combustion can actually help to prevent a start failure when the conditions are suitable. Apparently, once combustion was initiated in the 21 kPa reacting experiment, the projectile was accelerated ahead of the shock. It can therefore be surmised that in the 24 and 28 kPa experiments, the generated thrust was insufficient to allow the projectile to overtake the shock, and an unstart resulted. Additional experiments with even greater instrument density beyond the diaphragm and computational modeling could confirm this hypothesis.

3. Conclusions

Experiments were performed in the University of Washington's 38-mm-bore ram accelerator to investigate the shock dynamics in the gun launch tube. Experiments with the vent tube perforations closed showed that the shock reflections in the launch tube caused a start failure at 2.5 MPa ram accelerator fill pressure in a mixture of $2.8CH_4+2O_2+5.7N_2$, but did not prevent ventless operation at higher fill pressures. By strengthening the entrance diaphragm, this start failure was prevented. Additional experiments were performed with increased residual air pressure in the launch tube and with the vent perforations open. The ram accelerator test section was filled with 5.0 MPa of the same propellant mixture used in the ventless experiments. It was found that for residual gas pressures of 21 kPa or less the ram accelerator performed normally, however, for initial launch tube pressures of 24 kPa or greater, a start failure resulted. Examination of the pressure data from the launch tube showed that a shock struck the entrance diaphragm at approximately the same time as the projectile nose in the cases where a start failure occurred. This phenomenon was also observed for the successful start at 21 kPa residual air pressure, however, it was not present at 7 and 14 kPa. Thus, it can be concluded that the start failure was not solely a function of a shock striking the diaphragm coincident with the projectile nose. Data taken in subsequent experiments with pressure transducers beyond the entrance diaphragm showed that in the 21 kPa experiment, the projectile was able to overtake the transmitted shock wave and start successfully. A non-reacting experiment at 21 kPa, however, experienced a start failure. This indicated that the initiation of combustion was providing enough thrust to allow the projectile to overtake the shock and start successfully. Presumably, at higher launch tube pressures, insufficient thrust was generated to allow the transmitted shock to be overtaken by the projectile during the starting process.

Acknowledgement

This work was performed under Army Research Office grant number DAAL03-92-G-0100. Thanks are due to David Buckwalter, Chris Bundy, Josh Elvander, Andrew Higgins, Eric Schultz, and Ryan Schwab for their suggestions and for their help in the laboratory.

References

Bruckner AP, Burnham EA, Knowlen C, Hertzberg A, Bogdanoff DW (1992) Initiation of combustion in the thermally choked ram accelerator. In: Takayama K (ed), Shock Waves, Springer-Verlag, Heidelberg, 1:623-630

Burnham EA (1993) Investigation of starting and ignition transients in the thermally choked ram accelerator. PhD Thesis, Dept Aero Astro, Univ Washington, Seattle, WA, USA

Sasoh A, Hirakata S, Ujigawa Y, Takayama, K (1996) Operation tests of a 25-mm-bore ram accelerator. AIAA paper 96-2677

Schultz E (1997) The subdetonative ram accelerator starting process. MSAA Thesis, Dept Aero Astro, Univ Washington, Seattle, WA, USA

Stewart JF (1997) Effects of launch tube gases on starting of the ram accelerator. MSAA Thesis, Dept Aero Astro, Univ Washington, Seattle, WA, USA

Overview of the subdetonative ram accelerator starting process

E. Schultz, C. Knowlen, and A.P. Bruckner
Aerospace and Energetics Research Program, University of Washington, Seattle WA 98195, U.S.A.

Abstract. The ram accelerator requires a conventional gun to initially boost the projectile to supersonic entrance velocity. An experimental investigation has been undertaken to improve the understanding of transition from the conventional gun to the ram accelerator and initiation of the thermally choked propulsive mode, referred to as the starting process. Developing a robust starting process is instrumental for utilizing the ram accelerator in a variety of applications. Four possible outcomes of a start attempt have been identified. A successful start is achieved when supersonic flow is maintained throughout the diffuser, and the normal shock system is stabilized on the projectile body through propellant energy release. A sonic diffuser unstart is caused by conditions upstream of the throat resulting in subsonic flow in the diffuser. A wave fall-off occurs when insufficient energy is released from the propellant to keep the shock system on the body from receding behind the base. A wave unstart is caused by conditions downstream of the throat resulting in disgorgment of the shock system on the body into the diffuser. Experimental results are presented, along with a discussion of the factors involved that determine which of these outcomes actually occurs.

Key words: Initiation, Starting process

1. Introduction

A schematic of the idealized flowfield of the thermally choked ram accelerator propulsive mode is presented in Fig. 1. The propellant flowing over the projectile nose is ram compressed by a series of reflected shocks and expands supersonically relative to the projectile behind the throat. The propellant then encounters a normal shock followed by a subsonic combustion zone which releases the chemical energy and thermally chokes the flow at full tube area behind the projectile. The combustion-supported normal shock generates a high base pressure which accelerates the projectile down the tube. This mode of ram accelerator operation is observed at projectile velocities below the Chapman-Jouguet detonation velocity (V_{CJ}) of the combustible gas mixture (Bruckner et al. 1991, Hertzberg et al. 1991). The research presented here lies within the subdetonative velocity regime, and considers the process for establishing the flowfield depicted in Fig. 1.

The idealized starting process, illustrated in Fig. 2, begins with the initial launcher accelerating a perforated obturator (sealed with a backplate) and projectile from rest through the evacuated launch tube. A Mylar entrance diaphragm separates the evacuated launch tube from the high pressure, propellant-filled ram accelerator tube. The projectile pierces the entrance diaphragm and enters the propellant with the obturator at supersonic entrance velocity. The propellant is ram compressed up to the projectile throat and expands over the body where it encounters the obturator. A normal shock is driven onto the projectile body as the obturator and backplate separate from the projectile and rapidly decelerate. The propellant energy is released in a combustion zone behind the normal shock, which thermally chokes the flow and stabilizes the shock on the projectile body. The starting process is loosely defined as the period between projectile acceleration from rest to the point where the flowfield illustrated in Fig. 1 is stabilized. Experiments were conducted to investigate the possible outcomes of a start attempt and to determine the effects that some non-ideal factors have on the starting process.

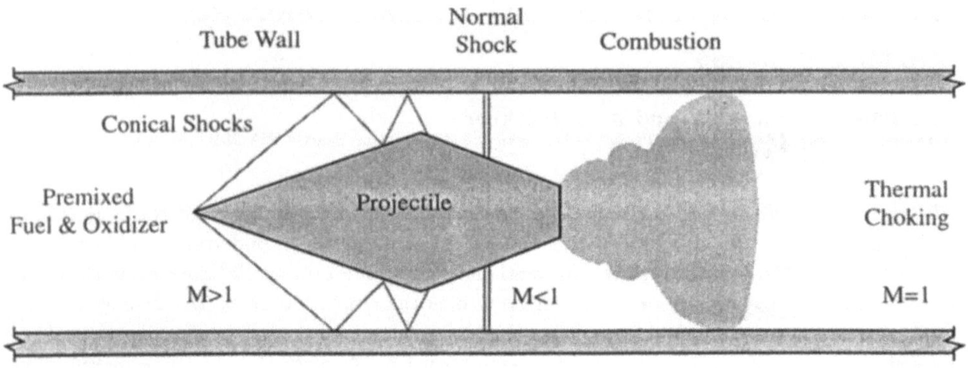

Fig. 1. Ram accelerator propulsive mechanism at subdetonative velocities

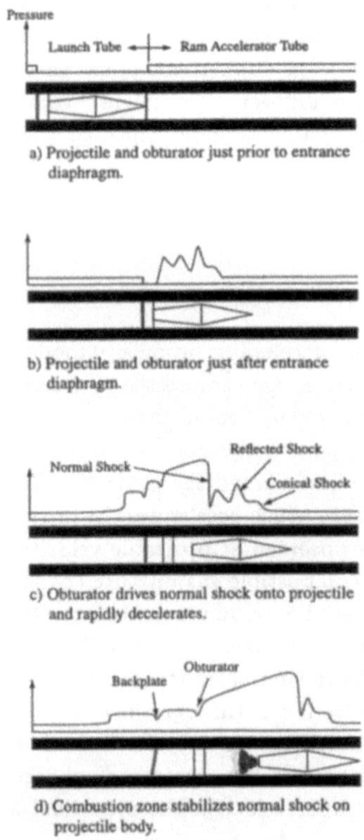

Fig. 2. Idealized starting process

2. Experimental Procedure

The experiments were conducted in the 38.1 mm bore University of Washington (UW) ram accelerator facility, presented in Fig. 3 (Knowlen et al. 1991a). It consists of a light gas gun initial launcher with a 6 m launch tube, a launch tube dump tank, 16 m of ram acceleration tube, a final dump tank, and a catcher tube. A schematic of the instrument station configuration near the entrance diaphragm employed in many of the experiments is presented in Fig. 4. There are two instrument stations in the launch tube, 12 and 139 mm before the entrance diaphragm. The first instrument station in the ram accelerator tube is located 12 mm after the diaphragm, followed by a station 227 mm down the tube. There is a 400 mm interval between subsequent stations along the remainder of the ram accelerator tube. All stations are instrumented with electromagnetic sensors and piezoelectric pressure transducers (PCB 119) at the tube wall. Electromagnetic sensors provide a time-distance history of the projectile by detecting the passage of a magnet ($\pm 1\mu s$) in the projectile throat. Differentiating these data resulted in velocity data with an uncertainty of 1%. Piezoelectric pressure transducers allowed measurement of the flowfield pressure at the tube wall with a $1\mu s$ response time.

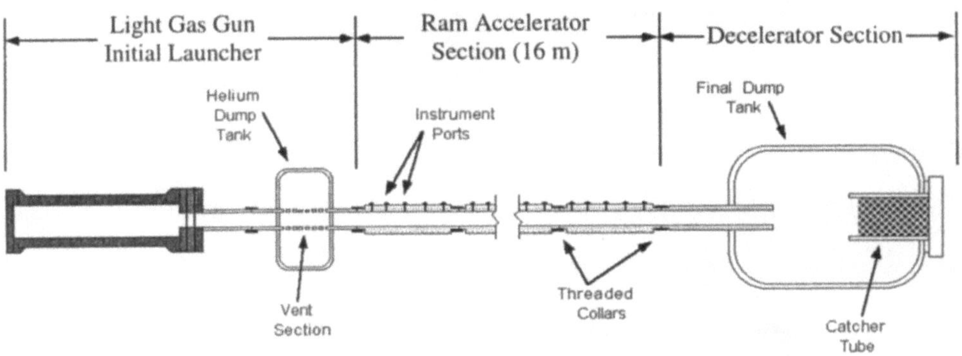

Fig. 3. Ram accelerator facility diagram

The initial launcher is a helium-filled double diaphragm light gas gun with a maximum breech pressure of 34 MPa. The Mach number of the projectile as it enters the ram accelerator test section is varied by adjusting the helium breech pressure, and through propellant chemistry changes to adjust the acoustic speed. The launch tube is 6.4 m long and has a series of thirty-six opposing 6.4 mm diameter vent holes, distributed over a 305 mm section of tube beginning 600 mm from the entrance diaphragm, to allow venting of gases in the launch tube to an evacuated dump tank. The nominal launch tube residual air pressure is $\sim$ 530 Pa (4 Torr), monitored by a piezoresistive transducer (Kistler 4043A1). The entrance diaphragms are made of Mylar, up to 0.72 mm thick for a 5.0 MPa propellant fill pressure.

The propellant characteristics are tailored by the type and concentration of fuel, oxidizer, and diluent in the mixture. A propellant class is characterized by its fuel-oxygen stoichiometry and its concentration of a specific diluent. Typical fuels used are methane and hydrogen, while diluents include nitrogen, argon, carbon dioxide, helium, or additional fuel. The oxidizer is oxygen in all cases, and propellant fill pressures are generally 2.5-5.0 MPa. The nominal UW propellant for starting at $\sim$1100 m/s is $2.8CH_4+2O_2+5.7N_2$, with small perturbations in the methane and nitrogen content. The propellant composition control is based on sonic orifice metering and is supported by gas chromatography. Absolute mixture accuracy is within 5%, and the relative precision when varying mixtures is within 1%. Propellant properties of interest include the acoustic speed and the heat release parameter $Q = \Delta q/c_pT$, where the equilibrium heat release, Δq, is normalized by the constant pressure specific heat capacity, c_p, and the temperature, T, of the quiescent gas mixture. The energy release parameter is usually evaluated at the thermally choked Chapman-Jouguet (CJ) state, in which case $Q = Q_{CJ}$. Chemical reaction rates, induction lengths,

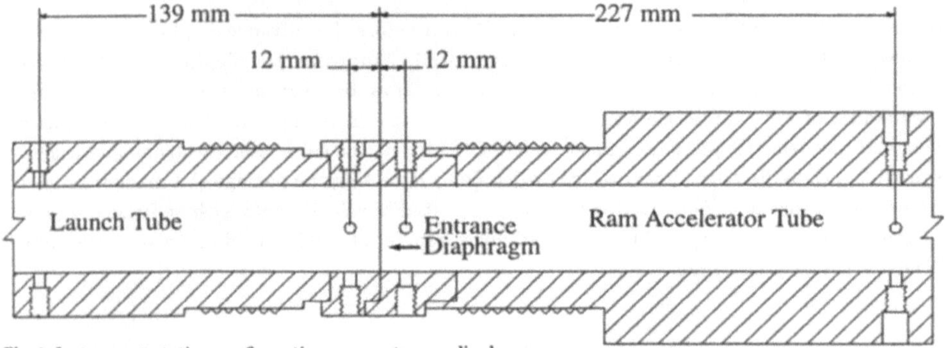

Fig. 4. Instrument station configuration near entrance diaphragm

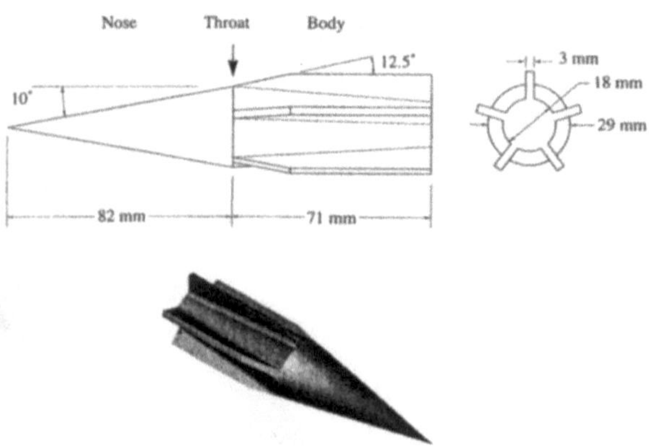

Fig. 5. Standard ram accelerator projectile

and activation energies, collectively referred to as "reactivity parameters," are also of interest but difficult to quantify under typical ram accelerator operating conditions.

The standard projectile shown in Fig. 5 is typically fabricated of 7075-T6 aluminum alloy and consists of two pieces, the nose and body, which screw together at the throat. Primary characteristics of the standard projectile include a 10° half-angle nose, 5 fins to stabilize the subcaliber projectile in the tube, a mass of approximately 75 gm, and a flow throat-to-tube area ratio (A_{throat}/A_{tube}) of 0.42. A magnet in the throat allows tracking of the projectile with the electromagnetic sensors. A polycarbonate obturator (Fig. 6), consisting of a 13 gm perforated piston glued to a 3 mm thick, 3 gm polycarbonate backplate, is attached to the projectile base. All obturators are 38.1 mm (±0.5 mm) in diameter, and in some cases a magnet is glued into a circumferential groove in the obturator to facilitate its tracking. Projectile firings into combustible mixtures are referred to as "hot shots" and given the designator HS, while experiments in non-combustible mixtures are referred to as "cold shots" and given the designator CS.

3. Outcomes of a Start Attempt

There are four possible outcomes of a ram accelerator start attempt at supersonic entrance velocity: a successful start, sonic diffuser unstart, wave fall-off, or wave unstart. Each of these phenomena, illustrated in Fig. 7, is described in the following paragraphs.

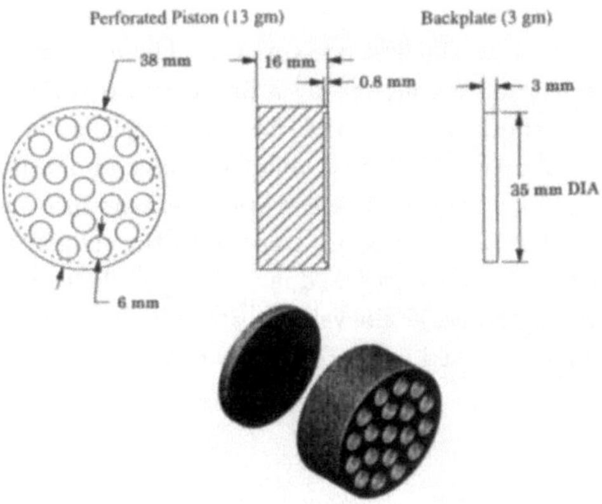

Fig. 6. Standard obturator configuration

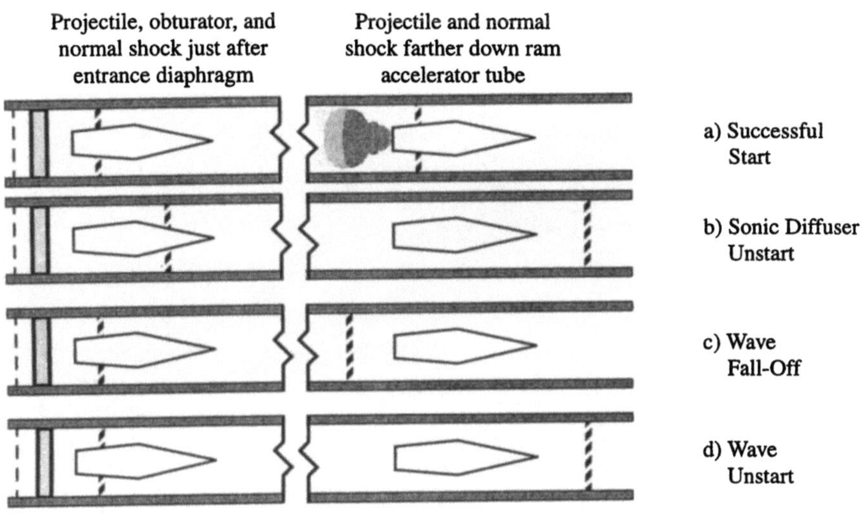

Fig. 7. Possible outcomes of an attempted start

3.1 Successful Start

A successful start (Fig. 7a) is accomplished by satisfying two criteria: obtaining supersonic flow past the projectile throat and stabilizing a high pressure shock system on the body by igniting the propellant in a combustion zone behind the shock system. A successful start is defined in the context of this work as a projectile which has accelerated for more than 2 m beyond the entrance diaphragm. At this point the obturator is considered decoupled from the starting process and a change in propellant via staging can be made at the 2 m tube joint (Bruckner et al. 1991 and Elvander et al. 1996). Plots of projectile velocity versus distance and tube wall pressure data for a successful ram accelerator start are presented in Fig. 8. A distance of 0 m is assigned to the location of the entrance diaphragm, and 0 μs corresponds to the projectile throat passing the instrument station. The tube wall pressure data, presented as measured pressure normalized by propellant fill pressure, are shown as a series of traces which originate from different instrumentation stations along the tube. A projectile outline is shown scaled relative to the average velocity. The lowest pressure trace, 12 mm after the entrance diaphragm, clearly shows the obturator-driven shock.

The conical shock from the projectile nose tip is evident as the first pressure rise. The next shock impacts the tube wall in the projectile throat region and is followed by a pressure drop arising from the subsequent flow expansion. The following shock is assumed to be the leading wave of the normal shock system which is supported by combustion. Reflected oblique shocks, appearing as periodic pressure fluctuations in the traces from the other instrument stations, extend into the full tube area behind the projectile (Hinkey et al. 1992). The pressure trace from the station at 1427 mm shows the tail of the signal decaying below the fill pressure, due to thermal drifting of the piezoelectric transducer. Upon successful starting, the projectile velocity steadily increases at a rate consistent with that predicted for the thermally choked ram accelerator propulsive mode.

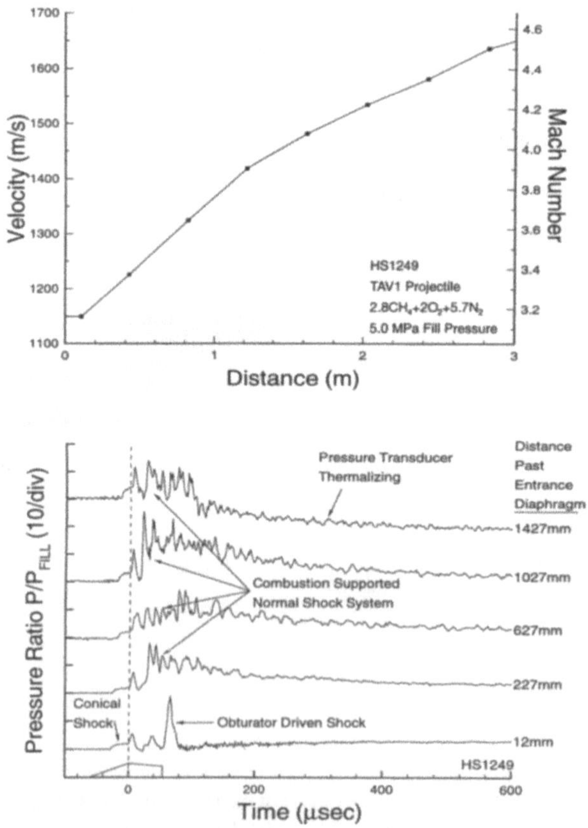

Fig. 8. Experiment resulting in a successful start

As previously noted, the standard propellant for the starting process is $2.8CH_4+2O_2+5.7N_2$, with small perturbations in the methane and nitrogen content. Other methane/oxygen-based propellants with carbon dioxide, argon, or helium as diluents have also produced successful starts. Fill pressures from 0.3 to 5.0 MPa have been used, with projectile entrance velocities typically around 1150 m/s (Mach 3.1 in the standard propellant). Successful starts, in different mixtures, have been achieved at a minimum and maximum entrance velocity of 715 m/s and 1361 m/s, respectively.

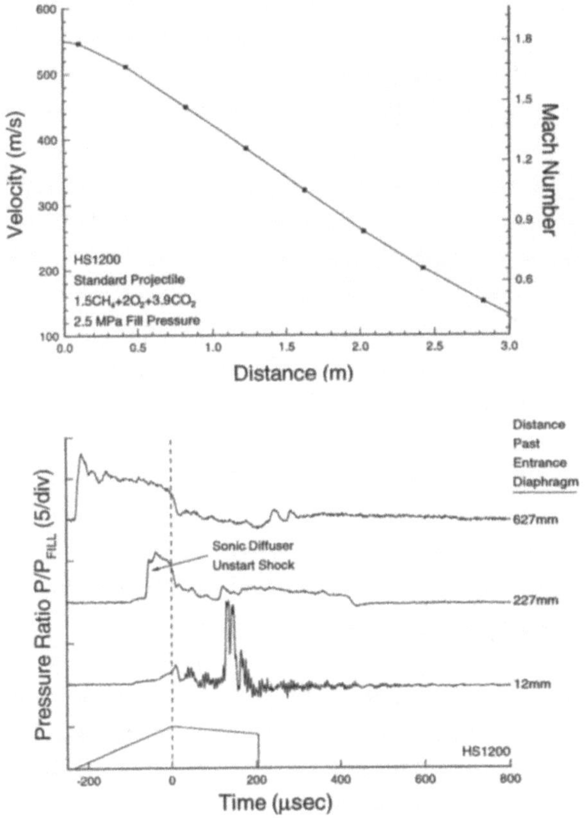

Fig. 9. Experiment resulting in a sonic diffuser unstart

There have been 1371 hot shots and 63 cold shots conducted at the UW between September 1985 and June 1997. From April 1990 to the present, the facility has been configured in the manner described above. The reliability of the starting process can be quite high when experiments are conducted with the same projectile and obturator geometries, propellant composition and fill pressure, launch tube evacuation, and entrance velocity. For example: 102 experiments with the standard projectile and obturator (Figs. 5 and 6) have been attempted in $2.8CH_4+2O_2+5.7N_2$ at or above 2.5 MPa, with entrance velocities greater than 1100 m/s and nominal launch tube evacuation. Of these experiments, 99 were successful starts and 3 failed, yielding a 97% success rate.

3.2 Sonic Diffuser Unstart

A sonic diffuser unstart (Fig. 7b) occurs when the flow is rendered sonic ahead of the throat due to upstream conditions. In one scenario, the flow chokes ahead of the throat when a projectile enters the propellant at too low a Mach number. A sonic diffuser unstart will also occur if the flow reaches sonic conditions ahead of the throat due to excessive upstream combustion or supersonic drag deceleration to a sufficiently low Mach number. Compression waves emanating from the sonic region coalesce into a shock and propagate ahead of the projectile, causing rapid deceleration in the resulting high pressure zone on the projectile nose. Figure 9 shows projectile velocity versus

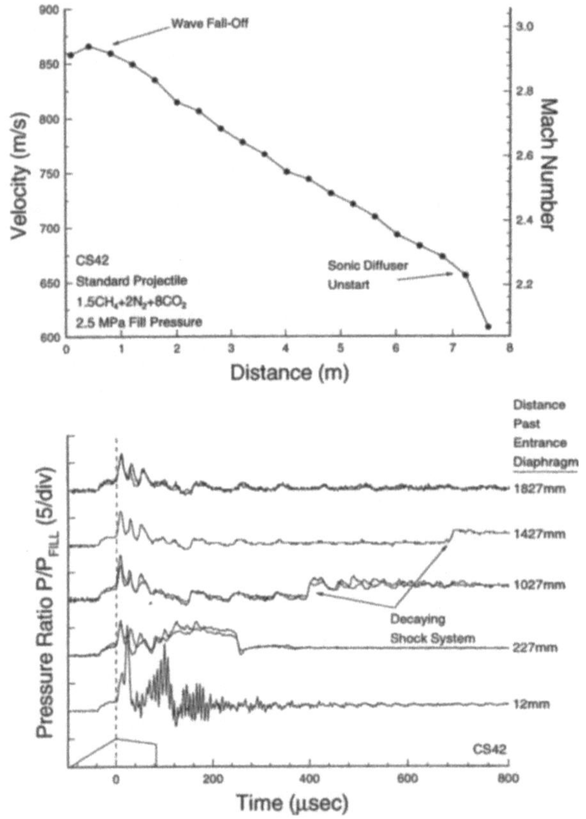

Fig. 10. Non-reactive experiment resulting in a wave fall-off

distance and tube wall pressure data from a shot in a CO_2-diluted propellant which resulted in a sonic diffuser unstart caused by a low entrance Mach number. The projectile velocity steadily decreases following entry of the projectile into the ram accelerator tube, and the pressure trace 227 mm past the entrance diaphragm shows the shock which arises after the sonic diffuser unstart.

The Fig. 9 pressure trace from the instrument station at 12 mm appears similar to that shown for the successful start of Fig. 8. During the sonic diffuser unstart experiments with a high spatial resolution test section of Burnham (1993), it was noted that pressure measurements immediately after the entrance diaphragm appeared just as those during a successful start. The unstart shock wave ahead of the throat was only observed after 70 mm of travel past the entrance diaphragm. Complex, unsteady gasdynamic processes govern the formation of the shock ahead of the throat, and the unstart shock in the data presented here formed between the first two instrument stations of the test section.

3.3 Wave Fall-Off

A wave fall-off (Fig. 7c) occurs when the combustion-supported shock system recedes from the projectile body because of insufficient energy release. The projectile outruns the shock system and decelerates due to supersonic drag as it travels down the tube. Figures 10 and 11 present projectile velocity versus distance and tube wall pressure data from shots which resulted in wave

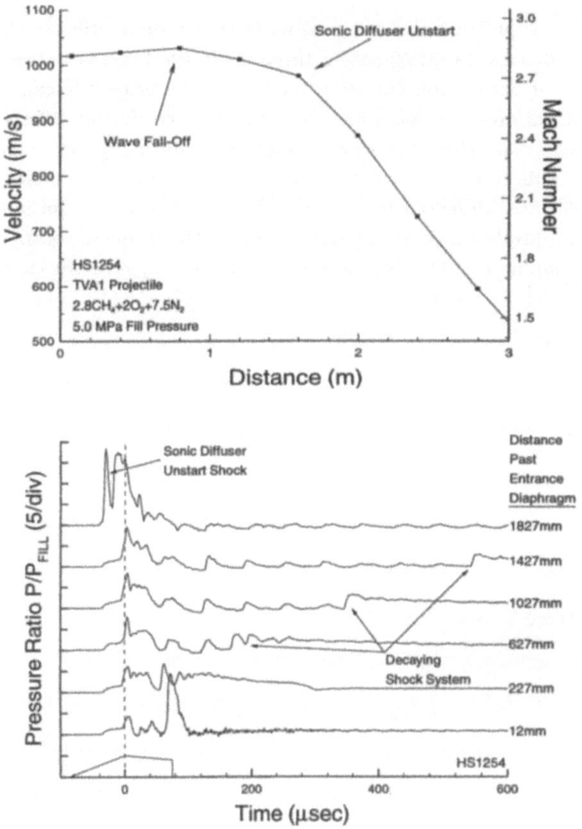

Fig. 11. Reactive experiment resulting in a wave fall-off

fall-offs in non-reactive and reactive mixtures, respectively. The double traces shown in Fig. 10 are from instrument stations that had opposing pressure transducers. The velocity increases after projectile entry due to the high base pressure generated by the obturator, and then decreases after the obturator-supported shock system recedes from the projectile body. As seen in the figures, both experiments eventually resulted in a sonic diffuser unstart after the projectile was sufficiently decelerated.

The similarity of the shock system recession from the projectile base in the pressure plots, regardless of whether the mixture is reactive or unreactive, supports the theory of insufficient energy release leading to a wave fall-off. A quasi-one-dimensional ram accelerator model employed by Higgins et al. (1993) also predicts recession of the normal shock from the projectile body at a minimum level of energy release.

3.4 Wave Unstart

A wave unstart (Fig. 7d) occurs when conditions downstream of the throat cause the shock system on the body to disgorge into the diffuser. The projectile then rapidly decelerates in the high pressure region behind the shock. Figure 12 shows projectile and shock velocity versus distance and tube wall pressure data from a shot which resulted in a wave unstart. The obturator-driven shock reaches the throat 227 mm after projectile entry and rapidly disgorges through the diffuser.

The wave velocity and pressure amplitude indicate that the unstart wave is a detonation in this case. Due to limited instrumentation density (approximately three projectile lengths between every instrument station) the location of detonation initiation relative to the projectile cannot be accurately discerned. Significant variation in pressure history data observed during different wave unstarts suggests that there exists more than one wave unstart mechanism. For example, projectile velocity versus distance and tube wall pressure data from a shot in which the unstart wave developed in the projectile wake are presented in Fig. 13. The pressure traces initially indicate a wave fall-off until a high amplitude shock of unknown origin in the projectile wake is observed 1027 mm past the entrance diaphragm. This waves surges past the projectile throat by 1427 mm.

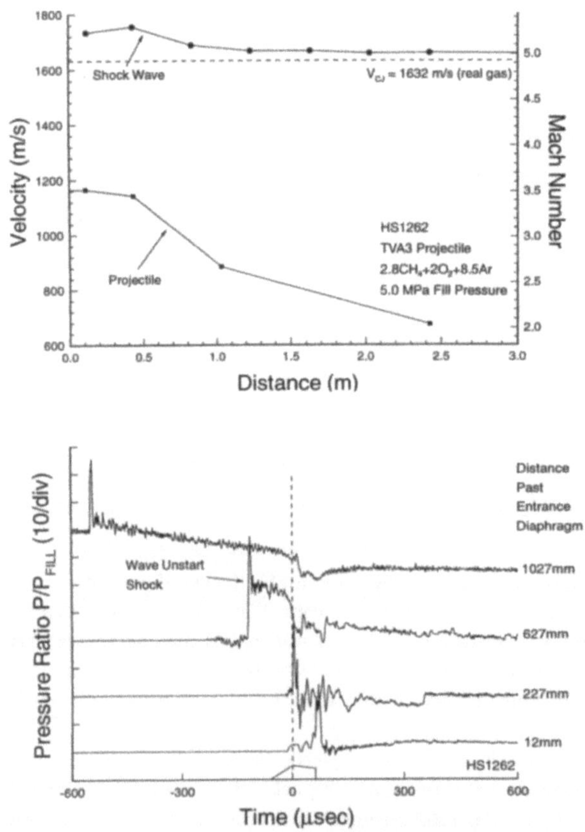

Fig. 12. Experiment resulting in a wave unstart originating on projectile body

The exact mechanism(s) by which a wave unstart occurs is unknown, although several hypotheses have been suggested. The following synopsis will be limited to the gasdynamic wave unstart hypotheses, as structural failure leading to a wave unstart has already been well documented by Hinkey et al. (1993). The obturator and/or compressed launch tube gases may drive a shock system onto the body with such a forward velocity relative to the projectile that it overruns the throat. Alternatively, the wave unstart could be due to the combustion region releasing too much energy for the projectile Mach number (Higgins et al. 1993). Another scenario involves conditions such that the mixture detonates behind the throat, resulting in a detonation wave propagating ahead of the throat (Schultz et al. 1997). Theories of shock-boundary layer interactions causing separation that leads to a wave unstart in high energy propellants have also been suggested by Higgins (1993) and Choi et al. (1997). A different concept is that as the projectile Mach number increases, the strength of the reflected shocks may grow to the point that propellant ignition takes place between the body and tube wall. Combustion in this region may propagate forward through the boundary layer and contribute to the wave unstart mechanism. Excessive energy release occurring on the projectile nose was once thought to play a major role in the wave unstart process but has been ruled out as a result of combustion stripping and re-ignition experiments by Knowlen et al. (1994).

4. Non-idealities of the Starting Process

The exact mechanism by which the propellant is ignited and combustion is stabilized behind the projectile throat is unclear, due to the numerous variables present which are unaccounted for in the ideal discussion of the starting process presented earlier. Ignition and stabilization of the combustion process are probably dependent in some way on all of these factors. The following sections give a summary description of each, along with a brief review of pertinent research at the UW. Some description of the starting process at ram accelerator facilities other than the UW is provided in the literature summarized by Schultz (1997).

4.1 Initial Launcher

The high pressure gas released in the initial launcher can cause structural failure of the obturator depending on the gas pressure, obturator geometry and material, and the projectile base area. This problem is suspected to have occurred during some start attempts with quasi-two-dimensional projectiles (Chang et al. 1997) and hollow projectiles (Sasoh et al. 1996a), and with a powder gun initial launcher (Sasoh et al. 1996b).

The launch tube is not perfectly evacuated, and some gas from the initial launcher does blow-by the obturator. The residual launch tube air and blow-by gas are compressed by a series of reflected shocks between the obturator and entrance diaphragm as the projectile accelerates down the launch tube (Stewart et al. 1997). A perforated section of the launch tube permits some venting of the compressed launch tube and initial launcher gases to an evacuated dump tank. The obturator backplate will separate when a sufficiently high pressure differential exists across it. Similarly, the obturator is glued to the base of the projectile and will separate when a sufficiently high pressure differential exists (Stewart et al. 1997). It is also possible for the entrance diaphragm to rupture prematurely if the compressed launch tube gas exceeds a certain pressure level.

Compression of the launch tube gases has been observed, during cold and hot shots, by pressure transducers and luminosity sensors (Burnham 1993, Sasoh et al. 1996b, Schultz et al. 1997, Stewart et al. 1997). Repeated shock reflections were tracked, and a radiating luminosity region was detected just prior to the entrance diaphragm. Burnham (1993) identified the compressed launch tube gas as a primary propellant ignition source through an experimental and computational investigation varying the residual air pressure, quantity of blow-by gas, and utilization of an inert buffer section between the launch tube and the ram accelerator section. Wave fall-offs occurred in the absence of compressed launch tube gas contacting the propellant. An effort to quantify the conditions for successful starting without launch tube venting and determining the maximum allowable residual air pressure has been undertaken by Stewart et al. (1997). These studies demonstrated sonic diffuser unstarts in the presence of excessive launch tube gas compression and showed that the timing of entrance diaphragm rupture relative to the projectile location does play a role in the starting process.

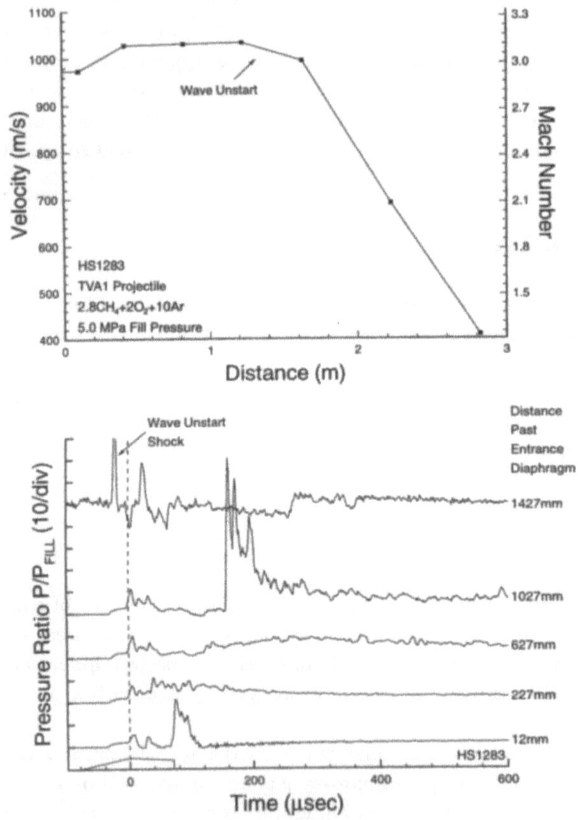

Fig. 13. Experiment resulting in a wave unstart originating in projectile wake

4.2 Entrance Diaphragm

If the compressed launch tube gas does not prematurely break the entrance diaphragm, then the projectile nose will. The dynamics of the diaphragm rupture will vary, depending on what time scale this event occurs relative to the motion of the projectile and launch tube shocks. Other potential effects from the entrance diaphragm impact are that the projectile and/or obturator may be damaged. Projectile condition upon entrance is of concern because a damaged or eroded geometry can significantly affect the propellant flowfield, especially in the case of severe projectile canting (Hinkey et al. 1993). Short obturator lengths may allow the obturator to tumble in the tube if struck by diaphragm fragments, influencing the area profile encountered by the flow.

No data have been formally presented regarding the effect of the diaphragm on the projectile, but several experiments have been conducted which demonstrated that minimal aluminum projectile damage is incurred by passing through Mylar diaphragms. For example, some of the experiments by Stewart et al. (1997) utilized a 3.6 mm thick Mylar entrance diaphragm and resulted in successful starts.

4.3 Ram Accelerator Section

Propellant variables governed by the chemical composition and ambient environment include the reactivity parameters (finite reaction rates, activation energy, and induction time), acoustic speed, energy release, temperature, and pressure. The propellant is initially quiescent at room temperature with a fill pressure up to 5.0 MPa. Variation of the type and relative amounts of fuel, oxidizer, and diluent allows tailoring of the propellant chemical properties. These variables are typically coupled in such a manner that it is difficult to control one without affecting the others.

The propellant is compressed as it passes through a conical shock generated by the projectile nose, and then through reflections of this shock between the projectile and the tube wall. The propellant expands as it encounters an increasing area behind the throat, but the fin leading edges in the throat area create high pressure stagnation regions and complex oblique shock patterns. The geometry-dependent, three-dimensional, turbulent flowfield on the projectile body contains mixed regions of subsonic and supersonic flow, shock reflections between the body and tube wall, shock-shock intersections (Hinkey et al. 1992), and shock interactions with the boundary layers on the projectile surface and the tube wall. The shock system supported by the obturator is typically idealized as a single normal shock, but is actually a complicated system of reflected oblique shocks and/or lambda shocks forming a shock train that renders the flow subsonic relative to the projectile (Waltrup et al. 1973, Billig 1993).

Due to the limited diagnostics available for use in the extreme environment found in the ram accelerator, there is no direct evidence of whether or not thermal choking actually occurs between the projectile and the obturator. The excellent agreement between experimental velocity versus distance data and the predictions of an end-state Hugoniot analysis performance code, however, provides strong support for the case that a thermally choked region does exist at full tube area behind the projectile travelling at subdetonative velocities (Knowlen et al. 1991b). Luminosity measurements for projectiles at subdetonative velocities indicate that the primary combustion zone is at or behind the base of the projectile (Burnham 1993, Hinkey et al. 1992).

Higgins et al. (1993) conducted experiments in which a driving projectile (already successfully started) was run through 4 m of inert gas to strip the combustion wave, and then re-entered a combustible mixture. For certain propellant energy release values and projectile Mach numbers, a new combustion wave formed in the projectile wake, propagated toward the projectile, attached itself to the base, and accelerated the projectile, while under other conditions no propellant ignition (similar to a wave fall-off in start attempts with an obturator) occurred or a wave unstart was observed. The combustion stripping and re-ignition experiments demonstrated that successful starting can occur without the presence of an obturator and residual launch tube gas.

4.4 Projectile and Obturator

Projectiles with variations in nose cone angle, nose length, throat diameter, body length, body taper angle, base diameter, number of fins, fin thickness, fin rake angle, and fin location have been used in experiments. Geometric variations such as these, along with the entrance Mach number, govern the flowfield about the projectile. The projectile is typically hollowed out to a wall thickness of 1.5 mm, and materials used include magnesium, aluminum, and titanium. The structural variables are impacted by the relatively high flowfield temperatures and pressures. In addition, projectile masses ranging from 50 to 110 gm experience accelerations up to 35,000 g during a successful start. These experiments have demonstrated that a relatively wide range of projectile geometries can be successfully started in the ram accelerator (Imrich 1995, Schultz 1997).

The obturator continues to affect the flowfield about the projectile as long as there exists a subsonic region between the two. The obturator dynamics are affected by its mass, geometry, and the pressure differential across it. Once the backplate has dislodged from a perforated obturator, the flow can expand through the exposed perforations and a significant pressure differential causes them to act as choked orifices. Flow occlusion is also altered by a short obturator if it tumbles in the tube when subjected to a non-uniform pressure distribution. Results of computational and experimental studies of the effects of the obturator on the starting process can be found in Burnham (1993), Giraud et al. (1995), Schultz et al. (1997), and Nusca (1997).

5. Conclusions

Experiments have been used to explore the four possible outcomes of a start attempt in the ram accelerator: successful start, sonic diffuser unstart, wave fall-off, and wave unstart. A successful start is achieved when supersonic flow is maintained throughout the diffuser, and a normal shock system is stabilized on the projectile body through propellant energy release. A sonic diffuser unstart is caused by conditions upstream of the throat resulting in subsonic flow in the diffuser. A wave fall-off occurs when insufficient energy is released from the propellant to keep the shock system on the projectile body from receding behind the base. A wave unstart is caused by conditions downstream of the throat resulting in disgorgment of the shock system on the body into the diffuser.

Development of a robust starting process is instrumental for utilizing the ram accelerator in a variety of applications. It is obvious that the starting process is far from the idealized model presented initially. Many variables attributed to the projectile, obturator, initial launcher, entrance diaphragm, and ram accelerator section have been identified as influential. Further research on the starting process will seek to quantify the effects of these variables, thereby generating envelopes of operating conditions under which successful starts occur.

Acknowledgement

This work summarizes the basic knowledge accumulated on the starting process over the 12 year operational lifetime of the UW ram accelerator. Therefore the efforts of all ram accelerator personnel at the UW and the support of USAF, AFOSR, ARO, ONR, and NASA are greatly appreciated.

References

Billig FS (1993) Research on supersonic combustion. J Prop Power 9:499-514

Bruckner AP, Knowlen C, Hertzberg A, Bogdanoff DW (1991) Operational characteristics of the thermally choked ram accelerator. J Prop Power 7:828-836

Burnham EA (1993) Investigation of starting and ignition transients in the thermally choked ram accelerator. Ph.D dissertation, Dept Aero Astro, Univ Washington, Seattle, WA

Chang X, Higgins AJ, Schultz E, Bruckner AP (1997) Operation of quasi-two-dimensional projectiles in a ram accelerator. J Prop Power 13:802-804

Choi J, Jeung I, Yoon Y (1997) Numerical study of scram accelerator starting characteristics. AIAA paper 97-0915

Elvander JE, Knowlen C, Bruckner AP (1996) High velocity performance of the ram accelerator. AIAA paper 96-2675

Giraud M, Legendre JF, Simon G, Henner M, Voisin D (1995) RAMAC in 90 mm caliber or RAMAC90. Starting process, control of the ignition location and performance in the thermally choked propulsion mode. Proc 2nd Int Workshop on Ram Accelerators, Seattle, WA, USA

Hertzberg A, Bruckner AP, Knowlen C (1991) Experimental investigation of ram accelerator propulsion modes. Shock Waves 1:17-25

Higgins AJ, Knowlen C, Bruckner AP (1993) An investigation of ram accelerator gas dynamic limits. AIAA paper 93-2181

Higgins AJ (1993) Gas dynamic limits of the ram accelerator. Master Thesis, Dept Aero Astro, Univ Washington, Seattle, WA

Hinkey JB, Burnham EA, Bruckner AP (1992) High spatial resolution measurements of ram accelerator gas dynamic phenomena. AIAA paper 92-3244,

Hinkey JB, Burnham EA, Bruckner AP (1993) Investigation of ram accelerator flow fields induced by canted projectiles. AIAA paper 93-2186

Imrich TS (1995) The impact of projectile geometry on ram accelerator performance. Master Thesis, Dept Aero Astro, Univ Washington, Seattle, WA

Knowlen C, Li JG, Hinkey J, Dunmire B (1991a) University of Washington ram accelerator facility. 42nd Meeting of the Aeroballistic Range Association, Adelaide, Australia

Knowlen C, Bruckner AP (1991b) A Hugoniot analysis of the ram accelerator. In: Takayama (ed) Shock Waves, Proc 18th Int Symp on Shock Waves, Vol I, pp 617-622

Knowlen C, Higgins AJ, Bruckner AP (1994) Investigation of operational limits to the ram accelerator, AIAA paper 94-2967

Nusca MJ (1997) Computational simulation of the ram accelerator using a coupled CFD/interior ballistic approach. AIAA paper 97-2653

Sasoh A, Higgins AJ, Knowlen C, Bruckner AP (1996a) Hollow projectile operation in the ram accelerator, J Prop Power 12:1183-1186.

Sasoh A, Hirakata S, Ujigawa Y, Takayama, K (1996b) Operation tests of a 25 mm bore ram accelerator. AIAA paper 96-2677

Schultz E, Knowlen C, Bruckner AP (1997) Detonation limits applied to the subdetonative ram accelerator starting process. AIAA paper 97-0807

Schultz E (1997) The subdetonative ram accelerator starting process. Master Thesis, Dept Aero Astro, Univ Washington, Seattle, WA

Stewart JF, Knowlen C, Bruckner AP (1997) Effects of launch tube gases on starting of the ram accelerator. AIAA paper 97-3175

Waltrup PJ, Billig FS (1973) Structure of shock waves in cylindrical ducts. AIAA J 11:1404-1408

Diaphragm rupturing processes by a ram accelerator projectile

J. Maemura[1], S. Hirakata[1], A. Sasoh[1], K. Takayama[1], J. Falcovitz[2]
[1]Shock Wave Research Center, Institute of Fluid Science, Tohoku University, 2-1-1 Katahira, Aoba-ku, Sendai 980-77, Japan
[2]Institute of Mathematics, Hebrew University, Jerusalem 91904, ISRAEL

Abstract. Fluid dynamic processes of diaphragm rupturing by a conically-nosed projectile were experimentally studied. The experiments were conducted using a 25-mm-bore ram accelerator. The projectile was launched by a powder gun which was the pre-accelerator of the ram accelerator. Either air or nitrogen was used as the test gas. Through framing photography using a high-speed image converter camera, it was observed that in the processes of the diaphragm rupturing a radiating region developed between the diaphragm and the approaching sabot. This region corresponded to the shock-heated gas which originated from a precursory shock wave driven by the accelerating projectile. With this precursory shock wave eliminated by a gasdynamic method, flow field around the projectile which was entering the test section by rupturing the diaphragm was visualized by holographic interferometer. During the rupturing, the system of oblique shock waves around the conical nose of the projectile was seen undisturbed on the fore side of the diaphragm. Under the same condition, numerical simulation was conducted using a GRP (Generalized Riemann Problem) scheme which was extended for computing fluid dynamic problems with moving boundaries. Two extreme models were examined; one assumed that the diaphragm deformed in the same way as the projectile piercing the diaphragm. The other assumed that the diaphragm was ruptured instantly at the moment that the tip of the nose hit the diaphragm. Comparing these models with the experimentally visualized result, the former was concluded to better express the diaphragm rupturing process under the present study.

Key words: Starting process, Diaphragm, Precursor shock wave

1. Introduction

Recently, studies on unsteady phenomena in chemically reacting supersonic flows, such as detonation initiation by spheres (Higgins and Bruckner 1995, Bélanger et al. 1995, Ju and Sasoh 1997), and ram accelerator starting processes (Hertzberg et al. 1988, Burnham 1993, Li et al. 1995, Leblanc et al. 1996, Choi et al. 1996) are intensively being conducted. In studying those subjects, an accurate model of diaphragm rupturing processes by moving bodies needs to be established. Quantitatively clarifying the processes will lead to progress in the understanding experimentally observed phenomena and to improved related numerical simulation. In particular, in ram accelerators the diaphragm rupturing process essentially affects whether or not the starting of ram acceleration becomes successful. Therefore, the development of an accurate model is important for identifying wide ranges of ram accelerator operation.

2. Experimental apparatus

Experiments were conducted using the 25-mm-bore ram accelerator (RAMAC25, Sasoh et al. 1996) at Shock Wave Research Center, Institute of Fluid Science, Tohoku University. A projectile made of A7075-T651 with a conical nose (half-apex angle 10°) was launched using 25-gram singly-based smokeless powder. The powder gun has a 3-m-long launch tube. In the acceleration process in the launch tube, the projectile was backed by a sabot made either of polycarbonate or

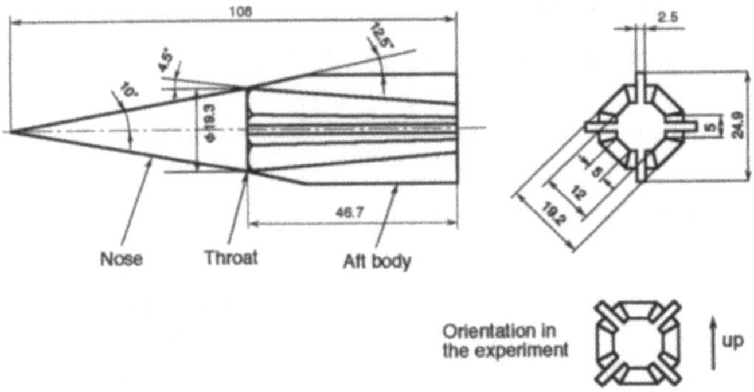

Fig. 1. Conically-nosed projectile and its orientation in the experiment

magnesium alloy, AZ31(F). From 0.05 m to 0.15 m upstream of the exit of the launch tube which is connected to the ram acceleration tubes, twenty four 8-mm-dia. perforation holes exist.

The passage moment of the projectile at each position was measured by sensing magnetic induction produced by the coasting projectile in which a ring of rubber magnet was inserted. The speed of the projectile was determined from the method of time of flight. In this study, the speed at the diaphragm was about 1100 m/s.

Figure 2 shows the schematic illustration of the test section. It is composed of two aspherical lenses made of acrylic which sandwich a single or two layers of Mylar diaphragm with the aid of ring holders made of brass. The aspherical lenses (Takayama and Onodera 1983) are designed so that an in-tube image is vertically magnified uniformly by 2.06 times whereas its horizontal image is not distorted. All through the present experiments, the orientation of the fins of the projectile was set, respectively to 45°, 135°, 225° and 315° measured from the vertical line, seen in Fig. 1. With this orientation, flow field around the aft body of the projectile in the test section was best observed.

3. Generation of precursory shock wave in projectile acceleration processes

No matter how low the initial pressure is, during the projectile acceleration process in the launch tube of the powder gun the projectile generates compression waves ahead of it. This phenomenon is well known as the classical piston problem (Courant and Friedrichs 1948, Shapiro 1983). In the present case, the projectile speed is so high that the front of the compression waves makes the transition to a shock wave early in the launch tube. This shock wave will hereafter be referred to as a "precursory shock wave." This precursory shock wave causes repeated reflections between the diaphragm and mainly the sabot – partially on the projectile (Burnham 1993, Sasoh et al. 1996). Immediately before the diaphragm rupturing, after several shock wave reflections, a region which is thus shock heated can ignite the combustible mixture which is initially loaded in the ram acceleration tubes.

In the present study, a precursory shock wave which experienced several reflections was observed by high-speed framing photography (Sasoh et al. 1996). The visualization was conducted using a high-speed image converter camera (ULTRANAC; framing rate; 2×10^7 frame/s max., exposure duration; 10 ns min). In this experiment, the framing rate was set 1×10^5 frame/s. The exposure duration was set to 200 ns for Tri-X negative film (Kodak, ASA 200). A Xenon stroboflash back-lighted the test section through a 3-mm-thick diffuser. The rise time of the stroboflash was about 130 μs, after which a constant intensity was maintained for longer than

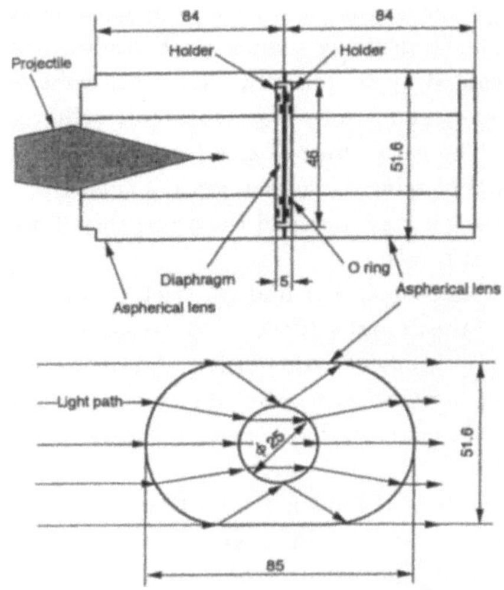

Fig. 2. Test section using aspherical lenses

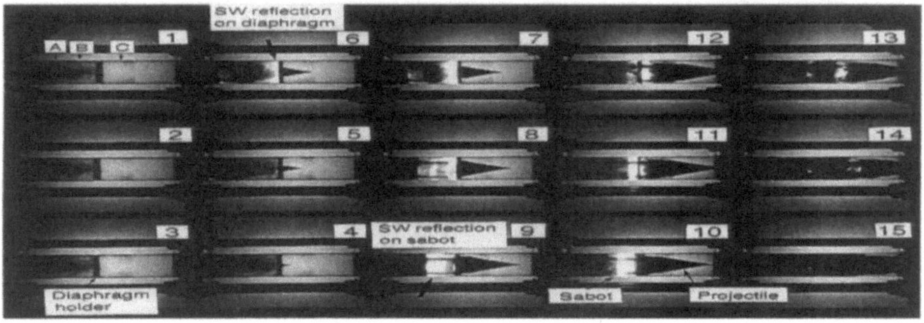

Fig. 3. High-speed framing photography of projectile entry processes. Frame interval; 10 μs. p_1 = 240 Pa (air), p_2=0.8 MPa (N_2), diaphragm 50-μm-thick Mylar diaphragm, U_p=1120 m/s

500 μs. The camera delay from the stroboflash was set to 150 μs, during another 140 μs fifteen consecutive frames of photograph were taken at time intervals of 10 μs. The pressure signal measured at 0.6 m upstream of the aspherical lenses was used for triggering the system. Out of the total length of six meters of the ram acceleration tube, the first 2-m segment was used as the extension of the launch tube of the powder gun, so that the effective length of the launch tube was 5 meters.

Figure 3 shows a series of high-speed framing photographs of the projectile which was rupturing the diaphragm. In the middle of each frame, the image of the whole aspherical lenses and the projectile is seen. The dark strip between the aspherical lenses is the shade of the metal holders in which the diaphragm was sandwiched. The speed of the projectile was 1120 m/s. In each frame interval, the projectile coasted 11.2 mm. Each frame is sequentially numbered. Hereafter, the time corresponding to a frame number, k, will be denoted by t_k. Initially, the launch tube of the powder gun was evacuated to 240 Pa. The precursory shock wave was repeatedly reflected between on the diaphragm and mainly on the sabot. The early stage of this precursory shock wave is not observable in this figure. The projectile starts piercing the diaphragm at $t = t_3$. Up to $t = t_5$ no radiation is noticeable. After several reflections the region behind the precursory shock wave is heated and starts emitting radiation as is seen at $t = t_6$.

The shock wave reflected from the diaphragm propagates upstream and is reflected again on the sabot. At $t = t_9$, the region of the highest intensity of the radiation moves onto the sabot. This illuminating gas slug is transported by the sabot, passing through the diaphragm. After entering the downstream side of the diaphragm, this hot gas slug is mixed with the cold nitrogen initially loaded there which has not experienced the shock reflections, thereby eventually the radiation quenches.

The present results are consistent with the conclusion of Burnham (1993) that this shock heated region acts as an igniter to the ram accelerator mixtures. Although the investigation on this precursory shock wave is of importance for studying ram accelerator starting processes, it disturbs visualization of fluid dynamic processes of diaphragm rupturing by the projectile. Therefore, in the next section the visualization conducted by using holographic interferometry after eliminating the precursory shock wave by a gasdynamic method is described.

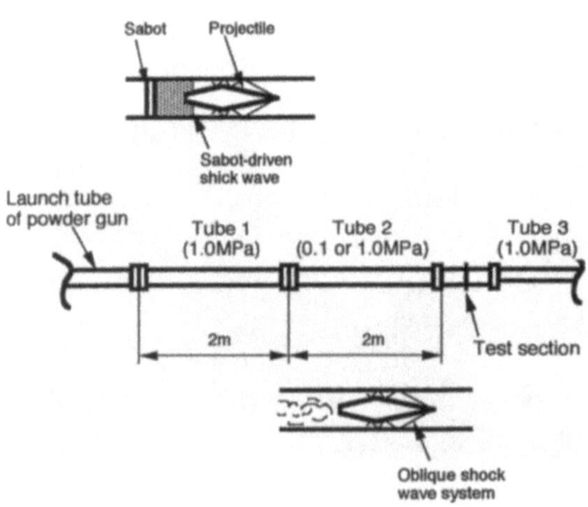

Fig. 4. Gasdynamic method of eliminating precursory shock wave

4. Visualization of fluid dynamic processes of diaphragm rupturing by a supersonically coasting projectile

In order to eliminate the precursory shock wave from the visualization, the sabot and then the precursory shock wave were separated from the projectile in the gasdynamic manner as follows (Fig. 4): After the projectile emerged from the launch tube of the powder gun, it enters Tube 1 which is filled with nitrogen at 1.0 MPa. At the moment of the entry, it is accompanied by the sabot. In Tube 1, the sabot is separated from the projectile due to the difference in aerodynamic drag coefficients between the projectile and the sabot. The projectile starts coasting supersonically,

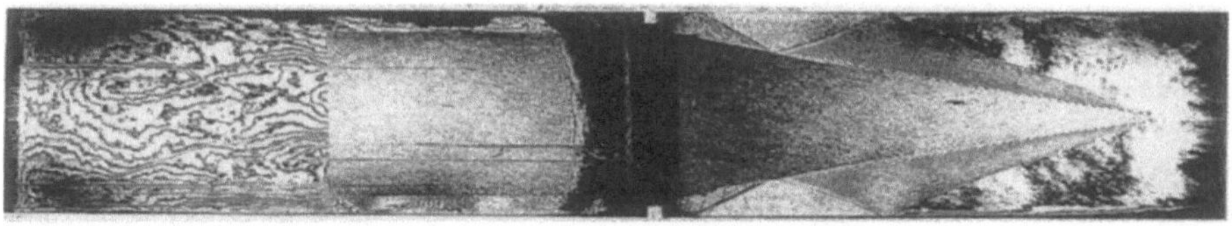

Fig. 5. Reconstructed interferogram of projectile entry process by impinging against Mylar diaphragm, p_2=0.1 MPa, p_3=1.0MPa, two-layer diaphragm (thickness = 188 μm), U_p = 1105m/s

while the separated sabot does not affect the flow field around the projectile. The test section was located between Tubes 2 and 3. If the pressure on the upstream side of the diaphragm is set lower than the pressure in Tube 1, Tube 2 is diaphragmed from Tube 1 and the pressure there is independently adjusted. Tube 3 is connected to the downstream side of the test section, where the pressure is also independently adjusted.

In the optical system for the flow visualization using double exposure holographic interferometer (Takayama 1983), a Q-switched ruby laser (Apollo Lasers Inc. 22DH, 2J/pulse and 25 ns pulse duration) was used as the light source. In order to minimize the distortion of the visualized image, the object beam path formed a 'Z' shape. The first exposure was done before firing the powder gun. The second exposure was triggered by the pressure signal measured at 0.7 m upstream of the diaphragm.

Figures 5 shows reconstructed interferograms of the flow field around the projectile piercing the diaphragm. The pressures in Tubes 2 and 3 were 0.1 MPa and 1.0 MPa respectively. The entry velocity was 1105 m/s; the Mach number was 3.25 – the real gas effects are not significant at this level of the Mach number. Two layers of thick diaphragm (188 μm thick each) were used. By the method described above, the projectile supersonically coasted in Tube 2 without the precursory shock wave and the radiation emission from the shock-heated region was thus eliminated from the visualization.

Even when the projectile enters deep downstream from the diaphragm, on the fore side only a conical shock wave system appears and remains undisturbed. This implies that on the fore side, the flow field remains supersonic with respect to the projectile. Neither fragments of the ruptured diaphragm nor disturbances generated by the fragments are observed in the figure. If the diaphragm was ruptured instantly when the tip of the projectile hit the diaphragm, expansion waves would be observed even on the fore side of the diaphragm. However, in the visualization they were not. Further discussion will be made in the next section.

In Fig 5, a wake region is observed on the aft side of the projectile. This region is not affected by the diaphragm rupturing process. The region between the aft body of the projectile and the tube wall in the process of the diaphragm rupturing must be affected by the diaphragm rupturing. However, the effects of the diaphragm rupturing on the flow field in the region are not well diagnosed solely from this interferogram.

5. Numerical simulation

In order to determine what kind of model is suitable for describing the experimentally visualized diaphragm rupturing process, numerical simulations were conducted. The method used for the simulation is the conservation laws scheme GRP (Generalized Riemann Problem), which has been extended to treat flows with moving boundaries. Under this scheme, the boundary is moving relative to an underlying 2-D Cartesian grid, in which the flow is computed by finite-difference (indeed, finite-volume) approximation. For the details of the numerical method, the readers should refer to Falcovitz and Ben-Artzi (1995).

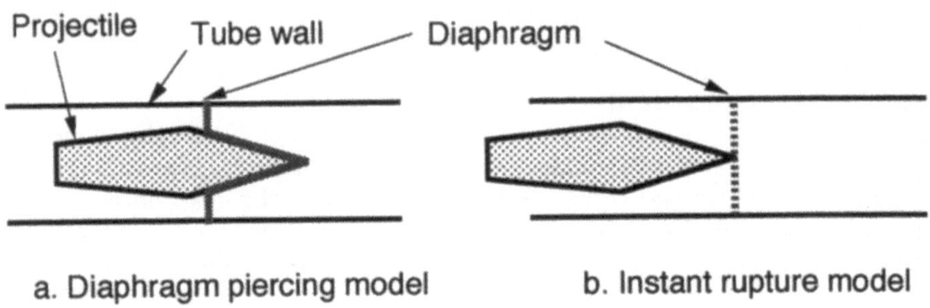

Fig. 6. Two models of diaphragm rupture: left, diaphragm piercing model (Model 1); right, instant rupture model (Model 2)

As previously mentioned, two extreme models which are schematically shown in Fig. 6 are used in the simulation of the piercing diaphragm flow process. In the first model, the diaphragm is pierced by the projectile cone in a "rigid" way. By that we mean that the perforated part of the diaphragm disappears, but the unperforated part remains intact. However, beyond the point of maximum projectile radius, the rate of growth of the perforated part of the diaphragm is maintained constant, so from that moment on, flow commences from the high pressure side to the lower pressure side. In the second model, the diaphragm instantaneously disappears at the moment that the cone starts piercing it. From that moment on, a kind of shock tube flow commences, and it is superposed on the advancing projectile flow field. We denote these two models by Model 1 (the diaphragm piercing model) and Model 2 (the instant rupture model), respectively.

The data of these computation is as follows. The equation of state is that of perfect gas with $\gamma = 1.4$. The initial temperature is 293 K and the molecular weight is assumed $W = 29$. The initial pressures are $p_2 = 0.1$ MPa on the left side of the diaphragm (prior to piercing) and $p_3 = 1.0$ MPa on the right side. The projectile velocity is 1.1 km/s and it is assumed constant throughout the process. The projectile is assumed to have an axisymmentric geometry and the same length and throat area as those in the experiment. The rectangular computation domain is of dimensions 0.0125 m $\times$ 0.150 m; it is divided into a grid of 40 $\times$ 480 square cells. The diaphragm is located in the middle of the computational rectangle, and the projectile is initially touching the leftmost boundary line, i.e., it is about to enter the computation domain.

Figures 7 and 8 show numerical isopycnics after the rupture corresponding to Model 1 and Model 2, respectively. The flow about the cone is well established when piercing diaphragm commences for Model 1 at about $t = 70\mu s$ (Fig. 7). The flowfield for Model 2 at that time (Fig. 8), clearly shows the beginning of "shock tube" flow, initiated by the instant diaphragm rupture.

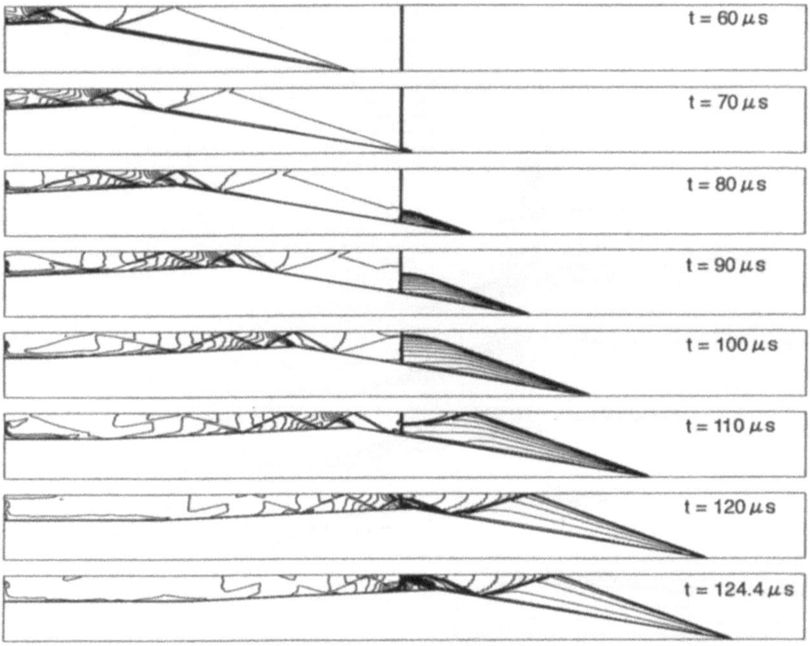

Fig. 7. Numerical isopycnics with the diaphragm piercing model (Model 1): p_2=0.1 MPa, p_3 = 1.0 MPa, U_p = 1.1 km/s; γ = 1.4

From this time on, the two model flows evolve in different ways, and we consider each one of them separately.

In Model 1 (Fig. 7), at $t = 100\mu s$ the conical shock has almost reached the tube wall, and the flow pattern is clearly analogous to that of the former time frame. This is a manifestation of the self-similar nature of the present flowfield, which persists up until the moment of first reflection (from the tube wall). That first reflection has occurred by $t = 110\mu s$, and the second reflection – from the cone – takes place at $t = 120\mu s$. At this time, the maximum-diameter point, i.e. the throat, has already passed the diaphragm plane, so that the diaphragm opening continues at the previous rate, and it is perceived that some flow starts through the small opening between the diaphragm and the projectile. The experimental visualization that we compare with is Fig. 5, taken at about $t = 124.4\mu s$). There, the second reflected shock is fully formed, and it is barely affected by the incipient rarefaction flow induced by the diaphragm opening. At this moment, there is clearly a good agreement between the simulation and the holographic interferogram with respect to all three shocks.

Turning to Model 2 (Fig. 8), the shock tube flow due to instant diaphragm rupture is clearly visible from $t = 80\mu s$ to 100 μs, consisting of a right rarefaction wave and a left shock wave, separated by a contact discontinuity. This flow interferes with the conical shock flow. However, the rightmost extent of this interference is only about one third of the distance covered by the cone tip, since the cone is moving at a Mach number of about 3.2. At $t = 100\mu s$, we clearly observe the conical shock "trace" as it passes through the rarefaction wave. Then at $t = 110\mu s$ the conical shock is about to reflect from the tube wall, and at $t = 120\mu s$ the reflected shock is seen curved through the rarefaction fan, barely touching the projectile. Finally, at the visualization time $t = 124.4\mu s$, we see the curved shock reflected from the tube wall, but no second reflected shock. From these results, it is well established that Model 2 does not agree with the visualization,

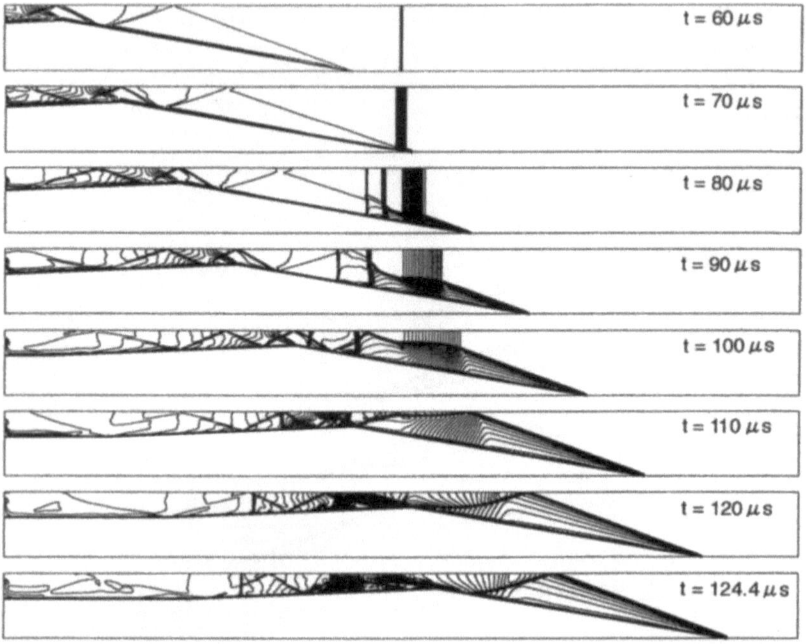

Fig. 8. Numerical ispycnics with the instant rupturing model (Model 2): p_2=0.1 MPa, p_3 = 1.0 MPa, U_p = 1.1 km/s; γ = 1.4

inasmuch as the second reflected shock is completely dissipated by the flow induced by the instant diaphragm rupture.

6. Conclusions

In this paper, the process of diaphragm rupturing by the conically-nosed projectile was experimentally visualized by high-speed framing photography and holographic interferometry. A shock-heated region generated by the precursory shock wave which was repeatedly reflected between the diaphragm and the sabot, was visualized as a luminous gaseous slug in the vicinity of the diaphragm. The observed phenomena are consistent with the pressure measurements and then with the conclusion made by Burnham that this shock heated slug can act as the igniter to the mixture in the ram accelerator operation. Eliminating this precursory shock wave by the gasdynamic method, flow fields around the supersonically coasting projectile which ruptured the diaphragm were visualized by holographic interferometry. On the fore side of the diaphragm, a system of conical shock waves was observed and no significant disturbance was observed in the system. The experimentally observed flow field agreed well with the numerically simulated one, using the diaphragm piercing model. This result suggests that within the framework of the present study, the diaphragm piercing model is better simulates the process of the diaphragm rupturing by the conically nosed projectile.

7. Acknowledgement

The authors would like to thank Dr. O. Onodera for his valuable suggestions and advices on the flow visualization techniques. Messrs. H. Ojima and T. Ogawa are also appreciated for their helps in conducting the experiment. The authors appreciate efforts for machining related items for the experiment contributed by Messrs. M. Kato, T. Watanabe, K. Asano, N. Ito, K. Takahashi, K. Kikuta and Y. Fushimi, all of who belong to the machine shop of Institute of Fluid Science. This study was conducted as an in-campus project for an intensified research subject funded by Tohoku University.

8. References

Bélanger J, Kaneshige M, Shepherd JE (1995) Detonation initiation by hypervelocity projectiles. In: Sturtevant B et al (eds) Proc 20th Int Symp on Shock Waves, World Scientific, Singapore, Vol 2, pp 1119-1124

Burnham E (1993) Investigation of starting and ignition transients in the thermally choked ram accelerator. Ph.D Dissertation, Univ Washington

Choi JY, Jeung IS, Yoon Y (1996) Transient simulation of superdetonative mode initiation process in SCRam-accelerator. 26th Int Symp Comb, The Combustion Inst, pp 2957-2963

Courant R and Friedrichs KO (1948) Supersonic Flow and Shock Waves. Interscience Publ Inc, New York

Falcovitz J and Ben-Artzi M (1995) Recent developments of the GRP method. JSME International Journal, Series B 38:497-517

Hertzberg A, Bruckner AP and Bogdanof DW (1988) Ram accelerator: A new chemical method for accelerating.projectiles to ultrahigh velocities. AIAA J 26:195-203

Higgins AJ, Bruckner AP (1995) Detonation initiation by supersonic blunt bodies. 15th ICDERS

Ju Y, Sasoh A (1997) Numerical study of detonation initiation by a supersonic sphere. Trans Jpn Soc Aero & Space Sci 40:19-29

Leblanc JE, Lefebvre MH, Fujiwara T (1996) Detailed flowfields of a RAMAC device in H_2-O_2 full chemistry. Shock Waves 6:85-92

Li C, Kailasanath K, Oran ES, Landsberg AM and Boris JP (1995) Dynamics of oblique detonations in ram accelerators. Shock Waves 5:97-101

Sasoh A, Hirakata S, Ujigawa Y and Takayama K (1996) Operation tests of a 25-mm-bore ram accelerator. AIAA paper 96-2677

Sasoh A, Hirakata S, and Takayama K (1997) Experimental studies on ram accelerator starting processes. J Jpn Soc Aeronautical Space Sci 46:37-45

Shapiro AH (1983) The Dynamics and Thermodynamics of Compressible Fluid Flow, Vol 2. Robert E Krieger Publ, Reprint Edition

Takayama K (1983) Application of holographic interferometry to shock wave research. In: Proc SPIE 398, pp. 174-181

Takayama K and Onodera O (1983) Shock wave propagation past circular cross sectional 90 degree bend. In: Archer RD, Milton BE (eds) Proc 14th Int Symp on Shock Tubes and Waves, New South Wales Univ, Sydney, pp. 205-212

Numerical simulation of the unsteady processes in starting period of ram accelerator

S. Taki, C. Zhang, X. Chang
Department of Mechanical Engineering, Hiroshima University, Higashi Hiroshima, 739 Japan

Abstract. The ram accelerator is a device to accelerate projectile up to hypervelocity in a tube filled with combustible gaseous mixtures. One of the most troublesome problems in operating facility is the difficulty to start. In order to improve the ram accelerator system in Hiroshima University, direct numerical simulations are carried using finite difference methods. A projectile and an igniter move at speed of 1200 m/s from an evacuated chamber into a ram acceleration tube, where hydrogen-oxygen-nitrogen gas mixtures are filled as combustible gases. When the ram acceleration tube is filled by nitrogen as a non-reactive gas, we got a numerical solution that the projectile fly without choking. When combustible gas mixtures are filled in the ram acceleration tube, we could not get a solution without detonation or quenching. Then, we introduce an ignition tube between the evacuated chamber and the ram acceleration tube. In the ram acceleration tube, undetonable but combustible gas mixtures are filled, while in the ignition tube, detonable gas mixtures are filled. The ignition tube works well in getting a very close solution to choking mode combustion.

Key words: Starting process, Detonation, Combustion, Shock waves

1. Introduction

The ram accelerator (RAMAC) is a device to accelerate a projectile up to hypervelocity in a tube filled with combustible gas mixtures. The shape of the projectile is similar to the center body of a ramjet, the nose of which is conical or wedge shaped. The thrust accelerating the projectile is generated by combustion of gaseous mixtures (Hertzberg et al. 1988). A few different propulsion mode are expected caused by the type of combustion. When the moving velocity of the projectile is below the Chapman-Jouguet detonation velocity, the thermal choked ram accelerator propulsive mode will be observed (Hertzberg et al. 1991). The numerical investigation described here is on the starting process to this subdetonative velocity mode. There exists a difficulty to start in this velocity regime to prevent the detonation initiation. If a detonation initiates around the projectile and overtakes it, a reverse thrust works to unstart.

The ram accelerator facility in Hiroshima University basically consists of a light gas gun for an initial launcher, an evacuation chamber, a ram acceleration tube, a dump tank and a catcher tube. A projectile and an igniter are accelerated by light gas gun using an obturator up to an entrance velocity of the ram acceleration tube (Chang et al. 1996). In the evacuation chamber, the launch tube gases and the obturator are evacuated in order to avoid the detonation initiation. Starting process in the present numerical simulation starts just at time the projectile enters the ram acceleration tube from the evacuated chamber (Zhang et al. 1996).

To overcome the starting difficulty, we introduce an ignition tube. Numerical result show that the ignition tube works well to establish a choking mode structure.

2. Model

The ram acceleration tube in Hiroshima University has a rectangular cross section of 20 mm high × 15 mm wide. The shape and the size of the projectile and the igniter are almost same as those of our experiments. The length of the projectile is 53 mm and the height between shoulders is 12 mm. The wedge angle of the head is 20 degrees. We assume the phenomena are two-dimensional and symmetric at the center plane.

Numerical simulations start just after the projectile enters the ram acceleration tube from evacuated chamber tube. An igniter follows the projectile to ignite the combustible gas mixtures in the ram-acceleration tube. After our experiments we tried to solve following three cases for the acceleration tube; (i) the case when the acceleration tube is filled with incombustible gas, (ii) when combustible gas mixtures are filled, and (iii) when an ignition tube, where detonable gases are filled, is inserted between the evacuation chamber and the acceleration tube. As the combustible mixtures, though methane-oxygen system is used in our experiments, we adopted hydrogen-oxygen system, for we only qualitatively inquire into the possibility of stable combustion in the conditions of sub-detonative mode.

3. Governing Equations

3.1 Chemical reaction kinetics The reaction mechanism of the hydrogen-oxygen system we used includes 13 chain elementary reactions as shown in Table 1, where the chemical species: O_2, H_2, O, OH, H_2O and HO_2 are taken into account, as well as the diluent gas of N_2 which is treated as inert gas. H_2O_2 is neglected to save the computer memory.

Table 1. Used kinetics of the reactions for H_2-O_2 system. $B_j = k_j T^{n_j} \exp(-E_j/T)$

j	Elementary Reaction	k_j [m, s, kmol]	E_j [K]	n_j	Third body factor
(1)	$H_2+O_2 \rightleftharpoons 2OH$	1.7×10^{10}	24,200	0	
(2)	$H+O_2 \rightleftharpoons OH+O$	2.2×10^{11}	8,455	0	
(3)	$O+H_2 \rightleftharpoons OH+H$	1.8×10^{7}	4,480	1	
(4)	$H+O_2+M \rightleftharpoons HO_2+M$	1.59×10^{9}	-500	0	18 for H_2O
(5)	$H+HO_2 \rightleftharpoons 2OH$	2.5×10^{11}	950	0	
(6)	$OH+H_2 \rightleftharpoons H+H_2O$	2.2×10^{10}	2,590	0	
(7)	$OH+OH \rightleftharpoons O+H_2O$	5.5×10^{10}	3,520	0	
(8)	$H+OH+M \rightleftharpoons H_2O+M$	1.6×10^{16}	0	-2	4 for H_2O
(9)	$H_2+HO_2 \rightleftharpoons OH+H_2O$	2.0×10^{8}	12,000	0	
(10)	$OH+HO_2 \rightleftharpoons H_2O+O_2$	1.6×10^{10}	0	0	
(11)	$O+HO_2 \rightleftharpoons OH+O_2$	4.8×10^{10}	500	0	
(12)	$H+H+M \rightleftharpoons H_2+M$	1.0×10^{11}	0	-1	
(13)	$O+O+M \rightleftharpoons O_2+M$	1.9×10^{7}	-900	0	3 for O

3.2 Equations of state It is assumed that all the species are thermally ideal. The specific heat of each species could not be treated as constant but depends on the temperature, so that it is approximated by polynomials of the temperature as follows.

$$e_t = \sum_{i=1}^{N} h_i Y_i - \frac{p}{\rho} + \frac{1}{2}\mathbf{v}^2 \tag{1}$$

$$\frac{h_i}{RT} = \left[a_1 + \frac{a_2}{2}T + \frac{a_3}{3}T^2 + \frac{a_4}{4}T^3 + \frac{a_5}{5}T^4 + \frac{a_6}{T}\right]_i, \quad (i = 1, 2, \ldots, N) \tag{2}$$

where e_t and h_i denote the specific total energy and the specific enthalpy of the species i, respectively.

3.3 Transport flux The molecular diffusion and the heat conduction are essential in the flame propagation. In the present calculations, the bimolecular diffusion and the heat conduction are taken into account, but the radiation is neglected. The viscosity is also taken into account, but at the surface of the projectile, boundary layer is neglected and slip wall boundary conditions are used. The molecular transport coefficients for each species are approximated as functions of temperature, which are basically estimated by Chapman-Enskog formula. All the walls are assumed as adiabatic and non-catalytic, as well as slip.

3.4 Conservation equations The conservation equations of mass, momentum, energy and chemical species are the fundamental differential equations, to be integrated.

$$\frac{\partial \rho}{\partial t} + \nabla \cdot (\rho \mathbf{v}) = 0 \tag{3}$$

$$\frac{\partial \rho \mathbf{v}}{\partial t} + \nabla \cdot (\rho \mathbf{v}\mathbf{v}) + \nabla p = -\nabla \cdot \tau \tag{4}$$

$$\frac{\partial \rho e_t}{\partial t} + \nabla \cdot \left[\rho \mathbf{v}\left(e_t + \frac{p}{\rho}\right)\right] = -\nabla \cdot \mathbf{q} - \nabla \cdot (\tau \mathbf{v}) \tag{5}$$

$$\frac{\partial \rho Y_i}{\partial t} + \nabla \cdot (\rho \mathbf{v} Y_i) = \rho W_i - \nabla \cdot (\rho Y_i \mathbf{V}_i), \quad (i = 1, 2, \ldots, N) \tag{6}$$

where τ denotes the stress tensor, $\mathbf{q}$ the heat flux vector and $\mathbf{V}_i$ the diffusion velocity of species i.

4. Computational Methods

The finite difference method is used to get the solutions of reactive gas flow fields. The square grids are used to keep the uniformity. To solve the high-speed flow, explicit methods are preferable, while an implicit method is necessary to integrate stiff system of chemical kinetics.

4.1 Time splitting method The governing equations include hyperbolic terms of high-speed flow, parabolic terms of dissipation, and source terms by fast chemical kinetics. Because of all the terms are important, time splitting method is used separated by their character.

4.2 Multilevel grid refinement method The flame propagation depends mainly on the diffusion and the heat conduction, which require the very fine mesh size to solve. We adopted the adaptive multilevel grid refinement method we have developed, which tremendously save the computer time and memory. The finest mesh size is about 12 μm around flames in the present calculations.

4.3 Moving boundary method Two lattice point systems are used, which are fixed on the projectile and the igniter, respectively. The igniter will separate and move away from the projectile because of the different drag forces. As the main coordinates are fixed on the projectile, the inflow gas velocity is the flight velocity of the projectile. The tail of the main grid system is connected with the head of the sub grid system fixed on the igniter. The space behind the igniter is not solved, for the igniter is expected as a role of the throat.

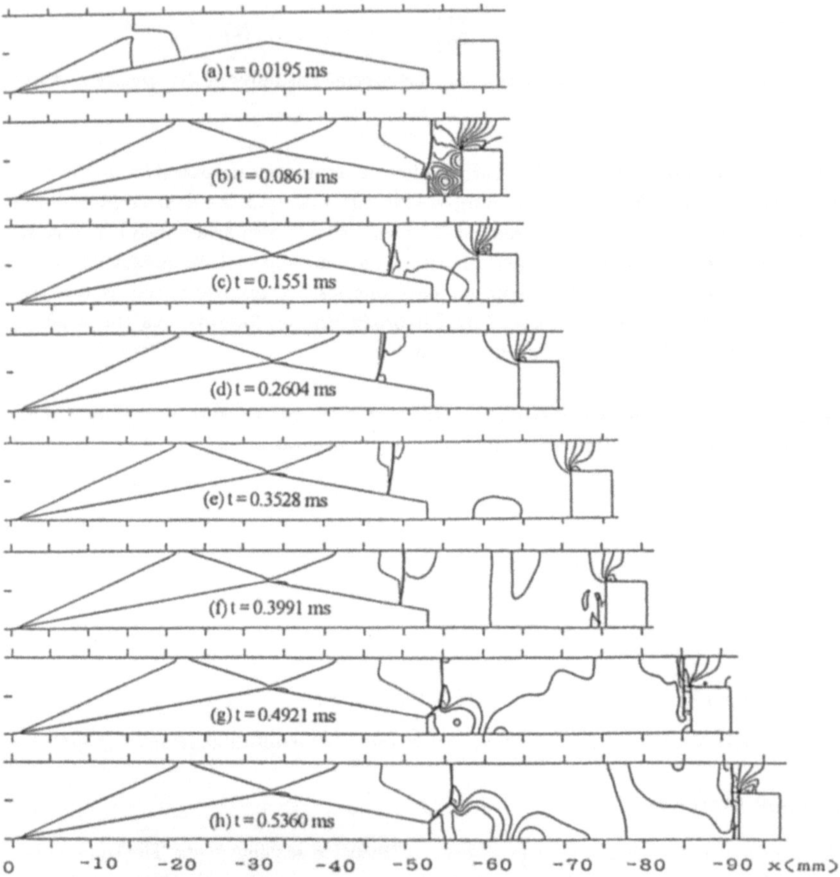

Fig. 1. Time sequence of the isobars for the cold shot case, where the nitrogen gas is filled in the ram acceleration tube at 0.1 MPa. Initial speed of the projectile and the igniter is 1200 m/s. Only upper half space is shown.

5. Numerical Results

In the ram accelerator, the projectile and the igniter move from the evacuated chamber tube (initial pressure is about 0.1 Pa) into the ram acceleration tube which is filled with gases of initial pressure 0.1013 MPa. The numerical simulations start at the time when the nose of the projectile arrives at the entrance of the ram acceleration tube or the ignition tube. Five cases are shown in Figs. 1 to 5. The initial velocity of the projectile and igniter is 1200 m/s for all the three cases. The first case is the one of the cold shot, i.e., an inert gas is filled in the ram acceleration tube. Fig. 1 shows the time sequence of isobars for the case of inert nitrogen gas. When the igniter following the projectile comes into the ram acceleration tube, a reflected normal shock is formed in front of the igniter. Because of high pressure behind the shock front, the igniter starts to leave from the projectile, while the reflected shock travels forwards to about $t = 0.26$ ms. After a while, the normal shock goes back behind the projectile as the igniter leaves away from it. This case shows the success of start for cold shot. If the igniter is bigger, e.g. the height of igniter is larger than 12 mm in the case shown in Fig. 1, or heavier, then the normal shock would go up to the forward of the projectile, called unstart.

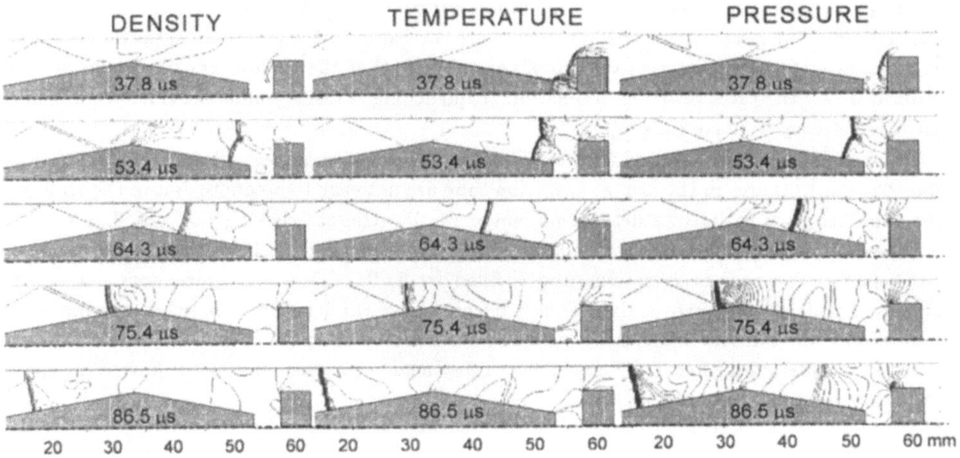

Fig. 2. Time sequence of the contours of the density, the temperature and the pressure for the case that the ram acceleration tube is filled with the mixture of $2H_2+O_2+2N_2$ at 0.1 MPa. Initial speed of the projectile and the igniter is 1200 m/s. The height of the igniter is 12 mm.

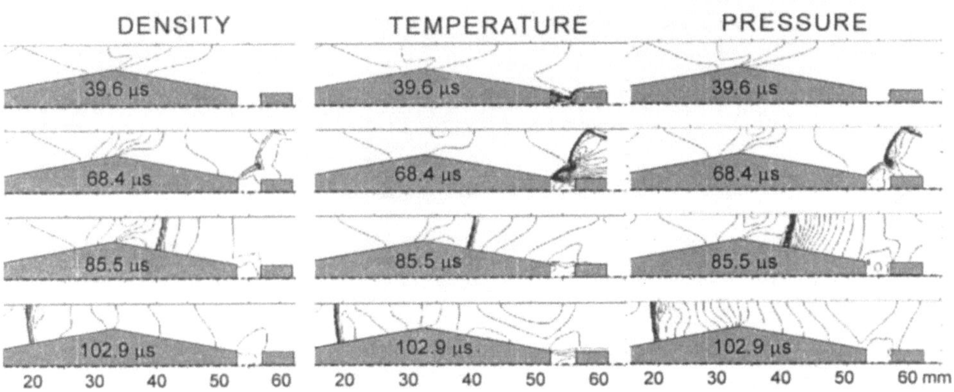

Fig. 3. Time sequence of the contours of the density, the temperature and the pressure. All the conditions are same as those of the case of Fig. 2 except the height of the igniter of 4 mm.

The second case is shown in Fig. 2, where the time sequences of the contours of the density, the pressure and the temperature are shown. The initial composition of the gas mixture in the ram acceleration tube in this case is $2H_2+O_2+2N_2$, which is easily detonable. Behind the normal shock wave caused by the igniter immediately generates a detonation. Once a detonation is formed, it travels forward and overtakes the projectile. In order to avoid the explosive ignition, we tried that the size of igniter is reduced from 12 mm high to 4 mm high as in the next numerical result, shown in Fig. 3. The results are almost same as previous one, except the delay of detonation generation. Then, to avoid the detonation, we tried the next case that the gas mixture is changed to $1/2H_2+O_2+2N_2$, which could burn but would not detonate. The results are shown in Fig. 4. Although the numerical simulations in this case are carried for much longer time, the flame would not propagate to the whole cross section of the tube, so that the thermal choking could not be established. All the cases result in detonation or quenching. Present numerical simulations show that it is not easy to make choking mode.

Then, we introduce an ignition tube between the evacuated chamber and the ram acceleration tube. The gas mixtures in the ram acceleration tube are not easy to detonate, but in the ignition tube they are easy to initiate detonation, those would generate choking hot gas. The results are shown in Fig. 5, where the ignition tube is 40 mm long and filled by the gas mixtures of $H_2+O_2+2N_2$. The ram acceleration tube is filled by $1/2H_2+O_2+2N_2$, which does not easily detonate. Until 78 μs in Fig. 5, the gases in burning is of the ignition tube, when the mode of combustion is detonation. After then, the flame is separated from the normal shock in the gases of the ram acceleration tube. The flame is not stably held yet in this case. But it looks very close to the choking mode combustion.

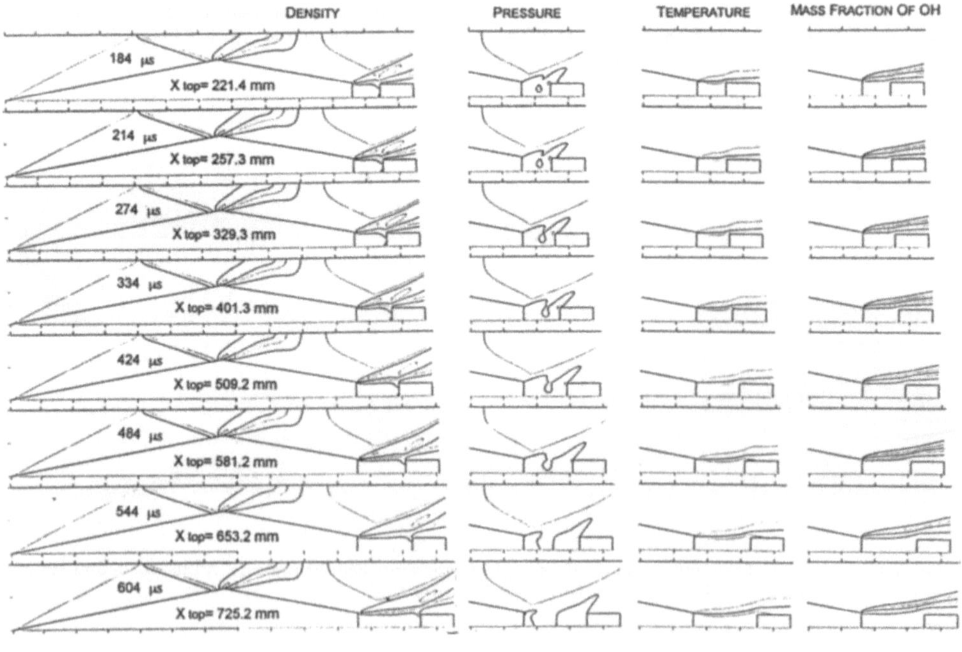

Fig. 4. Time sequence of the contours of the density, the temperature, the pressure and the mass fraction of OH radical for the case that the ram acceleration tube is filled with the mixture of $1/2H_2+O_2+2N_2$ at 0.1 MPa. Initial speed of the projectile and the igniter is 1200 m/s. The height of the igniter is 4 mm.

6. Conclusions

Numerical simulations are carried to get a choking mode combustion of a ram accelerator as a starting process. When the ram acceleration tube is filled by the nitrogen as a non-reactive gas, we got a numerical solution that the diffuser of the projectile works well without choking. When combustible gas mixtures are filled in the ram acceleration tube, we could not get a solution without detonation or quenching. Then, when we introduce an ignition tube between the evacuated chamber and the ram acceleration tube, we get a very close solution to choking mode combustion.

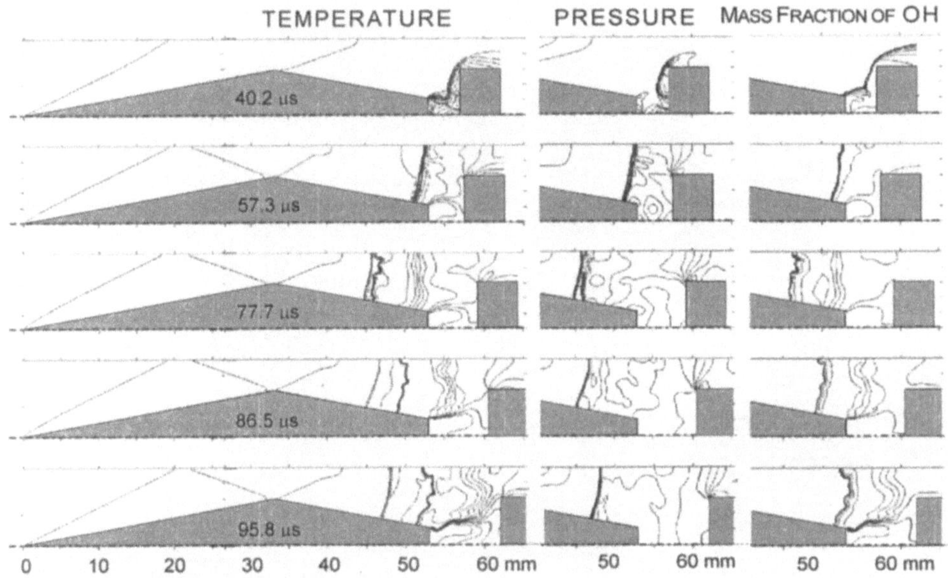

Fig. 5. Time sequences of the contours of the temperature, the pressure and the mass fraction of OH for the case when a short ignition tube (40 mm long) is inserted between the ram acceleration tube and the evacuated chamber. The initially filled gas mixtures are $H_2+O_2+2N_2$ for the ignition tube and $1/2H_2+O_2+2N_2$ for the ram acceleration tube, respectively. Initial speed of the projectile and the igniter is 1200 m/s. The height of the igniter is 12 mm.

References

Hertzberg A, Brruckner AP and Bogdanoff DW (1988) Ram accelerator: A new chemical method for accelerating projectiles to ultrahigh velocities. AIAA J 26:195-203

Hertzberg A, Brruckner AP and Knowlen C (1991) Experimental investigation of ram accelerator propulsion modes. Shock Waves 1:17-25

Chang X, Shimomura K, Matsuoka S, Watanabe T and Taki S (1996) Shock waves and combustion around a supersonic two-dimensional wedge-shaped projectile in a ram accelerator. 20th Int Symp on Space Tech & Sci (ISTS) 96-d-15

Zhang C, Inoue H and Taki S (1996) Direct numerical simulation of flow field with combustion in ram accelerator. 20th ISTS 96-a2-17

RAMAC 90: detonation initiation of insensitive dense methane-based mixtures by normal shock waves

J-F Legendre [1], P. Bauer [2], M. Giraud [1]
[1] ISL, French-German Research Institute of Saint-Louis, BP 34, F-68301, Saint-Louis Cedex, France
[2] LCD, Laboratory of Combustion and Detonics (UPR 9028-CNRS), ENSMA, BP 109, F-86960 Futuroscope Cedex, France

Abstract. Thorough investigations have been conducted so far at ISL on the sensitivity to detonation of dense gaseous explosive mixtures based on detonation tube experiments. Upper detonability limits have been determined in the 90mm caliber detonation tubes, using a direct initiation mode generated by a blasting cap. Fuel-rich methane-based mixtures diluted in various amounts of inert gas, respectively mono-atomic (Argon), di-atomic (Nitrogen) and tri-atomic (Carbon Dioxide) for an initial pressure of 2.5 MPa and initial temperature of 298 K were investigated. Typical Nitrogen-diluted mixtures currently used in different operational ram-accelerator facilities are situated outside, though at the vicinity, of the detonable area determined in detonation tube experiments. However, under particular experimental conditions, these mixtures could propagate a detonation during the so-called phenomenon of "unstart." In order to examine the validity of such upper detonable areas, as far as real initiation conditions of detonation around supersonic ram projectiles are concerned, different shock- generated initiation techniques have been investigated both at ISL and at the University of Washington. The aim of this paper is to make a special focus on the latest results of experiments carried out at ISL on the initiation of insensitive methane-oxygen-nitrogen mixture by means of normal shock waves. The technique used with ISL's 90mm caliber ram accelerator (RAMAC 90) involves the generation of a normal shock wave of known and adjustable characteristics in an intermediate ram section filled with an inert gas. This normal shock wave is then transmitted into a second test section filled with the dense explosive mixtures. The validity of the upper-detonable area determined in detonation tube with direct initiation by a blasting cap is finally discussed according to the initiation conditions found around a ram projectile at approximately Mach 3.

Key words: Starting process, Detonation, Normal shock wave, Sensitivity, Detonability

1. Introduction

Investigations are conducted at ISL on a 90mm caliber ram accelerator (noted RAMAC 90) (Giraud et al 1992, 1993, 1995, 1997) to develop a new laboratory hypervelocity launcher for ballistic applications. The key objective for this launcher is to accelerate a 1.5 to 2 kg projectile at velocities up to 3 km/s with an overall acceleration below 40 kG's. Experiments carried out so far along a 16.2-m-long ram section in the thermally choked propulsion mode allowed to reach a velocity increase from 1.3 to 2 km/s for a 1.34 kg projectile with a $2.95CH_4+2O_2+10N_2$ mixture at an initial pressure of 4.5 MPa (Giraud et al 1995). Besides projectile configuration (geometry, material), the composition of the dense combustible mixture plays a determinant role in the proper behavior of the thermally choked propulsion mode (Knowlen 1991). Accordingly, a thorough knowledge of the properties of these dense mixtures is required in terms of both detonability limits and Chapman-Jouguet equilibrium characteristics. Numerous experiments have been carried out so far at ISL on the detonation of fuel-rich methane-oxygen mixtures diluted in various amount of mono-atomic gases (i.e. argon and helium), di-atomic gases (nitrogen) or tri-atomic gases (carbon dioxide) at an initial pressure in the range 2.5 to 3.5 MPa (Legendre et al. 1995, Legendre 1996). These experiments have been carried out in two 90-mm-caliber smooth-bore detonation tubes (1.35-m-long, i.e., 15 calibers noted 90L15 and 3.15-m-long, i.e., 35 calibers noted 90L35 respectively) in order to provide geometrical conditions of confinement as close as possible to those encountered in the 90-mm-caliber ram accelerator (Legendre et al. 1993, 1995, 1998, Legendre 1996). Direct detonation initiation has been achieved through the spherical blast wave generated by the detonation of a charge of condensed explosive (# 8 detonator, 1g equivalent tetryl) (Bauer et al. 1994).

The experimental upper detonability envelope obtained in a detonation tube with direct pyrotechnic initiation technique at an initial pressure of 2.5 MPa has been compared to the observations of operational ram accelerator experiments carried out with methane-oxygen-nitrogen mixtures. Three basic mixture compositions currently used in three operational facilities in the thermally choked combustion mode have been considered :

- $2.7CH_4+2O_2+5.6N_2$: basic mixture used at the University of Washington (Seattle) in 38mm ram accelerator (Knowlen 1991)
- $3.2CH_4+2O_2+7.5N_2$: basic mixture used at ISL in 90-mm-caliber ram accelerator (Giraud et al 1992)
- $3.0CH_4+2O_2+10.0N_2$: basic mixture used at the US Army Research Laboratory in 120-mm-caliber ram accelerator (Kruczynski 1993) and at ISL in 90-mm-caliber ram accelerator (Giraud et al 1995, 1997)

The position of these three mixtures are plotted in a triangular-coordinate graph as well as the upper detonable area (Fig. 1).

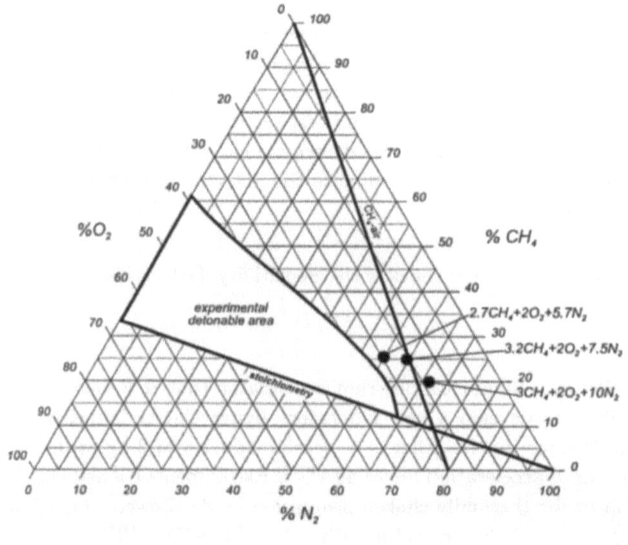

Fig. 1. Experimental detonable area: CH_4-O_2-N_2 mixtures, P_0 = 2.5 MPa, T_0 = 298 K, 90L15 detonation tube, # 8 detonator

This graph clearly shows that these three mixtures are situated outside the detonability area which means that no detonation could have been observed in our particular experimental conditions in a detonation tube. However, under particular experimental ramac conditions, these mixtures could propagate what looks like a detonation during the so-called phenomenon of "unstart" which occurs when a detonation front overtakes the projectile. Accordingly, a validation of this detonable area was required. Another series of experiments involving an initiation technique that would reproduce initiation conditions closer to those encountered during a real ram accelerator experiment was undertaken. Basically, only shock induced detonation initiation has been considered either on the body of the projectile or in the projectile wake on the normal shock detached from the sabot.

The aim of this paper is to check the applicability for ramac conditions of the results previously determined in detonation tubes in terms of detonability limits and to a lesser extent of prediction of the detonation velocity. This required (i) a better understanding and (ii) a better simulation tool for the real initiation conditions around an operational ram projectile.

2. Initiation conditions on ram-projectiles

The characteristics of the flow around a fin-guided supersonic ram projectile propagating in a tube filled with dense gas mixture are extremely difficult to accurately simulate. CFD computations are carried out at ISL in dense inert gas on the influence of the geometry of fin-guided projectile on the flow characteristics around and in the projectile wake (Henner 1996, Henner et al 1997a, 1997b). However, characteristics in terms of pressure and temperature are extremely difficult to experimentally validate in such highly three-dimensional flowfields. Accordingly, it has been decided to take into account only most simplified initiation conditions obtained by the observation of real signals of pressure transducers installed in the tube wall. Reference conditions of gas mixture and projectile geometry, respectively are : (i) initial density of ~30 kg/m^3, ambient temperature and sound velocity ~360m/s (ii) typical ISL RAMAC 90 projectile (Giraud et al 1995), (Legendre 1996) at a velocity of ~1300m/s ($M \sim 3.5$).

Reference shock characteristics have then been determined for a RAMAC 90 experiment with the ram section filled with nitrogen at an initial pressure of 2.8 MPa at a position of observation situated 1.05 m after the entrance of the ram section. Two shock initiation conditions have been considered: shock attached to the afterbody of the projectile and normal shock detached from the sabot and situated in the projectile wake, respectively (Fig. 2).

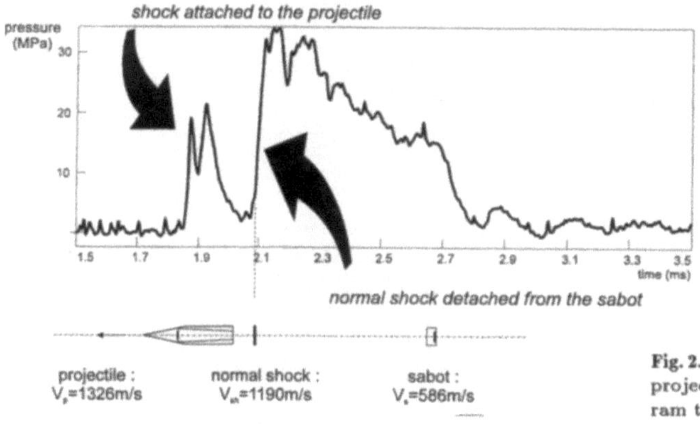

Fig. 2. Reference RAMAC 90 projectile: pressure transducer signal, ram tubes 1-5: N_2, P_0=2.8 MPa

In order to simulate these shock conditions, different experimental setups have been investigated, keeping the dense test gas mixture at rest. Two groups of experiments have been considered, either with the use of detonation tubes or with the use of ram-accelerators (Fig. 3). For the former facilities, initiation can occur either through the spherical blast wave generated by a charge of solid explosive, configuration which has been used during our previous experiments at ISL (Legendre 1996), or transmission in the test mixture of a normal detonation front propagating in a primary sensitive mixture (Bauer et al. 1998). Ram-accelerator facilities can also provide a valuable tool to study the shock induced initiation. In that case shock conditions are generated by a supersonic projectile either sub-calibrated (Higgins 1995) or a full-caliber piston directly injected into the test gas mixture (Bauer et al. 1996, 1997). A variant of the latter technique consists in injecting a full-caliber piston into an intermediate section filled with an inert gas. A normal shock is then generated and propagates ahead of the piston. The key feature of this technique is that it allows the transmission of a normal shock of adjustable characteristics with respect to inert gas nature, initial pressure and length of inert section, into the test section. Furthermore, this technique allows to unambiguously decouple the initiation location between the shock front and the forward face of the piston.

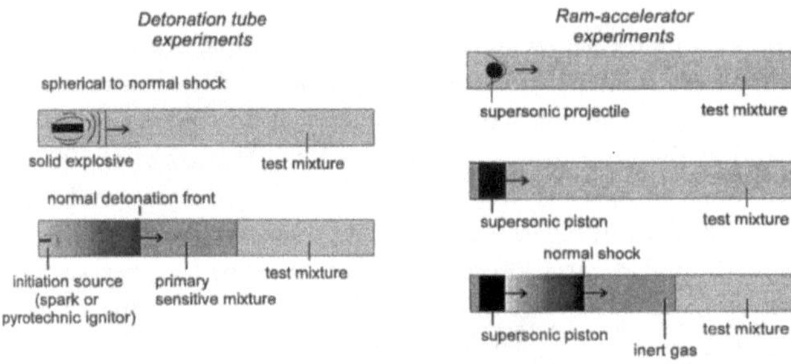

Fig. 3. Shock initiation techniques

This technique, also called normal shock initiation with intermediate inert gas, is that used at ISL (Legendre 1996) with the RAMAC 90 facility (Fig. 4). This experimental configuration is based on a test section composed of five 3-m-long ram tubes which can be filled either with an inert gas or test mixture. This 16.2-m-long test section is equipped with 99 instrument ports in order to reconstitute the trajectories of piston, shock and detonation fronts by means of respectively magnetic transducers, pressure transducer and light probes (Legendre 1996).

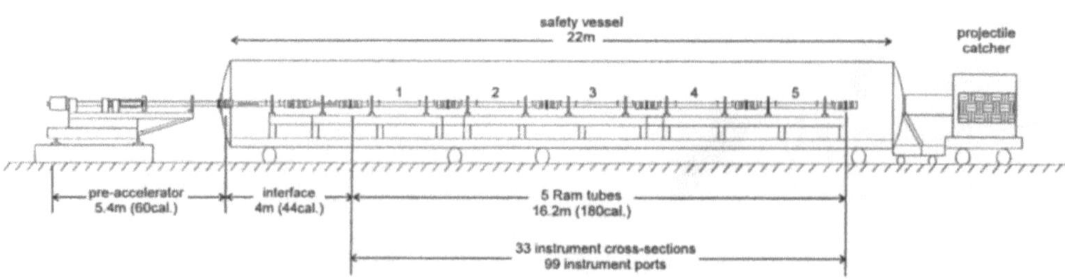

Fig. 4. RAMAC 90 facility

2.1 Simulation of initiation conditions on the body of the projectile The first shock initiation condition we have attempted to simulate is that found on the body of a RAMAC 90 projectile in the previously described reference conditions which yields a shock at Mach 3.5 (equal to the projectile Mach number) and a maximum peak pressure of 25 MPa. In that case, it must be emphasized that a normal shock is not the most representative situation to simulate the complex highly three dimensional oblique shock field around the undercalibrated projectile body. However we tried to reconstitute only the discontinuity of pressure through the shock front and absolutely not the flow conditions behind the shock. To simulate this pressure discontinuity, a 1.35 kg 90-mm-caliber magnesium piston has been injected at an initial velocity of 1.4 km/s at the entrance of the inert section composed of the first 3 ram tubes (9.9m) filled with nitrogen at an initial pressure of 1.5 MPa. The transmitted shock into the test section composed of the remaining 2 ram tubes (6.6m) filled with the test mixture at an initial pressure of 2.5 MPa is characterized by a shock Mach number of 3.05 (1.1 km/s) and a peak pressure of 25 MPa. With these experimental conditions, three mixtures have been investigated for an initial pressure of 2.5 MPa and ambient temperature. The first two mixtures currently used at ISL for RAMAC 90 experiments in the thermally choked propulsion mode (Giraud et al. 1995, 1997), i.e. $3.2CH_4+2O_2+7.5N_2$ and $3CH_4+2O_2+10N_2$, did not yield the initiation of a detonation. This is consistent with the observation that during operational RAMAC 90 experiments, when the diffuser is properly started, no initiation of a detonation occurs on the body of the projectile (at least at the considered Mach number of ~3.5).

In order to check the reliability of this normal shock initiation technique, a third mixture has been investigated, namely $\sim 2CH_4+2O_2+3N_2$, which has been chosen at purpose inside the detonable area determined in detonation tube. In the previous experimental configuration for normal shock initiation, this mixture has been initiated and a detonation propagated along the test section as presented on the trajectories plot on Fig. 5. It can be observed that the detonation initiation of the test gas mixture is achieved within the first meter.

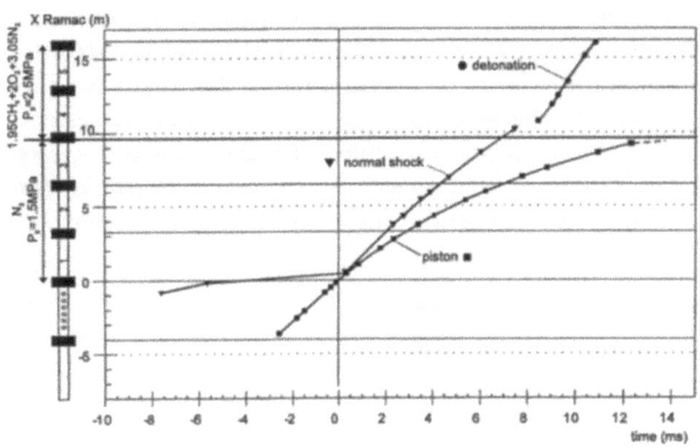

Fig. 5. *x-t* **trajectories: 90-mm-caliber piston, tubes 1-3: 1.5 MPa** N_2**, tubes 4-5: 2.5 MPa** $1.95CH_4+2O_2+3.05N_2$

2.2. Simulation of initiation conditions on the normal shock detached from the sabot In the reference conditions, the characteristics of the normal shock detached from the sabot and situated in the projectile wake are Mach number of 3.25 (1.2 km/s) and a maximum peak pressure of 35 MPa (Fig. 2). To simulate those conditions, a 1.35 kg magnesium piston was injected at a velocity of 1.4 km/s at the entrance of the inert section composed of the first 2 ram tubes (6.6 m) filled with nitrogen at an initial pressure of 2.5 MPa. The remaining three ram tubes (9.9 m) then contained the test gas mixture at an initial pressure of 2.5 MPa. For these conditions, the normal shock transmitted into the test section propagates at Mach number 3.2 (1.2 km/s) and its maximum peak pressure is 36 MPa. The normal shock initiation technique is, in this case, relatively consistent with the real initiation conditions ahead of the sabot. The test mixture investigated in that case was the current nitrogen-based mixture used at the University of Washington in the thermally choked and transdetonative propulsion mode, i.e., $2.8CH_4+2O_2+5.7N_2$ (Knowlen 1991). In our experimental conditions, no initiation either of detonation or even combustion could have been obtained as is shown on the trajectories plotted in Fig. 6.

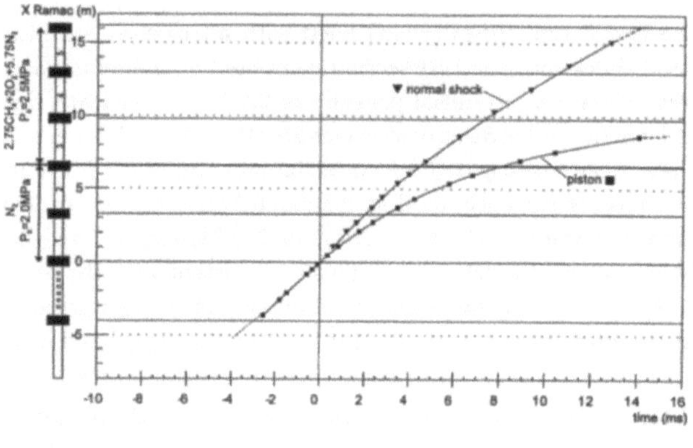

Fig. 6. x-t trajectories: 90-mm-caliber piston, tubes 1-2: 2.5 MPa N_2, tubes 3-5: 2.5 MPa $2.8CH_4+2O_2+5.7N_2$

It appears that although this test mixture could not be initiated with initiation conditions meant to be well representative of the real ramac conditions, it actually was initiated during ramac experiment. This was the reason why a more thorough investigation of the initiation phenomena found in the wake of a ramac projectile was requested. In order to unambiguously decouple the initiation conditions on the projectile and in the wake on the shock detached from the sabot, a so-called delayed initiation RAMAC 90 experiment has been carried out. In that case the RAMAC 90 reference projectile is injected at an initial velocity of 1.4 km/s into a first ram tube filled with nitrogen at an initial pressure of 2.0 MPa. This dense inert gas yielded a separation distance between projectile base and sabot of 1.6m at the time the projectile throat enters the combustible mixture situated in the next 2 ram tubes. This large separation distance allows to determine more accurately the position of initiation. Trajectories of projectile, sabot, shock front and combustion front are therefore represented on Fig. 7.

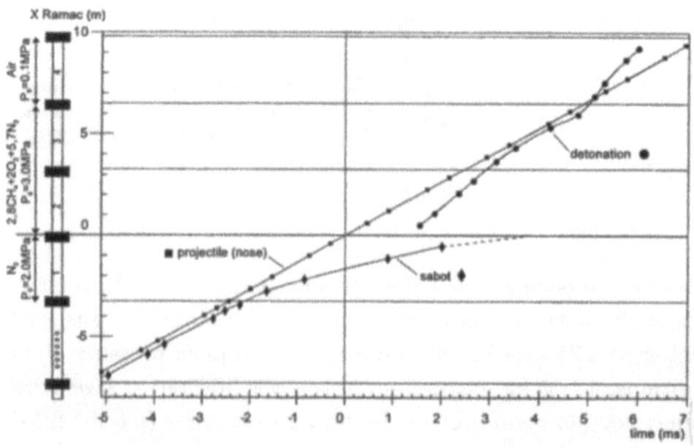

Fig. 7. x-t trajectories : 90-mm-caliber reference ram projectile, tube 1: 2.0 MPa N_2, tubes 2-3: 2.5 MPa $2.8CH_4+2O_2+5.7N_2$, tube 4: 0.1 MPa air

It appears that initiation of the combustion occurs in the projectile wake. A closer look at the trajectories around the entrance of the combustible mixture (Fig. 8) shows that initiation can be attributed to the normal shock detached from the sabot.

The detonation front propagates in the projectile wake at a quasi constant velocity of 2020m/s until it catches up the projectile base and decelerates down to the projectile velocity (Fig. 9). As soon as the projectile enters the last ram tube at ambient air, the combustion zone overtakes the projectile throat.

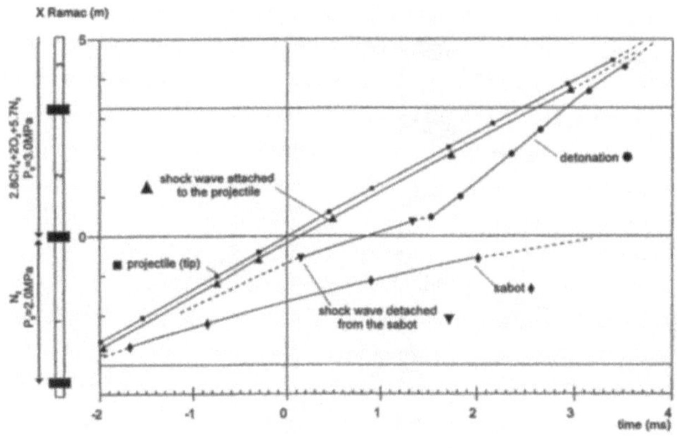

Fig. 8. x-t trajectories at the entrance of reactive mixture: 90-mm-caliber reference ram projectile, tube 1: 2.0 MPa N_2, tubes 2-3: 2.5 MPa $2.8CH_4+2O_2+5.7N_2$, tube 4: 0.1 MPa air

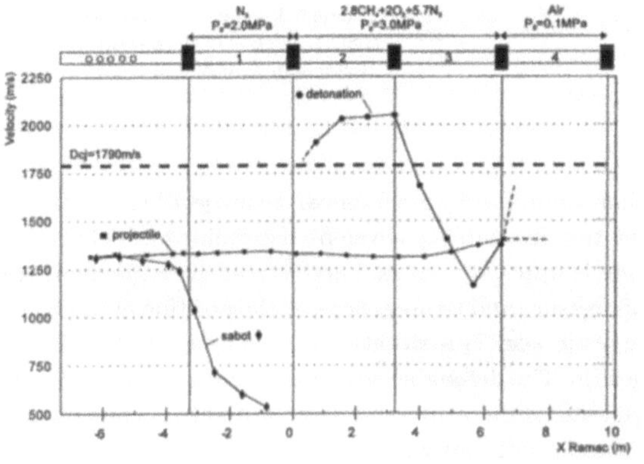

Fig. 9. Velocity profile: 90-mm-caliber reference ram projectile, tube 1: 2.0 MPa N_2, tubes 2-3: 2.5 MPa $2.8CH_4+2O_2+5.7N_2$, tube 4: 0.1 MPa air

The velocity profile shows that the quasi constant detonation velocity of 2020m/s is well above the computed Chapman-Jouguet detonation velocity of 1790m/s. Calculations have been undertaken by means of the Quartet code (Heuzé 1995) taking into account a real Gas equation of state (namely that of Boltzmann based on a virial development) to characterize the properties of the combustion products. The use of such real gas equation of state in this range of initial pressure for methane-oxygen-nitrogen has been validated through detonation tube experiments (Legendre 1996). The discrepancy between experimental and calculated detonation velocities can be explained by the values of pressure and temperature in the projectile wake which are actually higher than those of the unreacted mixture situated upstream the projectile as used for the calculations. In order to evaluate more thoroughly the combined effect of initial pressure and initial temperature, a contour plot of the Chapman-Jouguet detonation velocity has been computed by means of the Quartet thermochemical code and using the Boltzmann equation of state for the combustion products (Fig. 10). CFD computations of the flow characteristics in the wake in the reference conditions have been carried by means of the Tascflow code. For these given conditions, although the flow is highly three-dimensional, pressure and temperature mean values in the wake have been estimated at 10 MPa and 450 K respectively. On the basis of these values, a detonation velocity of 1950m/s was calculated. Finally the experimental detonation velocity of 2020m/s, which remains higher than the latter value can be accounted for a highly turbulent flow which, in turn, increases the detonation velocity.

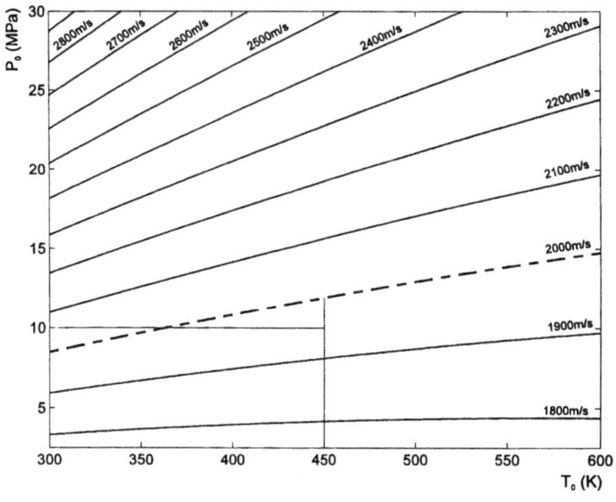

Fig. 10. Computed Chapman-Jouguet detonation velocities, $2.8CH_4+2O_2+5.7N_2$ mixtures, Boltzmann equation of state

3. Conclusion

This paper was intended to simulate initiation conditions by a reference ram projectile at a velocity around Mach 3.5. It can be concluded that the detonability envelope determined in detonation tube with direct pyrotechnic initiation mode is applicable to ram accelerator experiments if the initial conditions of the mixture, in terms of pressure and temperature, at the position of initiation are the same as those at rest upstream the projectile. This situation can be accounted for in the case of an initiation at the body of the projectile. The detonable envelope remains subdeterminant when the mixture exhibits higher pressure and temperature as well as a highly turbulent flow at the position of initiation. For instance, these conditions which yield an increase of detonation sensitivity, are found in case of an initiation on the normal shock detached from the sabot and situated in the wake of the projectile.

References

Bauer P, Legendre JF (1993) Detonability limits of propellant mixtures used in the Ramac. Proc 14th Int Symp on Ballistics, Québec, Canada, pp 389-398, ISL Rep ISL-PU361/93

Bauer P, Giraud M, Legendre JF, Catoire L (1994) Detonability limits of methane-oxygen mixtures at elevated initial pressures. Propellants, Explosives, Pyrotechnics 19:311-314, ISL Rep ISL-CO222/92

Bauer P, Presles H.N, Heuzé O, Legendre JF (1995) Prediction of detonation characteristics of dense gaseous explosives on the basis of virial equations of state. 20th Int Pyrotechnics Seminar, Colorado Springs, CO-USA, ISL Rep ISL-PU308/95

Bauer P, Legendre JF, Knowlen C, Higgins AJ (1996) Detonation of insensitive dense gaseous mixtures in tubes. AIAA paper 96-2682, ISL Rep ISL- PU304/97

Bauer P, Legendre JF, Knowlen C, Higgins AJ (1998) A review of detonation initiation techniques for insensitive dense methane-oxygen-nitrogen mixtures. J de Physique III, to be published

Bauer P, Knowlen C, Higgins AJ, Legendre JF (1997) Detonation initiation of insensitive dense gaseous mixtures by piston impact. Proc 21st Int Symp on Shock Waves, Great Keppel Island, Australia, Paper 1609, ISL Rep ISL-PU353/97

Giraud M, Legendre JF, Simon G, Catoire L (1992) Ram accelerator in 90mm caliber : first results concerning the scale effect in the thermally choked propulsion mode. 13th Int Symp on Ballistics, Stockholm, Sweden, ISL Rep ISL-CO210/92

Giraud M, Legendre JF, Simon G (1993) RAMAC 90: Experimental studies and results in 90mm caliber, length 108 calibers. Proc 1st Int Workshop on Ram Accelerators, Saint-Louis, France, ISL Rep ISL-PU360/93

Giraud M, Legendre JF, Simon G, Henner M, Voisin D (1995) Ramac in 90mm caliber: Starting process, control of the ignition location and performances in the thermally choked propulsion mode. Proc 2nd Int Workshop on Ram Accelerators, Seattle, WA, USA, ISL Rep ISL-PU349/95

Giraud M, Legendre JF, Henner M (1998) RAMAC in subdetonative propulsion mode - State of the ISL studies. In: Takayama K, Sasoh A (eds) Ram Accelerators, Springer-Verlag, Heidelberg, pp 65-77

Henner M (1996) Contribution to the design of a new Ramac projectile: Modelisation and experiments. 47th meeting of the Aeroballistic Range Association, Saint-Louis, France, ISL Rep ISL-PU361/96

Henner M, Legendre JF, Giraud M, Bauer P (1997a) Initiation of reactive mixtures in a ram accelerator. AIAA paper 97-3173

Henner M, Giraud M, Legendre JF, Berner C (1998) CFD computations of steady and non reactive flow around fin-guided RAM projectiles. In: Takayama K, Sasoh A (eds) Ram Accelerators, Springer-Verlag, Heidelberg, pp 325-332

Heuzé O (1985) Contribution au calcul des caractéristiques de détonation de substances explosives gazeuses ou condensées. Ph.D dissertation, University of Poitiers, Poitiers, France

Higgins AJ (1995) Detonation initiation by supersonic blunt bodies. 15th ICDERS Proceedings, Boulder, CO-USA

Knowlen C (1991) Theoretical and experimental investigation of the thermodynamics of the thermally choked ram accelerator. Ph.D dissertation, Univ Washington, Seattle, WA, USA

Kruczynski D (1993) Analysis of firings in a 120mm ram accelerator at high pressures. Proc 1st Int Workshop on Ram Accelerators, Saint-Louis, France

Legendre JF, Giraud M, Bauer P (1993) Effect of inert additives on the detonation properties of dense gaseous explosives. Proc 1st Int Workshop on Ram Accelerators, Saint-Louis, France, ISL Rep ISL-PU359/93

Legendre JF, Giraud M, Bauer P, Voisin D (1995) 90L35 detonation tube experiments: influence of diluent nature on the detonation characteristics of dense methane-based gaseous explosive mixtures. Proc 2nd Int Workshop on Ram Accelerators, Seattle, WA, USA, ISL Rep ISL-PU348/95

Legendre JF (1996) Contribution à l'étude de la sensibilité et des caractéristiques de détonation de mélanges explosifs gazeux denses à base de méthane utilisés pour la propulsion dans les accélérateurs à effet stato. Ph.D dissertation, University of Poitiers, Poitiers, France, ISL Rep ISL-R101/97

Legendre JF, Giraud M, Bauer P (1998) Detonation properties of dense methane-based gaseous explosive mixtures: Application to ram accelerators. Shock Waves, to be published

Hypersonic blunt body in chemically reacting flows

Comparison of numerical simulations and PLIF imaging results of hypersonic inert and reactive flows around blunt projectiles

K. Toshimitsu[1], A. Matsuo[2], M. R. Kamel[3], C. I. Morris[3], R. K. Hanson[3]
[1]Department of Aeronautics and Astronautics, Kyushu University, Fukuoka 812-81, Japan
[2]Department of Mechanical Engineering, Keio University, Yokohama 223, Japan
[3]Department of Mechanical Engineering, Stanford University, Stanford, CA 94305-3032, U.S.A.

Abstract. This paper shows comparisons between computational fluid dynamics (CFD) calculations and PLIF and Schlieren measurements of inert and reactive hypersonic flows around 2D and axisymmetric bodies. In particular, both hydrogen-oxygen and methane-oxygen chemical reactions are considered for the shock-induced combustion in hypersonic flows. The hydrogen-oxidation mechanism consists of an existing mechanism of 8 reacting species and 19 elementary reactions. The reduced model of the methane-oxidation mechanism is newly derived from the GRI-Mech 1.2 optimized detailed chemical reaction mechanism, and consists of 14 species and 19 chemical reaction steps. Both chemical reaction mechanisms are combined with a point-implicit Euler CFD code. The OH species density distributions of the present numerical calculations and imaging experiments for both mixtures are found to be in qualitative agreement.

Key words: Hypersonic flows, Shock-induced combustion, Hydrogen-oxygen, Methane-oxygen, CFD, PLIF, Blunt projectiles

1. Introduction

It is very important for the optimal design of ram accelerators that the flow characteristics around hypervelocity projectiles be well predicted by computational calculations. The characteristics of hydrogen-oxidation and methane-oxidation reactive flows around supersonic and hypersonic projectiles have been previously investigated. (Ahuja et al. 1996, Nusca 1993, 1995, Li et al. 1994, Matsuo et al. 1995, 1996, Soetrisno et al. 1992, Wilson et al. 1992, 1993 and Yungster et al. 1991, 1994) These works have made clear the important role of the combustion mechanism, unsteadiness, viscous effects and chemical kinetics of reactive flows. However, the comparison of CFD results with experimental data for species distribution profiles has not been carried out to investigate the reactive flow fields.

The Stanford expansion tube facility was used to obtain spatially and temporally resolved optical measurements of hypersonic reactive flow fields around several projectiles. Qualitative (Kamel et al. 1995, 1997) and quantitative (Houwing et al. 1996) planar laser-induced fluorescence (PLIF) imaging experiments of hypersonic inert and reactive flows over projectiles have been performed for comparison with the CFD calculations.

In this paper, the flow fields around a two-dimensional rod and axisymmetric hemisphere projectiles are investigated using a CFD program based on Matsuo's code (Matsuo et al. 1995, 1996). A hydrogen-oxidation mechanism of 8 species and 19 chemical reaction steps is used for the hydrogen-oxygen-nitrogen mixtures. A reduced chemical reaction mechanism of 14 species and 19 chemical reaction steps is proposed for the CFD simulations of methane-oxygen-nitrogen mixtures. Through the comparison of CFD calculations and PLIF imaging results, the code accuracy, chemical kinetics and the flow field characteristics are investigated.

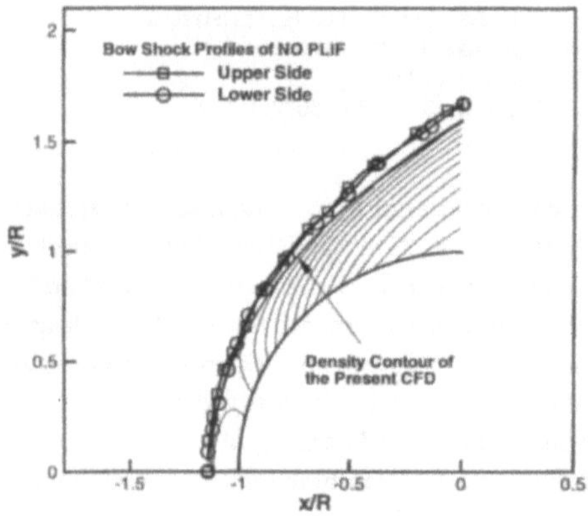

Fig. 1. Comparison of bow shock profiles by CFD calculation and NO PLIF measurement around the hemispheric body in condition 1

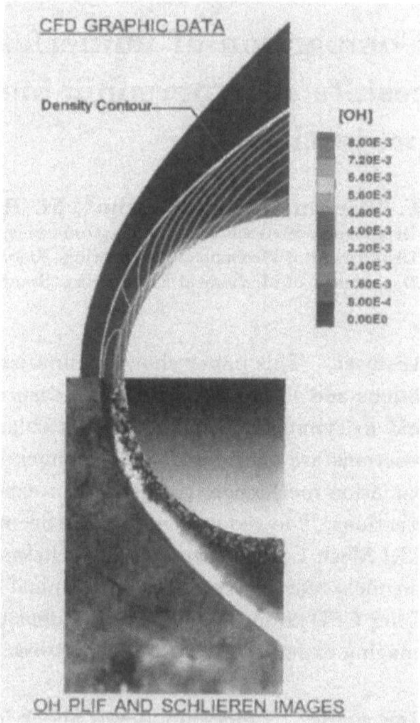

Fig. 2. Comparison of OH species density distributions and bow shock profiles by CFD calculation, OH PLIF and Schlieren imaging results around the hemispheric body in condition 2

2. Numerical formulation

The computational simulations are conducted on the following Euler equation for two-dimensional and axisymmetric flows with the species conservation equations,

$$\frac{\partial \hat{\mathbf{Q}}}{\partial \tau} + \frac{\partial \hat{\mathbf{E}}}{\partial \xi} + \frac{\partial \hat{\mathbf{F}}}{\partial \eta} = \hat{\mathbf{S}} + \hat{\mathbf{H}} \tag{1}$$

where $\hat{\mathbf{E}}$ and $\hat{\mathbf{F}}$ are inviscid flux vector in ξ and η coordinates, respectively. The term $\hat{\mathbf{S}}$ means the chemical reaction source vector, and $\hat{\mathbf{H}}$ is the axisymmetric source term. For a two-dimensional projectile, $\hat{\mathbf{H}}$ is equal to $\mathbf{0}$. The numerical analyses are based on non-MUSCL type total variation diminishing (TVD, Yee 1987) upwind explicit scheme for the conservation equation and an implicit integration of the chemical reaction processes, i.e. so-called point-implicit scheme.

3. Numerical results and reaction models

In the following three sections, the numerical results are presented according to the type of gas used in the calculations - air, hydrogen-oxygen-nitrogen and methane-oxygen-nitrogen mixtures. The numerical and experimental conditions investigated in this work are listed in Table 1. In order to remove the dependence of mesh size on the numerical simulations, various mesh sizes are used in similar simulations and their results compared. Convergence was satisfied when the

Table 1. Investigated conditions

No.	U_∞ [m/s(M)]	p_∞ [kPa (psi)]	T_∞ [K]	Mix.(mole %)	Mesh Size	Projectile	Experiment
				O_2/N_2			
1	2230 (7.0)	6.4 (0.94)	280	21.0/79.0 (inert)	76 × 51 151 × 151	hemispere (d=19mm)	NO PLIF
				$N_2/O_2/N_2$			
2	1960 (5.2)	11.2 (1.65)	350	10.0/5.0/85.0	101 × 101 151 × 151	hemispere (d=19mm)	OH PLIF
				$CH_4/O_2/N_2$			
3	2330 (6.8)	50 (7.11)	295	9.5/19.0/71.5	76 × 51 101 × 81 151 × 101	2-D rod (d=1, 3 and 7mm)	
4	2150 (6.8)	20 (2.94)	260	16.7/33.3/50.0	151 × 151	hemispere (d=19mm)	
5	ibid.	10 (1.47)	ibid.	ibid.	ibid.	ibid.	
6	ibid.	6.4 (0.91)	ibid.	ibid.	ibid.	ibid.	OH PLIF

The U_∞, p_∞ and T_∞ denote velocity at free stream, static pressure and static temperature, respectively.

value of the numerical residual decreased below at the end of an iteration step. The convergence condition is given by

$$\frac{\sum_{i,j}^{N} |\rho_{i,j}^{t} - \rho_{i,j}^{t+\Delta t}|}{N\rho_\infty} < 10^{-4.5} \tag{2}$$

where i and j are node number, N is total node number of the meshes, t is the time level, and ρ_∞ is the density of incoming flow. For the inert gas experiments, NO PLIF and Schlieren imaging measurements have been applied to obtain the shock profiles for hypersonic flows around a hemispherical cylinder. OH PLIF and Schlieren imaging measurements were used for the reactive cases, with the OH PLIF showing the regions of combustion and Schlieren imaging used to visualize the shock wave.

3.1 Inert gas (air) For the hemispherical body of condition 1, a 151 × 151 mesh size was used to calculate the density contour and to show the bow shock shape and position. The result was compared to NO PLIF measurements as shown in Fig. 1. The agreement between the calculated shock position and the imaged one is good on the center line (y/R=0). The described example indicates that the CFD calculation is useful to predict the inert-gas flow fields around the axisymmetric hemisphere projectile.

3.2 H_2-O_2 shock-induced combustion

The Hydrogen-Oxidation mechanism used in this work consists of 8 species (H_2, O_2, H, O, OH, H_2O, HO_2 and H_2O_2) and 19 chemical reaction steps, which are shown in Matsuo et al. 1995. It is reduced from the mechanisms of Jachimowski (1988) and Wilson and MacComack (1992) since the nitrogen reactions are not important at Mach numbers less than 5.

Figure 2 shows the OH species density distributions and bow shock profiles calculated using CFD and OH PLIF and Schlieren measurements (Kamel et al. 1997) around a hemispheric body in

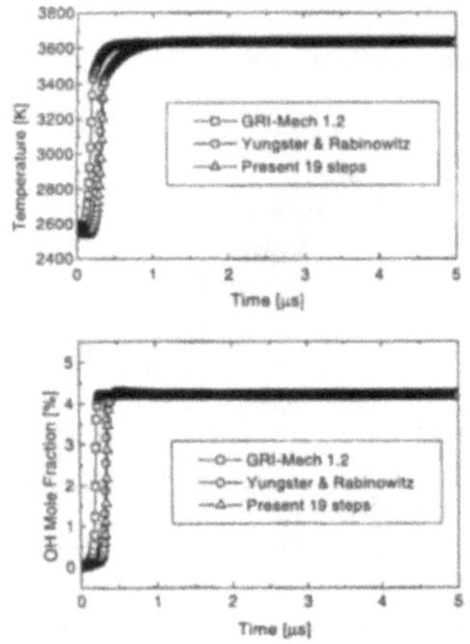

Fig. 3. Temperatures and OH mole fractions of GRI-Mech 1.2, Yungster and Rabinowitz and present chemical model at 2600 K and 3.0 MPa

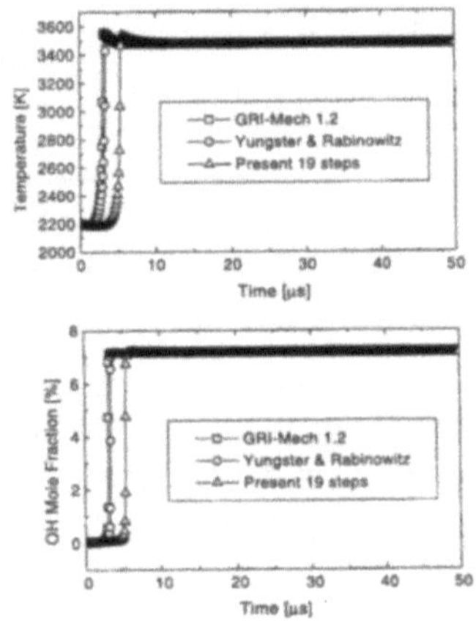

Fig. 4. Temperatures and OH mole fractions of GRI-Mech 1.2, Yungster and Rabinowitz and present chemical model at 2200 K and 0.4 MPa

condition 2. Both the bow shock and OH species profiles are in good agreement. This demonstrate that the CFD code can correctly simulate the hypersonic reactive flow field experiments (shock-induced combustion of a hydrogen-oxygen-nitrogen mixture) generated in the Stanford expansion tube facility.

3.3 CH_4-O_2 shock-induced combustion

3.3.1 Methane-oxygen reaction mechanism Many detailed and reduced mechanisms of methane oxidation chemical reaction have been suggested. For example, the 19 species 52 reactions mechanism proposed by Yungster and Rabinowitz (1994) has a good accuracy however, it is still complex and requires large computational power, for example a CRAY parallel supercomputer, to calculate the hypersonic reactive flow field. In this section, a new reduced chemical kinetic model is proposed in order to calculate (using a standard computer, for example a Hewlett Packard HP9000/712) the characteristics of methane-oxidation reactive flows under 3.0 MPa pressure. The chemical reaction model is reduced from the full mechanism of GRI-Mech 1.2 (Frenklach et al. 1995) through the sensitivity analyses of GRI-Mech 1.2 (Petersen et at 1995) and the Senkin program (Lutz et al. 1988). The mechanism consists of 14 species (H_2, O_2, H, O, OH, H_2O, CH_3, CH_4, CO, CO_2, HCO, CH_2O, CH_3O and C_2H_6) and 19 chemical reaction steps. Its rate coefficients are listed in Table 2. In order to compare the accuracy of the different chemical kinetic models, temperature and OH mole fraction profiles are calculated using Chemkin at two different conditions: (a) 2600 K and 3.0 MPa, and (b) 2200 K and 0.4 MPa, under constant internal energy and volume. The three different chemical kinetic models used in this comparative study are: the GRI-Mech 1.2 mechanism, the Yungster and Rabinowitz mechanism and the present 14 species 19 chemical reaction model with pressure dependence of rate coefficients of the Troe formulas. Those are derived from the data at the stagnation point of the O_2-N_2 non reactive flow in conditions 3 and 6 (Table 1) by the present CFD analyses. Figures 3 and 4 show temperature and OH mole fraction in conditions (a) and (b) respectively. Those of the GRI-Mech 1.2 mechanism, the Yungster and Rabinowitz mechanism and the present 19 steps one are in good agreement in condition (a). In condition (b), the ignition times of the present 19 steps model exhibit delay of $2\mu s$ from the ones of both the GRI-Mech 1.2 and Yungster and Rabinowitz mechanisms. However, in

Table 2. CH_4-O_2 Reaction Mechanism

k	Reaction	A	n	E
1	$H + O_2 = O + OH$	8.300×10^{13}	0.0	14413
2	$O + H_2 = H + OH$	5.000×10^{4}	2.670	6290
3	$OH + H_2 = H + H_2O$	2.160×10^{8}	1.510	3430
4	$H + OH + M = H_2O + M$	2.200×10^{22}	-2.000	0
5	$OH + CO = H + CO_2$	4.760×10^{7}	1.228	70
6	$H + HCO = H_2 + CO$	7.340×10^{13}	0.0	0
7	$OH + HCO = H_2O + CO$	5.000×10^{13}	0.0	0
8	$HCO + H_2O = H + CO + H_2O$	2.244×10^{18}	-1.000	17000
9	$H + CH_2O = HCO + H_2$	2.300×10^{10}	1.050	3275
10	$O + CH_2O = OH + HCO$	3.900×10^{13}	0.0	3540
11	$OH + CH_2O = HCO + H_2O$	3.430×10^{9}	1.180	-447
12	$O + CH_3 = H + CH_2O$	8.430×10^{13}	0.0	0
13	$CH_3 + O_2 = O + CH_3O$	2.675×10^{13}	0.0	28800
14	$H + CH_3$ (+M) $= CH_4$ (+M)			
	k_∞	1.270×10^{16}	-0.630	383
	k_0	2.477×10^{33}	-4.760	2440
	Troe/ 0.7830, 74.00, 2941.00, 6964.00/			
15	$H + CH_4 = CH_3 + H_2$	6.600×10^{8}	1.620	10840
16	$O + CH_4 = OH + CH_3$	1.020×10^{9}	1.500	8600
17	$OH + CH_4 = CH_3 + H_2O$	1.000×10^{8}	1.600	3120
18	$H + CH_2O$ (+M) $= CH_3O$ (+M)			
	k_∞	5.400×10^{11}	0.454	2600
	k_0	2.200×10^{30}	-4.800	5560
	Troe/ 0.7580, 94.00, 1555.00, 4200.00/			
19	$2CH_3$ (+M) $= C_2H_6$ (+M)			
	k_∞	2.120×10^{16}	-0.970	620
	k_0	1.770×10^{50}	-9.670	6220
	Troe/ 0.5325, 151.00, 1038.00, 4970.00/			

$k_f = AT^n exp(-E_k/RT)$; units are in second, moles, centimeters, calories, and Kelvins. Third body efficiencies: (4) H_2 = 0.73, H_2O = 3.65, CH_4 = 2.00, C_2H_6 = 3.00, (14) H_2 = 2.00, H_2O = 6.00, CH_4 = 2.00, CO = 1.50, CO_2 = 2.00, C_2H_6 = 3.00, (18) H_2 = 2.00, H_2O = 6.00, CH_4 = 2.00, CO = 1.50, CO_2 = 2.00, C_2H_6 = 3.00, (19) H_2 = 2.00, H_2O = 6.00, CH_4 = 2.00, CO = 1.50, CO_2 = 2.00, C_2H_6 = 3.00.

condition (b), i.e. in condition 6, the delay time may cause approximately 1mm error of reaction front stand-off distance. That means the delay effect on the reaction front position is small. The effects on the CFD simulations are discussed in the following sub-sections 3.3.2 and 3.3.3.

3.3.2 Rods Numerical simulations of the experiments on two-dimensional rods conducted by Srulijes et al. (1992) at the French-German ISL Institute are presented by Yungster and Rabinowitz (1994), Soetrisno et al. (1992) and ourselves. In our calculations, the 14 species and 19 steps reaction model with Troe formulas is used. The numerical results obtained using three different mesh sizes (76 × 51, 101 × 81 and 151 × 101) are almost the same. This implies that the numerical simulations are not dependent on the mesh sizes in these steady combustion cases.

The two-dimensional rods were placed in a stoichiometric methane-air mixture at super-detonative velocity 2330 m/s in condition 3. The computed Chapman-Jouguet detonation speed is 1800 m/s. Figure 5 shows the computational results for three cylindrical rods of the various

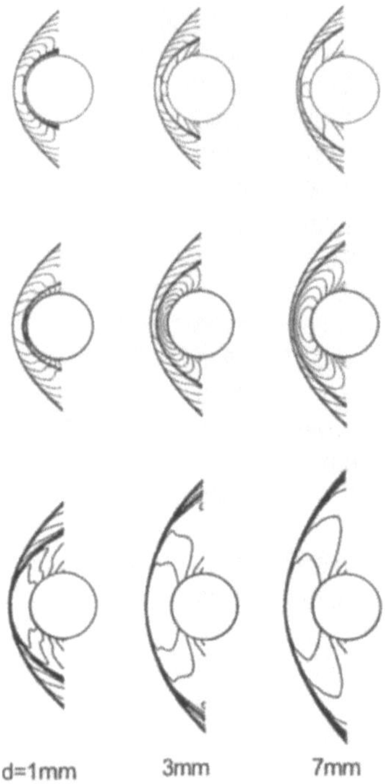

Fig. 5. Temperature contours around the two-dimensional rods by the present analyses (upper rank and mesh size 151 × 101), Yungster and Robinowitz (middle rank and mesh size 91 × 91) and Soetrisno et al. (lower rank and mesh size 72 × 65) in condition 3. The rod diameters are various as 1mm, 3mm and 7mm

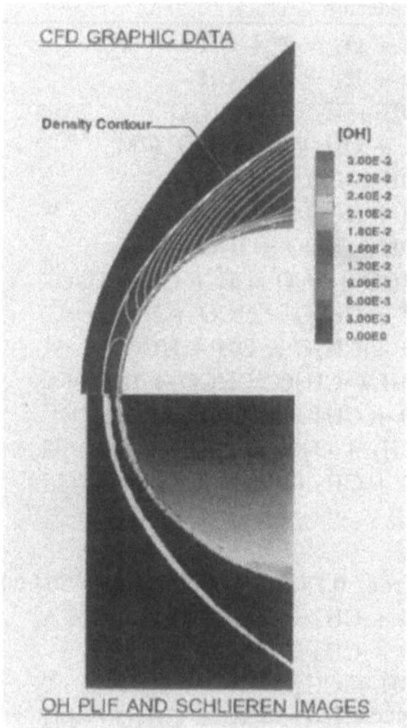

Fig. 7. Comparison of OH species density distributions and bow shock profiles by CFD calculation, OH PLIF and Schlieren imaging results around the hemispheric body in condition 6

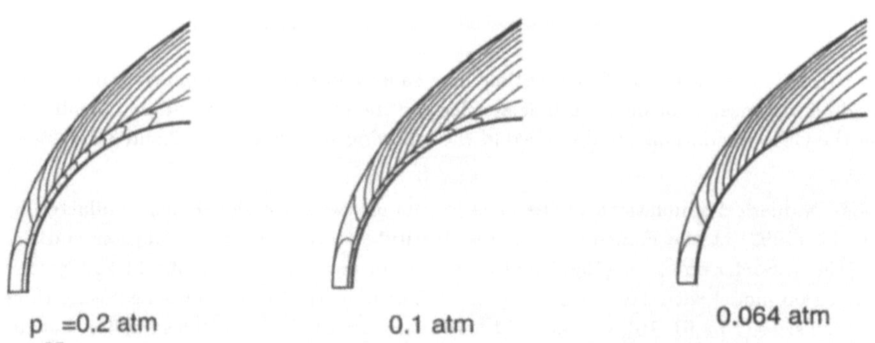

Fig. 6. Comparison of density contours around the hemispheric bodies in condition 4, 5 and 6. The free stream pressures are various as 20 kPa, 10 kPa and 6.4 kPa

diameters of 1 mm, 3 mm, and 7 mm. Yungster and Robinowitz and our computations predict the de-coupled bow shock and reaction front profiles; however, the conducted computations by Soetrisno et al. using the 9 species 12 steps model show the fully coupled shock-deflagration wave. This demonstrates the ability of the reduced 19 steps model to qualitatively predict the de-coupled bow shock and reaction front profiles for the three rods. However the temperature profiles behind the deflagration wave calculated using the 19 step model are different from the ones obtained using the Yungster and Robinowitz model. This is because the 14 species and 19 steps model predicted lower heat release and temperature distribution than Yungsters' one.

3.3.3 Hemispheric projectiles Figure 6 shows the density counters of the three numerical simulations various free stream pressures 20 kPa, 10 kPa and 6.4 kPa in condition 4, 5 and 6, respectively. Because the PLIF experimental condition 6 is eight times lower than one of the previous rod simulation (condition 3). The numerical simulations at $p_\infty = 20$ and 10 kPa show reaction fronts however, the one at 6.4 kPa does not indicate the reaction front. It is because that the Troe formulas may not activate the recombination reaction steps 14, 18 and 19 at the low pressure. It means that the pressure-dependent reaction coefficients in the reduced mechanism cause an incorrect ignition condition at the very low pressure. In order to avoid the incorrect ignition, we will omit the Troe formulas from the reduced mechanism under pressure 10 kPa.

Figure 7 shows the comparison of experimental results (Kamel et al. 1995) with the numerical simulation using the reduced model of Table 2 without the pressure dependence of rate coefficients of the Troe formulas at condition 6 of the superdetonative mode. The computed C-J velocity is 1958 m/s. Both OH species distributions and bow shock profiles are in qualitative agreement.

Here, these calculations required 6920 iterations and approximately 10 hours CPU time on a Hewlett Packard HP 9000/712. It means that this calculation need no large computer power.

4. Conclusions

The comparisons between computational fluid dynamics calculations and PLIF and Schlieren measurements of air and shock-induced combustion around blunt projectiles are presented. The hydrogen-oxygen-nitrogen and methane-oxygen-nitrogen mixtures are considered for the combustible flows. In particular, a new reduced model consisting of 14 reaction species and 19 chemical steps is proposed to simulate the methane combustion for the expansion tube experiments. The numerical calculations and experimental imaging results for both mixtures are in qualitative agreement. It is concluded that the CFD code can be satisfactorily used for estimating the characteristics of bow shock and reaction front profiles of flow fields around hypersonic projectiles.

Acknowledgement

The authors gratefully acknowledge the contributions of Dr. D. F. Davidson (Stanford Univ.), and Dr. Michael Nusca (ARL) to this investigation. Financial support for the first author's participation in the research was obtained through the Ministry of Education in Japan. This work was supported in part by a grant of HPC time from the DoD HPC center, CEWES MSRC C90.

References

Ahuja JK, Kumar A, Singh DJ, Tiwari SN (1996) Simulation of shock-induced combustion past blunt projectiles using shock-fitting technique. J Prop Power 12:518-526

Frenklach M, Wang H, Goldenberg M, Smith GP, Golden DM, Bowman CT, Hanson RK, Gardiner WC, Lissianski V (1995) GRI-Mech - an optimized detailed chemical reaction mechanism for methane combustion. GRI-95/0058, Topical Rep

Gilbert RG, Luther K, Troe J (1983) Ber Bunsenges. Phys Chem 87:169

Houwing AFP, Kamel MR, Morris CI, Wehe SD, Boyce RR, Thurber MC, Hanson RK (1996) PLIF imaging and thermometry of NO/N_2 shock layer flows in an expansion Tube. AIAA paper 96-0537

Jachimowski CJ (1988) An analytical study of the hydrogen-air reaction mechanism with application to scramjet combustion. NASA TP-2791

Kamel MR, Morris CI, Stouklov IG, Hanson RK (1995) Imaging of hypersonic reactive flow around cylinders and wedges. Paper No 95F-196, Western State Section/The combustion Institute 1995 Fall Meeting in Stanford University, October 30-31

Kamel MR, Morris CI, Hanson RK (1997) Simulation PLIF and schlieren imaging of hypersonic reactive flows around blunted cylinders. AIAA paper 97-0913

Kee RJ, Rupley FM, Miller JA (1989) Chemkin-II: A fortran chemical kinetics package for the analysis of gas-phase chemical kinetics. Sandia Rep SAND89-8009, UC-4001, Sandia Nat Lab, Livermore, CA

Li C, Kailasanath K, Oran ES (1994) Detonation structures on ram-accelerator projectiles. AIAA paper 94-0551

Lutz AE, Rupley FM, Miller JA (1988) SENKIN: A fortran program for predicting homogeneous gas phase chemical kinetics with sensitivity analysis. Rep SAND87-8248, Sandia Nat Lab, Livermore, CA

Matsuo A, Fujii K, Fujiwara T (1995) Flow features of shock-induced combustion around projectile traveling at hypervelocities. AIAA J 33:1056-1063

Matsuo A, Fujii K (1996) Detailed mechanism of the unsteady combustion around hypersonic projectiles. AIAA J 34:2082-2089

Nusca MJ (1993) Numerical simulation of fluid dynamics with finite-rate and equilibrium combustion kinetics for the 120-mm ram accelerator. AIAA paper 93-2182

Nusca MJ (1995) Investigation of ram accelerator flows for high pressure mixtures of various chemical compositions. AIAA paper 96-2946

Petersen EL, Davidson DF, Röhrig M, Hanson RK, Bowman CT (1995) A shock tube study of high-pressure methane oxidation. Paper No. 95F-153, Western State Section/The combustion Inst 1995 Fall Meeting in Stanford University, October 30-31

Soetrisno M, Imlay ST, Roberts DW (1992) Numerical simulations of the transdetonative ram accelerator combusting flow field on a parallel computer. AIAA paper 92-3249

Srulijes J, Smeets G, Seiler F (1992) Expansion tube experiments for the investigation of ram-accelerator-related combustion and gasdynamic problems. AIAA paper 92-3246

Wilson GJ, MacComack RW (1992) Modeling supersonic combustion using a fully implicit numerical method. AIAA J 30:1008-1015

Wilson GJ, Sussman MA (1993) Computation of unsteady shock-induced combustion using logarithmic species conservation equations. AIAA J 31:294-301

Yee HC (1987) Upwind and symmetric shock capturing schemes. NASA TM 89464

Yungster S, Eberhardt S, Bruckner AP (1991) Numerical simulation of hypervelocity projectiles in detonable gases. AIAA J 29:187-199

Yungster S, Rabinowitz MJ (1994) Computation of shock-induced combustion using a detailed methane-air mechanism. AIAA J 10:609-617

Comparisons of numerical methods for the analysis of unsteady shock-induced combustion

J. Y. Choi, I. S. Jeung, Y. Yoon
Department of Aerospace Engineering, Seoul National University, Seoul 151-742, Korea

Abstract. Numerical experiments were performed to investigate the characteristics of the various numerical approaches in the analysis of periodically unstable shock-induced combustion around a blunt body. Inviscid Euler equations and species conservation equations are used as the governing equations with a detailed chemistry mechanism of hydrogen-air combustion. The base-line numerical method is composed of the third order accurate spatial discretization scheme based on Roe's FDS method and the second order time accurate LU-SGS scheme with exact flux Jacobian splitting and Newton sub-iteration. As a first step of the numerical experiments, simulations of experimental results were conducted to confirm the reliability of base-line method. Secondly, some numerical experiments were conducted for a selected experimental case to compare time integration strategies. The effects of the order of time integration, the number of sub-iterations and the use of approximate flux Jacobian splitting were examined. The point-implicit method and Crank-Nicolson type fully implicit method were compared with base-line time integration method through time step refinement studies. In view point of spatial discretization method, the use of various limiter functions and the use of AUSM based flux splitting methods were examined. Grid refinement study was also performed in addition.

Keyword: Unstable shock-induced combustion, CFD, Numerical experiment

1. Introduction

After the experimental observations in 1960s and 1970s, many researchers showed their interest on shock-induced combustion over the last 30 years. Recently, the shock-induced combustion gains public interests as a promising combustion mechanism for hypersonic propulsion devices such as ODWE (Oblique Detonation Wave Engine, see Shepherd 1994) and ram accelerator (Hertzberg et al. 1988). In 1980s and 1990s, the development in numerical method and computational power make it possible to simulate such kind of combustion phenomena numerically, and the computational fluid dynamics has been used as a major research tool of the phenomena.

The shock-induced combustion is the self-ignited combustion phenomena of premixed gas induced by the flight of hypervelocity projectile in a combustible gas mixture. Thus, the shock-induced combustion flow field is characterized by the hypersonic range flow speed and the finite rate exothermic chemistry behind the bow shock. And, the coupling and the interaction between the bow shock and the reaction wave show the various and distinctive flow features according to the chemical and fluid dynamic conditions. Among the various features of shock-induced combustion, periodically unstable regime would be the most interesting one due to its naturally oscillating characters, as notably shown by Lehr (1972). The oscillating mechanism of shock-induced combustion was suggested by Toong in the 1970s and proved by Matsuo et al.(1993, 1995) by numerical analysis in 1990s. That is, the instability mechanism is a wave interaction mechanism between shock wave, combustion wave, compression waves, expansion wave and projectile surface.

Since the phenomena is the result of wave interactions between the bow shock, combustion wave and body surface, the analysis of the unsteady combustion phenomena requires a stable and time accurate numerical approach that can capture the strong bow shock and weak pressure waves without numerical instability or distortion. In addition, a detailed combustion mechanism that include sufficient number of chemical species and elementary reaction steps is needed to accurately reproduce the experimental results of shock stand-off distance, induction distance and oscillation frequency because these characteristics are determined by finite rate reaction mechanisms. However, the inclusion of more species results an increase of the variables proportional to the number of species and an increase of the matrix inversion time proportional to cubic number of the equations. Moreover, huge grid system is also required to resolve the wave interactions, which results in a severe increase in computational time. Therefore, the analysis of the oscillating combustion phenomena requires enormous memory capacity and computing time.

Thus, the reliable results of simulation begins to appear in early 1990's (Wilson et al. 1993), and the approach of numerical simulation is used at present in the analysis of the unknown shock-induced combustion phenomena for which the experimental or analytic methods cannot be applied adequately. During the previous several years, many researchers performed the simulation of experimentally known shock-induced combustion and reproduced the experimental result with sufficiently good accuracy. However, there are some discrepancies between their simulations and the experimental result, since they used different numerical methods and computational conditions. They explained the discrepancies by the inaccuracy of used chemistry mechanism, numerical method and etc. However, the huge computational expense needed in the analysis prevented the sufficient validational study that is needed in case of using the computational methods in the analysis of unknown physical problems.

In this study, a comparative study of numerical methods is made as a developing process of an efficient computational fluid dynamics code that capable of analyzing unsteady reacting flow. The computational code uses a time accurate fully implicit algorithm and a third order spatially accurate upwind scheme. The validation of the base line method is made from the simulation of experimental result and the results from different choices of numerical approaches are compared with that from the base line method. Grid and time step refinement studies are also presented.

2. Governing equation and chemistry model

The oscillatory phenomena of the shock-induced combustion are basically due to the interaction between shock and combustion wave, and the effects of transport properties are known to be negligible. Therefore, species conservation equations and inviscid Euler equations are used as governing equations of the reacting flow around an axisymmetric blunt body. The conservation form of these governing equation set involving N number of species is written in curvilinear coordinate as follows.

$$\frac{\partial \mathbf{Q}}{\partial t} + \frac{\partial \mathbf{F}}{\partial x} + \frac{\partial \mathbf{G}}{\partial y} + \mathbf{H} = \mathbf{W} \quad \text{where} \quad \mathbf{Q} = \begin{pmatrix} \rho_k \\ \rho u \\ \rho v \\ e \end{pmatrix} \tag{1}$$

Here, $\mathbf{Q}$ is conservative variable vector, $\mathbf{F}$ and $\mathbf{G}$ are flux vectors, $\mathbf{H}$ is axisymmetric term and $\mathbf{W}$ is reaction source term. ρ_k is partial density of k-th species and ρ is total density the sum of the partial densities. u and v are the velocity components and e is the total energy per unit volume. Pressure is evaluated from the ideal gas law for a mixture of thermally perfect gases and temperature is evaluated implicitly from the definition of the total energy with specific heat data. The specific heats are obtained as functions of temperature from NASA thermochemical polynomial data (Gardiner 1984).

In this study, the reduced form of Jachimowski's detailed chemistry mechanism is used for hydrogen/air combustion. Because the nitrogen oxidation process does not have a significant effect on the fluid dynamics of shock-induced combustion, the reaction steps are neglected and nitrogen is assumed as an inert species. This combustion model has been successfully used and validated in a number of previous studies(Matsuo et al. 1995, Yungster et al. 1994).

3. Time integration methods

A fully implicit second order accurate time integration method is used for the analysis of unsteady supersonic reacting flow. Newton-Raphson sub-iteration method is used to ensure the second order time accuracy and stability in case of using a large time step equivalent to CFL number greater than one. The governing equation (1) could be written in discretized form as follows.

$$[c_o\mathbf{I} - \mathbf{R_Q}]_{i,j}^{n+1,m} \Delta\mathbf{Q}_{i,j}^{n+1,m} = -c_o\mathbf{Q}_{i,j}^{n+1,m} - c_1\mathbf{Q}_{i,j}^{n} - c_2\mathbf{Q}_{i,j}^{n-1} + c_o\mathbf{R}\left(\mathbf{Q}_{i,j}^{n+1,m}\right) \tag{2}$$

where $\mathbf{R}$ is residual vector and $\mathbf{R_Q}$ is Jacobian matrix of the residual vector.

$$\mathbf{R}_{i,j} = \mathbf{W}_{i,j} - \mathbf{H}_{i,j} - \tilde{\mathbf{F}}_{i+1/2,j} + \tilde{\mathbf{F}}_{i-1/2,j} - \tilde{\mathbf{G}}_{i,j+1/2} + \tilde{\mathbf{G}}_{i,j-1/2} \tag{3}$$

Here, n is time integration level and m is sub-iteration index. Time marching is performed by assuming the initial guess of conservative variable at n+1 time level $Q^{n+1,0}$ as Q^n. The coefficients c_0, c_1 and c_2 are the functions of time step size defined from the discretization of time derivative. The scheme also can be the first order accurate method or Crank-Nicolson type second order scheme by the appropriate selection of the coefficients. The implicit part of equation (2) constitutes a block penta diagonal systems of equations. Normally, ADI scheme is often used for solving these kind of penta diagonal systems based on approximate factorization but its efficiency degrades severely in case of the number of equations are very large. Thus the LU relaxation scheme (see Shuen et al 1989) is used in these study, which involves just point-wise matrix inversion. By applying the upwind discretization of flux Jacobian matrices, Equation (2) can be rewritten in the following factorized form.

$$\begin{aligned}
&\mathbf{D}_{i,j}\Delta\mathbf{Q}_{i,j}^{*} = \mathbf{RHS}_{i,j} + \mathbf{A}_{i-1,j}^{+}\Delta\mathbf{Q}_{i-1,j}^{*} + \mathbf{B}_{i,j-1}^{+}\Delta\mathbf{Q}_{i,j-1}^{*} \\
&\mathbf{D}_{i,j}(\Delta\mathbf{Q}_{i,j}^{n+1,m} - \Delta\mathbf{Q}_{i,j}^{*}) = -\mathbf{A}_{i+1,j}^{-}\Delta\mathbf{Q}_{i+1,j}^{n+1,m} - \mathbf{B}_{i,j-1}^{-}\Delta\mathbf{Q}_{i,j-1}^{n+1,m}
\end{aligned} \tag{4}$$

Here, $\mathbf{A}$ and $\mathbf{B}$ are Jacobian matrix of flux vector $\mathbf{F}$ and $\mathbf{G}$, and $\mathbf{D}$ is the diagonal element of the left hand side of the equation (2).

Among several choices available in evaluating splitted flux Jacobian matrix, Steger-Warming flux splitting methods and approximate splitting method are used in this study. As these methods involve just point-wise Jacobian matrix without neighboring point, the evaluation of the left and right hand sides in equation (4) could be done in much more efficiently using a simplified formulation for the evaluation of diagonal term and Jacobian-variable vector product (see Choi 1997). Since these formulation is an exact one and gives a vector form of the matrix-vector product, the so-called 'frozen' matrix technique is not needed, which requires the huge memory capacity and the matrix vector product with the frozen matrix at initial value of sub-iteration. In case of using approximate splitting approach, the difference term of Jacobian matrix is just a scalar value and matrix inversion can be done much more efficiently, even though this procedure could have some problems in unsteady problems due to the approximation. The upper part matrix inversion could be scalar inversion by including the chemical source Jacobian matrix only in lower part of equation and the lower could be done by $N \times N$ matrix inversion by partitioning the souce Jacobian matrix. In addition, the matrix and vector product in right-hand-side of the equation can be evaluated using a simplified formulation similar to that used in Steger-Warming formulation. Because these kinds of simplified formula take only 10 % computational time of the matrix evaluation and matrix vector product, the total computing time is saved up to 50 %.

4. Spatial discretization methods

The finite volume cell-vertex scheme is used for spatial discretization of governing equations. The convective terms in residual vector are expressed as differences of numerical fluxes at cell interface. The numerical fluxes containing artificial dissipation are formulated using Roe's FDS (Flux Difference Splitting) method. The complete formulation of Roe's FDS method for multi-species chemically reacting flow is based on the primitive variable approach of Grossman and Cinnella (1990) and extended to two-dimensional curvilinear coordinates (Choi 1997). Since Roe's FDS method does not satisfy the second law, its solution sometimes shows unphysical solutions such as carbuncle phenomena. As a remedy of such problems, the entropy fixing function by Montagne et al. (1988) is used in the present study. The entropy fixing parameter, which is known to affect artificial dissipation, shock thickness and convergence rate, is used in the range of 0.01 to 0.4.

As a higher order extension of the upwind scheme, MUSCL (Monotone Upstream Method for Scalar Conservation Law) scheme is used for the extrapolation of primitive variables at cell interface. In this study, the third order scheme is used for all cases. In addition, some limiter functions are used to overcome the severe dispersion error introduced by the higher order extrapolation and to preserve the TVD (Total Variation Diminishing) property. In this study, minmod, Koren and van Albada limiters are considered. The minmod limiter shows different characteristics according to parameter β, and the minmod limiter preserves TVD property in the range of $1 \leq \beta \leq 4$ in case of the third order spatial accuracy. Smaller value of β preserves the monotonicity but introduces the larger numerical dissipation. The value of β =1.25 is used as a standard value for present study, since prior one-dimensional shock tube study showed stable shock capturing characteristic in Mach number range 4 to 5. On the other hand, the Koren and van Albada differentiable limiter functions are slightly efficient than minmod function since they do not need logical operations such as 'min' and 'max'. They showed the shock capturing characteristics between minmod with β =1.0 and β =4.0, although the Koren's limiter showed somewhat diffusive solution.

As an alternative to Roe's FDS method, AUSM (Advection Upstream Splitting Method, Liou et al. 1993) variants were also considered in this study. The flux splitting methods are based on the splitting of cell interface Mach number and the difference of the variants mainly owes to the cell interface Mach number. Thus, the formulations involve only vector manipulation and are simple and easily extendable to reactive flow. So, the details are not included here. In this study, the newly developed AUSM+W (Liou 1995) is considered and used in connection with minmod or pressure limiter (Edwards 1995).

5. Results of numerical investigations

To find accurate and efficient numerical methods for the analysis of unsteady shock-induced combustion, some comparative numerical experiments are performed in the following way. At first, simulations of experimental results were conducted to validate the performance of base-line method. The base-line method is composed with fully implicit second order accurate time integration method and third order spatially accurate upwind scheme. The implicit part is constructed with the Steger-Warming splitting and inverted by LU-SGS scheme. A time step equivalent to CFL number of 3 and 4 sub-iterations are used. The minmod limiter is used in spatial discretization with β =1.25. Computational domain is composed with 200 × 300 grid system normal to the projectile surface and tangential to body surface. Inflow boundary is set just before the oscillating bow shock. As a next step, some investigations were made for time integration method. The effect of the order of integration accuracy is examined and the optimum number of sub-iteration is pursued. The use of the approximate splitting is compared with that of the Steger-Warming splitting. In addition, the point-implicit method and Crank-Nicolson method are compared with base line method through time step refinement study. Finally, the spatial discretization related considerations are investigated. The effect of limiter functions and numerical diffusivity is examined and the use of an AUSM based flux splitting method is considered. The grid refinement study is also conducted in this criterion.

5.1. Simulation of experimental results using base line method

As a validational purpose of the developed computational fluid dynamics code, Lehr's (1972) experiments of periodically oscillating shock-induced combustion are numerically simulated. Stoichiometric hydrogen/air mixture is used in the experiment at 320mmHg and 403m/s of sonic velocity, which corresponds to the temperature of 293K. Diameter of the hemispherical projectile used in the experiment is 15mm. The cases show oscillating results at Mach number 4.18, 4.48, and 4.79. Experimentally observed oscillation frequencies 148, 425, and 712kHz, respectively. For the efficiency of computation, 2000 steps of time integration is performed by first order scheme without sub-iteration before second order accurate time integration. After 1000 steps of the time integration, the oscillation begins to appear and goes to regular manner. The first order solution after the 2000 steps is used as an initial condition of second order integration and this solution is used as an initial condition for the all the cases in this paper for the clear comparison.

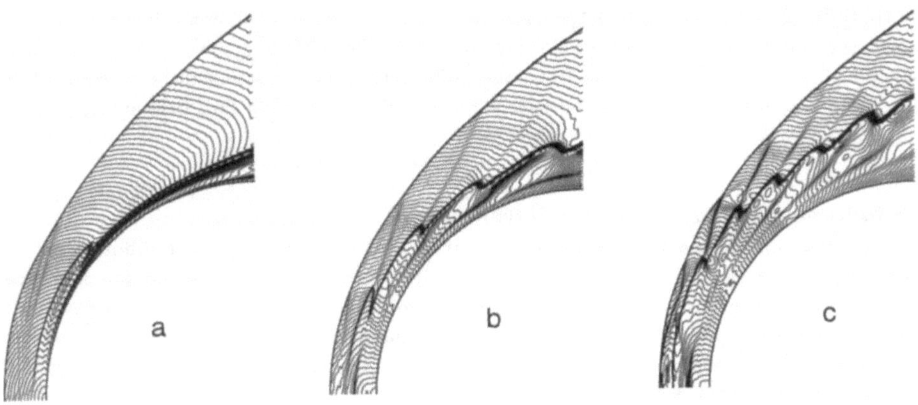

Fig. 1. Local Mach number distributions for three experimental cases. **a**, $M = 4.18$; **b**, $M = 4.48$; **c**, $M = 4.79$

Final solution of second order integration is obtained after the final 1000 steps with 4 sub-iterations. The error norm of sub-iteration is less than 1 % of initial error after 4 sub-iterations in every time steps. Figure 1 and 2 show the local Mach number distributions and the temporal variation of density along stagnation streamline for three cases. The plots of Mach number distribution agrees well with the experimental results and the density history show all the details of the instability mechanism predicted theoretically. The discussion on instability mechanism is beyond the scope of this paper and will not be described any more. The frequency of oscillation is obtained from these results and compared in Table 1 with experimental and previous numerical values. The results of present simulation show satisfactory correspondence with experimental results even though not exactly agree. The reliability of chemistry model has known to have a responsibility for large differences of frequencies by Wilson et al. (1993) and Yungster et al.

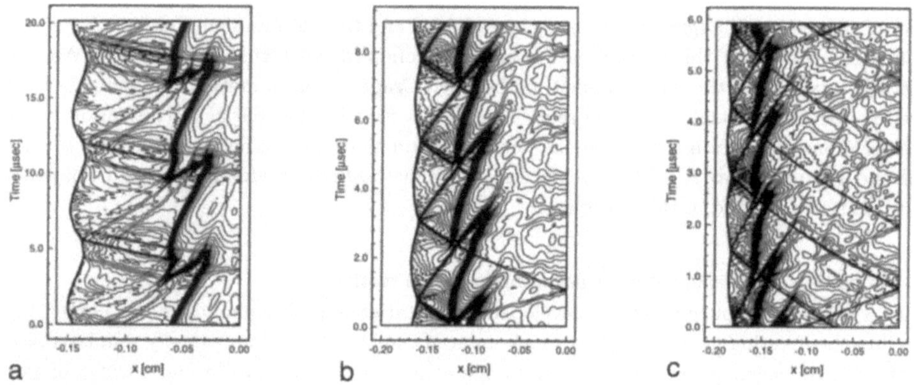

Fig. 2. Temporal variation of density along stagnation line for the experimental cases. a, $M = 4.18$; b, $M = 4.48$; c, $M = 4.79$

(1994). However, since the chemistry mechanism used in this study is same to the one used by Yungster et al. (1994) and Matsuo et al. (1995) and the chemistry mechanism shows satisfactory results, the discussion on the effect of chemistry mechanism would be a recapitulation. Anyway, the base-line method could be considered as having reliable solution accuracy from the above comparisons of the results.

On the basis of the reliability of the base-line method, the effect of different choices of numerical approaches will be discussed in the following. Since the case of Mach number 4.48 shows sufficient number of wiggles in the reaction front and sufficiently long induction distance between bow shock and reaction front, the case is regarded as showing most clear flow field among the three and used as a reference case.

5.2. Order of integration and number of sub-iteration

The effect of the order of temporal integration and the convergence of sub-iteration were examined for the case of Mach number 4.48. A first order time accurate solution and second order accurate solutions with 0, 2, 4 and 10 sub-iterations were obtained using a same initial condition described previously. All other computational conditions are fixed to that of the base-line method. The pressure variations at stagnation point are plotted for comparison in Fig. 3. The first order solution show smooth variation of pressure at stagnation point while the second order solutions show some overlaid high frequency peaks on the main pressure oscillation. However, the period of oscillation seems to be unchanged in all cases except the case of a second order solution with zero sub-iteration. In cases sub-iteration is performed, the solutions show converging manner as the number of sub-iteration is increased, although the solution is nearly unchanged after the two sub-iterations. After two sub-iterations, the sub-iteration error norm was less than 10 % of initial variation for all the time integration steps. As a reference, the error norms were less than 1 % and 0.01 % for four sub-iterations and 10 sub-iterations, respectively. Thus, two sub-iterations or the convergence criterion of 10 % is considered as sufficient in this case since the computational time increases nearly proportional to the number of sub-iterations. The flow field solution of the first order time integration shows relatively smooth variation in comparison with second order time accurate solution although not presented here. This smoothness smears the sharpness of the waves and some contact surfaces between the normal shock and reaction front. Thus, the first order solution seems to smear some waves but shows very periodic oscillation, while the second order solution shows sharp but numerically oscillating behavior in the post shock region.

Table 1. Comparison of the oscillation frequency of unstable shock-induced combustion

kHz	$M = 4.18$	$M = 4.48$	$M = 4.79$
Experiment	148	425	712
Present	155	426	707
Yungster et al.	163	431	701
Matsuo et al.	160	-	725
Wilson et al.	-	-	530

5.3 Flux Jacobian splitting methods

The use of approximate flux Jacobian splitting could be an efficient way of matrix inversion of LU factored implicit formulation. Because it needs just one $N \times N$ matrix inversion in lower sweep and a scalar diagonal inversion in upper sweep while the use of exactly splitted Jacobian matrix needs two $(N+3) \times (N+3)$ matrix inversions in lower and upper sweep. The result with the approximate Jacobian is plotted in Fig. 4 for the first and the second order time integration. In the case of the first order time integration, the time step equivalent to CFL number of two is used because the CFL number of three shows an instability problem. Except the time step, all the computational conditions are fixed to that of base line method. The Mach number distribution shows diffusive solution and the worse is that the periodicity begins to be broken in density variation. Temporally lagged solution is recognized from the Fig. 4. However, the solution is improved by using second order time integration with four sub-iterations. A time step equivalent to CFL number 3.0 was available in this case and the solution show nearly same quality to that obtained by using Steger-Warming flux Jacobian splitting method as shown in the figure. Thus it is understood from this result that the approximation of left-hand-side matrix would not be a significant matter if the sufficient convergence could be attained. However, it is hard to say that approximate splitting is more efficient than exact splitting method since two sub-iterations seems to be sufficient in case of using exact Steger-Warming Jacobian splitting method.

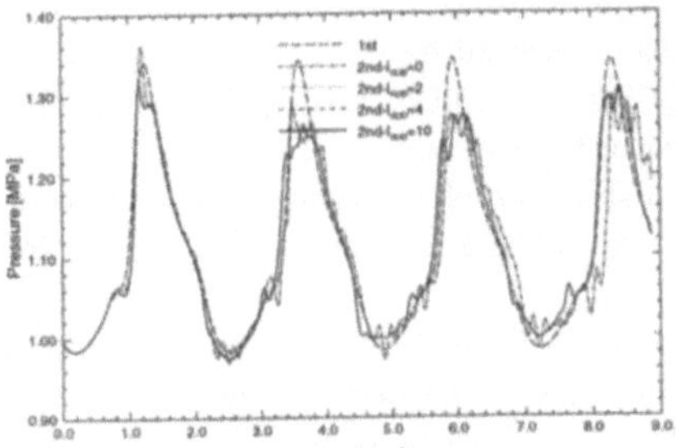

Fig. 3. Temporal variations of pressure at stagnation point by different orders of time integration and numbers of sub-iteration

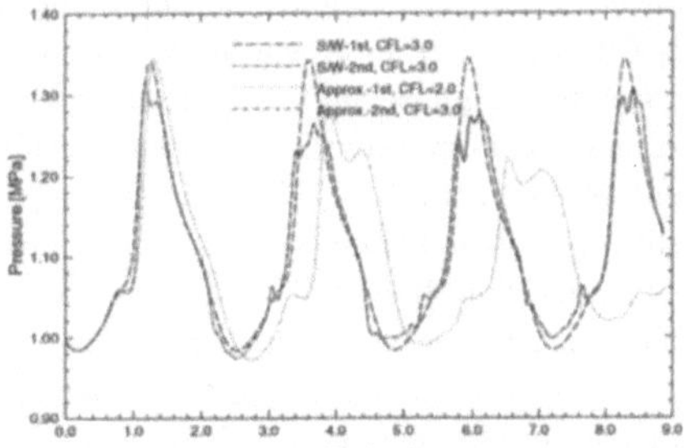

Fig. 4. Temporal variations of pressure at stagnation point by different flux Jacobian splitting methods and orders of time integration

5.4. Time step refinement studies for point-implicit, Crank-Nicolson and sub-iteration method

As alternatives to fully implicit time integration approaches, well-known point-implicit method and Crank- Nicolson type fully implicit method is considered. The point-implicit method is used in single matrix inversion step where chemistry source term is treated implicitly while the convective terms are treated explicitly. Although this approach has just a first order time accuracy, it is known to produce a reasonable solution when the time step is sufficiently small. In case of point-implicit method, time steps equivalent to CFL numbers of 0.03 to 0.5 were used. Though not presented in detail, the results at large time steps show severe high frequency oscillation that could not be considered as a physical result. However the amplitude of oscillation is mitigated and tends to converge to certain amplitude, although the period of the high frequency oscillation is nearly unchanged. From this time step refinement study, it is considered that the CFL number much less than 0.1 should be used in the analysis of shock-induced combustion with the present computational conditions. Even though the method takes just an $N \times N$ matrix inversion for a time step, the limitation of time step size is a severe efficiency restriction of the point-implicit method.

In case of Crank-Nicolson type fully implicit method, CFL number ranging from 0.3 to 3.0 can be used without divergence. The solution with different time step size is plotted in Fig. 5 as an example of time step refinement study. The solutions also show high frequency oscillation when large step is used and the solution also tends to converge to a solution with small amplitude as using small time step size. The oscillation of nearly converged solution seems to have nearly same frequency and amplitude to those from point-implicit method. From this examination, see Crank-Nicolson method is considered to be used with CFL number less than 1.0 since it shows a stability problem of high frequency oscillation at large time step.

The time step refinement study is also conducted for present base-line method with CFL number ranging from 0.3 to 6.0. Sub-iteration convergence criterion of 1 % error was used in this study rather than using the restriction on maximum sub-iteration number. With this convergence criterion, about 6 sub-iterations are needed in most of the time steps for the case of CFL number of 6 to satisfy the convergence criterion, and about 2 sub-iterations for CFL number of 1.0. Surprisingly, just one sub-iteration is needed for CFL number of 0.3, that means actually no sub-iteration needed. This result says that the efficiency increase is mitigated with increased number of sub-iterations needed for the convergence. However, even with this mitigation of efficiency increase, the solution with large time step seems to be still efficient since the solution is not oscillatory and looks stable. This solution stability is a comparable result to that for the point-implicit method and Crank-Nicolson method, which have severe oscillation problem at large time step size. However, an actual limitation of time step size is also appeared for this sub-iteration method. Slight phase shift is noticed for the case of CFL number of 6 and larger time of CFL number of 12 showed solution divergence problem even with the large number of sub-iterations. On the other hand, the LU-SGS sub-iteration approach also showed high frequency small amplitude oscillation at small time step, and the solution showed the converging trend similar to that for point implicit method and Crank-Nicolson method.

For the comparison of the converging trends, the solutions from the above three integration methods with smallest time step cases are summarized in Fig. 6. The three solutions show nearly same pressure history with high frequency small amplitude oscillation. The oscillatory solution is considered as reliable since the results from different method shows nearly same converging solution at small time step. The oscillation frequency deduced from the Fig. 6 is about 12MHz. However, authors cannot be sure that the high frequency oscillation is a physical one or a numerical one originated from the nonlinear shock capturing scheme, because such a high frequency oscillation has not been observed in the previous researches although it is a converged solution with small time step size.

5.5. Effect of spatial limiter function

The most distinctive characteristics of different limiter functions would be the shock capturing characteristics originated from generic numerical diffusivity. As an example, minmod limiter shows steepest solution with β =4.0 and most diffusive one with β =1.0 in TVD region. Thus,

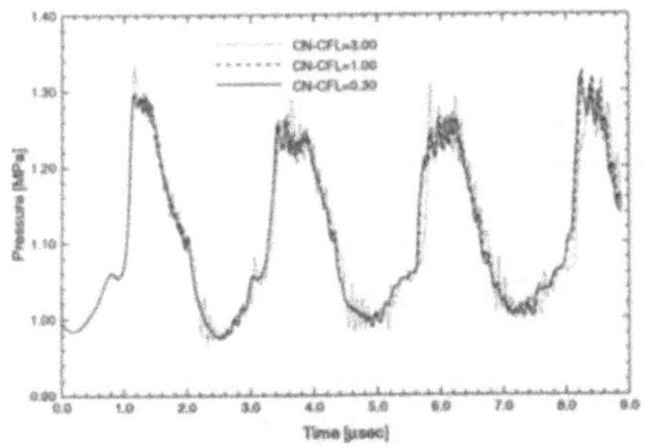

Fig. 5. Temporal variations of pressure at stagnation point by Crank-Nicolson method with different time steps

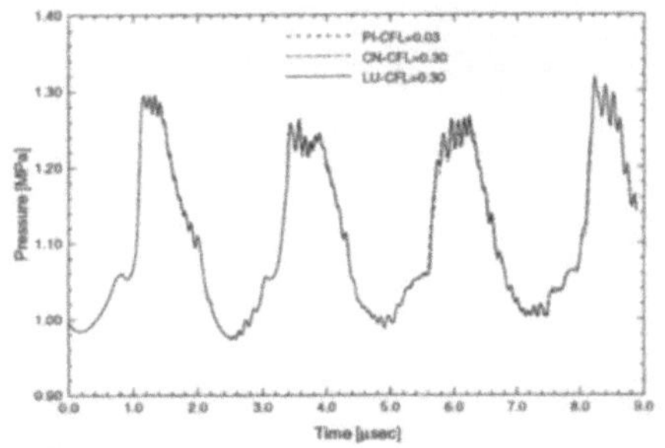

Fig. 6. Comparison of temporal variations of pressure at stagnation point by different time integration methods

the results with different limiter functions are compared for the understanding of the effect of spatial discretization and numerical diffusivity. In this examination, first order time integration was carried without sub-iteration for the computational efficiency and for the clear comparison without the corruption of pressure history from high frequency oscillation. All the computational conditions are same as the base-line method. In Fig. 7, the most distinguished feature of the solution is the phase shift of the oscillatory pressure wave. The steepest limiter (minmod with β =4.0) shows largest period of oscillation and most diffusive limiter (minmod with β =1.0) shows smallest period. Other limiter functions show intermediate solutions since they have intermediate shock capturing characteristic between the above two limiter functions.

To find the change of oscillation frequency more precisely, the time integration has been done for a longer time of 4000 iterations to include more oscillation periods. This iteration number includes 15 or 16 periods of oscillation. The oscillation frequency is evaluated from the average of the first peak to last peak time and the first valley to last valley time those divided by the number of periods. Table 2 is the resulting oscillation frequencies for different limiter functions. The steepest limiter shows smallest frequency and diffusive limiter shows largest frequency among the calculations of present study. The maximum deviation of 11kHz corresponds to 2.5 % error of frequency and the experimental result is listed within the deviation. From this result in Table 2, it is considered that the result of numerical simulation could have some deviation which may originates from the choice of spatial discretization method that possesses its generic numerical dissipation implicitly. Nearly same results of the phase shift were still observed when the time integration has been done with second order accuracy, even though the pressure histories show some high frequency oscillation.

5.6. An application of AUSM based method

Recently, AUSM variants flux splitting methods begin to be used widely for the problem in which the stable shock capturing is the major concern and Roe's FDS sometimes fails. However, the previous studies showed that the method has a problem of post shock oscillation where strong shock wave is involved. The AUSM based method is considered as still under developing and not yet validated for sufficient number of problems. In this study, AUSM+W, an advanced version of AUSM, is adopted with MUSCL approach. The minmod and pressure limiters (Edward 1995) are used for stable shock capturing. The pressure limiter has been showed somewhat diffusive characters and often used with AUSM based method as a remedy of post-shock oscillation problem that AUSM could have.

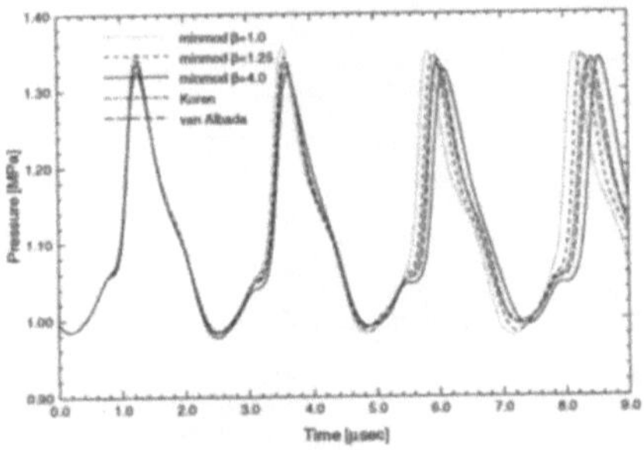

Fig. 7. Temporal variations of pressure at stagnation point for different limiter functions and first order time integration of 1000 time steps

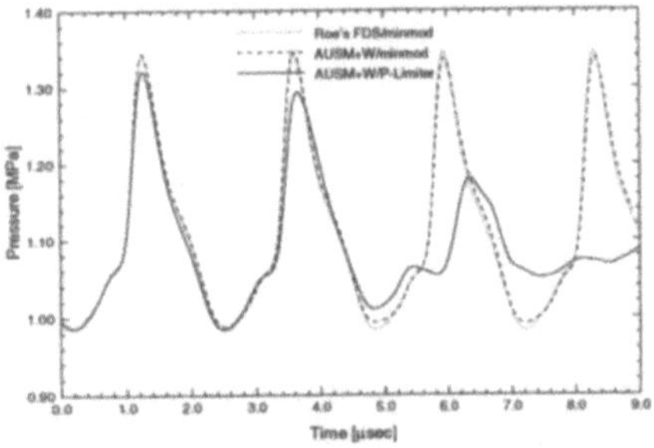

Fig. 8. Comparison of temporal variations of pressure at stagnation point for Roe's FDS and AUSM+W with first and second order time integration of 4000 time steps

Table 2. Comparison of the oscillation frequency for different limiter functions

Limiter Function	Frequency [kHz]
Experiment	425
Yungster	431
minmod, β =1.00	429.48
minmod, β =1.25	426.13
minmod, β =4.00	418.73
Koren	423.19
van Albada	421.78

Figure 8 is the pressure history by using the AUSM+W method. The most notable one is that the use of pressure limiter shows a decay of oscillation and nearly steady solution after sufficiently long time. The decay of oscillation could be obtained by both the first and the second order time integration while the use of minmod limiter shows similar oscillating solution to that of Roe's FDS method. This unphysical behavior of AUSM+W with pressure limiter could be a result of the excessive generic numerical diffusivity of pressure limiter. Thus, we can understand that the shock-capturing scheme and its generic numerical diffusivity are very important to capture the physically correct solution of these kind of unstable shock-induced combustion phenomena.

5.7. Grid refinement study

A grid refinement Study was carried out for 5 grid systems as an extension of the considerations on spatial discretization. The computations were performed by the first order time integration with same time step size that used in the base line method. All other conditions were fixed at the base-line method. The resulting pressure history is plotted in Fig. 9 for comparison. Because it is not possible to use the same initial condition for different grid system, the pressure curves are translated to match the time of second pressure peaks. The periodicity of the solution is reasonably captured in all grid systems. The frequency of oscillation is smallest for 400 $\times$ 600 grid and other results show the trend of convergence to the results by 400 $\times$ 600 grid system. The deviation of oscillation is within 2 % error. This trend could be considered in relation with results of limiter function studies in which the steepest method shows smallest frequency. Even though the result by 100 $\times$ 150 grid system looks to presenting a reasonable solution, 150 $\times$ 200 grid system is a smallest that adequate for capturing the oscillatory behavior of shock-induced combustion because the periodicity of solution breaks down in 100 $\times$ 150 grid system if more time passes.

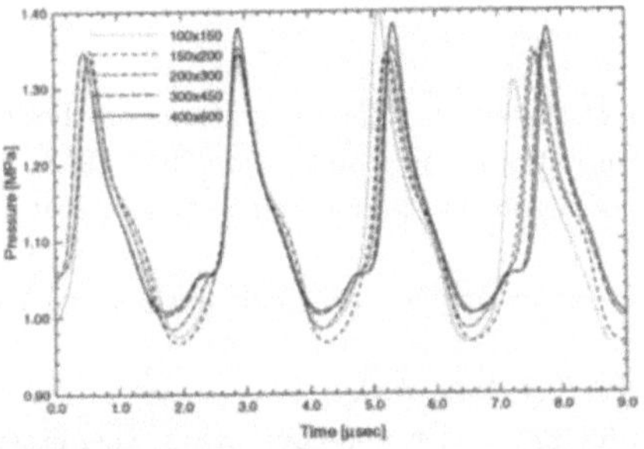

Fig. 9. Comparison of temporal variations of pressure at stagnation point for different grid systems. Reference time is translated to match the time of second pressure peaks

6. Conclusions

A comparative study of numerical methods is made as a developing process of an efficient computational fluid dynamics code capable of analyzing unsteady reacting flow. As a base line method, an iterative fully implicit algorithm is used for time integration and a third order accurate upwind scheme is used for spatial discretization. The simulation of experimental results shows the validity of the present base line method. The additional studies on time integration methods and spatial discretization methods present some important points that should be taken into account in the analysis of these kinds of unsteady problem.

The first order accurate time integration method results in a spatially diffusive solution while the periodicity of the solution is preserved well. For the iterative solution of second order accuracy, two sub- iterations seem to be sufficient with Steger-Warming formulation of flux Jacobian matrix. However, the approximation of implicit part would not be a matter of importance if the sufficient number of sub-iteration is carried out for convergence. A high frequency oscillation involved in second order solutions showed converging manner to the case of small time step size even though it cannot be assured as a physical one or a numerical one at the present status. The sub-iterative second order fully implicit method shows a stable solution with large time step size while the point-implicit method and Crank-Nicolson method fails due to the severe oscillation.

Although the deviation is not so significant, the choice of different spatial limiter function showed the different oscillation frequency that explained the discrepancy of the results from the previous researches. The combination of AUSM+W method with pressure limiter function showed an unphysical solution, and it is understood that the generic numerical diffusivity of each numerical method is important for the physically correct simulation of shock-induced combustion. From the grid refinement study, 100×150 grid system seems not to be sufficient for the reproduction of the periodic solution. 150×200 grid system would be the smallest one for the stable periodic solution, though the use of finer grid exhibits a converging trend to a solution with smaller frequency.

7. Acknowledgement

This research is supported by Turbo and Power Machinery Research Center and Korea Science and Engineering Foundation with contract number 971-1005-031-2.

8. References

Choi JY, Jeung IS, Lee S (1996) Dimensional analysis of the effect of flow conditions on shock-induced combustion, Twenty-Sixth Symposium (International) on Combustion 2957-2963

Choi JY (1997) Superdetonative mode starting process of supersonic combustion ram accelerator, Ph.D dissertation, Seoul Nat Univ, Korea

Edwards JR (1995) A low-diffusion flux-splitting schemes for navier-stokes calculations. AIAA paper 95-1713

Gardiner WC (1984) Combustion Chemistry. Springer-Verlag, New York

Grossman B, Cinnella P (1990) Flux split algorithms for flows with non-equilibrium chemistry and vibrational relaxation. J Comp. Phys 88:131-168

Hertzberg A, Bruckner AP, Bogdanoff DW (1988) Ram accelerator: A new chemical method for accelerating projectiles to ultrahigh velocities. AIAA J 26:195-203

Hirsch C (1990) Numerical Computation of Internal and External Flows. John Wiely & Sons, New York, Vol 2, p 1990

Jachimowski CJ (1988) An analytical study of the hydrogen-air reaction mechanism with application to scramjet combustion. NASA TP-2791

Lehr HF (1972) Experiment on shock-induced combustion. Astro Acta 17:589-597

Liou MS (1995) Progress towards an improved CFD method: AUSM+. AIAA paper 95-1701

Liou MS, Steffen CJ (1993) A new flux splitting scheme. J Comp Phys 107:23-39

Matsuo A, Fujii K, Fujiwara T (1995) Flow features of shock-induced combustion around projectile traveling at hypervelocities. AIAA J 33:1056-1063

Matsuo A, Fujiwara T (1993) Numerical investigation of oscillatory instability in shock-induced combustion around a blunt body. AIAA J 31:1835-1841

Montagne JL, Yee HC, Klopfer GH, Vinokur M (1988) Hypersonic blunt body computation including real gas effects. NASA TM 10074

Shepherd JE (1994) Detonation waves and propulsion. In: Buckmaster J et al (eds) Combustion in High-Speed Flows, Kluwer Academic Publ, Dordrecht, Netherlands, pp 373-420

Shuen S, Yoon S (1989) Numerical study of chemically reacting flows using a lower-upper symmetric successive overrelaxation scheme. AIAA J 27:1752-1760

Toong TY (1983) Combustion Dynamics: The Dynamic of Chemically Reacting Fluids, McGraw-Hill, New York

Wilson GJ, Sussman MA (1993) Computation of unsteady shock-induced combustion using logarithmic species conservation equations. AIAA J 31:294-301

Yungster S, Radhakrishnan K (1994) A fully implicit time accurate method for hypersonic combustion: application to shock-induced combustion instability. AIAA paper 94-2965

On the detonation initiation by a supersonic sphere

Y. Ju[1], A. Sasoh[2] and G. Masuya[1]
[1] Dept. of Astronautics and Space Engr., Tohoku University, Sendai 980-77, Japan
[2] Institute of Fluid Science, Tohoku University, Sendai 980-77, Japan

Abstract. Conditions of detonation initiation induced by a supersonic sphere in a stoichiometric hydrogen/oxygen mixture with 70% argon dilution, are investigated numerically and theoretically. The physical model includes a detailed full chemistry and is solved using an LU-SGS TVD scheme. Transition of the two extremes, shock induced combustion and detonation initiation, are examined over pressures ranging between 0.02 MPa and 1 MPa. Numerical results show a very distinct detonation initiation boundary which separates the detonatable and undetonatable regions. A good agreement is shown through the comparison of the numerical results with the recent experimental data. Additionally, a theoretical investigation on the determination of detonation initiation boundary is carried out by introducing a concept of the kinetic limit. The kinetic limit is defined by the ignition Damköhler number. A joint theory for the determination of the detonation initiation boundary is presented by relating the present kinetic limit defined by the unity ignition Damköhler number and the energy limit given by Lee. Comparison between the theory and the experiment over a wide pressure range shows that the detonation initiation boundary can be well defined by the present joint theory.

Key words: Detonation initiation, Supersonic projectile, High pressure

1. Introduction

Detonation initiation by a supersonic projectile has a renewed interest due to its fundamental significance in the theory of detonation and its close relevance to the operation of Ram-accelerators (Hertzberg et al 1988). It is well known that firing of a supersonic projectile into a detonatable gaseous mixture has two extremes, shock induced combustion and detonation. Earlier studies concerning this issue were conducted by Zeldovich and Leipunsky (1943). In the past twenty years, detailed flow regimes of supersonic blunt projectiles in combustible gases were examined by experiment and numerical simulation (Lee 1984, Behrens 1965, McVey et al 1971, Lehr 1972, Ahuja et al 1995, Matsuo et al 1995, Lefebvre and Fujiwara 1995). Efforts of these studies have been paid to the steady and periodic regular regimes of combustion at projectile speeds higher or slightly lower than the CJ velocity, particularly to the unstable oscillating phenomenon. Detailed mechanisms of this instability have been made clear through the above studies.

Compared to the study of the steady and regular combustion regimes, few attempts have been made to estimate the requirement for direct initiation. Although a limited number of experiments on detonation initiation were conducted by Zeldovich and Leipunsky (1943) and Lehr (1972), quantitative conditions for detonation initiation are poorly understood, particularly at high pressures and sub-CJ velocities.

Recent development of the supersonic propulsive device and control of the desired and undesired detonation initiation requires a clear understanding of both the mechanism of detonation initiation and the relation between the dynamic parameters such as the projectile velocity, size as well as the mixture pressure. An energy criterion defining the detonation initiation boundary is theoretically developed by Lee (1994) and Vasiljev (1994). The theory equates the minimum energy required to initiate a cylindrical detonation wave to the work done by the drag force of the projectile per unit length. Recently, two experimental work on detonation initiation are systematically performed by Belanger et al. (1995) and Higgins and Bruckner (1995), respectively. The former emphasizes the phenomena of overdriven detonation for projectile speed higher than CJ velocity. The latter was conducted to examine the detonation initiation conditions for projectile speed lower than CJ velocity at a pressure range between 0.04 and 0.75 MPa. The experimental data are compared with Lee's theory. Unfortunately, they reported that the theory fails at high mixture pressures. In order to examine the effect of chemistry on detonation initiation, a numerical simulation with the same pressure conditions as the experiment is conducted by the present authors (Ju and Sasoh 1997) using a detailed chemistry. Although the numerical simulation has a better agreement with the experiment than the theory, the experiment still show a much lower initiation Mach number than the simulation at pressure 0.75MPa. This large difference makes these authors question the effect of the diaphram in the experiment. The reason is that a large pressure jump in the experiment across the diaphragm may result in a rapid gas expansion from the test chamber to the dump tank when the projectile passes through the diaphragm. The induced gas velocity relative to the projectile may considerably affect the experimental data. Recently, Higgins and Bruckner improved their experimental facility by adding a buffer of argon ahead of the diaphragm on the dump tank side and matched the pressures on the both sides of the diaphragm (Higgins 1996).

The object of the present work is to seek a theoretical determination of the detonation initiation boundary. First, numerical results of detonation initiation by a supersonic sphere in a stoichiometric hydrogen/oxygen mixture diluted by 70% of argon is reported. The results are compared with the new experimental data. Then a kinetic limit is theoretically defined by introducing the ignition Damköhler number. A joint theory for the detonation initiation is presented by combining the kinetic limit with the energy limit presented by Lee (1994). Finally, a comparison between the present joint theory and the experiment is made.

2. Physical model and governing equations

As shown in Fig.1, detonation initiation by firing a supersonic sphere into a detonatable gaseous mixture is modeled by considering the shock induced combustion or detonation with a blunt sphere fixed in an infinite supersonic flow. The inflow gas is a stoichiometric hydrogen/oxygen mixture diluted by 70% argon. Sphere size is 12.7 mm in diameter, which is the same as that in the experiments (Higgins and Bruckner 1995, Higgins 1996). The inflow Mach number varies between 2 and 5. In supersonic flow, a bow shock is formed around the blunt sphere. The bow shock induces chemical reaction and leads to an ignition behind. If the combustion wave is strong enough, the combustion wave will propagate upstream and begins to couple with the bow shock, and finally results in a propagating detonation wave. If the combustion wave is not strong enough, the interaction between the two waves will be decoupled and results in a detonation failure. The interest of the present study is to determine the detonation initiation boundary over a wide range of initial pressure (0.02 MPa-1 MPa).

In the previous studies of detonation phenomena, viscosity is usually ignored in numerical simulation. The two-dimensional, axisymmetric, time-dependent, chemically reactive Euler equation with n species and finite rate chemistry in a generalized curvilinear coordinate (ξ, η) can be written as

$$\frac{\partial \hat{U}}{\partial t} + \frac{\partial \hat{E}}{\partial \xi} + \frac{\partial \hat{F}}{\partial \eta} = \hat{S} + \hat{H} \tag{1}$$

where t is time. ξ and η are transformed coordinates along and normal to the sphere surface. $\hat{U}$ is the vector of conversation variables; $\hat{E}$ and $\hat{F}$ are respectively the vectors of convective fluxes in ξ and η directions. $\hat{S}$ is the vector of the source of chemical reaction and $\hat{H}$ is the vector of axisymmetric source term. For illustration, vectors $\hat{U}$, $\hat{E}$, $\hat{S}$ and $\hat{H}$ are written as

$$\hat{U} = J \begin{pmatrix} \rho_1 \\ \rho_2 \\ \vdots \\ \rho_n \\ \rho u \\ \rho v \\ E \end{pmatrix}, \qquad \hat{E} = J \begin{pmatrix} \rho_1 U \\ \rho_2 U \\ \vdots \\ \rho_n U \\ \rho u U + \xi_x p \\ \rho v U + \xi_r p \\ (E+p)U \end{pmatrix} \tag{2}$$

$$\hat{H} = \frac{J}{r} \begin{pmatrix} \rho_1 v \\ \rho_2 v \\ \vdots \\ \rho_n v \\ \rho u v \\ \rho v^2 \\ (E+p)v \end{pmatrix}, \qquad \hat{S} = J \begin{pmatrix} \dot{\omega}_1 \\ \dot{\omega}_2 \\ \vdots \\ \dot{\omega}_n \\ 0 \\ 0 \\ 0 \end{pmatrix} \tag{3}$$

where x and r are respectively axial and radial coordinates. u and v are the velocities in x and r directions. ρ_i and ρ are respectively the species density and total density. p and E denote the pressure and the total energy. $\dot{\omega}_i$ is the reaction rate of each species. J is the Jacobian of coordinate transformation and U is the contravariant velocity in ξ direction

$$U = \xi_x u + \xi_r v \tag{4}$$

Detailed description of these vectors are given in Ref. 16. The gas mixture is considered thermally ideal. The specific heat of each species is computed by a fourth-order polynomial of temperature that fits the data tabulated in JANAF.

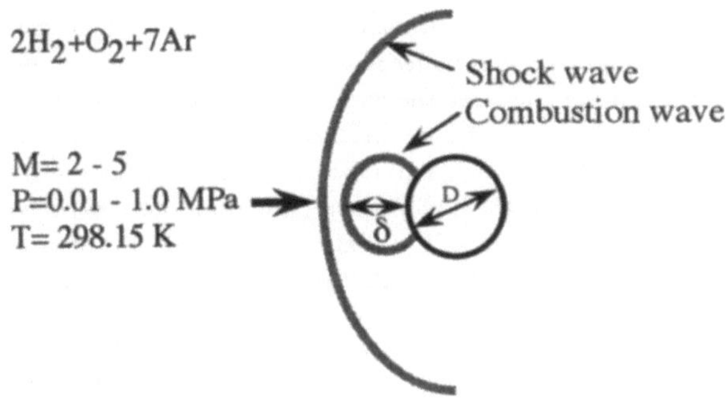

Fig. 1 Schematic illustration of the computational domain.

In the present study, a full chemistry, which was originally given by Stahl and Warnatz (1991) and revised and tested by the first author (Ju and Niioka 1994) in the supersonic mixing layer, is employed. In order to exactly calculate the ignition at low temperature, where H_2O_2 is an important species, reaction H_2O_2+H=H_2O+OH is supplemented. The mechanism includes 35 elementary reactions (see Refs.17 and 18) and 9 species (Ar, O_2, H_2, H_2O, H, HO_2, OH, O, H_2O_2). Here, Ar is considered reactively inert.

The governing equation is implicitly solved using the LU-SGS algorithm developed by Ju (1995). The Crank-Nicholson method is employed to achieve a second-order time accuracy. These numerical code has been tested for hydrogen/air ignition in a supersonic mixing layer and for combustion in a ramped duct. Since increase of the inflow pressure makes the chemical source term much stiff, a CFL number of 0.2 is used in the current studies. 141 × 101 grid points are used in ξ and η directions, respectively.

The upstream boundary condition is determined by specifying inflow temperature, pressure, velocity and mole fraction of each species. At the outflow boundary, variables are extrapolated from their upstream values. On the outer boundary, variables are extrapolated from their values at inner grids. At inner boundary, sphere surface and axisymmetrical axis, slip and adiabatic conditions are adopted.

In all the calculations, the converged flow field is firstly calculated by frozen the chemical reaction. This converged solution is then used as a start for the computation of detonation initiation. Calculation is terminated when the leading shock wave reaches the upstream boundary or the calculation time is longer than 1 millisecond, which is the largest traveling time of sphere in the detonatable mixture in the experiment of Higgins and Bruckner (1995) .

3. Numerical results and discussions

Pressure contours for an inflow pressure of 0.15MPa at 0.007 ms and 0.018ms are plotted in Figs.2 and 3, respectively. The inflow velocity is 1260 m/s, which is the same as that in the experimental study (Higgins and Bruckner 1995). At t=0.007 ms, it can be seen that there is a pressure jump just at the front nose of the sphere. This jump is attributed to the ignition induced by the shock wave. The resulting reaction zone locates far behind the standing off shock wave. At t=0.018 ms, Figure 3 clearly shows that the shock wave which is overdriven by the combustion wave propagate far ahead of the original location of the standing off bow shock. A subsequent calculation shows that the propagating shock wave is coupled with the combustion wave and runs out the left boundary. This phenomenon indicates that the supersonic sphere sucessfully initiated a detonation wave. Therefore, a detonation initiation boundary can be determined from the shock wave trajectory.

A comparison between the numerical results and the experimental data obtained with (Higgins 1996) and without (Higgins and Bruckner 1995) an argon buffer is shown in Fig.4. It can be seen that the initial pressure jump across the diaphragm between the dump tank and the test chamber dramatically affects the measured critical Mach number for detonation initiation. The improved experiment with an argon buffer gives a lower detonation initiation Mach number on the low pressure side but a higher initiation Mach number on the high pressure side than the experiment without buffer. This discrepancy increases with an increase of the mixture pressure. At the mixture pressure of 0.75 MPa, the experiment without buffer indicates that detonation can be initiated even when the projectile is as slow as Mach 2, which is too slow to ignite the mixture based on the stagnation temperature. However, the experiment with an argon buffer shows that the projectile below Mach 3.2 is unable to initiate detonation. This large difference is produced by the occurrence of the unsteady shock wave traveling from the test chamber to the dump tank when the projectile passes through the diaphragm at the experiment without buffer.

Since both the present numerical simulation and the theory exclude the occurrence of this unsteady shock wave, the experimental data with an argon buffer should be used when a comparison between the experiment and numerical simulation is made. As shown in Fig.4, the present numerical results agrees reasonably well with the experimental data. On the high pressure side, numerical simulation can exactly reproduce the experimental data at pressures 0.15 and 0.75

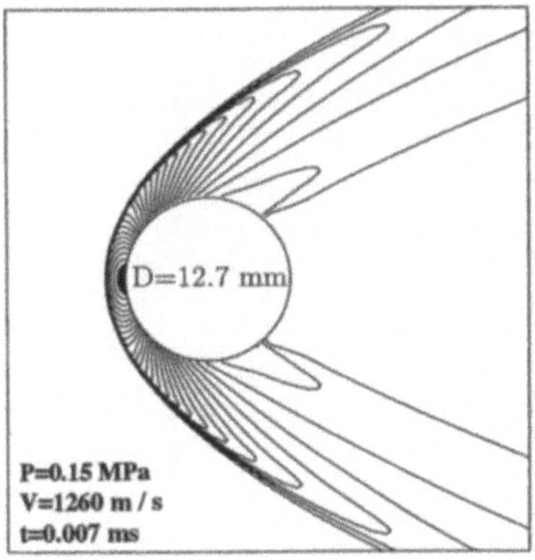

Fig. 2. Pressure contours at 0.007ms for initial pressure of 0.15 MPa and inflow velocity of 1260 m/s

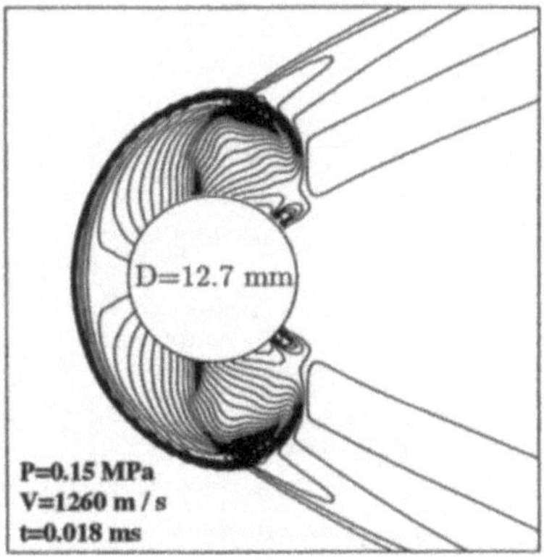

Fig. 3. Pressure contours at 0.018ms for initial pressure of 0.15 MPa and inflow velocity of 1260 m/s

MPa. However, for mixture pressure higher than 0.4 MPa, experimental data shows an slightly increasing tendency of the critical Mach number with the pressure while the numerical simulation exhibits a decreasing tendency. As stated in our previous study (Ju and Sasoh 1997), the decreasing dependence of the critical Mach number can be explained by the ignition kinetics. However, the increasing dependence given by the experiment is difficult to be explained. Further examination of this critical Mach number using a kinetic mechanism which is suitable for high pressure is necessary. On the low pressure side, numerical results show a detonation limit of 0.065 MPa, which agrees well with the limit of 0.08 MPa obtained by the experiment. Furthermore, numerical data shows there is a minimum critical Mach number near the mixture pressure of 0.08 MPa. For mixture pressure below it, the predicted critical Mach number increases rapidly. This minimum critical Mach number corresponds to the second explosion limit of hydrogen/air mixture and the rapid increase of it is determined by the energy limit which will be discussed in the next section.

4. Joint theory for detonation initiation

4.1 Energy limit A theoretical study to estimate the energy limit required for direct detonation initiation of detonation by a supersonic projectile is conducted by Lee (1994) using the hypersonic blast wave analogy. The theory equates the critical energy per unit length required for a cylindrical detonation initiation to the work done by the drag per unit length. Using the ideal blast solution of Taylor (1950) to give the blast trajectory and assuming that the blast radius must be at least equal to a critical radius (3.2 λ), the critical energy per unit length for the initiation of a cylindrical detonation is given by Lee as

$$E_0 = 14.5 p_0 M_{CJ}^2 \lambda^2 \tag{5}$$

where p_0 is the mixture pressure and M_{CJ} is the Mach number of CJ velocity normalized by the sound speed c_0 of the undisturbed mixture. λ is the detonation cell size.

On the other hand, the work per unit length done by the projectile at the critical Mach number M_{cr} is

$$E_d = \frac{1}{2}\rho_0 c_0^2 M_{cr}^2 (\frac{\pi d^2}{4}) C_D \tag{6}$$

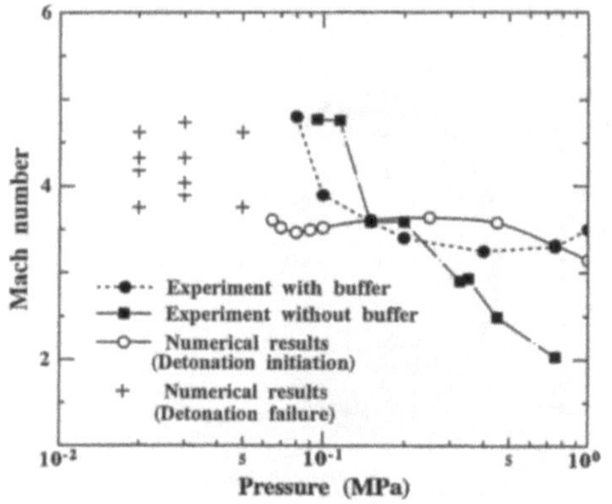

Fig. 4. Comparison of numerical results with experimental data with and without buffer

where C_D is the drag coefficient and takes on a numerical value of 0.92 for a hypersonic sphere. ρ_0 is the mixture density and d is the diameter of the sphere. By equating the work done by the drag force per unit length to the critical energy for direct initiation of cylindrical detonation, we obtain

$$M_{cr} = M_{CJ}\sqrt{\frac{116}{\gamma C_D \pi}\frac{\lambda}{d}} \tag{7}$$

where γ is the ratio of the specific heats. Therefore, for given projectile diameter and mixture pressure, detonation initiation Mach number can be calculated directly from Eq.(7).

A comparison between the theory and the experiment is shown in Fig.5. It can be seen that the theory agrees well with the experiment on the low pressure side. However, for mixture pressure higher than 0.1 MPar, the theory gives a much lower value than the experimental data. The reason for this discrepancy has been discussed in details in our previous study (Ju and Sasoh 1997). Our conclusion is that the energy limit give by Eq.(7) does not include the requirement for auto-ignition. Therefore, quantitative determination of the detonation initiation needs another limit, the kinetic limit, which states the requirement of the shock compression for auto-ignition.

4.2 Kinetic limit defined by the unity ignition Damköhler number In experimental determination of the detonation boundary, the test chamber has a finite dimension and thus the traveling time of the supersonic sphere in the test chamber will be limited both by the projectile speed and by the dimension of the test chamber. Therefore, a success of detonation initiation observed in experiment must requires that the mixture should be ignited by the projectile at least within the traveling time. This requirement provides a natural definition of the kinetic limit (ignition limit) using the unity ignition Damköhler number as

$$D_{aig} = \frac{\tau_{ig}}{\tau_{tr}} = 1 \tag{8}$$

in which τ_{tr} is the traveling time of the supersonic sphere in the test chamber. τ_{ig} is chosen to be the auto-ignition time at the stagnation temperature T^* and at the pressure p^* behind the standing bow shock. Thus, the detonation initiation Mach number can be determined by the follow relation

$$\tau_{ig}(p^*, T^*) = \frac{L}{M_{cr}c_0} \tag{9}$$

$$p^* = p_0[1 + \frac{2\gamma}{\gamma+1}(M_{cr}^2 - 1)], \quad T^* = T_0(1 + \frac{\gamma-1}{2}M_{cr}^2) \tag{10}$$

where L is the length of the test chamber. p_0 and T_0 are respectively the initial pressure and temperature of the mixture. c_0 is the sound speed. With the initial values of T_0 and p_0 and a presumed value of M_{cr}, a new M_{cr} can be calculated from Eq.(9). After several times of iterations, the critical Mach number for a given mixture pressure can be easily obtained. It should be noted that τ_{ig} can be obtained both asymptotic and numerically. In the present study, numerical method is used and an asymptotically expression of τ_{ig} will be given in the forthcoming study.

We have shown that the energy limit defined by Lee (1994) gives a reasonable detonation boundary at low pressure while fails at high pressure. On the other hand, we also shown that the kinetic limit can well define the detonation initiation boundary at high pressure but fails at low pressure. These two facts suggest that a joint theory for detonation initiation can be built by combining the energy limit requirement.

The comparison between the present joint theory with the experiment is given in Fig.5. The upper-right region of curve AOB corresponds to the detonable region and the lower-left part of curve AOB is the undetonatable region. It can be seen that the detonation boundary AOB defined by the present joint theory agrees well with that measured by the experiment.

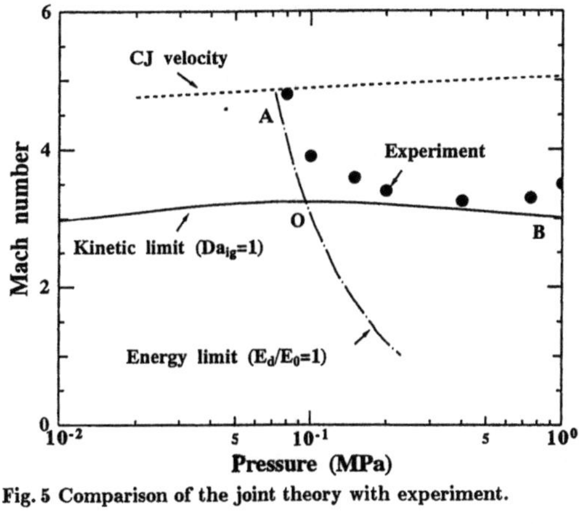

Fig. 5 Comparison of the joint theory with experiment.

5. Conclusions

The problem of detonation initiation by a supersonic sphere is investigated numerically and theoretically. Comparison between the numerical results and the experimental data obtained with and without buffer shows that the pressure jump across the diaphragm between the test chamber and the dump tank in experiment has a dramatic effect on the measured results, particularly at high pressure. The numerical results reasonably agrees with the experimental data using an argon buffer. The theory presented by Lee correctly predicts the detonation initiation condition at pressures below 0.1 MPa but no longer gives appropriate values as the pressure increases. A condition for kinetic limit of auto-ignition is shown to be required to quantitatively determine the detonation initiation boundary. A kinetic limit defined by the unity ignition Damköhler number reproduces well the experimental data. A joint theory presented by combining the kinetic limit and the energy limit agrees very well with the experiment. This agreement shows that a successful detonation initiation requires not only that the auto-ignition time at the stagnation temperature is shorter than the traveling time of the projectile in the test chamber, but also that the work done by the drag force of the projectile is larger than the minimum energy for direct initiation of a cylindrical detonation wave to keep from the decoupling of the shock wave and the combustion wave. Future research is needed to address how the pressure dependent reactions affects the critical Mach number at high pressure.

References

Ahuja JK, Kumar A, Tiwari SN (1995) Numerical investigation of shock-induced combustion past blunt projectiles in regular and large disturbance regimes. AIAA paper 95-0153

Behrens H, Struth W, Wecken F (1965) Studies of hypervelocity firings into mixtures of hydrogen with air or with oxygen. In: 10th Symp (Int) on Comb, pp 245-252

Belanger J, Kaneshige M, Shepherd JE (1995) Detonation initiation by hypervelocity projectile. In: Sturtevant et al (eds) Proc 20th Int Symp on Shock Waves, pp 1119-1124

Matsuo A, Fujii K, Fujiwara T (1995) Flow features of shock-induced combustion past blunt projectiles in regular and large disturbance regimes. AIAA J 33:1056-1063

Hertzberg A, Bruckner AP, Bogdanoff DW (1988), Ram accelerator: A new chemical method for accelerating projectiles to ultrahigh velocities. AIAA J 26:195-203

Higgins AJ, Bruckner AP (1995) Detonation initiation by supersonic blunt bodies. In: Proc 15th ICDERS, Boulder

Higgins AJ (1996) Investigation of detonation initiation by supersonic blunt bodies. Ph.D thesis, Univ Washington

Ju Y, Niioka T (1994) Reduced kinetic mechanism of ignition for non-premixed hydrogen/air in a supersonic mixing layer. Comb Flame 99:240

Ju Y (1995) Lower-upper scheme for chemically reacting flow with finite rate chemistry. AIAA J 33:1418-1425

Ju Y, Sasoh A (1997) Numerical study of detonation initiation by a supersonic sphere. Trans Jpn Soc Aero Space Sci 40:19-29

Lefebvre MH, Fujiwara T (1995) Numerical modeling of combustion processes induced by a supersonic conical blunt bodies. Comb and Flame 100:85-93

Lee JHS (1984) Dynamic parameters of gaseous detonations. Ann Rev Fluid Mech 16:311-336

Lee JHS (1994) On the initiation of detonation by a hypervelocity projectile. Zelodovich Memorial Conf on Comb, Voronovo, Russia

Lehr HF (1972) Experiments on shock-induced combustion. Astr Acta 17:589-597

McVey JB, Toong TY (1971) Mechanism of instabilities of exothermic hypersonic blunt-body flows. Comb Sci and Tech 3:63-76

Stahl G, Warnatz J (1991) Numerical investigation of time dependent properties and extinction of structures of methane and propane air flamelets. Comb Flame 85:285

Taylor GI (1950) Proc Roy Soc London A201:159-164

Trevino C (1990) Ignition phenomena in H_2-O_2 mixtures. In: Prog in Astro Aero, AIAA, Vol 131, pp 19-43

Vasiljev AA (1994) Initiation of gaseous detonation by a high speed body. Shock Waves 3:321-326

Zeldovich J, Leipunsky O (1943) A study of chemical reactions in shock waves. Acto Physicochimica USSR 18:167-171

Experimental observation of oblique detonation waves around hypersonic free projectiles

J. Kasahara[1], A. Takeishi[2], H. Kuroda[1], M. Horiba[1], K. Matsukawa[2], J. E. Leblanc[2], T. Endo[3], T. Fujiwara[2]

[1]Department of Aerospace Engineering, Nagoya University
[2]Department of Microsystem Engineering, Nagoya University
[3]Center for Integrated Research in Science and Engineering, Nagoya University , Furo-cho, Chikusa-ku, Nagoya, Aichi 464-8603, Japan

Abstract. We studied the oblique detonation waves around hypersonic projectiles. Projectiles (10 mm diameter, conical nose shape) were fired at hypersonic speeds (2.8 ± 0.1 km/s) into stoichiometric hydrogen-oxygen mixtures (pressure between 10 kPa and 50kPa, at room temperature T = 300.2 ± 1.4 K). The flowfields around the projectiles were visualized using a multi-frame schlieren technique. We made a comprehensive study of the oblique detonation wave phenomena around the hypersonic projectiles with variations in two parameters: the projectile nose shape and the initial gas pressure. The projectile nose cone open-angle was varied from 60° to 180°. The initial gas pressure was varied from 10 kPa to 50 kPa. Four types of combustion were observed. At the lowest initial gas pressure 10 kPa or at the smallest open angle 60°. two non-ignition types: a detached bow shock wave type and an attached bow shock wave type were observed. At the upper initial gas pressure and bigger opened-angle, an oblique detonation type was observed. The presence of an intermediate type (which we label a straw hat type, consisting of an oblique detonation + a shock induced bow combustion wave) was observed at 20 - 33 kPa and 90° - 120°. The shock shape of this type is similar to the shape of a typical straw hat. Using multi-frame schlieren pictures and comparing the pictures shot in two observation windows separated by a distance of 260 mm, we confirmed that the oblique detonation waves were steady within 92.5 μs. By detonation polar analysis, the heat release behind the detonation wave front was close to that of the theoretical Chapman-Jouguet detonation, (within 20 %, and in most cases within 10 %). It should be noted that the minimum curvature radius in the detonation wave of the straw hat type was a radius of 40 - 50 times the induction length of the detonation wave. We think this is a significant parameter in the phenomena.

Key words: Oblique detonation waves, hypersonic free projectiles

1. Introduction

Experimental studies of oblique detonation waves are not only interesting in themselves, but also important for the stable operation of the new hypersonic propulsion applications, such as the ram accelerator (Hertzberg et al. 1988) and the oblique detonation wave engine (Powers 1994). Oblique detonation waves of hydrogen-oxygen or hydrogen-air mixtures were investigated using ballistic range projectiles (Lehr 1972, Kaneshige and Shephered 1996, Higgins 1997).

We made a comprehensive study of the oblique detonation wave phenomena around the hypersonic projectiles with variations in two parameters: the projectile nose shape and the initial gas pressure.

Table 1. Experimental conditions

Shot No.	Cone Angle [deg.]	Initial Pres. [kPa]	Initial Temp. [K]	Proj. Velocity [km/s]
119	120	33.3	295.1	2.92
128	120	33.0	301.4	2.86
130	120	50.5	301.0	2.73
134	120	21.3	301.0	2.83
137	120	40.5	301.6	2.91
138	120	9.5	301.1	2.83
140	60	33.6	300.4	2.96
141	180	33.6	299.4	2.89
142	90	33.3	299.9	2.81
143	150	33.4	300.4	2.82
146	60	50.3	300.1	2.98
147	90	51.2	300.0	2.80
148	180	10.0	300.0	2.84
149	180	51.3	300.0	2.75
151	120	34.2	299.6	2.81
153	90	21.1	298.9	2.93
173	120	32.9	296.1	2.84

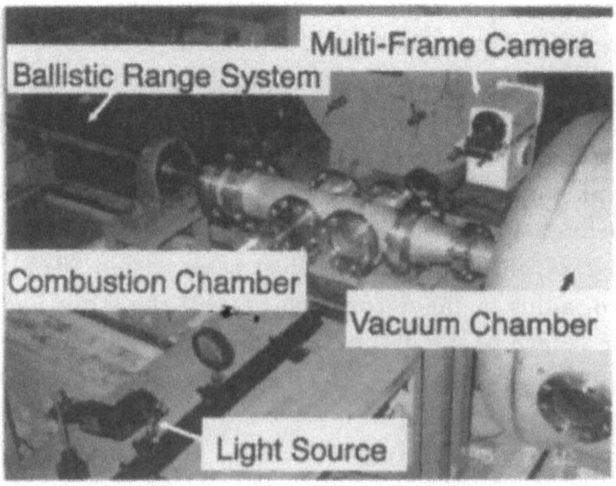

Fig. 1. Experimental facility.

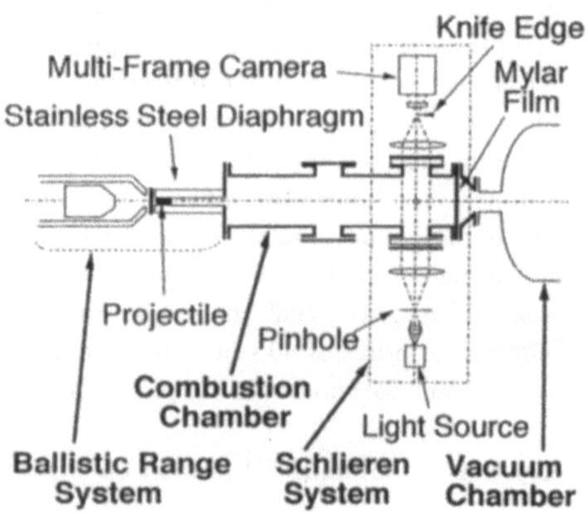

Fig. 2. Experimental arrangement (top view).

2. Experimental facility and conditions

The experimental facility is shown in Fig. 1 and a schematic diagram of the experimental arrangement (top view) is shown in Fig. 2. The experimental equipment is composed of four elements; (1) a two-stage, light-gas gun to inject the hypersonic projectiles, (2) a combustion chamber with the gas mixture, (3) a schlieren system for observation, and (4) a vacuum chamber for the exhausted projectile and burned gas. The two-stage, light-gas gun uses high-pressure air, at 5 MPa, to drive the piston. The driver light gas is helium gas. The maximum projectile speed is 5 km/s.

A stainless steel, cylindrical tube of 140-mm-internal diameter and 10-mm thickness was adopted as the combustion chamber containing the hydrogen-oxygen mixture. The combustion chamber had a 90 mm observation window on the horizontal plane, perpendicular to the center line of the tube. The position of the bow shock wave was determined by the piezo pressure gages Kistler 603B (Kistler Japan) attached to the wall of the combustion chamber. About 100 μs after detection, the projectile arrived at the observation-window section and schlieren system recorded the flowfield around the projectile.

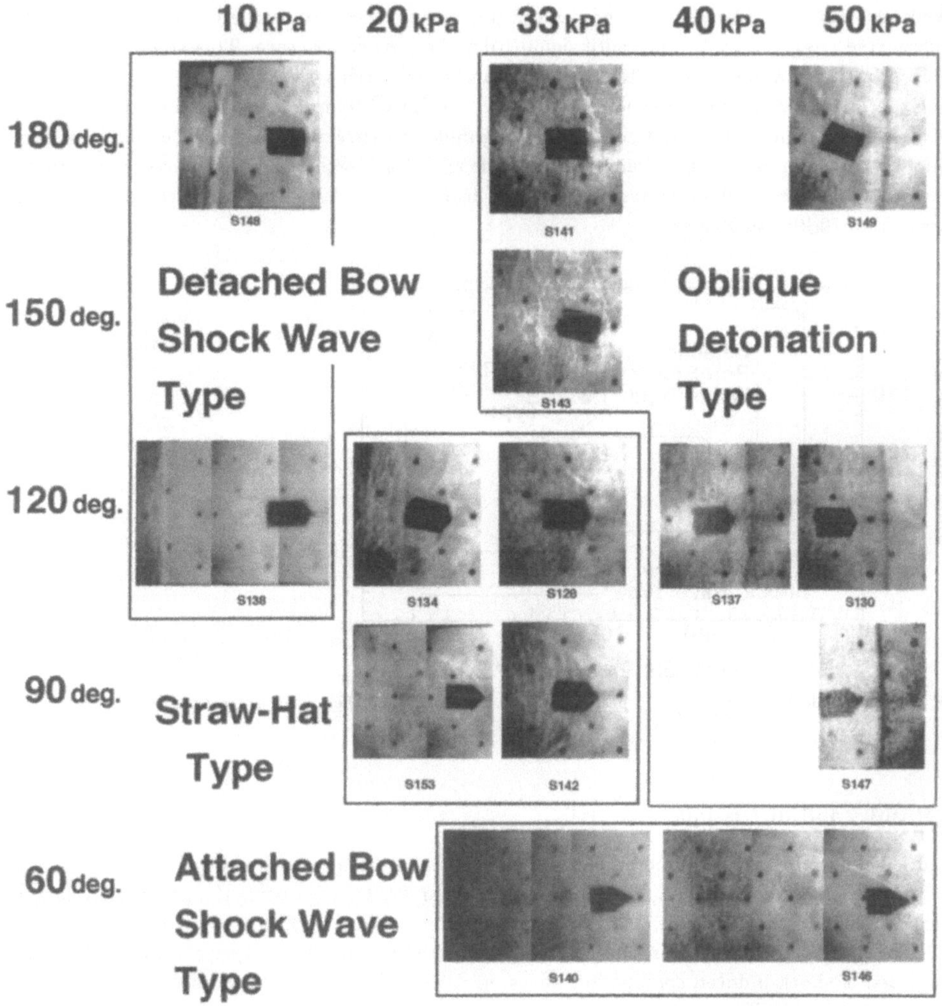

Fig. 3. Four combustion types categorized by the initial gas pressure and the nose cone angle.

After the combustion chamber, the projectile breaks a Mylar film and it is recovered in a vacuum chamber. The vacuum chamber is ready for injecting when the pressure drops to 100 Pa. As soon as the projectile breaks the diaphragm, the high-pressure burned gas is inhaled into the vacuum chamber, so the combustion chamber was not exposed to high pressure for long time. The total time for the operation, from releasing high pressure air in the gas gun, to recovering projectile in the vacuum chamber, was about 10 ms.

The schlieren method was used for visualizing the flowfield around projectile, using a stroboscope PS-240 (SUGAWARA Lab.) and SL540EZ (Canon) with the spark duration of 0.1 ms and

1.2 ms, respectively. The pinhole diameter was 2 mm. The objective lens was a spherical convex lens of 150 mm in diameter. The knife edge was placed of one side of the optical path, as shown in Fig. 2. Two kinds of high-speed multi-frame cameras (image converter camera) were used: an IMACON 790 (HADLAND PHOTONICS), and a FS501 ULTRANAC (NAC Inc.). The spatial resolution was approximately 0.2 mm at the objective plane. The frame camera and the light source were triggered by the pressure signal from the bow shock wave. This trigger signal was delayed by a digital delay generator DG535 (Stanford research systems).

Experimental conditions are shown in Table 1. The initial temperature was room temperature, 300.2 ± 1.4 K, which was almost constant throughout the experiment. All of the experiment were done with a cone-nose, 10-mm-diameter projectile and stoichiometric hydrogen-oxygen mixtures. The projectile nose cone open-angle was varied from 60° to 180°. The initial gas pressure was varied from 10 kPa to 50 kPa.

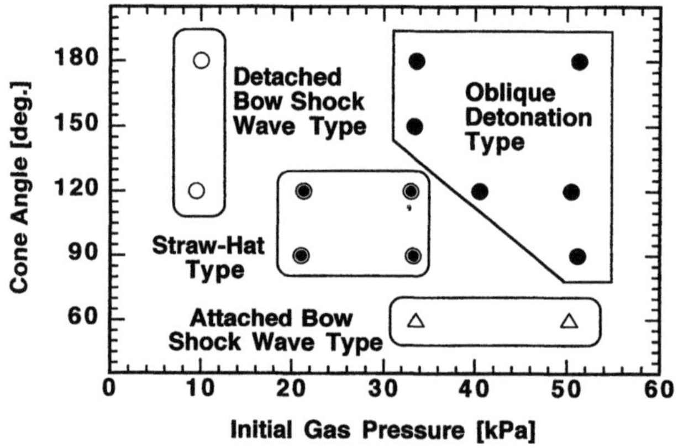

Fig. 4. The dependence of shock wave types on the nose cone angle and the initial gas pressure.

3. Results and discussions

Figure 3 shows the typical schlieren pictures against two parameters, the initial gas pressure and the projectile nose cone angles. The black dots are markers for measuring the projectile velocity. Each dot is 2 mm wide and is separated 20 mm from the next dot in the same horizontal or vertical line. The marker are the same in all schlieren pictures. The combustion types observed, which are composed of shock-induced combustion waves or oblique detonation waves, are categorized into four types shown in Fig. 3.

We shows the distribution of these types in Fig. 4 as a function of the projectile nose cone angle and the initial gas pressure. Figure 5 shows schematic pictures of four combustion types.

(1) Detached bow shock wave type is shown in Fig. 5a. This type was observed at 10 kPa, the lowest pressure. This combustion type generated a detached bow shock wave with no reaction and a normal detonation wave trailing behind and propagating with the same velocity as the projectile.

(2) Straw hat type were observed at the initial pressure, 22 - 33 kPa, and the projectile cone nose angle, 90°-120°. The conditions for the straw hat type were between (1) the detached bow shock wave type and (2) the oblique detonation type, which we will discuss next. Figure 5b shows that the waves generated around the projectiles in the straw hat type were made of two types of shock waves; a detached bow shock wave generated near the projectile tip and a detonation wave which

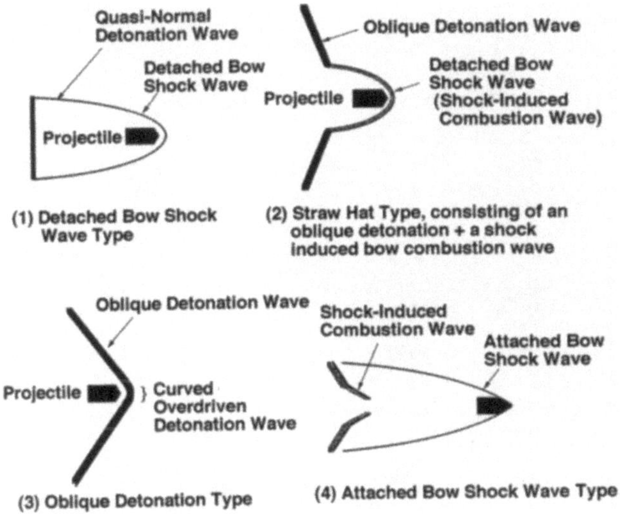

Fig. 5. Schematic pictures of four combustion types.

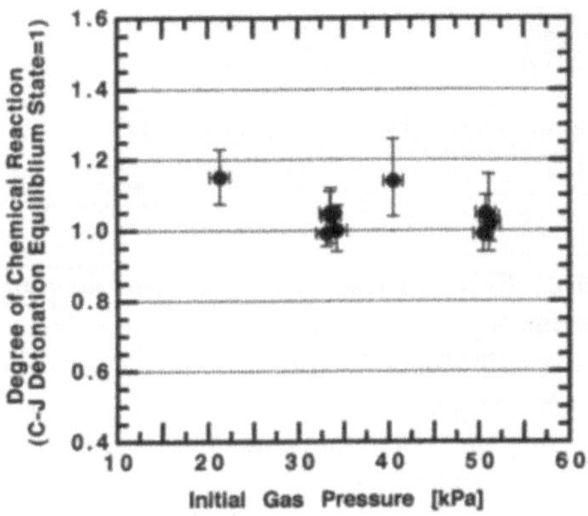

Fig. 6. The degree of chemical reaction of the oblique detonation as a function of the initial gas pressure.

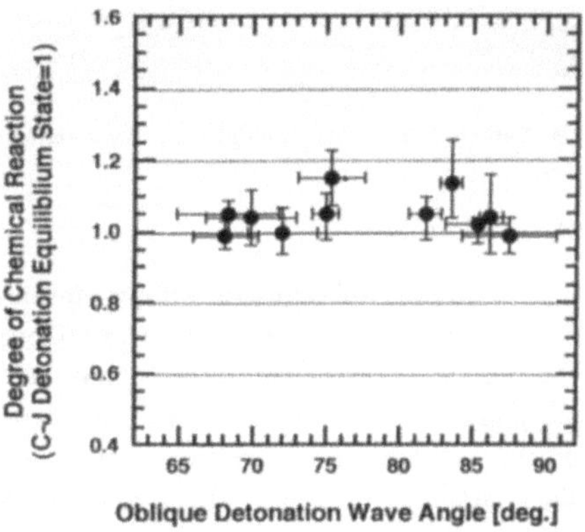

Fig. 7. The degree of chemical reaction of the oblique detonation as a function of the oblique detonation wave angle.

intersected the detached bow shock wave and propagated oblique to the projectile flight axis. It seems that the detonation wave was part of a cone. The propagation velocity of the detonation wave were approximately Chapman-Jouguet. The reason for our calling "Straw hat type" is that the shock wave shape of this type looks like a straw hat hanging on a hat rack.

(3) Oblique detonation type was observed for the initial pressure higher than 33 kPa and the cone angles greater than 90°. except that some conditions overlap with the straw hat type conditions. Figure 5c shows that in this type the curved overdriven detonation wave were generated around the projectiles and connected smoothly to the oblique detonation waves propagating at the Chapman-Jouguet velocity, forming a continuous shock front.

(4) Attached bow shock wave type was observed at the nose cone angle lower than 60°. As shown Fig. 5d, this combustion type generated an attached bow shock wave with no reaction behind the shock wave but a reaction zone generated at a distance of 50 mm behind the projectile.

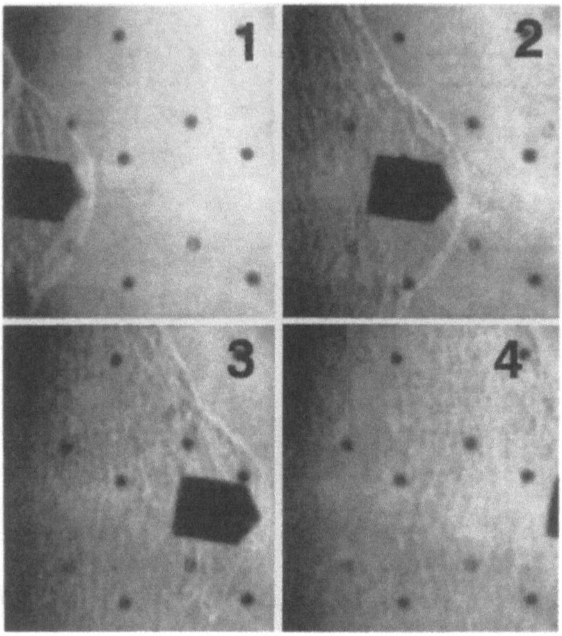

Fig. 8. A multi-framing schlieren picture of the straw hat type (SHOT119) . This type maintained steady state for 16 μs.

3.1 Heat release behind the oblique detonation wave By detonation polar analysis, we can estimate the heat release behind the oblique detonation waves. By assuming that the oblique detonation waves were two dimensional and the flow couldn't be affected by the flow after the sonic point, because of the small curvature of the detonation wave and the thermal choking condition, we can use detonation polar analysis. By using the detonation angle observed in the oblique detonation type or the straw hat type, we can calculate the heat release behind the oblique detonation wave, q_{od}. We define the degree of chemical reaction, $k = q_{od}/q_{cj}$; q_{cj} is the heat release behind the theoretical Chapman-Jouguet detonation wave. The degree of chemical reaction, k, is plotted as a function of the initial gas pressure in Fig. 6, and as a function of the oblique detonation wave angle in Fig. 7. From Figs. 6 and 7 it can be stated that the heat release behind the detonation wave front was close to that of the theoretical Chapman-Jouguet detonation, within 20 % and in most cases within 10 %. The degree of chemical reaction does not depend on either the initial gas pressure or the oblique detonation wave angle. It may be concluded that the oblique detonation waves observed in the present study were in the Chapman-Jouguet detonation state, and the oblique detonation waves were straight and did not diverge.

3.2 Measuring the steady-state duration of the straw hat type and the oblique detonation type Using multi-frame schlieren pictures and comparing the pictures shot in two observation windows separated by a distance of 260 mm, we can estimate the steady-state duration of the straw hat type and the oblique detonation type. Figure 8 shows that the straw hat type maintained steady state for 16 μs. Figure 9 shows the pictures shot at the two observation windows. From this figure the projectile caused detonation as soon as it injected combustion chamber, and the oblique detonation type had been steady state for 92.5 μs.

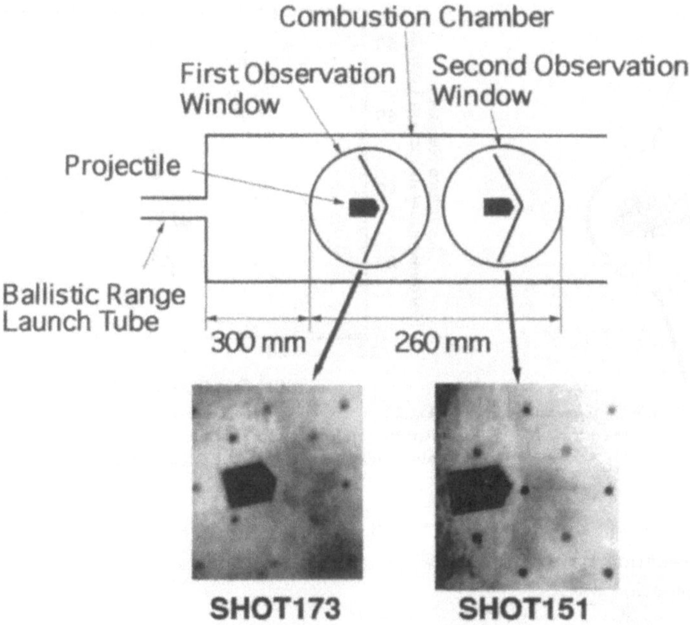

Fig. 9. The oblique detonation type had been steady state for 92.5 μs, comparing the pictures shot in two observation windows separated by a distance of 260 mm.

3.3 Minimum curvature radius in the detonation wave Figure 10 also shows the definition of the minimum curvature radius R of the detonation wave at the initiation point. R is equal to the radius of a cylinder attached to the initiation point. Figure 11 shows the curvature radius normalized by induction length, l_i, as a function of the initial gas pressure. l_i is defined as the induction length of steady two-dimensional C-J oblique detonation waves. We use Westbrook's chemical reaction model (Westbrook et al. 1982). R_{max} is the upper limit because of the finite size of the observation window. R_{min} is the lower limit because R cannot be smaller than the projectile size. It should be noted that the minimum curvature radius for any shot was about 40-50 times the induction length of the detonation wave.

It is assumed that the normalized curvature radius is constant and that the detonation ignites. Figure 11 shows that as the initial pressure increase or the projectile Mach number decrease, which is able to generate smaller R and smooth contact from the oblique detonation to overdriven bow detonation waves, the combustion type will change gradually from the straw hat type to the oblique detonation wave type.

4. Conclusions

We made a comprehensive study of the oblique detonation wave phenomena around the hypersonic projectiles with variations in two parameters: the projectile nose shape and the initial gas pressure. At the lowest initial gas pressure a detached bow shock wave type and an attached bow shock wave type were observed. At the upper initial gas pressure and bigger opened-angle, an oblique detonation type was observed. The presence of an intermediate type, a straw hat type, was observed at 20 - 33 kPa and 90° - 120°.

We confirmed that the oblique detonation waves were steady within 92.5 μs. The heat release behind the detonation wave front was close to that of the theoretical Chapman-Jouguet deto-

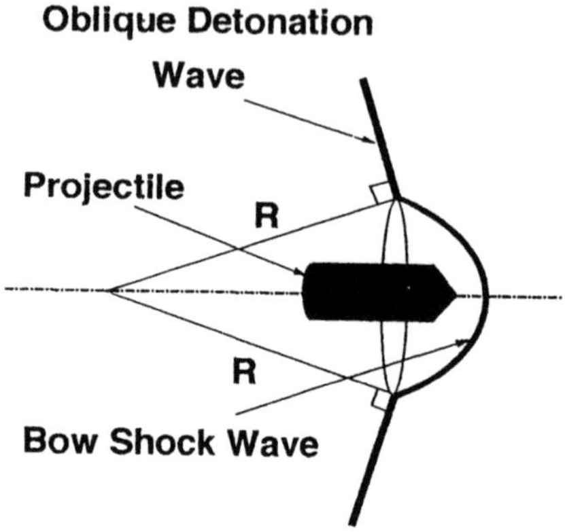

Fig. 10. The definition of the minimum curvature radius of the detonation wave, R, in the straw hat type.

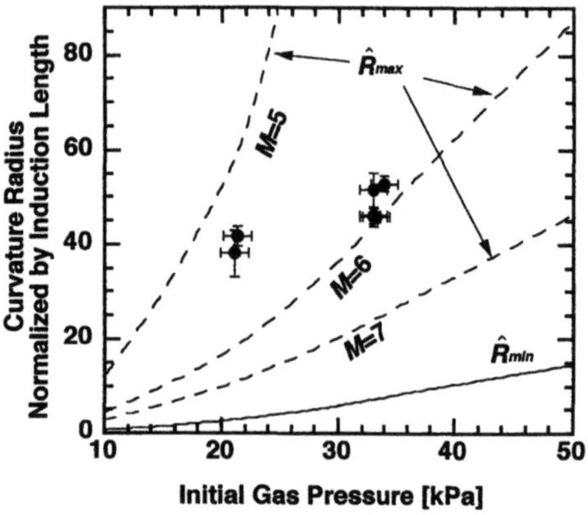

Fig. 11. Curvature radius normalized by induction length as a function of the initial gas pressure. It was about 40-50 in the present experimental conditions.

nation, (within 20 %, and in most cases within 10 %). It should be noted that the minimum curvature radius in the detonation wave of the straw hat type was a radius of 40-50 times the induction length of the detonation wave.

Acknowledgement

The authors are grateful to Kistler Japan Co. Ltd., J. Osawa Co. Ltd., and NAC inc. for the provision of laboratory facilities. We thank A. Saito, S. Takahashi, and K. Tachibana of Nagoya University for supporting this present work. The authors express their sincere thanks to Prof. N. Yoshikawa of Nagoya University, Dr. A. Matsuo of Keio University and Dr. H. Sakakita of Electrotechnical Laboratory. This work was made possible by a grant from JSPS Research Fellowship for Young Scientists.

References

Hertzberg A, Bruckner AP and Bogdanoff DW (1988) Ram accelerator: A new chemical method for accelerating projectiles to ultrahigh velocities. AIAA J 26:195-203

Higgins AJ (1997) The effect of confinement on detonation initiation by blunt projectiles. 33rd Joint Prop Conf and Exhibit, AIAA paper 97-3179

Kaneshige MJ, Shephered JE (1996) Oblique detonation stabilized on a hypervelocity projectile. 26th Int Symp on Combustion, Combustion Inst, Naples, Italy

Lehr HF (1972) Experiments on shock induced combustion. Astronautica Acta 17:589-597

Powers JM (1994) Oblique detonations: Theory and propulsion applications. In: Combustion in High-Speed Flows, Kluwer Academic Publishers, pp 345-371

Westbrook CK, Dryer FL, Schug KP (1982) A comprehensive mechanism for the pyrolysis and oxidation of ethylene. 19th Int Symp on Combustion, Combustion Inst, Haifa, Israel, pp 153-166

Numerical prediction of envelope oscillation phenomena of shock-induced combustion

A. Matsuo
Department of Mechanical Engineering, Keio University, 3-14-1, Hiyoshi, Kohoku-ku, Yokohama 223, Japan

Abstract. The envelope oscillation phenomena of the shock-induced combustion is numerically investigated by a series of the projectile velocities. The zero-dimensional analysis shows that the period of the envelope oscillation is essentially the low frequency oscillation coupled with the high frequency oscillations. As for the flowfield of the envelope oscillation, the numerical simulation reveals the detailed oscillation mechanism. The reaction front is always away from the bow shock wave, and the detonable reaction front is never established. Therefore, the corrugated pattern of the reaction boundary moves with fluid like a flowfield of the regular regime of the unsteady shock-induced combustion.

Key words: Shock-induced combustion, computational fluid dynamics, Detonation

1. Introduction

Unsteady shock-induced combustion has been experimentally investigated in 1960s and 1970s. A number of ballistic range experiments (Alpert and Toong 1972, Ruegg and Dorsey 1962, Lehr 1972, McVey and Toong 1971, Chernyi 1968, Behrens et al. 1965) with quiescent combustible gases were conducted, and they revealed the presence of a remarkably periodic manner over the whole region within the bow shock of the projectiles in the experimental photographs with spark light sources. These periodic density variations appear in two distinct regimes (Alpert and Toong 1972). One is referred to as the regular regime whose density variations are highly regular and low in amplitude, and the other is referred to as the large-disturbance regime whose oscillations are less regular and low in frequency but far more pronounced. The regular and large-disturbance regimes are not steady flowfields with respect to the projectile. The period of the large-disturbance regime is several times longer than that of the regular regime. The period in both the regular and the large-disturbance regimes depends on the projectile speed and is also a function of the induction time derived from the projectile velocity. The period was normalized by the induction time, τ_s, in the nearly uniform region behind the normal segment of the bow shock. The experimental results (Alpert and Toong 1972) suggested that the range of the period of the large-disturbance regime is $3 < t/t_{ind} < 12$, and the averaged period of the large-disturbance regime derived by the method of the least square was 5.23. The period of the regular regime oscillations has the same order of the induction time.

The mechanisms of the regular and the large-disturbance regimes have been numerically clarified by the recent investigations of author's group (Matsuo and Fujiwara 1993, Matsuo and Fujii 1995, 1996, Matsuo et al. 1995). A wave-interaction model on the stagnation streamline was proposed based on the simulation results. The recent simulation technique allows us to discuss the detail physics of the coupling between gas dynamics and chemical kinetics quantitatively. According to the simulation results (Mastuo and Fujii 1996, Kasahara et al. 1996) of the large-disturbance regime of the shock-induced combustion around the spherical projectiles, the typical flow feature of the large-disturbance regime has the cellular structure of the detonable reaction front along the bow shock in one cycle of the oscillations. On the other hand, the flow features of the regular regime has no detonable structure.

Recently, Kasahara et al. (1996) have investigated the unsteady shock-induced combustion using ballistic range projectiles and have categorized the combustion cells into five modes composed

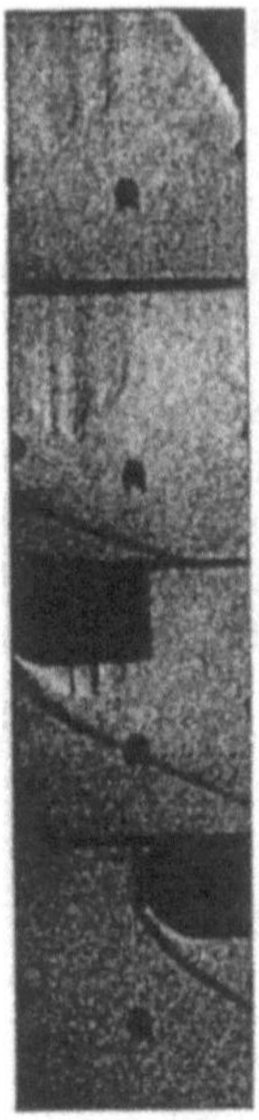

Fig. 1. Experimental output by Kasahara et al. (1996): Normal cell oscillation + Envelope oscillation modes

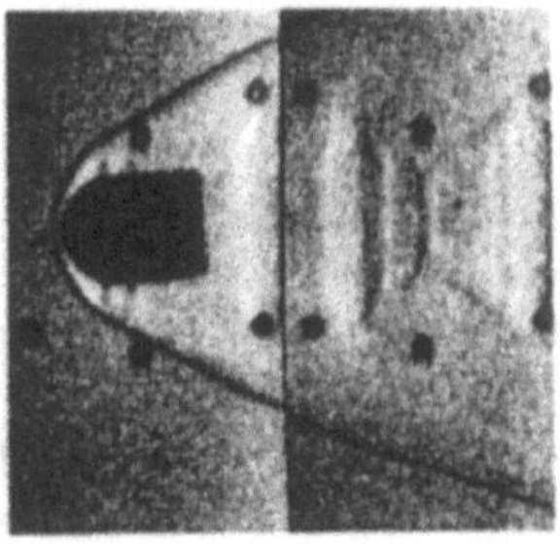

Fig. 2. Experimental output by Kasahara et al. (1996): High-frequency oscillation + Envelope oscillation modes

of three fundamental modes (Normal cell oscillation, High-frequency cell oscillation, and Envelope oscillation). The envelope oscillation has been newly reported in their work, and they explain that the envelope oscillation is superimposed on the high-frequency cell oscillation mode. In the present work, the numerical study is carried out to clarify the envelope oscillation phenomena revealed by Kasahara et al.

2. Envelope oscillation

Multi-frame schlieren photographs showing the envelope oscillation taken by Kasahara et al. (1996) are shown in Figs. 1 and 2. The combustion feature in Fig. 1 is categorized into (Normal cell oscillation + Envelope oscillation). In Fig. 2, the feature is categorized into(High-frequency cell oscillation + Envelope oscillation). In the work of Kasahara et al., the normal cell oscillation correspond to the feature of the large-disturbance regime, and the high-frequency cell oscillation correspond to the feature of the regular regime (Alpert and Toong 1972). However, Figs. 1 and 2 do not show the detonable reaction front which have the triple point merged by the bow shock and the reaction front. The high frequency striations with the corrugated pattern appear in low frequency mode, and the low frequency mode is called "Envelope oscillation." As observed in the experimental result in Fig. 3, the envelope oscillation appears only in the lower projectile velocity case. In other words, the envelope oscillation is observed around the lower limit of the unsteady shock-induced combustion.

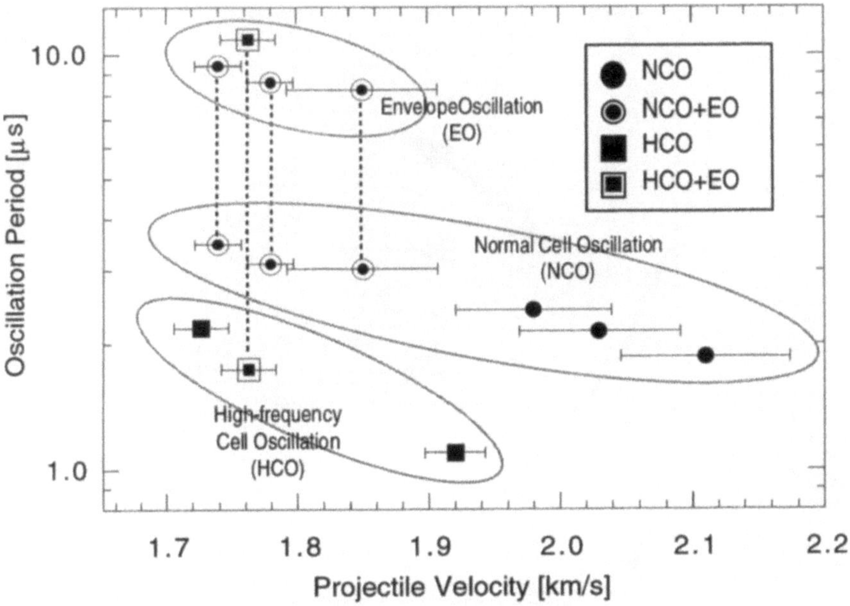

Fig. 3. The dependance of oscillation on the projectile velocity of the experimental results by Kasahara et al. (1996)

3. Period of oscillation

The period of the large-disturbance regime (Normal cell oscillation) is $3 < t/t_{ind} < 12$, and the averaged period derived by the method of the least square was 5.23. The period of the regular regime (High-frequency cell oscillation) has the same order of the induction time. Generally, the oscillation period is normalized by the induction time for general discussion of the period, but in Fig. 3 the period is not normalized. It must be more careful to classify the unsteady mode, and then the oscillation period should be normalized by the induction time for qualitative discussion of the unsteady shock-induced combustion. The induction time t_{ind} is derived by the zero-dimensional analysis, which is carried out by the time-integration of species equations using the hydrogen-oxygen reaction mechanism (Jachimowski 1988) consisting of 9 species and 19 reactions in zero-dimension in space under the constant volume mode of the thermally perfect gas mixture. For an initial condition, all the variables are given by the normal shock relation ahead of the projectile body under the thermally perfect gas. The temperature profile is obtained by the analysis, and the induction time is defined as the time where the temperature increase per unit time indicates a maximum value, $(dT/dt)_{max}$.

The induction times are calculated under the experimental conditions, pressure 75 kPa, temperature 300 K and gas mixture $2H_2+O_2+3.76N_2$. The induction time for each case is indicated as the dashed bold line in Fig. 4. Based on the previous experimental observations (Alpert and Toong 1972), the averaged period of the large-disturbance regime of the shock-induced combustion is reported as 5.23 and overlaps the experimental results as the bold line in Fig. 4. As mentioned above, the period of the large-disturbance regime (Normal cell oscillation) is $3 < t/t_{ind} < 12$, and two lines, $3t/t_{ind}$ and $12t/t_{ind}$, are also indicated by the dashed lines in Fig. 4. Therefore, the period of the large-disturbance regime (Normal cell oscillation) should locate between the lower and the upper dashed lines, at least. However, the area of the envelope oscillation in Fig.

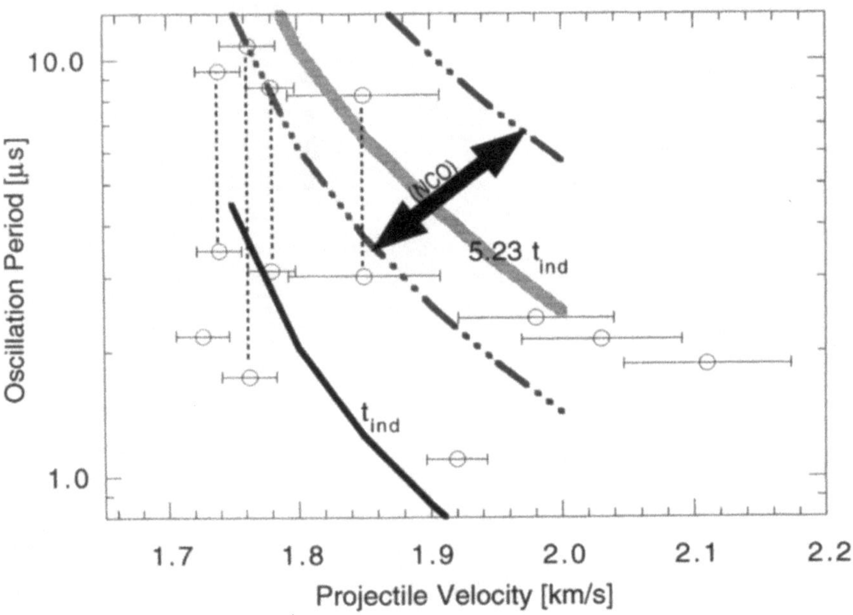

Fig. 4. Classification of the experimental results by Kasahara et al. (1996) by the induction time

3 correspond to that of the large-disturbance regime (Normal cell oscillation). The period coupled with the envelope oscillation in the area of the normal cell oscillation is obviously shorter than that of the normal cell oscillation, and it is almost the induction time. Therefore, such high frequency oscillations coupled with the envelope oscillation should be classified into one of the high-frequency cell oscillation.

4. Computational setup

To clarify the envelope oscillation phenomena, a series of computations of the flowfields are carried out. The governing equations are Euler equation under the axisymmetric assumption, and the chemical reaction is considered as hydrogen-oxygen mechanism consisting of 8 species and 19 elementary reactions (Jachimowski 1988) omitting nitrogen reactions. The reproduction of the physical unsteadiness by our computational code is confirmed by the previous computational works (Matsuo et al. 1995, Matsuo and Fujii 1996), and the detailed computational setup is referred in them.

A computational domain is limited to the region in front of the hemispherical body, and the flow is assumed to be axisymmetric based on the experimental observations. The number of grid points is 401 $\times$ 401, which are equally distributed in η-direction.

5. Results and discussions

A series of computations under the experimental conditions of Kasahara et al. (Kasahara et al. 1996) are carried out. The computed conditions are that the test gas pressure is 75 kPa and the projectile diameter is 10.0 mm, and three kinds of the projectile velocity (1,850, 1,800, 1,758.7 m/s) are simulated. Those conditions correspond to both the envelope oscillation and the normal cell oscillation in the experimental outputs.

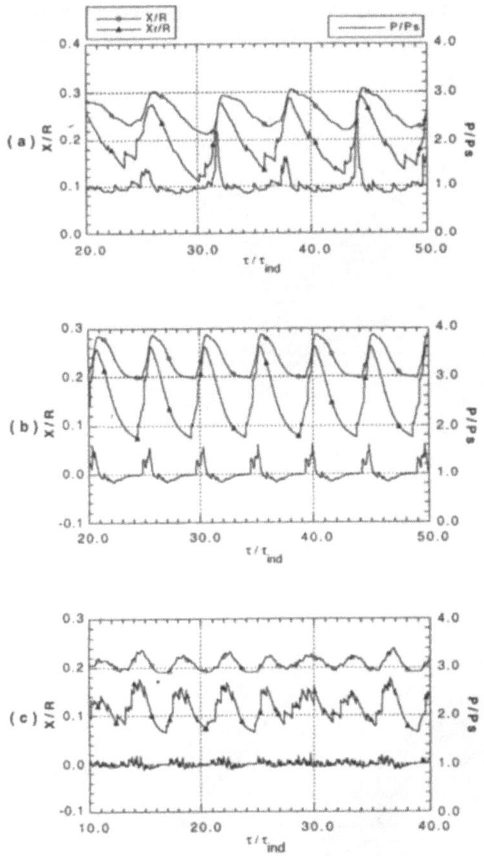

Fig. 5. Histories of the shock standoff distance (sphere), the location of the reaction front (triangle) and the shock front pressure on the stagnation streamline: Projectile Velocity: (a) 1850.0 m/s. (b) 1800.0 m/s and (c) 1758.7 m/s

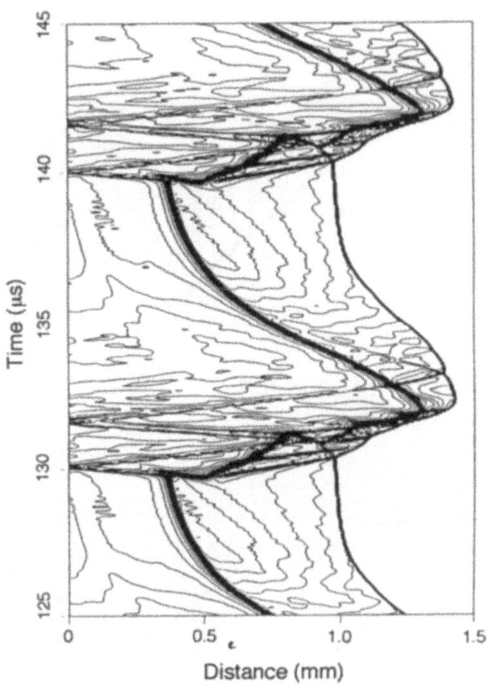

Fig. 6. X-t diagram of the density distribution on the stagnation streamline: Projectile velocity 1800.0 m/s, pressure 75 kPa, diameter 10.0mm

6. Histories of stagnation streamline

A series of simulations are conducted by changing of the projectile velocity. Figures 5 show the histories of the shock standoff distance, the location of the reaction front and the shock strength of the bow shock on the stagnation streamline. The line with spheres symbol is the history of the shock standoff distance, and the line with shaded triangle symbol is the history of the location of the reaction front. The symbols are plotted at every 4,000 iteration steps of the computations. The location of the shock standoff distance and the reaction front is indicated by the distance from the stagnation point, and the distance is normalized by the projectile radius. The shock strength normalized by the steady shock strength is represented by the pressure level of immediately behind the bow shock wave. The shock strength of the steady state solution is calculated by the normal shock relation with real gas effects.

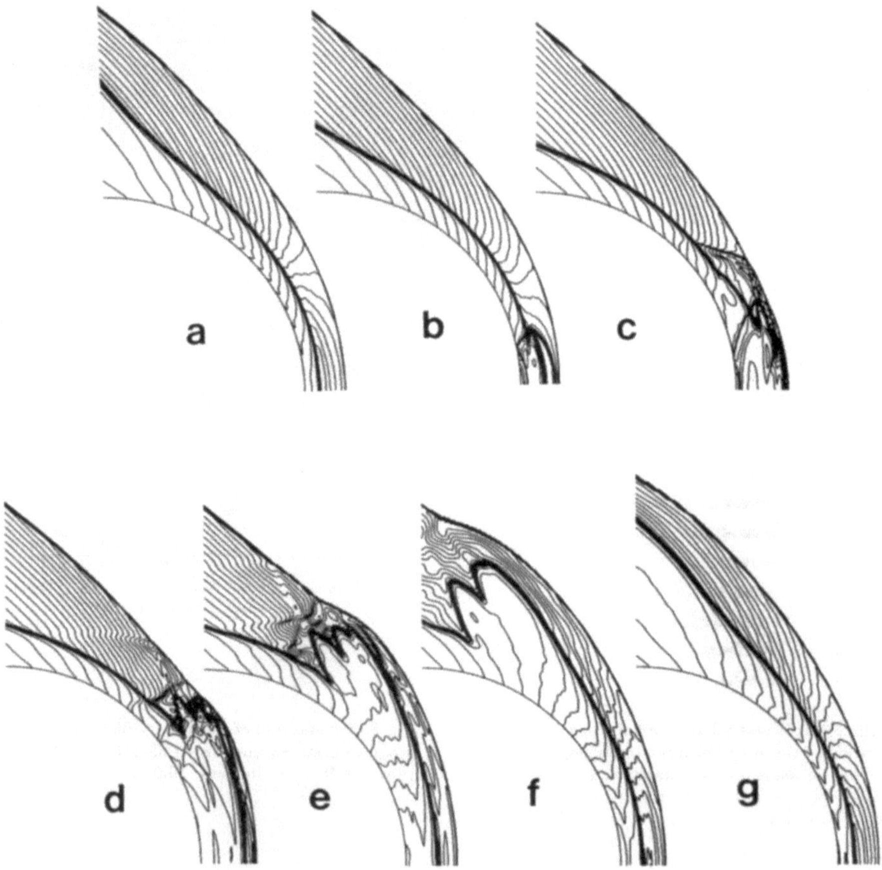

Fig. 7. Density contour plots: Projectile velocity 1800.0 m/s, pressure 75 kPa, diameter 10.0mm

All of results mainly show that the low frequency mode although the reaction front always away from the bow shock and never penetrates it in Fig. 5c, and the shock pressure history does not show the low frequency mode. The shock pressure history in Fig. 5a shows the strong peak in one cycle, and its level is more than 2.0. The flowfield of Fig. 5a shows the typical features of the large-disturbance regime in the simulation results. See Fig. 5b, the low frequency oscillation is regularly repeated, and each shock pressure shows completely the same pattern. The x-t diagram of the density distribution on the stagnation streamline is shown in Fig. 6. The bow shock, the reaction front and the wave interaction are completely repeated on the stagnation streamline. The wave interaction for the low frequency mode consists of the self-enforced explosion sub-mechanism. Figures 7 show the time-evolving density contour plots of one cycle. In Fig. 7a, no corrugated patterns are observed on the reaction boundary, and in Fig. 7b, the new reaction region in front of the projectile is generated. From Fig. 7b to 7c, the wave interaction of the self-enforced explosion sub-mechanism must be occurred. After Fig. 7c, the corrugated pattern created by the self-enforced explosion moves downstream. At Figs. 7e and 7f, the corrugated pattern is clearly separated from the bow shock, and two peninsulas of the reaction region appear in the flowfield. The regularly corrugated high-frequency oscillation in Figs. 1 and 2 is considered as such peninsulas of the reaction region in Fig. 7f. After Fig. 7g, the new reaction occurs in front of the projectile as well as Fig. 7b.

Hence, the envelope oscillation can be explained by the simulation results as follows: since the reaction boundary does not merge the bow shock wave, the detonable reaction front consisting of the bow shock and the reaction front is not established. If the reaction front is connected with the bow shock wave, the flow feature is recognized as the large-disturbance regime The detonable reaction front propagates along the bow shock wave. Then, the corrugated reaction boundary is erased from the flowfield around the projectile. In the case of the envelope oscillation, the reaction front is always away from the bow shock wave, and the corrugated pattern of the reaction boundary moves with fluid like a flowfield of the regular regime of the unsteady shock-induced combustion.

7. Conclusions

The envelope oscillation phenomena of the shock-induced combustion was numerically investigated by a series of the projectile velocities under the experimental condition of Kasahara et al. The zero-dimensional analysis was carried out to clarify the oscillation mode of the shock-induced combustion. The analysis showed that the period of the envelope oscillation is essentially the low frequency oscillation coupled with the high frequency oscillations. The flowfield of the envelope oscillation was also simulated under the axisymmetric condition, and the detailed oscillation mechanism was revealed. The reaction front is always away from the bow shock wave, and the detonable reaction front was never established. Therefore, the corrugated pattern of the reaction boundary moves with fluid like a flowfield of the regular regime of the unsteady shock-induced combustion.

References

Alpert RL and Toong TY (1972) Periodicity in exothermic hypersonic flow about blunt projectiles. Astronautica acta 17:539-560

Behrens H, Struth W, Wecken F (1965) Studies of hypervelocity firings into mixtures of hydrogen with air or with oxygen. In: Proc 10th Int Symp on Combustion, pp 245-252

Chernyi GG (1968) Supersonic flow past bodies with formation of detonation and combustion fronts. Astronautica acta 13:467-480

Jachimowski CJ (1988) An analytical study of the hydrogen-air reaction mechanism with application of scramjet combustion. NASA TP-2791

Kasahara J, Horii T, Endo T, Fujiwara T (1996) Experimental observation of unsteady H_2-O_2 combustion phenomena around hypersonic projectiles using a multi-frame aamera. In: Proc 26th Int Symp on Combustion, pp 2903-2908

Lehr HF (1972) Experiments on shock-induced combustion. Astronautica Acta 17:589-597

Matsuo A, Fujiwara T (1993) Numerical investigation of oscillatory instability mechanism in shock-induced combustion around an axisymmetric blunt body. AIAA J 31:1835-1841

Matsuo A, Fujii K, Fujiwara T (1995) Flow features of shock-induced combustion around projectile traveling at hypervelocities. AIAA J 33:1056-1063

Matsuo A, Fujii K (1995) Computational atudy of large-disturbance oscillations in unsteady supersonic combustion around projectiles. AIAA J 33:1828-1835

Matsuo A, Fujii K (1996) Detailed mechanism of the unsteady combustion around hypersonic projectiles. AIAA J 34:2082-2089

McVey JB, Toong TY (1971) Mechanism of instabilities of exothermic hypersonic blunt-body flow. Combustion Sci Tech 3:63-76

Ruegg FW, Dorsey W (1962) A missile technique for the study of detonation wave. J Res Nat Bureau Standards 66C:51-58

Diagnostics

Experimental investigation of ram accelerator flow fields and combustion kinetics

M. R. Kamel, C. I. Morris, A. Ben-Yakar, E. L. Petersen, R. K. Hanson
High Temperature Gasdynamics Laboratory Stanford University, California 94305

Abstract. Ram accelerator-related research at Stanford in the areas of hypersonic reactive flows and high-pressure combustion kinetics is presented. Research on reactive flows includes investigation of the combustion modes observed in hypersonic reactive flows over blunt cylinders and 2D bodies. In the experiments reported herein, simultaneous OH PLIF and schlieren imaging experiments of hypersonic reactive flow fields around spherical-nosed and flat-faced cylinders, 19 and 25 mm in diameter, have been performed. Stagnation pressure histories were recorded using a pressure transducer embedded in the cylinders. Methane-, ethylene-, and hydrogen-based fuel-oxidizer mixtures were used at different free stream conditions. Three different combustion modes were observed as flow velocity was varied relative to the Chapman-Jouget speed, consistent with previous work. These experiments represent the first time pressure disturbances in the unsteady combustion modes have been directly measured. Gas-phase combustion kinetics research involves the use of a high-pressure shock tube facility for ignition delay time measurements and detailed kinetics modeling for ram accelerator mixtures and conditions. Ignition time measurements are presented, and a detailed kinetics mechanism developed to model CH_4/O_2 ram accelerator ignition is reviewed.

1. Introduction

Crucial to the development of the ram accelerator and related CFD codes is the availability of data bases for hypersonic reactive flowfield structure and high-pressure chemical kinetics. The research at Stanford covers two areas: (1) hypersonic reactive flows around 2D and axisymmetric bodies, and (2) high pressure, gas-phase combustion kinetics. An expansion tube facility (Kamel et al. 1995) is utilized for the fundamental experiments on hypersonic reactive flowfields. Flowfield measurements in the expansion tube are based primarily on planar laser-induced fluorescence (PLIF) of key species, together with conventional schlieren imaging, pressure sensing, and absorption/emission spectroscopy. The expansion tube facility has been used to accelerate inert and reactive mixtures (hydrogen, methane and ethylene-based fuels) to velocities ranging from 1700 to 2200 m/s, with Mach numbers in the range of 4 to 7. In the reactive experiments, simultaneous schlieren and OH PLIF imaging is performed to obtain information on the location of the shock wave (schlieren) and the regions of combustion (PLIF) in the flowfield during the same test.

Work on high pressure gas-phase kinetics is performed in a high-pressure shock tube facility (Petersen et al. 1996a). The facility, with test pressure capability to 100 MPa, is used to measure chemical reaction rate parameters in fuel-oxidizer mixtures at high pressures and temperatures. The diagnostic methods utilized in the high-pressure shock tube include cw laser absorption techniques developed at Stanford (UV, visible, and IR wavelengths), conventional emission spectroscopy, and pressure gauges. This paper is divided into two sections. Investigation of hypersonic reactive flows using the expansion tube facility is presented first, and the second section covers the shock tube studies of combustion kinetics at high pressures.

2. Expansion tube measurements of supersonic reactive flowfields

Research in the expansion tube facility is divided into two categories; (1) the study of combustion instabilities in hypersonic reactive flows around blunt projectiles; and (2) the investigation of shock-induced combustion and oblique detonations on 2D wedges. Only the work on blunted projectiles is presented here, interested readers can refer to previous publications (Morris et al. 1995) for discussions of experiments on 2D wedges. The flowfield optical diagnostics consist of simultaneous schlieren and PLIF imaging to obtain information on the reaction front and the location of the shock. The application of these two imaging techniques allows for the measurement of shock and reaction front stand-off distances during the same test.

2.1 PLIF description Combustion information was acquired using PLIF imaging (Hanson et al. 1990), which provided temporally and spatially-resolved pictures of the reactive expansion tube flows. This PLIF work relied on OH, a naturally-occurring combustion radical, as the fluorescent tracer. The resulting images, in which regions of high signal correspond qualitatively to regions of high OH concentration, are therefore useful for purposes of combustion visualization. Quantitative measurements of flow field quantities such as rotational and vibrational temperature are also realizable (Houwing et al. 1996, Palmer et al. 1993).

The basis of PLIF as applied here is linear laser excitation of tracer molecules in the flow followed by broadband collection of the fluorescence from the radiative decay of these excited molecules. Based on considerations of separation from neighboring lines and measurement sensitivity at the temperatures in question, the $Q_1(7)$ transition of the $A^2\Sigma^+ \leftarrow X^2\Pi$ (1,0) band of OH, located at 283.31 nm, was selected for laser excitation. Radiation at this wavelength was provided by the frequency-doubled output of a dye laser pumped by a pulsed Nd:YAG laser. Rhodamine 590 dye is used for OH PLIF transitions near 283 nm, with pulse energies of approximately 9 mJ. The sheet is roughly 0.5 mm thick $\times$ 30 mm wide at the viewing section. The fluorescence signal is collected through the same exit window as that of the schlieren system. To separate both light signals, a 50 mm diameter dichroic mirror is mounted at 45° to the optical axis perpendicular to the exit window. The dichroic, designed for larger than 99 % reflectivity between 300 and 320 nm, reflects the OH fluorescence but is transparent to the schlieren beam. The reflected fluorescence is collected onto the 578 $\times$ 364 pixel array of an ICCD camera. UG11 and WG305 Schott glass filters were placed in front of the f/4.5, 105 mm UV lens in order to block elastically scattered laser light while still passing the majority of the OH fluorescence. For each shot, the CCD array accumulated a fluorescence signal given by:

$$S_f = C_{opt} V_c n_0 f_{J''} B E g \phi, \tag{1}$$

where C_{opt} represents the overall efficiency of the optical setup at converting photons from fluorescence into photoelectrons incident on the CCD, V_c is the collection volume, n_0 is the number density of the absorbing species (OH in this case), $f_{J''}$ is the Boltzmann fraction of the tracer species molecules that are in the absorbing state, B is the Einstein coefficient for stimulated absorption, E is the laser energy fluence (energy per unit area), and g is the spectral convolution of the laser line and the absorption transition. The fluorescence yield ϕ , defined as $\frac{A}{(A+Q)}$, where A is the Einstein coefficient for spontaneous emission from all the populated upper states and Q is the rate of electronic quenching to the lower state, represents the fraction of the molecules pumped to the upper state that decay radiatively. Note that the given equation (1) is valid only for the case of weak excitation, which is the scenario considered here.

As examination of the above fluorescence equation shows, rigorous interpretation of a PLIF image is complicated by the multiple dependencies of the fluorescence signal. Nevertheless, for the purposes of qualitatively evaluating the flowfield, some conclusions can be drawn from the fluorescence images acquired here, which have been post-processed only to subtract the camera dark noise. Namely, it can be seen that the parameter in the fluorescence equation that will vary most significantly in the imaged region is the OH number density. Other parameters are less responsible for the variation in signal observed. For instance, the Boltzmann fraction $f_{J''}$ of molecules in the N"=7, v"=0 level of the ground electronic state does not vary by more than a factor of two across the temperature range encountered behind the bow shock, while the overall signal varies by more than a factor of 30 in that region. Variations in other relevant quantities across an image are similarly small contributors to the signal relative to the OH concentration gradients. These variations would include those due to the temperature and overall number density dependencies of the quenching term in the fluorescence yield, as well as those attributable to the particulars of the collection optics and to non-uniformities in the laser sheet profile. Straightforward corrections to account for the latter two effects (Seitzman et al. 1993) will be incorporated into future PLIF work in the expansion tube. Given, then, that the OH PLIF images constitute rough maps of OH concentration, they can be used to identify regions of combustion in a way that was impossible with schlieren photography.

2.2 Hypersonic flows around blunt projectiles The problem of hypersonic reactive flow over blunt bodies has been investigated experimentally and numerically over the past thirty years (Alpert and Toong 1972, Ruegg and Dorsey 1962). Most experimental investigations consisted of firing blunt projectiles into a chamber filled with a quiescent combustible mixture, typically hydrogen-based. The observed combustion modes have been characterized into three regimes: the smooth flame front regime, the regular disturbance regime, and the large disturbance regime.

In the smooth flame front regime, the combustion mode is that of an adiabatic shock wave followed some time later by a flame front, whose boundaries are roughly parallel to those of the blunt body. The regular regime is distinguished by the observation of disturbances traveling between the flame front and the shock wave. These disturbances perturb the flame front and cause it to have a corrugated boundary. In the large disturbance regime, the disturbances are strong enough to perturb the shock wave, and can lead to the formation of a cellular detonation. The criteria determining which of these regimes occur at different conditions remain under investigation.

In this paper, results of imaging experiments of hypersonic reactive flows around blunt bodies are presented. In addition to the optical diagnostics discussed above, a pressure transducer is embedded inside the blunted body to record the time history of the pressure in the stagnation region. Pressure disturbances in the stagnation area are apparent in the stagnation pressure traces, and their frequency and amplitude can be measured. Such quantitative measurements of the pressure oscillations in the stagnation region are important for the validation of CFD codes used to simulate the unsteady combustion modes. These experiments represent the first time that the pressure oscillations in the unsteady combustion modes have been directly measured.

The experiments performed at Stanford are aimed at reproducing, in an expansion tube facility, the different combustion modes observed previously in ballistic ranges. Two kinds of bodies were used in the current experiments: spherical-nosed cylinders and flat-faced cylinders, with two body diameters for each cylinder type: 25.4 mm and 19 mm. The mixtures used are stoichiometric hydrogen-oxygen diluted in 80 % nitrogen, stoichiometric methane-oxygen diluted in 50 % nitrogen, and stoichiometric ethylene-oxygen mixtures diluted in 60 - 80 % nitrogen.

2.2.1 Smooth flame front regime The smooth flame front regime is distinguished by an adiabatic shock, followed by a distinct induction zone and flame front. Figures 1a and 1b are examples of this case. This condition is observed to occur in cases where the ratio of free stream velocity, V, to the mixture's CJ velocity, V_{CJ}, is larger than unity. The stagnation pressure traces for these conditions do not exhibit any obvious oscillations within the frequency response range of the pressure transducer ($<$ 500 kHz).

As observed in the images, the combustion is closest to the shock wave along the stagnation line where the temperature increase is the largest. In the radial direction away from the stagnation streamline, the curvature of the bow shock increases leading to a decrease in the post shock temperatures and an increase in the corresponding induction time. The increase in induction time, and the subsequent cooling of the shocked gas particles as they expand around the cylinder, cause the quenching of the flame front at some radial distance away from the stagnation line. The last particle streamline to burn thus defines the boundary of the burnt region (Chernyi 1968, Lee 1994). This is apparent in the OH images where the boundary of the OH PLIF signals seem to follow a streamline parallel to the body, Fig. 1.

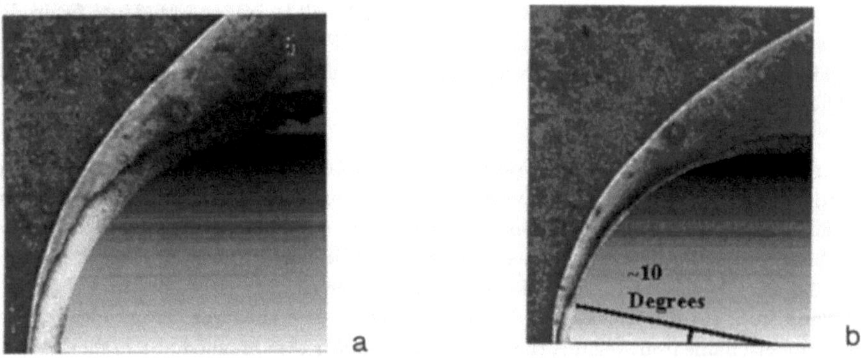

Fig. 1. Simultaneous OH PLIF and schlieren images for flow over a 25 mm spherical-nosed cylinder of (a) a hydrogen-based mixture with $V_\infty = 1780$ m/s, $P_\infty = 23.0$ kPa, and $T_\infty = 280$K with a V/V_{CJ} value of 1.3; and, (b) a methane-based mixture with $V_\infty = 2150$ m/s, $P_\infty = 7.1$ kPa, and $T_\infty = 260$ K with a V/V_{CJ} value of 1.13.

Upon inspection of the PLIF and schlieren images, the angle θ_c at which the flame front is extinguished is measured to be around 40° for the hydrogen-based mixture and around 10° for the methane-based mixture. Here θ is the angle between the flow axis and different points along the bow shock, with the vertex at the center of the hemisphere ($\theta = 0$ is the stagnation line). It is expected that the value of θ_c depends on the flow conditions, the post-shock chemical induction time of the mixture used, and the specific geometry of the blunt body. A non-dimensional parameter is proposed to estimate the value of θ_c for the different mixtures around spherical-nosed bodies. The parameter is defined as the ratio of the chemical induction time, τ_{ign} , over a characteristic flow time, τ_{flow}. The characteristic flow time used is the ratio of body diameter over the post-shock velocity. Note that the velocity is normal to the shock only along the stagnation streamline. The chemical induction times along the bow shock are calculated at conditions immediately downstream the shock. For the images shown in Fig. 1, the shock shape is divided into short segments which are then treated as oblique shocks. The post-shock conditions behind each oblique segment are then calculated using the 1D shock equations, which allows for the calculation of τ_{ign}/τ_{flow}.

Figure 2 shows plots of the post-shock temperature ratio, $T_y/T_{y,\theta=0}$, and the non-dimensional parameter τ_{ign}/τ_{flow}, just downstream of the bow shock as a function of θ. It can be seen that the angle θ_c for both cases occurs around values of $\tau_{ign}/\tau_{flow} \sim$ O[1]. Figure 2 also shows the values of the non-dimensional parameter calculated using experimental fits for the bow shock shape obtained from the literature (Billig 1967).

The hydrogen-based mixture of Fig. 2a has a lower activation energy and shorter induction times than the methane-based one of Fig. 2b. Hence, the combustion is sustained behind the shock wave over a wider region than for the methane-based mixture.

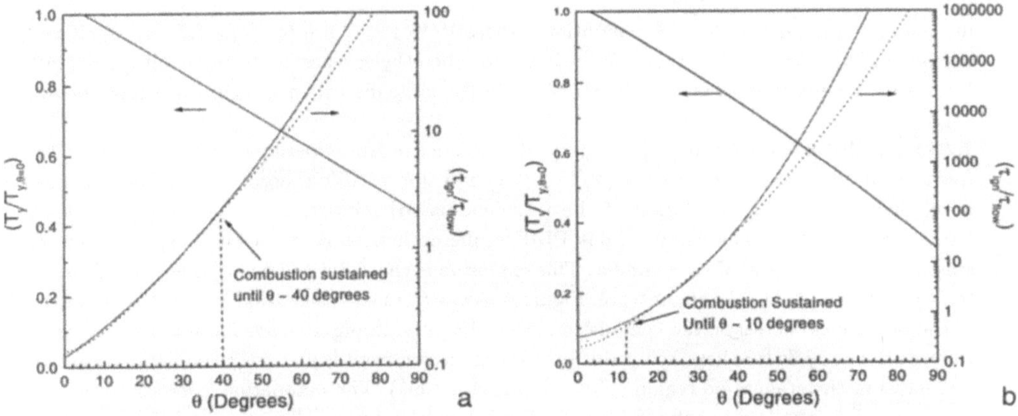

Fig. 2. Plots of temperature and characteristic times ratios immediately downstream the bow shock evaluated for flow over a 25mm spherical-nosed cylinder of (a) 13.3%H_2 + 6.7%O_2 + 80%N_2 at free stream conditions of M=5, P=23 kPa, T=280K, and (b) 16.7%CH_4 + 33.3%O_2 + 50%N_2 at free stream conditions of M=6.7, P = 6.6 kPa and T=260K. Post-shock temperatures are calculated using 1D shock equations. Induction times for the hydrogen-based mixture are calculated using a correlation given by Cheng & Oppenheim (1984) and those for methane are calculated using a correlation given by Lifshitz and Skinner (1971). θ is the angle formed between the stagnation line (θ=0) and points along the bow shock. Solid lines are calculations behind the shock shapes traced from the images of Fig. 1. Dashed lines are calculations behind shock shapes given by experimental fits obtained from the literature (Billig 1967).

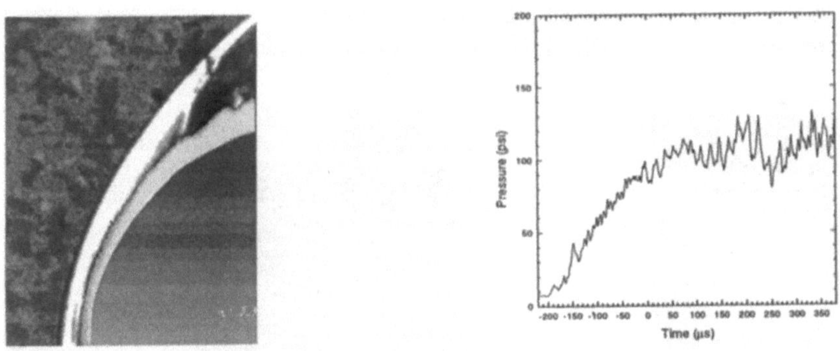

Fig. 3. Simultaneous OH PLIF and schlieren image for M=5.2 flow of a 7.5%C_2H_4 + 22.5%O_2 + 70%N_2 over a 25.4 mm spherical-nosed cylinder. The trace on the right represents the stagnation pressure history measured by a pressure transducer embedded inside the cylinder, with t = 0 denoting the beginning of the 200 μs steady test time. For this case, the pressure trace exhibited regular oscillations at a frequency of 53 kHz. The free stream conditions are V_∞ = 1730 m/s, P_∞ = 23 kPa, and T_∞ = 280K, with a V/V_{CJ} value of 0.94.

2.2.2 Regular regime In this regime, regular oscillations are observed in the stagnation pressure traces, and corrugations are observed in the flame front. The bow shock, however, remains unaffected by the oscillations and maintains its smooth front. Figure 3 shows a simultaneous PLIF and schlieren image for the regular regime case, along with the stagnation pressure trace for that test. Time $t = 0$ on the pressure plot denotes the beginning of the steady test time, which lasts about 200 ms after which the free stream velocity decreases and the pressures (static and stagnation) increase. The frequency of the oscillations in the regular regime typically vary between 50 and 55 kHz for the tests performed.

Regular regime oscillations are only observed with the spherical-nosed cylinders and only for the ethylene-based mixture for cases where values of V/V_{CJ} are nearly 1. The induction times for the methane mixture for flow conditions where $V/V_{CJ} \leq 1$ are too long for any significant combustion to occur in the stagnation region. For the ethylene-based mixtures, all cases (with $V/V_{CJ} \leq 1$) using the flat-faced cylinder result in the large disturbance regime discussed below.

2.2.3 Large disturbance regime In the large disturbance regime experiments performed here, the free stream velocity is kept constant ($V/V_{CJ} \leq 1$) and the conditions are varied for each cylinder by changing the nitrogen dilution of the (ethylene-based) mixtures, and increasing the flow's free stream pressure. The simultaneous PLIF/schlieren images reveal a local explosion in the stagnation region ahead of the cylinder. This explosion is characterized by an intense combustion front closely coupled to the shock wave, Figure 4. Away from the stagnation line, as the curvature of the bow shock increases, the combustion front abruptly decouples from the shock and recedes towards the body. It is believed that this primary explosion front is an overdriven detonation that originates in the stagnation region (Matsuo and Fuji 1995). The detonation loses strength as it propagates into the bow shock, and eventually decouples into an adiabatic shock followed by a reaction zone.

The pressure traces for cases of the large disturbance regime exhibit very sharp peaks at frequencies between 16 and 25 kHz, as shown in Figure 5. Low amplitude high frequency oscillations between the high pressure peaks are also apparent in the pressure traces. Similar features in the shock front pressure history have been observed by Matsuo and Fuji (1996) in their CFD simulations of the large disturbance regime. In their paper, Matsuo and Fuji show through a series of test simulations how oscillations in the pressure vary from small amplitude high frequency ones (regular regime) to large amplitude low frequency oscillations (large disturbance regime), as the body diameter is increased keeping all other parameters constant. This is consistent with our measured pressure traces for the regular and large disturbance regimes.

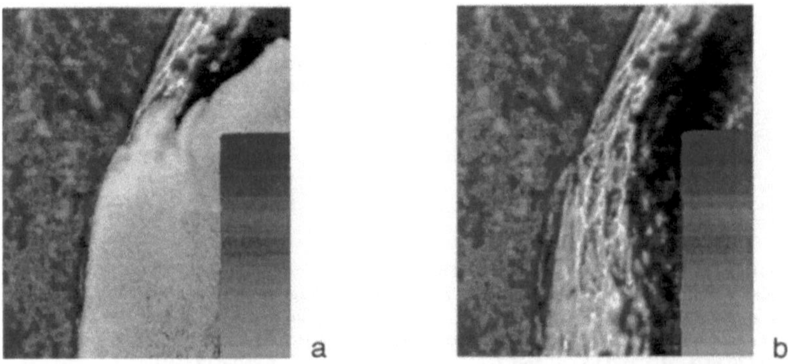

Fig. 4. (a) Schlieren and OH PLIF image for $M = 5.2$ flow of $5\%C_2H_4 + 15\%O_2 + 80\%N_2$ ($V/V_{CJ} \sim 1$) over a 25.4 mm cylinder, showing the initiation of a detonation front in the stagnation region in front of the cylinder. (b) The same schlieren image is shown separately to illustrate the non-uniformity in the shock front corresponding to the regions of intense combustion. The free stream conditions are $V_\infty = 1730$ m/s, $P_\infty = 34$ kPa , and $T_\infty = 280$K, with a V/V_{CJ} value of 1.

Based on their results, Matuso and Fuji proposed a Damköhler parameter to predict the occurrence of the large disturbances. The parameter D_I is expressed as

$$D_I = t_f/t_c \text{ where } t_{flow} = D/a_2 \text{ and } t_c = T_2/(dT/dt)_{max}$$

Here D = projectile diameter, a_2 = post-shock sound speed, T_2 = post-shock temperature, and $(dT/dt)_{max}$ = maximum value of temperature increase per unit time. Their simulations show that $D_I = 80$ is the critical Damköhler number distinguishing the high frequency modes from the lower frequency ones. Cases with $D_I < 80$ resulted in the steady solution or the high frequency mode, while cases exhibiting the large disturbances had $D_I > 80$. It is important to note that their simulations involved spherical-nosed blunted cylinders only, whereas flat-faced cylinders are used in the present experiments.

In some cases, for the same mixture and free stream conditions, the large disturbance regime was observed with the flat-faced cylinder whereas regular regime oscillations were observed using the spherical-nosed cylinder. Hence, in the unsteady regimes investigated here, the combustion characteristics in the near field in front and around the body are also dependent on the body geometry, and not just its diameter. A different value for the critical Damköhler number is thus expected for the flat-faced cylinders than the one given by Matsuo and Fuji. Indeed, the experiments with the flat-faced cylinder resulting in the large disturbance regime had a Damköhler number three orders of magnitude less than the critical number given by Matsuo and Fuji. In calculating the Damköhler number, post-shock conditions for the evaluation of the a_2 and T_2 were obtained using an (ARL) modified version of the NASA CET89 (Liberatore 1994) equilibrium code, and $(dT/dt)_{max}$ was based on a chemical kinetics calculation using the RAMEC (Petersen et al. 1997a) kinetics mechanism.

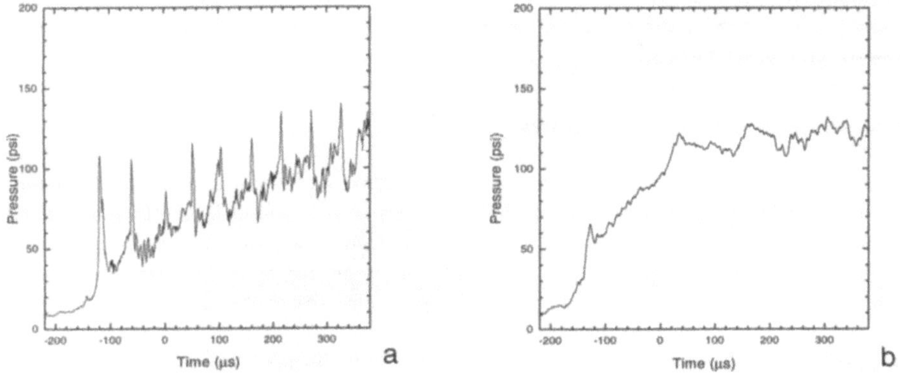

Fig. 5. (a) Stagnation pressure history for the large disturbance regime for M=5.2 flow of a mixture of 7.5%C_2H_4 + 22.5%O_2 + 70%N_2 over a 25.4 mm flat-faced cylinder. The free stream conditions are V_∞ = 1730 m/s, P_∞ = 23 kPa, and T_∞ = 280K, with a V/V_{CJ} value of 0.94. The frequency of the oscillations is approximately 19 kHz. (b) The pressure trace of the corresponding inert case, a mixture of 7.5%C_2H_4 + 92.5%N_2

In an effort to investigate the parameters which determine the period of the large disturbance oscillations, the measured period is plotted vs a characteristic flow time, τ_{flow}, as shown in Figure 6. Here the flow time is taken as $\tau_{flow} = 2\delta/C_{ps}$, where δ is the stand-off distance of a reactive shock, and C_{ps} is the post shock sound speed. Since the shock in front of the body is oscillating, the characteristic stand-off distance is obtained from an experimental correlation (Serbin 1958) of stand-off distance as a function of density ratio across the shock. However, to estimate the reactive stand-off distance, the density ratio across a reactive shock wave is used in the correlation. The choice of τ_{flow} is based on the qualitative interpretation of the oscillations as being due to pressure waves traveling between the bow shock and the flat-faced body. The peaks in the measured pressure traces registered over-pressures between 40 and 80 %. Such pressure jumps correspond to waves with Mach numbers between 1.1 and 1.3. Hence, the sound speed in the stagnation region is used as an approximation of the pressure wave velocity.

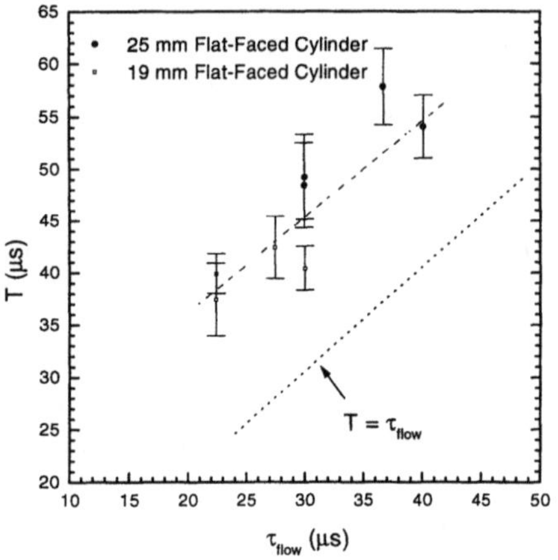

Fig. 6. Plot of the measured period of oscillations for the large disturbance regime vs. a characteristic flow time, $\tau_{flow} = (2\delta)/C_{ps}$. δ is the shock stand-off distance calculated from experimental fits given in the literature (Serbin 1958) and using the density ratio across a reactive shock. C_{ps} is the post-shock speed of sound.

In the experiments performed, the period of oscillations increased with decreasing nitrogen dilution, increasing body diameter and increasing free stream pressure. Thus, factors leading to larger amounts of heat release in the stagnation region in front of the cylinder, resulted in longer periods between successive pressure peaks. Increased heat release in the stagnation region decreases the local density there causing larger stand-off distances; the pressure waves therefore have to travel a longer distance.

3. Ram accelerator ignition chemistry

In an ideal ram accelerator, all combustion and heat release occurs downstream of the oblique shock/detonation located near the throat. However, premature ignition at the forebody of the projectile can produce a negative thrust component and possible unstart. Therefore, adequate knowledge of ignition chemistry at realistic conditions is needed to predict when and where the propellant ignites, making chemical kinetics a significant part of the ram accelerator design process.

Until recently, no ignition data were available for the mixtures and harsh conditions that exist inside a ram accelerator. Of the propellants currently in use, methane is the most frequently used fuel, and typical mixtures are fuel-rich ($\phi \geq 2$) with less than 80 % diluent gas (N_2, He, CO_2, etc.). With fill pressures of 5.0 MPa or more, pre-combustion (post-bow-shock) pressures reach hundreds of atmospheres where the mixture can easily ignite at temperatures less than 1400 K. At these high-pressure, fuel-rich conditions, methane oxidation kinetics are not well known.

Shock tube ignition delay time measurements and detailed kinetics modeling are therefore being performed in our laboratory as part of an ongoing study to advance the knowledge base for ram accelerator chemistry. Provided below is a summary of the ignition time (τ_{ign}) measurements accomplished thus far, followed by a review of the detailed kinetics mechanism developed to model CH_4/O_2 ram accelerator ignition. A reduced mechanism for inclusion in the ARL numerical flow field simulations is also presented.

3.1 Shock tube measurements Ignition delay time measurements were conducted behind reflected shock waves in the Stanford High Pressure Shock Tube at pre-ignition ram accelerator temperatures (1040 - 1600 K) and pressures (3.5 - 26 MPa) (Petersen et al. 1996b). The shock tube, rated for reflected-shock pressures up to 100 MPa, has a 50-mm inner diameter and uses helium as the driver gas. Pressure and emission (ir and visible) were used to monitor the onset of ignition. The mixtures investigated thus far cover the entire scope of propellant combinations utilized by the Army Research Laboratory (ARL). Table 1 summarizes the mixtures, which include a wide range of diluent gases (N_2, Ar, He) and equivalence ratios (0.4 - 6.0).

Table 1. Mixtures used in shock tube ignition delay time study

	Mixture	ϕ
1	$0.038CH_4 + 0.192O_2 + 0.77Ar$	0.5
2	$0.20CH_4 + 0.133O_2 + 0.667Ar$	3.0
3	$0.20CH_4 + 0.133O_2 + 0.667N_2$	3.0
4	$0.273CH_4 + 0.182O_2 + 0.545Ar$	3.0
5	$0.273CH_4 + 0.182O_2 + 0.545N_2$	3.0
6	$0.50CH_4 + 0.167O_2 + 0.333He$	6.0

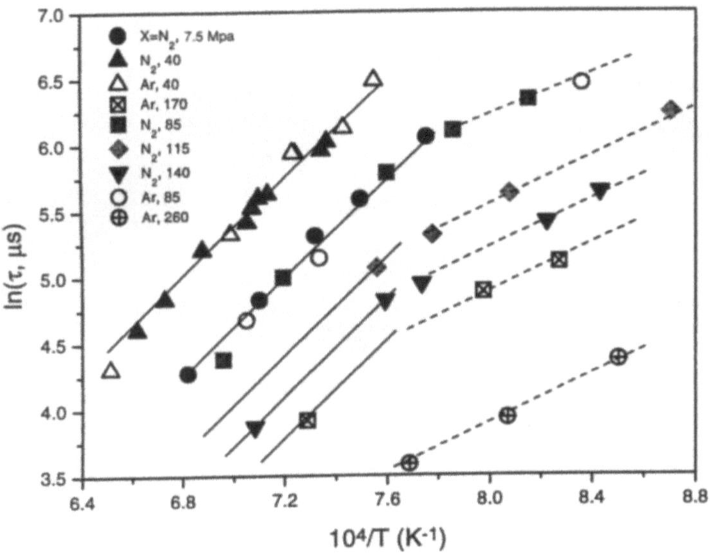

Fig. 7. Representative ignition delay time data. Standard ARL mixture, $3CH_4 + 2O_2 + 10X$ (X = N_2, Ar).

Presented in Fig. 7 are representative τ_{ign} results for the standard ARL mixture, $3CH_4 + 2O_2 + 10X$ (X = N_2, Ar). Three conclusions evident in Fig. 7 are typical of those seen in the entire data set. First, the ignition delay times exhibit a strong pressure dependence, with τ_{ign} decreasing with increasing pressure, as expected. Second, there is no discernible difference between the N_2- and Ar-based results, implying the methane/oxygen ignition time is independent of the diluent gas type. Finally, the slope (i.e., activation energy) at lower temperatures and higher pressures is noticeably different from the slope at higher temperatures and lower pressures. This accelerated ignition trend occurs at higher temperatures for increasing pressures and represents some interesting chemistry that is not predicted by existing detailed kinetics models designed primarily for high-temperature, low-pressure flames. Further details on the chemistry are provided in a later section.

Individual ignition delay time correlations for each Table 1 mixture were developed, and two overall correlations (one for each temperature regime) were derived. The correlation for the higher temperature data (typically above 1300 K) is

$$\tau_{ign,High-T} = 2.10 \times 10^{-15}[CH_4]^{0.04}[O_2]^{-1.35} \exp(35,000/RT)$$

while the lower temperature and higher pressure correlation is

$$\tau_{ign,Low-T} = 2.86 \times 10^{-15}[CH_4]^{-0.50}[O_2]^{-1.41} \exp(21,400/RT)$$

For each expression, τ_{ign} is the ignition delay time in seconds, [X] is the concentration of species X in mol/cm^3, R is the universal gas constant (1.987 cal/mol-K), and T is the temperature in degrees Kelvin. The high-temperature expression is valid for both fuel-rich and -lean mixtures, while only the fuel-rich mixtures follow the low-temperature expression. Both correlations have coefficients of multiple determination (i.e., r^2) greater than 0.96. As expected from Fig. 7, the high-temperature activation energy of 35.0 kcal/mol is larger than that of the low-temperature expression, 21.4 kcal/mol. The pressure dependence for the former expression, however, is somewhat lower (-1.31 compared to -1.91).

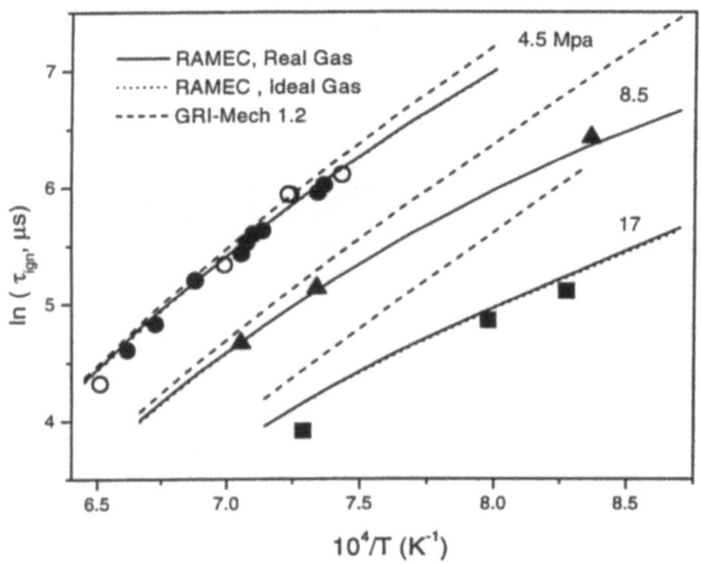

Fig. 8. Comparison between kinetics models and ignition data for $3CH_4 + 2O_2 + 10X$ (X = N_2, Ar) mixture. ○ - Ar, 4.0 MPa; • - N2, 4.0 MPa; △ - Ar, 8.5 MPa; ◇ - Ar, 17 MPa.

3.2 Kinetics Modeling Initial attempts to compare the shock tube ignition data with a detailed kinetics mechanism were done using the methane oxidation mechanism from the Gas Research Institute, GRI-Mech 1.2 (Frenklach et al. 1995). The GRI kinetics model, developed for a wide range of flame speed, ignition, and species profile data, contains 23 species and 175 elementary reaction steps. Although GRI-Mech 1.2 was shown to agree well with near-stoichiometric, high-pressure, high-temperature shock tube data in a previous study (Petersen et al. 1996c), it was unable to adequately predict the ignition delay times of ram accelerator mixtures, particularly at the fuel-rich, high-pressure, lower temperature conditions. As shown by the dashed line in Fig. 8, GRI-Mech 1.2 overpredicts the ignition delay time at the highest pressures and lowest temperatures by a factor of 4 or more. This result is not surprising, however, since the GRI mechanism was not designed for the lower temperature chemistry that becomes important at the high-pressure conditions of the ram accelerator.

To obtain better agreement between theory and experiment, reaction mechanisms designed for the lower temperatures commonly seen in flow reactor experiments were investigated to determine which additional reactions were required to model the ram accelerator ignition data. Using GRI-Mech 1.2 as the base mechanism, reactions from recent low-temperature methane oxidation and C_2 mechanisms were added until good agreement was found between the mechanism and the data. The final extended mechanism that best fits the ram accelerator ignition data contains 279 reactions and 47 species. This ram accelerator chemistry model is referred to as RAMEC (Petersen et al. 1997a).

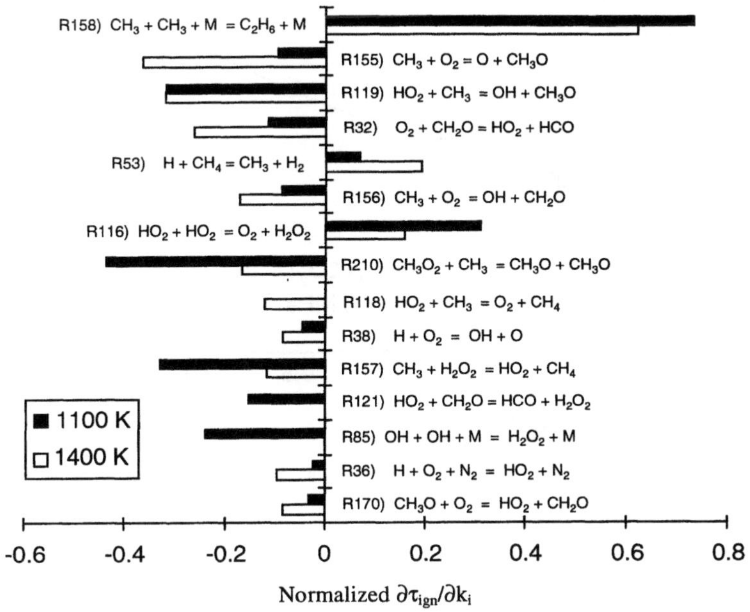

Fig. 9. Sensitivity of ignition delay time to key reactions at 1100 K and 1400K; P = 10 MPa, $3CH_4 + 2O_2 + 10N_2$ mixture.

Presented in Fig. 8 are the predictions of RAMEC for the standard ARL mixture at pressures up to 17 MPa. The agreement between the improved mechanism and the shock tube data is now much better than for the original mechanism, particularly at higher pressures and lower temperatures. Nonetheless, further improvement to RAMEC is required at temperatures below 1100 K and pressures greater than 30 MPa. Also shown in Fig. 8 is a comparison between the predicted ignition delay times assuming both ideal gas and real gas thermodynamics (Peng-Robinson equation of state); for ram accelerator ignition, there is little difference between either assumption.

With good agreement between the full mechanism (RAMEC) and the experimental data, the mechanism was used with confidence to elucidate the key reactions and pathways for ram accelerator ignition and oxidation. A sensitivity analysis using ignition delay time as the response variable is presented in Fig. 9 for 10 MPa and two different temperatures: 1400 K and 1100 K (representing the two different activation energy regimes in Fig. 7). In general, ram accelerator ignition chemistry is dominated by the CH_3 and HO_2 radicals, with methyl recombination to ethane as the controlling radical sink/inhibitor. Evident in Fig. 9 is the shift in primary promoters between 1400 K and 1100 K. At the lower temperature, hydrogen peroxide (H_2O_2) and methyl peroxy (CH_3O_2) have increased significance. The reaction pathways involving H_2O_2 and CH_3O_2 lead to a quicker buildup of the reactive H and OH radicals, hence the accelerated ignition trends seen at lower temperatures and higher pressures.

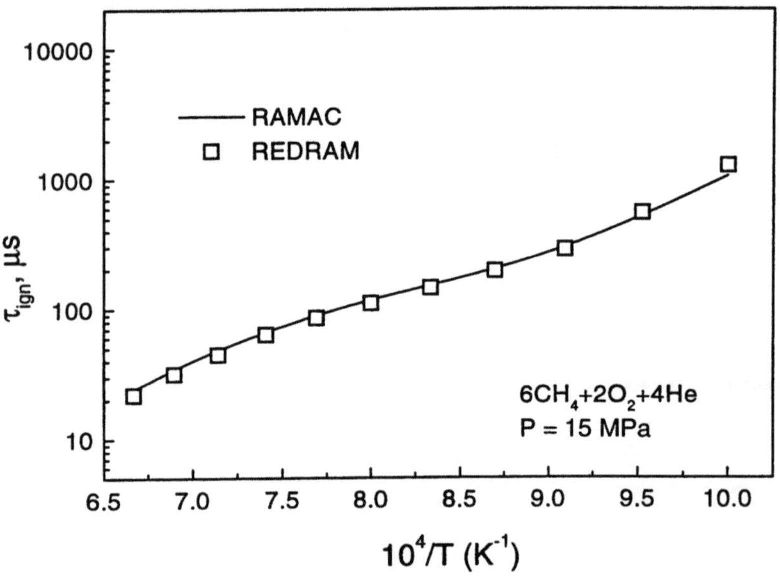

Fig. 10. Comparison between reduced mechanism (REDRAM) and the full kinetics mechanism (RAMEC) for $6CH_4 + 2O_2 + 4He$ mixture at 15 MPa.

3.3 Reduced mechanism While RAMEC successfully reproduces some key aspects of ram accelerator ignition and oxidation, the large number of reactions (279) and species (47) limit its usefulness, due to size and time constraints, as a chemistry module in present-day numerical flow solvers. A reduced mechanism was therefore needed. Employing a detailed reduction procedure, a 34-reaction, 22-species skeletal mechanism (REDRAM, Petersen et al. 1997b) was developed using ignition delay time and heat release as the selection criteria. REDRAM faithfully reproduces ignition delay times within 5 % and combustion product temperatures within 10 K of the values predicted by the full mechanism for the entire range of mixtures and conditions of the present study. Fig. 10 demonstrates the accuracy of REDRAM versus the full mechanism for the $6CH_4 + 2O_2 + 4He$ mixture at 15 MPa.

4. Summary

Two aspects of the ram-accelerator research conducted at Stanford have been presented. In the investigation of hypersonic reactive flow around blunt bodies, the Stanford expansion tube was used to generate the different combustion modes observed in previous ballistic range experiments. Simultaneous OH PLIF and schlieren imaging was performed to obtain information on the shock and flame front stand-off distances. The observed modes qualitatively agreed with results of previous works. In the shock tube studies of combustion kinetics at high pressures, ignition time measurements of methane-based mixtures at pre-ignition ram accelerator temperatures and pressures have been conducted. Future studies will continue to be both experimental and analytical in nature. For the experimental work, additional ignition delay time measurements will be performed in support of the ARL effort. Using advanced laser diagnostics, the high-pressure shock tube will also be employed to make quantitative measurements of kinetic rate coefficients of importance to ram accelerator chemistry. Analytical efforts will concentrate on improving RAMEC and on the design of an optimum ram accelerator mixture which has the best combination of ignition time and heat release.

Acknowledgments

The work has been supported by the U.S. Army Research Office, with Dr. David Mann as technical monitor, and the Office of Naval Research, with Dr. Richard Miller as technical monitor.

References

Alpert RL, Toong TY (1972) Periodicity in exothermic hypersonic flows about blunt bodies. Astronautica Acta 17:539-560

Billig FS (1967) Shock-wave around spherical- and cylindrical-nosed bodies,J Spacecraft 4:822-823

Cheng RK, Oppenheim AK (1984) Autoignition in methane-hydrogen mixtures. Combustion and Flame 58:125-139

Chernyi GG (1968) Supersonic flow past bodies with formation of detonation and combustion fronts. In: Problems of Hydrodynamic and Continuum Mechanics, Society Ind and App Math, Philadelphia, pp. 145-169.

Frenklach M, Wang H, Goldenberg M, Smith GP, Golden DM, Bowman CT, Hanson RK, Gardiner WC, Lissianski V (1995) GRI-Mech-an optimized detailed chemical reaction mechanism for methane combustion. GRI Topical Rep GRI-95/0058

Hanson RK, Seitzman JM, Paul PH (1990) Planar laser-fluorescence imaging of combustion gases. Appl Phys B 50:441-454

Houwing AFP, Kamel MR, Morris CI, Wehe SD, Boyce RR, Thurber MC, Hanson RK (1996) PLIF imaging and thermometry of NO/N_2 shock layer flows in an expansion tube. AIAA paper 96-0537

Kamel MR, Morris CI, Thurber MC, Wehe SD, Hanson RK (1995) New expansion tube facility for the investigation of hypersonic reactive flow. AIAA paper 95-0233

Lee JHS (1994) On the initiation of detonation by a hypervelocity projectile. Zeldovich Memorial on Combustion, Voronovo, Russia

Lehr HF (1972) Experiments on shock-induced combustion. Astronautica Acta 17:589-597

Liberatore F (1994) Ram accelerator performance calculations using a modified version of the NASA CET89 equilibrium chemistry code. ARL-TR-647

Lifshitz A, Scheler K, Burcat A, Skinner GB (1971) Shock-tube investigation of ignition in methane-oxygen-argon mixtures. Combustion and Flame 16:311-321

Matsuo A, Fuji K (1995) Examination of the improved model for the unsteady combustion around hypersonic projectiles. AIAA paper 95-2565

Matsuo A, Fuji K (1996) First Damköhler parameter for prediction and classification of unsteady combustions around hypersonic projectiles. AIAA paper 96-3137

Morris CI, Kamel MR, Hanson RK (1996) Investigation of ram-accelerator projectile flowfields in an expansion tube. 33rd JANNAF Combustion Subcommittee Meeting

Palmer JL, Hanson RK (1993) Planar laser-induced fluorescence imaging in free jet flows with vibrational nonequilibrium. AIAA paper 93-0046

Petersen EL, Davidson DF and Hanson RK (1996a) Ignition delay times of ram accelerator CH_4/O_2 mixtures, submitted to J Prop and Power, also AIAA paper 96-2681

Petersen EL, Davidson DF, Hanson RK (1996b) Ram accelerator mixture chemistry: Kinetics modeling and ignition measurements. JANNAF Combustion Subcommittee Meeting, Vol. I, CPIA Pub. 653, pp 395-407

Petersen EL, Röhrig M, Davidson DF, Hanson RK, Bowman CT (1996c) High-pressure methane oxidation behind reflected shock waves. Twenty-Sixth Symposium (Int) on Combustion, pp. 799-806

Petersen EL, Davidson DF, Hanson RK (1997a) Kinetics modeling of shock-induced ignition in low-dilution CH_4/O_2 mixtures at high pressures and intermediate temperatures, submitted to Combustion and Flame, also Paper 97S-066, Western States Section of the Combustion Institute

Petersen EL, Davidson DF, Hanson RK (1997b) Reduced kinetics mechanisms for ram accelerator combustion. AIAA paper 97-2892

Ruegg FW, Dorsey W (1962) A missile technique for the study of detonation wave. J Res Nat Bureau of Standards. 66C:51-58

Seitzman JM, Palmer JL, Antonio AL, Hanson RK, DeBarber PA, Hess CF (1993) Instantaneous planar thermometry of shock-heated flows using PLIF of OH. AIAA paper 93-0802

Serbin H (1958) Supersonic flow around blunt bodies. J Aeronautical Sci 25:58

Accelerating hydrogen/air mixtures to superdetonative speeds using an expansion tube

J. Srulijes, G. Smeets, G. Patz, F. Seiler
French-German Research Institute of Saint-Louis (ISL), F-68301 Saint-Louis, France

Abstract. A new concept for accelerating stoichiometric hydrogen-air mixtures to superdetonative velocities in an expansion tube using a Ludwieg tube as hydrogen reservoir is the subject of this paper. With this special set-up hydrogen/air gas mixtures can be successfully accelerated to a flow having a superdetonative speed with a static pressure of about 200 kPa at 350 K. The experiments in the expansion tube are carried out for basic research purposes on ram-accelerator-related combustion phenomena and also for supporting and optimizing the operation of the 30 mm and 90 mm ram accelerator facilities at ISL. Problems encountered while using an expansion tube for this purpose and some preliminary experiments performed with this new facility will be discussed.

Key words: Expansion tube, Ludwieg tube, Combustion, Stoichiometric hydrogen/air mixtures, Hypervelocity, Detonation

1. Introduction

The worldwide first ram accelerator was devised and successfully operated by Hertzberg et al (1986) and Bruckner et al (1987). The ram accelerator is based on the ramjet principle. A sharp-nosed projectile, which resembles the centrebody of a conventional ramjet runs inside a tube filled with a combustible gas mixture. The tube acts as the outer cowling of the ramjet and the energy release by combustion produces high pressure at the base of the projectile providing thrust. The reason for investigating the gasdynamic ram-accelerator-related phenomena in an expansion tube, is the difficulty in obtaining reliable information on the combustion processes taking place around a fast moving projectile inside the ram accelerator. Inversely, by using a fixed model in a moving gas, as is the case in the expansion tube, information can be obtained much more easily. As described by Srulijes et al (1992,1995), stoichiometric methane/air and ethylene/air combustible gas mixtures were accelerated under well-known conditions, without autoignition in the tube, to the superdetonative flow velocities corresponding to the projectile velocity in the 30-mm-caliber ram accelerator RAMAC 30 of ISL. A hydrogen-based stoichiometric gas mixture is not only of paramount interest for RAMAC applications. Due to its very small detonation cell size this gas mixture is also of interest for many other research subjects. Several attempts to accelerate stoichiometric hydrogen/air gas mixtures to superdetonative speeds having static pressures higher than 25 kPa were carried out without success at different institutes. At the ISL Shock Tube Laboratory we also tried to accelerate stoichiometric hydrogen/air gas mixtures to superdetonative speeds in the classical way, that is with the above mentioned expansion tube without success. The result was always an autoignition of the gas mixture inside the tube. The main problem is that this gas mixture autoignites very easily. Due to the incident shock wave the combustible gas mixture is heated in the driven tube and ignites spontaneously. Even weak incident shock waves produce, by shock reflection, a heated zone in front of the diaphragm between driven tube and expansion tube. This causes the gas mixture to autoignite at the beginning of the expansion tube. Only hydrogen/oxygen mixtures highly diluted with nitrogen at very small pressures were recently accelerated by this method by Morris et al (1996).

2. General considerations

Several years ago Smeets and Mathieu (1987) investigating turbulent boundary layers in our shock tunnel made a remarkable discovery. They used for Doppler velocity measurements soot particles generated by pyrolysis of a hydrocarbon gas. These particles are very small having a mean diameter well below 0.2 μm. It turned out that soot particles generated by pyrolysis are partly deposited on the tube wall serving as a store for quite a number of subsequent experiments. On each run - presumably by means of a strong ultrasonic field in the turbulent shock boundary layer - a sufficient quantity of particles is detached and mixed into the flow. They thus studied the transition to turbulent tube flows. From this investigation it turned out that after the thickness of the boundary layer has reached approximately 20 % of the tube diameter, the influence of the curved tube wall appears, resulting in an accelerated expansion of turbulence into the flow core.

In Fig. 1 the arrival of the boundary layer border as indicated by the arrival of soot particles and by the onset of the turbulent fluctuations of the recorded signals is plotted as a function of time t or distance to the shock x, respectively. Up to a distance to the wall of about $y = 1.5$ cm the data points almost coincide with the dotted line representing the calculated boundary layer border (defined by a 1 % deviation of the mean velocity from the free stream condition) for the two-dimensional case. Already 400 μs after shock passage, when the flat plate boundary would have reached a thickness of only 2 cm the transition into a turbulent flow in the 100 mm diameter tube is completed.

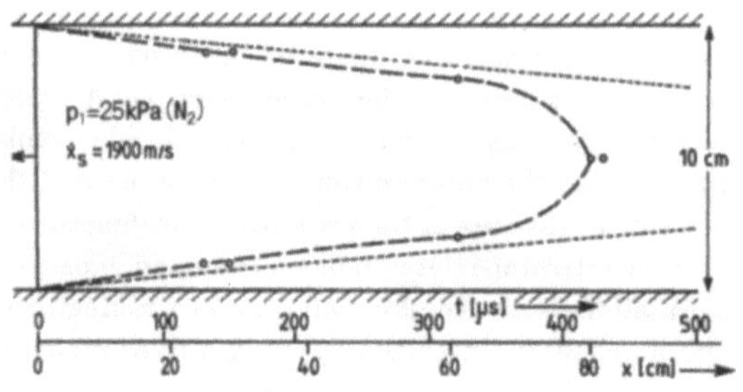

Fig. 1. Migration of the soot particles from the wall

Figure 2 shows an example of velocity fluctuations recorded at the tube axis after transition. The peak-to-peak values are of the order 60 m/s corresponding to a root-mean-square of about $u^* = 20$ m/s. When dividing this quantity u^* by the flow velocity behind the shock, u_2, the so defined turbulence degree is $u^*/u_2 = 0.013$. From this work it can be concluded that in shock tubes of more than 50 diameters length - except for rarefied gases forming laminar boundary layers - the flow behind the shock consists of two different phases. The first phase is free of turbulence in the core of the tube and the second - after a sudden transition - shows a finite degree of turbulence. Considering this behavior of the turbulent boundary layer in our shock tube and the above mentioned problems in trying to accelerate stoichiometric hydrogen/air mixtures to superdetonative speeds, Smeets proposed a new method to achieve this goal.

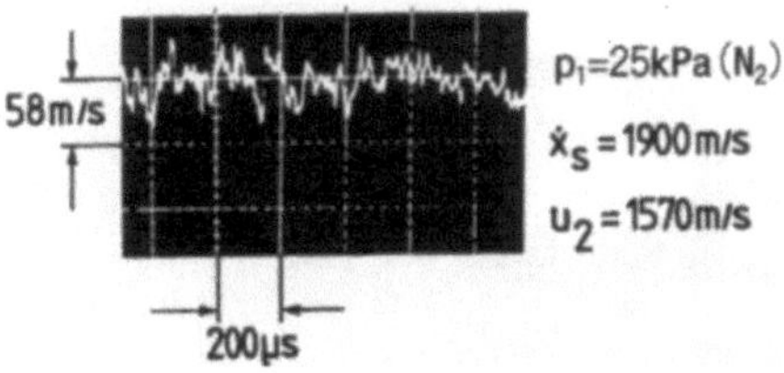

Fig. 2. Velocity fluctuations after transition

3. Coupling of expansion tube and Ludwieg tube

We used a special shock tube with a third section at a very low pressure filled with helium, called expansion section, beyond the driven section. Because of the non-stationary expansion, this set-up provides high gas velocities with very high stagnation temperatures in the expansion section. The shock tube has an additional section filled with nitrogen, named buffer section, placed between the driver tube containing hydrogen and the driven tube filled with air. The geometry of the expansion shock tube used for the experiments presented in this paper is as follows: inner diameter = 0.1 m, driver length = 1.7 m, buffer length = 3.5 m, driven tube length = 6.8 m and expansion tube length = 8.7 m.

Contrary to our previous method of accelerating the various combustible gas mixtures the expansion tube is now used to accelerate only the air to the required velocity. The basic idea is to inject the hydrogen into the expansion section parallel to the tube wall to the already accelerated cold air. The injected mass flux of hydrogen must be calculated to reach the stoichiometric level wanted. The hydrogen inflow must have almost the same velocity and pressure as the air flow passing the injection zone. The hydrogen temperature is much lower than the air flow temperature. Our calculations showed that also the temperature maxima inside the boundary layer in both, the hydrogen flow as well as the upstream air flow are below the autoignition temperature of the hydrogen/air mixture.

Knowing the behavior of the turbulent boundary layer in the shock tube flow described above, the hydrogen is expected to mix with the air after some tube diameters. To prove this assumption was the aim of our research work.

To inject the hydrogen we decided to use a Ludwieg tube of 1.5 m length as a gas reservoir (see Ludwieg 1955). The advantage of this facility compared to a classical reservoir is a constant pressure at the injection zone while the expansion wave travels up and down the Ludwieg tube after diaphragm bursting. The hydrogen is injected nearly parallel to the tube wall through a toroidal nozzle placed 4.2 m from the end of the expansion section (test section). The initial hydrogen pressure in the Ludwieg tube is 3.0 MPa. For this pressure level the hydrogen mass flux must be determined by an adapted nozzle design to produce a stoichiometric hydrogen/air gas mixture.

Now everything depends on the timing of the hydrogen gas injection. For this purpose we used a double diaphragm to trigger the Ludwieg tube with the incident shock wave. It is indispensable to start the injection some μs before the arrival of the air at the injection zone. If the injection starts after the arrival of the air flow, the air, having a higher pressure, enters the nozzle and after the bursting of the diaphragms the gas mixture ignites in the nozzle and consequently inside the expansion tube. We used Helium in the expansion section at a pressure of about 20 kPa. With hydrogen instead of helium we had a much higher failure rate. Figure 3 shows schematically the expansion-Ludwieg tube test facility

The calculated flow parameters in the hydrogen/air gas mixture were: velocity = 2200 m/s, static pressure = 240 kPa and static temperature = 340 K, corresponding to a Mach number M

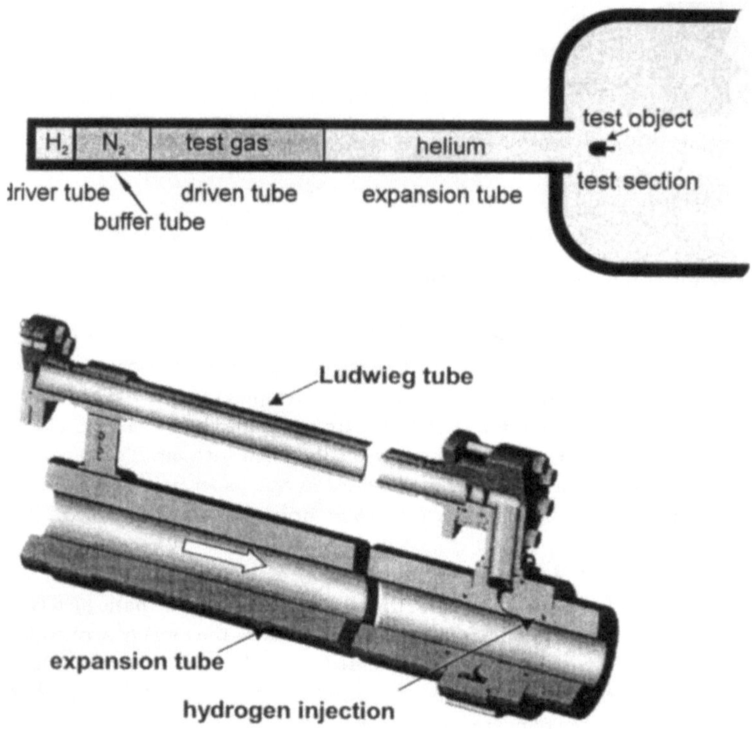

Fig. 3. Principle sketch of the expansion-Ludwieg tube facility

= 5.0. Both, the calculated and the measured test time for our expansion tube (constant flow parameters) is about 600 μs before the flow begins to decay.

The experiments shown in the paper were carried out using a nearly stoichiometric gas mixture of hydrogen/air and for the comparative experiments, without combustion, we used nitrogen instead of air. The Chapman-Jouguet velocity of the gas mixture is 1960 m/s, calculated with the real gas code MEGEC of Gatau (1979).

Flow visualization was done by means of differential interferometry as described by Smeets and George (1973). This optical measuring technique allows the visualization of density gradient fields. We visualized the flow by a series of framing pictures with a rotating drum camera. For these framing pictures the flow is focused on the rotating film and eight successive air sparks (Δt = 200 μs) were used as light source.

The calculated velocity was checked by measuring the pressure history on the surface of a wedge with a wedge angle = 15° shown in Fig. 4. The test flow was also checked by measuring the angle of the attached oblique shock wave on the wedge and also by Pitot pressure measurements. Both, wedge surface as well as Pitot probes, were in the Mach cone flow. The end of the Mach cone flow is visible by the curvature of the oblique shock wave in the interferogram.

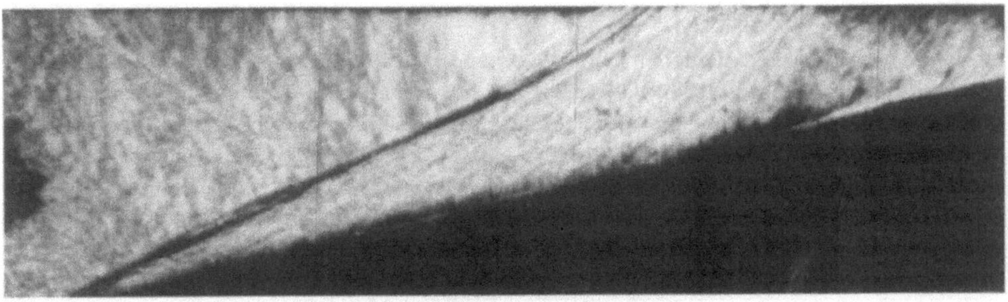

Fig. 4. Differential interferogram of the flow over a 15° wedge

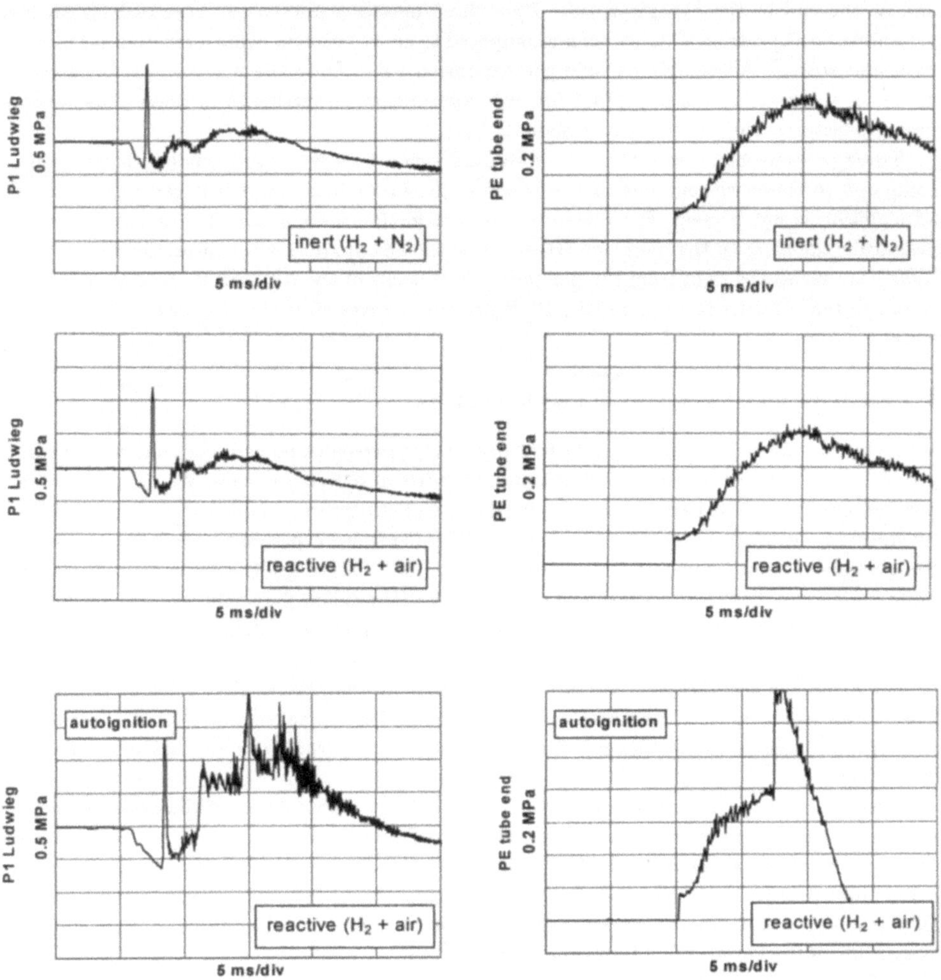

Fig. 5. Comparison of pressure histories inert-reactive with and without autoignition inside the expansion tube

4. Pressure measurements

Pressure measurements at different critical points inside the tube were extremely helpful to understand the phenomena taking place inside the expansion-Ludwieg tube. A pressure gauge at the end of the expansion tube provided the evidence for the absence of gas combustion (autoignition) inside the tube . The static pressure is measured at the end of the expansion tube with a PCB gauge type 113A21. Using PCB pressure gauges type 113A24 we measured Pitot and wedge surface pressures and the wall pressure at different tube locations. The pressure gauges were quasi-dynamically calibrated. We used a special device that generates pressure pulses of about 1 ms, giving calibration values better than one percent. All pressure gauges exposed to high temperature are protected with a 0.2 mm wax layer.

Figure 5 shows the pressure histories P1 measured 405 mm downstream of the injection zone and at the end of the expansion tube PE. These measurements were performed for an inert experiment using nitrogen instead of air compared to an experiment using a reactive gas mixture, with and without autoignition inside the expansion tube. Only the first 600 μs following the arrival of the test gas are important for our experiments. Nevertheless in these plots we show pressure histories with a duration of about 2.5 ms.

These pressure histories clearly show the difference between experiments with autoignition compared to those without during the whole cycle. From these records it can be seen that if autoignition is not present, the pressure development for the reactive experiment (H_2+air) is practically identical to the inert one (H_2+N_2). If the gas mixture autoignites, different pressure signals are measured right from the beginning. The decay of the pressure at the end of the cycle is due to temperature increase on the PCB pressure gauges after the wax protection is melted away.

5. Preliminary test experiments with a sphere

With this new facility a stoichiometric hydrogen/air hypersonic and superdetonative flow can be produced and used for testing the flow around models of different shapes. We repeated some of our former experiments carried out with methane and ethylene and added some new ones.

As an interesting example for experimental study using our expansion-Ludwieg tube, we chose some spherical shapes (see Lehr 1972 and Bélanger et al. 1996) to test the detonation initiation by blunt bodies based on the Lee-Vasiljev (1994) theory that predicts for a sphere of a given diameter travelling at the Chapman-Jouguet velocity direct initiation of detonation. A comprehensive work on this subject is given by Higgins (1996).

For a stoichiometric H_2/air mixture at a pressure of 100 kPa Vasiljev gives a cell size of λ = 9.93 mm requiring that the diameter d of the sphere for detonation initiation be at least 4.52 times the cell size. We used this experiment to test if our hydrogen/air gas flow was sufficiently well mixed.

These very preliminary experiments were carried out with spheres of diameter 40 mm, 60 mm and a half sphere with a diameter of 48 mm. All models were placed slightly after the end cross-section of the expansion tube but inside of the Mach cone to allow the low Mach number helium flow to leave the tube before the test gas arrives. Some experiments were carried out with off-axis position of the model, having a displacement of 25 mm from the flow axis, to use as much as possible of the core of the Mach cone flow.

In Fig. 6 we see the differential interferograms corresponding to the models mentioned above. The interferograms at the left show the control experiments using nitrogen instead of air. On the right we see the interferograms corresponding to the reactive flow experiments. Using the relation $d = 4.52\lambda$ and the values given above, a diameter for detonation initiation of 45 mm is calculated. Taking into account that the flow pressure in our experiments is higher than that

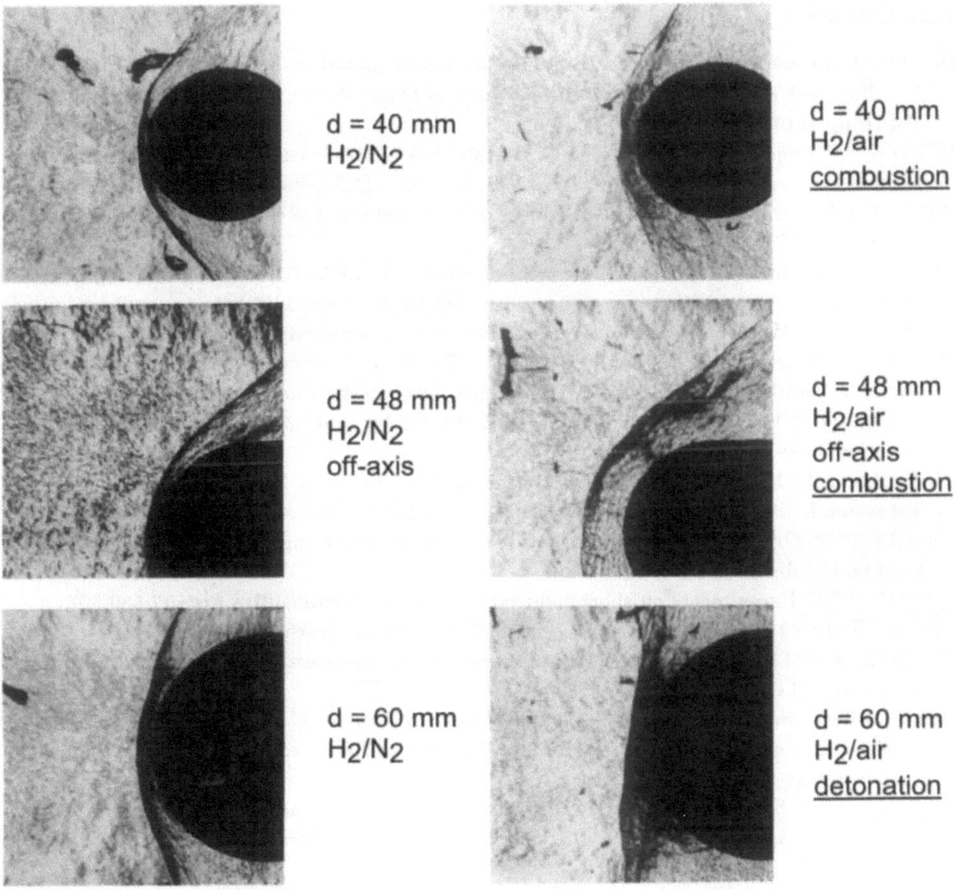

Fig. 6. Differential interferograms for different spheres showing the reaction fronts and the corresponding control experiments without combustion

given by Vasiljev and consequently the detonation cells are smaller, the detonation initiation should occur at a diameter smaller than that shown here for $d = 60$ mm (see Ju et al. 1997). Even if we do not have more detailed experiments yet because we measured spheres of only these three diameters the aim of the experiments to prove an adequate hydrogen/air flow mixing was achieved.

6. Summary and conclusions

A new method for accelerating stoichiometric hydrogen/air gas mixtures to superdetonative velocities with static pressures of the order of 200 kPa was tested. It combines the classical expansion tube with a Ludwieg tube used to inject the hydrogen into the already accelerated and expanded high speed air flow.

Besides some triggering difficulties and the unsolved problem of the simultaneous arrival of the test flow with diaphragm particles it was possible to show that this concept can be used for investigating general phenomena in supersonic flows of near stoichiometric hydrogen/air mixtures including research in the field of ram acceleration.

7. References

Bélanger J, Kaneshige M, Shepherd JE (1996) Detonation initiation by hypervelocity projectiles. In: .Stutevant B et al (ed) Proc 20th Int Symp on Shock Waves, World Scientific, Singapore, Vol II, pp 1119-1124

Bruckner AP, Bogdanoff DW, Knowlen C, Hertzberg A (1987) Investigation of gasdynamic phenomena associated with the ram accelerator concept. AIAA paper 87-1327

Gatau F (1979) Propriétés thermodynamiques d'un mélange de gaz en équilibre thermochimique Programme MEGEC, ISL, R 109/79

Giraud M, Legendre JF, Simon G, Henner M, Voisin D (1995) RAMAC 90 starting process: Control of the ignition location and performance in the thermally choked propulsion mode. Proc 2nd Int Workshop on Ram Accelerators, Univ Washington, Seattle, USA

Hertzberg A, Bruckner AP, Bogdanoff DW (1986) The Ram accelerator: A new chemical method for achieving ultrahigh velocities. 37th Meeting Aeroballistic Range Association, Québec

Higgins AJ (1996) Investigation of detonation initiation by supersonic blunt bodies, Ph.D. dissertation, Univ Washington, USA

Ju Y, Sasoh A, Masuya G (1998) On the detonation initiation by a supersonic sphere. In: Takayama K, Sasoh A (eds) Ram Accelerators, Springer-Verlag, Heidelberg, pp 255-262

Lee JHS (1994) On the initiation of detonation by a hypervelocity projectile. Zeldovich Memorial Conf on Combustion, Voronovo

Lehr HF (1972) Experiments on shock-induced combustion. Astronautica Acta 17:589-597

Ludwieg H (1955) Der Rohrwindkanal, Zeitschrift für Flugwissenschaften 3, Heft 7

Morris CI, Kamel MR, Hanson RK (1996) Expansion tube investigations of ram-accelerator projectile flowfields AIAA paper 96-2680

Patz G, Seiler F, Smeets G, Srulijes J (1995) Status of ISL's RAMAC 30 with fin guided projectiles accelerated in a smooth bore. Proc 2nd Int Workshop on Ram Accelerators, Univ Washington, Seattle, USA

Seiler F, Patz G, Smeets G, Srulijes J (1995) The rail tube in ram acceleration: Feasibility study with ISL's RAMAC 30. Proc 2nd Int Workshop on Ram Accelerators, Univ Washington, Seattle, USA

Smeets G, George A (1973) Anwendungen des Laserdifferentialinterferometers in de Gasdynamik, ISL, R 28/73

Smeets G, Mathieu G (1987) Investigation of turbulent boundary layers and turbulence in shock tubes by means of laser doppler velocimetry. Proc 16th Int Symp on Shock Waves, also as ISL, CO 219/87

Srulijes J, Smeets G, Seiler F (1992) Expansion tube experiments for the investigation of ram-accelerator-related combustion and gasdynamic problems. AIAA paper 92-3246

Srulijes J, Eichhorn A, Nusca MJ, Smeets G, Seiler F (1995) Shock tube experiments for modeling ram-accelerator-related phenomena. Proc 2nd Int Workshop on Ram Accelerators, Seattle, USA

Vasiljev AA (1994) Initiation of gaseous detonation by a high speed body. Shock Waves 3:321-326

Computational fluid dynamics

Numerical simulations of unsteady ram accelerator flow phenomena

M. J. Nusca
US ARL, Aberdeen Proving Ground, MD, 21005, USA

Abstract. Computational fluid dynamics (CFD) solutions of the full Navier-Stokes equations along with finite-rate chemical kinetics, are used to numerically simulate the reacting in-bore flowfield for a 120mm ram accelerator projectile propulsion system. Various unsteady flow phenomena are investigated, including projectile starting, obturator discard, and high-velocity unstarting. These simulations illustrate the importance of obturator discard dynamics in achieving a successful starting of the ram acceleration process. However, once started, high projectile velocities can induce an unstart as the combustion wave precedes the projectile under certain conditions. In addition, a CFD code validation case is presented for low pressure, steady flow, hydrogen/oxygen combustion around a wedge as compared to Schlieren/PLIF data from Stanford.

Key words: Starting process, Obturator, Unstart, CFD

1. Introduction

Experimental testing and gasdynamic modeling of the ram acceleration technique for in-bore projectile propulsion has been investigated at the U.S. ARL under the Hybrid Inbore RAM (HI-RAM) propulsion program (Nusca and Kruczynski 1996). This research program seeks to provide a highly efficient method of achieving hypervelocity ($\geq$ 3 km/s) projectile gun-launch for use in high speed impact testing applications. The ARL ram accelerator system uses a 120-mm (bore diameter) tube which is modeled after the 38-mm system at the University of Washington. Numerical solutions of the Navier-Stokes equations have been obtained at the ARL via computational fluid dynamics (CFD) codes for non-reacting and reacting, two-and three-dimensional flows. These codes are being used to investigate the complex gas dynamic physics of ram accelerator projectile propulsion. A variety of CFD techniques have been brought to bear on this problem, including models for chemically frozen (non-reacting) gas, finite-rate global and multiple step chemical kinetics, and equilibrium chemical processes. Accurate numerical simulation of hydrocarbon-based reacting flow creates a very great demand on computational resources since the number of intermediate species and the number of kinetic steps for typical hydrocarbon fuels are prohibitively large. Global reaction mechanisms based on up to thirty-two steps have therefore been investigated for use in preliminary design studies. Within several hours on supercomputers, viscous and chemically reacting gas dynamic simulations are used to assess the influence of projectile velocity, tube fill pressure and mixture composition, and projectile geometry on species consumption, tube wall pressure and projectile thrust. These studies are being used to seek optimum performance for the ARL ram accelerator with minimal gun firings. In particular, significant advances in CFD modeling for ram acceleration have been made at the ARL (Nusca 1994, 1995, 1996, 1997).

Recently, ARL CFD models have been modified to simulate the unsteady coupling between the launch gun and the ram accelerator (Nusca 1997). In the ARL experimental gun, projectile transition from the conventional (solid propellant) launcher to the accelerator is made through a transition/vent section. This section decouples the launch gun movement from the accelerator and vents combustion gases from the conventional launcher. Redesign of the ram accelerator may require that the ram accelerator operate without the vent section. The removal of the vent section requires that the CFD code be linked to an interior ballistics (IB) simulation of the launch gun in order to supply the pressure behind the obturator. The IB code must be able to adjust to the pressure on the forward face of the obturator, supplied from the CFD code simulating ram combustion. Simulation of ventless ram acceleration has been accomplished (Nusca 1997).

2. Reacting Flow Model

The chemically reacting hypervelocity flow field around a body is numerically simulated using a computer code based on computational fluid dynamics. The ARL-NSRG2 code solves the 2D/axisymmetric, unsteady, real-gas Navier-Stokes equations including equations for chemical kinetics and diffusion. These partial differential equations are cast in conservation form and converted to algebraic equations using a finite-volume formulation. The conservation law form of the equations assures that the end states of regions of discontinuity (e.g., shocks, flames and deflagrations) are physically correct even when smeared over a few computational cells. Solution takes place on a mesh of nodes distributed in a zonal fashion around the projectile and throughout the flow field such that sharp geometric corners and other details are accurately represented.

The Navier-Stokes equations for 2D/axisymmetric reacting and unsteady flow are written in the following conservation form with variables nondimensionalized in a conventional fashion (Nusca 1998).

$$\frac{\partial \mathbf{U}}{\partial t} + \frac{\partial \mathbf{F}}{\partial x} + \frac{\partial \mathbf{G}}{\partial y} + \mathbf{H} = 0 \tag{1}$$

$$\mathbf{U} = \begin{pmatrix} \rho y^\alpha \\ \rho u y^\alpha \\ \rho v y^\alpha \\ \rho e y^\alpha \\ \rho c_1 y^\alpha \\ \vdots \\ \rho c_N y^\alpha \end{pmatrix}, \quad \mathbf{F} = \begin{pmatrix} \rho u y^\alpha \\ (\rho u^2 + \sigma_{xx}) y^\alpha \\ (\rho u v + \tau_{xy}) y^\alpha \\ (\rho u e + u\sigma_{xx} + v\tau_{yx} + q_x) y^\alpha \\ (\rho u c_1 - \Lambda (c_1)_x - \Gamma_1 (\ln T)_x) y^\alpha \\ \vdots \\ (\rho u c_N - \Lambda (c_N)_x - \Gamma_N (\ln T)_x) y^\alpha \end{pmatrix}, \quad \mathbf{H} = J \begin{pmatrix} 0 \\ 0 \\ \alpha\sigma_+ \\ 0 \\ -\dot{w}_1 y^\alpha \\ \vdots \\ -\dot{w}_N y^\alpha \end{pmatrix} \tag{2}$$

where the vectors arrays **F** and **G** contain shear (τ) and normal (σ) stress terms as well as mass (Λ) and thermal (Γ) diffusion coefficients and heat transfer (q) terms. The dependent variables e, ρ, u, v represent the gas energy, density and velocity components (in Cartesian coordinates x, y), respectively. The coefficient/exponent α is used to customize the equations to either two-dimensional (α=0) or axisymmetric (α=1) problems (see Nusca and Kruczynski 1996, for three-dimensional simulations). The mass fraction and chemical production term for the i-th species are represented by c_i and $\dot{w}_i$, respectively.

The Peng-Robinson equation of state (Benedek and Olti 1985) is used since pressures in the ram accelerator system are very high (i.e., 5-50 MPa).

$$p = \frac{\overline{\Re} T}{\overline{v} - b} - \frac{a}{\overline{v}(\overline{v} + b) + b(\overline{v} - b)}$$

$$a \equiv 0.45724 \frac{\overline{\Re}^2 T_c^2}{p_c} \left(1 + 0.37464(1 - T_r^{0.5})\right)^2$$

$$b \equiv 0.0778 \frac{\overline{\Re} T_c}{p_c} \tag{3}$$

where a is dependent on the critical temperature and critical pressure for the mixture, b is the mixture co-volume, $\overline{v}$ is the specific volume of the mixture, and $\overline{\Re}$ is the universal gas constant.

Assuming N chemical species, chemical reactions can be expressed in a general reaction equation given by (X_i represents the symbol for species i),

$$\sum_{i=1}^{N} \nu_i' X_i \;\; {}_{k_b}\!\rightleftharpoons^{k_f} \;\; \sum_{i=1}^{N} \nu_i'' X_i \tag{4}$$

With a general reaction rate equation given by,

$$\frac{dC_i}{dt} = \frac{\dot{w}_i}{M_i} = (\nu_i'' - \nu_i') \left(k_f \prod_i^N C_i^{\nu_i'} - k_b \prod_i^N C_i^{\nu_i''} \right) \tag{5}$$

where k_f and k_b are the forward and backward reaction rates, respectively. The dimensional chemical production rate for species i is $\dot{w}_i$ (C is the concentration), with ν_i' and ν_i'' as the reactant and product stoichiometric coefficients for species i, respectively, and M is the molecular weight. The forward rates are specified,

$$k_f = \mathrm{C}_f T^{\beta} \mathrm{e}^{-E_f/kT}, \quad (\mathrm{cm}^3 \ \mathrm{mole}^{-1} \ \mathrm{s}^{-1}) \tag{6}$$

while the backward rates are obtained from the equilibrium constant, $K_{eq} = k_f/k_b$. Entropy, S, and enthalpy, h, are computed using curve-fits and then for each reaction,

$$K_{eq} = \exp\left(-\frac{\Delta g}{\Re T}\right)$$

$$\Delta g = g_{\mathrm{products}} - g_{\mathrm{reactants}}, \quad g = h - TS \tag{7}$$

Given an appropriate chemical kinetics mechanism, defining the species, reactions, and reaction rates, then the chemical production terms, $\dot{w}$, can be computed for each species conservation equation. Chemical kinetics mechanisms are discussed in the next section.

The integral form of the Navier-Stokes equations is convenient for the finite-volume solution scheme.

$$\frac{\partial}{\partial t} \int_{\Omega}\!\!\int U dx dy + \int_{d\Omega} (F dy - G dx) + \int_{\Omega}\!\!\int H dx dy = 0 \tag{8}$$

where, Ω is the region or computational cell, $d\Omega$ is the boundary or the cell sides. The semi-discrete form of this equation is given by,

$$\frac{d}{dt}(S_{ij} U_{ij}) + L U_{ij} - D U_{ij} + S_{ij} H_{ij} = 0 \tag{9}$$

where LU_{ij} is the spatial discretization operator and DU_{ij} is the smoothing or dissipative operator for each computational cell center located at grid index i, j. The effective flux through a cell face can be written as, $LU_{ij} - DU_{ij}$. An explicit fifth-order Runge-Kutta algorithm is utilized for time-integration of the semi-discrete equation. The convective and transport terms are resolved using upwind-differences along with flux-limiting to yield a second-order spatial resolution. Numerical stiffness due to chemical source terms is mitigated by convergence acceleration techniques. Flow turbulence is modeled using a mixing length approach. See Nusca (1997, 1998) for further details.

3. Chemical kinetics mechanisms

Hydrocarbon mixtures (e.g., $3CH_4+2O_2+10N_2$), pressurized to 5-10 MPa and above, are commonly used in the ram accelerator. Fuel rich mixtures are used with fuel equivalence ratios of about 3. For this value a complete understanding of CH_2/O_2 chemical kinetics, especially for $p >$ 1 MPa, is not available. Accurate numerical simulation of hydrocarbon-based reacting flow is very demanding in terms of computational resources since the number of intermediate species is prohibitively large (i.e., > 45). In order to decrease the computer time required of CFD simulations, which increases with N in a greater than linear fashion, skeletal or reduced reaction mechanisms (i.e., decreasing N by neglecting some intermediate reaction steps and/or combining elementary steps) have been developed.

A detailed kinetics mechanism was developed by Petersen (Petersen et al. 1996) to model shock tube ignition delay times for CH_4/O_2 mixtures at pressures from 4 - 26 MPa, temperatures from 1040-1500K, and fuel-rich stoichiometry. Using the shock tube facility at Stanford, this study produced ignition delay time data for mixtures used in the ram accelerator as well as curve-fit formula for these data. Petersen (Petersen et al. 1996) shows significant discrepancy between this data and the detailed mechanism GRIMECH (Frenklach et al. 1995) consisting of 23 species and 177 reactions, and produces a new detailed mechanism called RAMEC (47 species and 279 reactions). This new mechanism agrees well with the shock tube data. In addition, Petersen (Petersen et al. 1997) has produced a skeletal mechanism REDRAM (22 species and 32 reactions) that essentially reproduces the shock tube data and results of the detailed mechanism, RAMEC. The ARL-NSRG2 code has been linked with REDRAM and therefore currently includes chemical species: CH_4, CH_3, CH_3O_2, C_2H_2, C_2H_3, C_2H_4, C_2H_5, C_2H_6, CO, CO_2, CH_2O, CH_3O, H, H_2, O, O_2, OH, H_2O, HO_2, H_2O_2, HCO, and N_2.

4. Reacting flow model validation

Stanford has embarked on an experimental investigation of the shock wave structure and combustion fronts around geometries similar to the ram accelerator projectile. One such geometry is a 40 degree half-angle wedge, 38.1mm in length with a 23mm constant thickness (25.4mm) afterbody section. This model was mounted in the test section of an expansion tube described by Morris et al. (Morris, Kamel, and Hanson 1996). The test gas mixture was $2H_2+O_2+17N_2$ and the freestream flow conditions for the wedge were 2140 m/s, 18 kPa, 290 K, for velocity, pressure and temperature, respectively. The freestream Mach number was 5.8. Although the pressure in the expansion tube test section is much smaller than mixture fill pressures tested in the ram accelerator (e.g., 5.0 MPa), this case is a good test case for at least two reasons: 1) comparison with measured onset of combustion on the wedge tests the coupling of the flow code and the kinetics mechanism with respect to predicting flow-time/ignition-time interactions, 2) the kinetics mechanism REDRAM has been validated for low as well as high pressure so that it will be the same mechanism used for the high pressure ram accelerator computations. While ram acceleration uses principally hydrocarbon (e.g., CH_4) based mixtures, the H_2/O_2 reaction mechanism forms a subset of the CH_4/O_2 mechanism, REDRAM, and thus should be tested in conjunction with the CFD code. Figure 1 shows the computed OH mass fraction contours, using the Stanford mechanism, and illustrate the principle release of OH at about one-quarter to one-third of the wedge length downstream of the leading edge. Figure 1 also shows the results of simultaneous Schlieren/PLIF photography for the wedge flow produced by Stanford (Morris et al. 1996). Qualitative agreement between computations and experiment are good.

5. Results

Time-accurate CFD simulations are reported for conditions corresponding to shot 27 of the ARL 120-mm ram accelerator. The projectile was injected at approximately 1250 m/s ($Ma = 3.5$) into the accelerator tube filled with a gaseous mixture ($3CH_4+2O_2+10N_2$) at 300 K and pressurized to 5.0 MPa. The mixture has a sound speed of 361 m/s, a Chapman-Jouget detonation speed of 1448 m/s, and a pre-combustion γ of 1.379. Pressures on the accelerator tube wall were collected at 11 ports located from 0.3 to 9.1 meters from the entrance diaphragm. Doppler radar was used to obtain a continuous in-bore velocity-time history of the projectile. The CFD simulation was started with entrance of the projectile/obturator combination into the accelerator tube at t = 20.05 ms with specified velocity (1256.4 m/s) and obturator back pressure (27 MPa). The

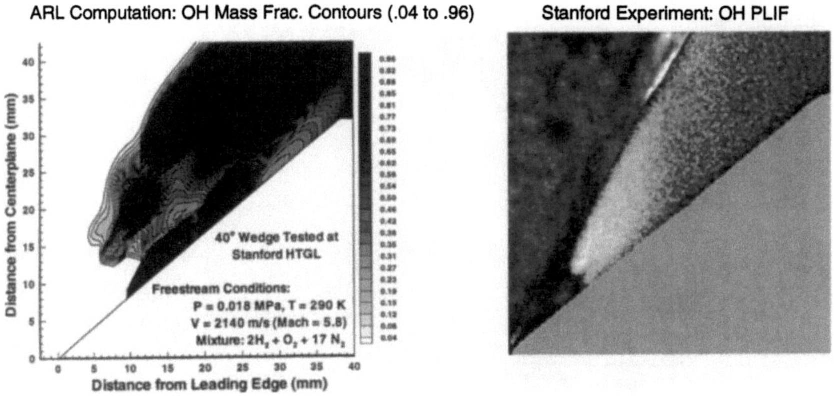

Fig. 1. Computed OH mass fraction contours compared to OH-PLIF results.

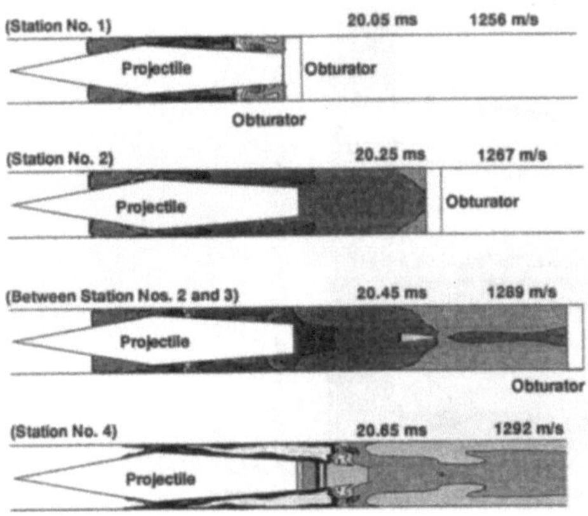

Fig. 2. Computed H_2O mass fraction contours (0 to 0.1). Times and projectile velocities are given.

projectile fins were neglected so that axisymmetric simulations could be used. The flowfield equations (Eq. 1) were then solved in a time-accurate fashion along with a force computation for the projectile and obturator, individually. The velocity of the projectile and the relative separation between the projectile base and obturator were updated for each time step (0.2 ms) using the computed, time-dependent force and given masses. The backpressure on the obturator due to the conventional charge was assumed to be ambient after 20.45 ms. When the obturator is greater than five projectile lengths downstream of the projectile base, its influence on the projectile flow field is negligible. A downstream outflow condition is then prescribed one projectile body length behind the projectile.

Figure 2 shows the computed time sequence of projectile obturator separation in terms of gas H_2O mass fraction contours. Water is a major product for the reaction, thus illustrating regions of significant combustion activity. For Fig. 2, white corresponds to the absence of H_2O in the flowfield, while dark-grey corresponds to a water mass fraction of about 0.1. Flow stagnation on the obturator, when in close proximity to the projectile, causes both a normal shock and a combustion front to occur on the projectile. The combustion front is located on the projectile forebody while the normal shock is located on the afterbody. Flow between the combustion front and the normal shock is slightly supersonic. As the obturator is pushed further back from the projectile the normal shock moves rearward and is eventually positioned behind the projectile. At this point the projectile is in fully supersonic flight. Combustion then occurs at the projectile-

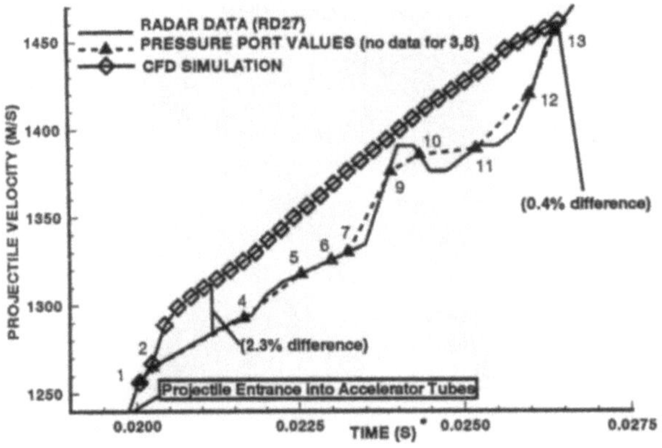

Fig. 3. Computed and measured projectile velocity vs time of flight.

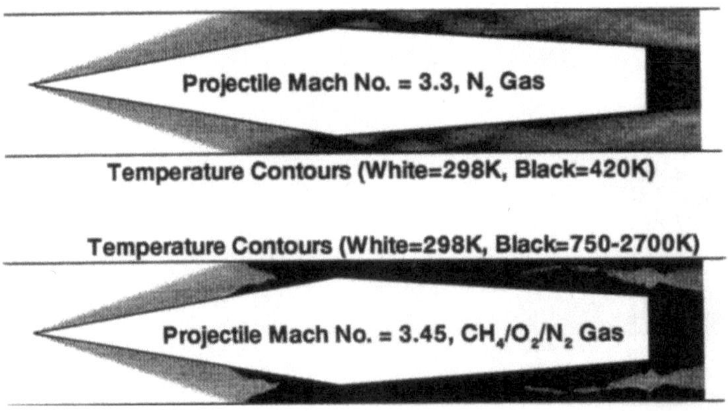

Fig. 4. Computed temperature contours for inert (upper view) and reacting (lower view) flows.

afterbody junction due to shock heating, in the projectile and tube wall boundary layers due to stagnation heating, and downstream of the normal shock traveling behind the projectile. This normal shock, which also initiates combustion, is now driven by the obturator (i.e., a piston). Due to the high combustion pressures on the side facing the projectile as well as the relieved backpressure, the obturator is moving slower than the projectile.

Figure 3 shows the corresponding projectile velocity-time history. The projectile velocity was measured using a Doppler radar positioned outside the accelerator and beamed down the tube bore using a sacrificial mirror. Tube-mounted transducers were used to measured the projectile pressure signature as it passed each station. The computer simulation shows overprediction with respect to measurements, beyond station 2. Thereafter the computed projectile velocity was fairly constant at 2.3% above measurements. The large projectile thrust computed after station 2 is due to a higher than measured pressure on the projectile afterbody. It is thought that perhaps the obturator is tilted (i.e., the obturator length is such that the obturator can spin within the tube cross-section), thus relieving pressure from behind the projectile.

Figure 4 corresponds to conditions for shot 14 station 5 ($M_a = 3.3$) and shot 15 station 4 ($M_a = 3.45$). These stations are located in the accelerator such that the obturator has been discarded and ram combustion is started. In these gray-scale figures, the lighter colors correspond to cool temperatures. The upper view in Fig. 4, for the inert shot, shows the projectile nose shock reflecting from the tube wall, impinging near the projectile forebody/afterbody junction,

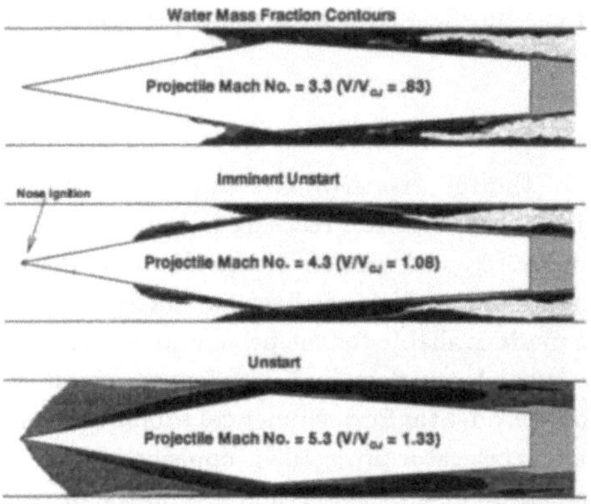

Fig. 5. Computed H_2O mass fractions for three different projectile Mach numbers. Projectile velocities normalized by the Chapman-Jouget detonation velocity for the mixture, are also shown.

and subsequently reflecting and weakening downstream. High temperature boundary layers are observed on the projectile and tube walls. The lower view of Fig. 4, for the combustion shot, shows wide regions of high temperature gas caused by shocks and combustion, superimposed. Note combustion occurs in the projectile and tube wall boundary layers as well; rapid boundary layer thickening caused by shock wave impingement is evident. Some localized combustion occurs in the forebody boundary layer. Results for higher pressure shots (see Nusca and Kruczynski 1996) shows that extensive combustion between the projectile forebody and tube wall becomes more predominant with increasing freestream pressure. In addition, increased projectile acceleration causes the nose shock to reflect further rearward on the tube wall, combustion to move from the second shock reflection to the first, and combustion in the forebody boundary layer to move forward. These effects shift high pressure to the forebody where drag is generated, counteracting thrust generated by high pressure on the afterbody.

Figures 5 shows the computed H_2O mass fractions (white corresponds to the absence of H_2O in the flowfield, while black corresponds to a water mass fraction of 0.1). Ignition of the mixture is indicated by the sudden appearance of H_2O. The standard mixture was used along with conditions: $3CH_4$+$2O_2$+$10N_2$, $p = 5.0$ MPa, $T = 300$K, projectile velocity = 1195 m/s ($M_a = 3.3$, $V/V_{CJ} = 0.83$). The projectile is assumed to be started in the accelerator tube for the first view in the Fig. 5 (which corresponds to the lower view in Fig. 4). Here chemical reactions start both in the forebody boundary layer and at the point of shock reflection from the tube wall. Reactions engulf the entire projectile afterbody surface and continue in the tube wall boundary layer downstream of the projectile. The projectile wake contains reaction products as well. Expansion (cooling) of the mixture over the projectile afterbody reduces the reaction progress and thus smaller amounts of H_2O are found between the tube wall and projectile wall boundary layers. For the second and third views in Fig. 5, computed results for projectile velocities corresponding to Mach numbers of 4.3 and 5.3 are shown with all other conditions held constant. The shock and boundary layer induced reactions produce more water near the projectile, which begins to diffuse further out into the flowfield on the projectile forebody. A small region of ignition on the projectile nose occurs at the flow stagnation point. Figure 5 illustrates the "high Mach number unstart" phenomena as a wave of combustion now travels ahead of the projectile, starting at the nosetip. The computed projectile thrust values (force normalized by fill pressure and tube cross sectional area, F/PA) corresponding to the three views of Fig. 5 are 2.5, 2.3, and -.21, respectively. A negative value indicates a projectile drag force. Thrust was determined by numerical integration of pressure and shear stress on the projectile surface along with the flowfield simulated using the present CFD code.

6. Conclusions

Computational fluid dynamics (CFD) solutions of the full Navier-Stokes equations along with finite-rate chemical kinetics, are used to numerically simulate the reacting in-bore flowfield for the ram accelerator projectile propulsion system. A reduced version of the Stanford RAMEC chemical kinetics mechanism (22 species and 32 reactions), which has been validated for ram accelerator conditions using ignition time data, was made available for the present investigations. A CFD code validation case is presented for low pressure, steady flow, hydrogen/oxygen combustion around a wedge as compared to Schlieren/PLIF data from Stanford. Numerical simulations of the reacting in-bore flowfield for the 120-mm ARL ram accelerator projectile propulsion system have been performed. These simulations show that stagnation of the combustible gas on the projectile obturator causes the formation of both a normal shock and a combustion wave in the flowfield. As the obturator is gasdynamically discarded, the normal shock trails behind the projectile and the combustion wave collapses into shock-induced combustion on the projectile. With the obturator sufficiently downstream, the projectile accelerates at supersonic speeds. Projectile velocity through the accelerator can be adequately predicted, but is dependent on accurate thermodynamic state equations and chemical kinetics models. Sufficiently high projectile velocities can induce an "unstart" as the combustion wave precedes the projectile. This phenomenon has also been numerically simulated.

References

Benedek P, Olti F (1985) Computer Aided Chemical Thermodynamics of Gases and Liquids, Wiley, New York

Frenklach M, Wang H, Goldenberg M, Smith GP, Golden DM, Bowman CT, Hanson RK, Gardiner WC, Lissianski V (1995) GRI-Mech – an optimized detailed chemical reaction mechanism for methane combustion. GRI Topical Rep No. GRI-95/0058

Morris C, Kamel M, Hanson R (1996) Investigation of ram accelerator projectile flowfields in an expansion tube. Proc 33rd JANNAF Combustion Subcommittee Meeting, Naval Postgraduate School, Monterrey, CA, Nov. 4-8

Nusca MJ (1994) Reacting flow simulation for a large scale ram accelerator. AIAA paper 94-2963

Nusca MJ (1995) Reacting flow simulation of transient multi-stage ram accelerator operation and design studies. AIAA paper 95-2494

Nusca MJ (1996) Investigation of ram accelerator flows for high pressure mixtures of various chemical compositions. AIAA paper 96-2946

Nusca MJ (1997) Computational simulation of the ram accelerator using a coupled CFD/interior-ballistics approach. AIAA paper 97-2653

Nusca MJ (1998) Computational fluid dynamics code for the real-gas Navier-Stokes equations – NSRG2. US ARL Techl Rep

Nusca MJ, Kruczynski DL (1996) Reacting flow simulation for a large-scale ram accelerator. J Prop Power 12:61-69

Petersen E, Davidson D, Hanson RK (1996) Ram accelerator mixture chemistry: Kinetics modeling and ignition measurements. Proc 33rd JANNAF Combustion Subcommittee Meeting, Naval Postgraduate School, Monterrey, CA, Nov. 4-8

Petersen E, Davidson D, Hanson RK (1997) Reduced kinetics mechanisms for ram accelerator combustion. AIAA paper 97-2892

Numerical investigation of ram accelerator flow field in expansion tube

J. Y. Choi, I. S. Jeung, Y. Yoon
Department of Aerospace Engineering, Seoul National University, Seoul 151-742, Korea

Abstract. Steady and unsteady numerical simulations are conducted for the experiments performed to investigate the ram accelerator flow field by using the expansion tube facility in Stanford University. Navier-Stokes equations for chemically reacting flows are analyzed by fully implicit time accurate numerical method with Jachimowski's detailed chemistry mechanism for hydrogen-air combustion involving 9 species and 19 reaction steps. Although the steady state assumption shows a good agreement with the experimental schlieren and OH PLIF images for the case of $2H_2+O_2+17N_2$, it fails in reproducing the combustion region behind the shock intersection point shown in the case of $2H_2+O_2+12N_2$ mixture. Therefore, an unsteady numerical simulation is conducted for this case and the result shows all the detailed flow stabilization process. The experimental result is revealed to be an instantaneous result during the flow stabilization process. The combustion behind the shock intersection point is the result of a normal detonation formed by the intersection of strong oblique shocks that exist at early stage of the stabilization process. At final stage, the combustion region behind the shock intersection point disappears and the steady state result is retained. The time required for stabilization of the reacting flow in the model ram accelerator is found to be very long in comparison with the experimental test time.

Key words: Expansion tube, Numerical simulation, Ignition mechanisms

1. Introduction

A ram accelerator is a concept using a principle of ramjet propulsion as a propelling mechanism of a projectile in a barrel (Hertzberg et al. 1988, Bogdanoff 1992); a combustible mixture gas is compressed by a series of shocks and then generates thrust force by high speed combustion mechanism such as shock-induced combustion or oblique detonation. The ram accelerator projectile can reach the higher velocity with relatively low maximum pressure than that in the conventional gun since the rear part of the projectile is pressurized at a certain level during the entire acceleration process. The understanding of internal supersonic combustion mechanism does not thought to be sufficiently or controllably understood yet although it is essential for the successful and efficient operation of ram accelerator.

For the past years, there have been a number of researches on shock-induced combustion and oblique detonation those thought as the essential feature of the reacting flow field in ram accelerator. However, the direct visualization or measurement of the internal combustion flow field in ram accelerator is very difficult due to the very short flight time and long flight distance. As alternatives to the direct visualization of ram accelerator flow field, some other experimental techniques have been used for the investigation of the phenomena using ballistic range facility (Kaneshige et al. 1996), detonation tube (Viguier et al. 1994) or expansion tube facility (Srulijes et al. 1992). In addition, CFD technique has been widely used for the understanding of the combustion flow field (Yungster 1992, Nusca et al. 1996, Li et al. 1996, Choi et al. 1998a).

Recently, Morris et al. (1996a) tried to visualize and measure the ram accelerator flow field quantitatively. Expansion tube facility that capable of accelerating combustible gas up to hypersonic velocity was used to model the stationary ram accelerator flow field and schlieren/PLIF technique was used for the visualization of flow field and burned gas region. Although their experiments did not suggest reliably quantitative data yet, their result is considered as demonstrating the importance of viscous effect in internal combustion flow of ram accelerator. Therefore, the steady and unsteady characteristics of the internal combustion flow in model ram accelerator are analyzed in this study by computational simulation.

2. Expansion tube facility and research motivation

Expansion tube is a two stage shock tube device that capable of accelerating a test gas up to hypersonic velocity. The difference in sonic velocities of test gas and acceleration gas is the key of the acceleration up to hypersonic range since the both gases are accelerated at same velocity induced by the incident normal shock. The principle of expansion tube could be explained by the schematic of the device and x-t wave diagram in Fig. 1. The result of analysis of one-dimensional compressible flow is plotted in Fig. 2. It is easily recognizable that hypersonic flow is formed after the contact surface of two different gases having the same pressure. While the initial pressure of the driver gas should be determined by the pressure limit of device, the initial condition of test gas is determined by the needed test condition. In order to make high Mach number test gas flow, the light gases such as hydrogen or helium are used as acceleration gas at extremely low pressure. The operation time of this device is the hypersonic steady flow time at test section, that is the time between the arrival of contact surface to the arrival of expansion waves. Generally, the test time is limited to below one msec and it is considered as sufficient for the flow stabilization and the steady state experiment.

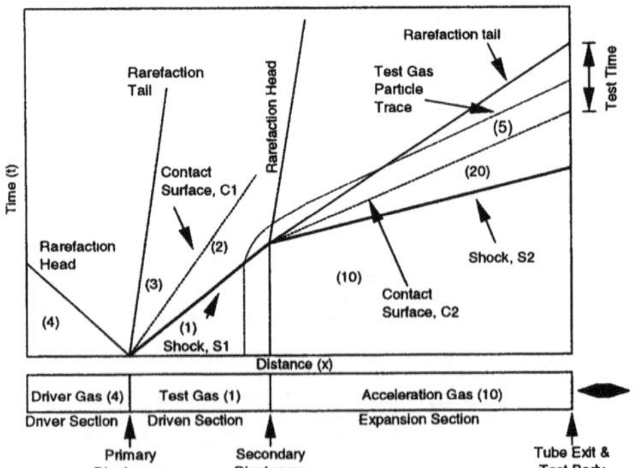

Fig. 1. $x-t$ wave diagram and schematic of expansion tube

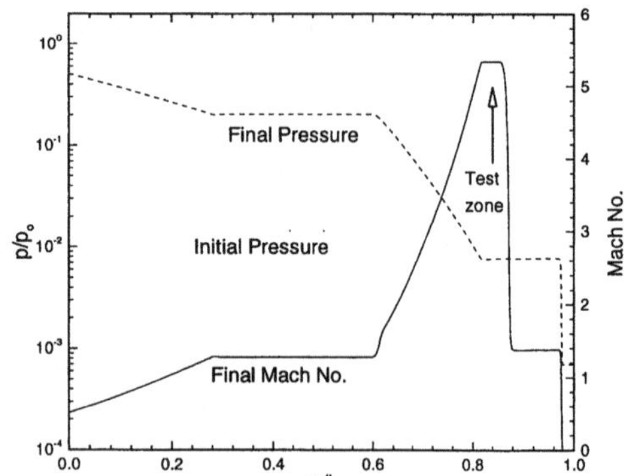

Fig. 2. Mach number and pressure distribution from one-dimensional simulation of expansion tube

Figure 3 is a schematic configuration of model ram accelerator used in experiment by Morris et al. (1996a). It was configured symmetrically in upper-lower side in order to get rid of the boundary layer effect on tube wall surface that is very small in real ram accelerator configuration. The flow conditions of test gases arrived at model ram accelerator are listed in Table 1.

Figure 4 is overlaid image of schlieren picture and OH PLIF image captured at almost same time. In case of (a) $2H_2+O_2+17N_2$ mixture, OH concentration is high in the region where shock and boundary layer interaction occurs. One can see the combustion occurring around the wall. However, ignition delay calculated from chemical kinetic theory in this region is much longer than other characteristic times of the flow (Morris et al 1996a), and the viscous effect is considered as a major reason of the formation of the combustion region along the body surface (Choi et al. 1998a), although there is no obvious experimental evidence. In case of (b) $2H_2+O_2+12N_2$

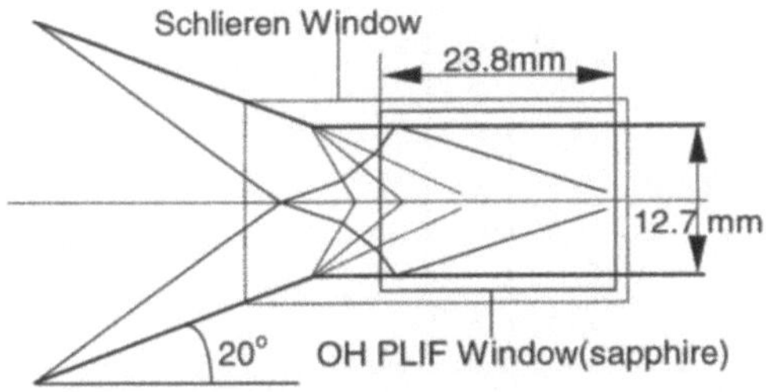

Fig. 3. Schematic configuration of model ram accelerator

Table 1. Experimental conditions of test gases at tube exit

Case	$2H_2+O_2+17N_2$	$2H_2+O_2+12N_2$
M_∞	5.2	5.2
T_∞(K)	350	350
P_∞(kPa)	11.2	11.2

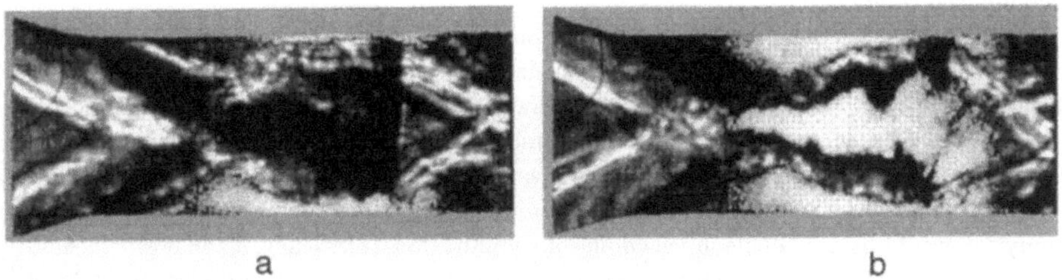

Fig. 4. Overlaid schlieren and OH PLIF images of combustion in model ram accelerator; **a**, $2H_2+O_2+17N_2$; **b**, $2H_2+O_2+12N_2$

mixture, the combustion region is formed not only around both wall but also in the center of flow region. However, according to the calculations of Morris et al.(1996a), the flow condition of the center of flow region generated by the intersection of normal shock waves is not sufficient to induce the combustion directly after the intersection point. Here, we can deduce that the two combustion mechanisms, viscous effect and unsteady shock-induced combustion could be a reason of the ignition, even though those have not been explained clearly yet. Therefore, in this study, numerical simulations of above experiments were carried out to clarify the combustion mechanisms.

3. Governing equations and chemistry model

For the simulation of the above expansion tube experiment, the coupled forms of species conservation equations and Reynolds averaged Navier-Stokes equations are employed to analyze the chemically reacting supersonic viscous flow in two dimensional coordinate. The conservation form of the equations is written as equation (1).

$$\frac{1}{J}\frac{\partial \mathbf{Q}}{\partial t}+\frac{\partial \mathbf{F}}{\partial \xi}+\frac{\partial \mathbf{G}}{\partial \eta}=\frac{1}{J}\mathbf{W}+\frac{\partial \mathbf{F}_v}{\partial \xi}+\frac{\partial \mathbf{G}_v}{\partial \eta} \qquad \text{where} \quad \mathbf{Q}=\begin{pmatrix}\rho_1\\ \rho_2\\ \vdots\\ \rho_N\\ \rho u\\ \rho v\\ e\end{pmatrix} \tag{1}$$

Since the details of the governing equations are described in the references (Choi et al. 1998a), that will be omitted in this paper. The specific heat data are obtained by NASA polynomial data (Gardiner 1984) that are valid up to temperature of 6000K. The laminar values of mixture viscosity and thermal conductivity are calculated using Wilke's mixing rule from the value of each species (Bird 1960). The binary mass diffusivity is obtained using the Chapman-Enskog theory and the diffusion velocities are found by Fick's Law for convenience (Bird 1960). The present hydrogen-air combustion mechanism is obtained from Jachimowski (1988) by ignoring the nitrogen dissociation mechanisms that has negligible effect on flow field characteristics. The model consists of nine-species and nineteen reaction-steps and has been validated by the simulation of shock-induced combustion.

For the ram accelerator model combustor considered here, Reynolds number based on projectile length is about 1×10^6 due to the small size of the test section size even though the flow velocity is very high. Since, there is not an evidence of the strong effect of turbulence and boundary layer transition occurs around Reynolds number of 1×10^6 for flat plat according to Shapiro (1954), laminar assumption is used for whole flow field. Although the test model is not a flat plat and shock wave/boundary layer interaction would make the flow being turbulent, the laminar assumption would be applicable in viewpoint of computational efficiency, if the flow field solution is not distorted by the assumption and the turbulence has no great influences on the global combustion mechanism in the experiment. Moreover, there is not an adequate turbulence model that can predict the interaction between the chemistry and turbulence in the region of strong shock wave/boundary layer interaction.

4. Numerical formulation

The finite volume approach is used for the spatial discretization of the governing equations. The convective fluxes are formulated using Roe's FDS method derived for multi-species reacting flow. MUSCL type variable extrapolation approach is used to get a high order spatial accuracy and differentiable limiter function is used to preserve the TVD property (Hirsh 1990). The general central difference scheme is used for viscous fluxes. The detailed description of the spatial discretization is described in the reference (Choi et al. 1998a).

For the case of steady state solution, the discretized governing equations are temporally integrated by first order accurate fully implicit method. LU relaxation scheme (Shuen et al. 1989) is used for the implicit analysis with approximate splitting method of flux Jacobian matrix. The detailed description of this formulation is also found in the reference (Choi et al. 1998a).

In case of unsteady simulation, second order accurate fully implicit time integration method is used with Newton-Raphson sub-iteration method to preserve the time accuracy and solution stability at large time step. Steger-Warming flux splitting approach is used in the splitting of Jacobian matrix to avoid the large diffusion error involved in the approximate splitting approach. The implicit part of the discretized equations is inverted using LU relaxation scheme with compact formulation of Jacobian matrix-variable vector product for the computational efficiency. The numerical method used in this study is described more precisely in the reference (Choi et al. 1998b) and has been validated by the simulation of unsteady shock-induced combustion mechanism.

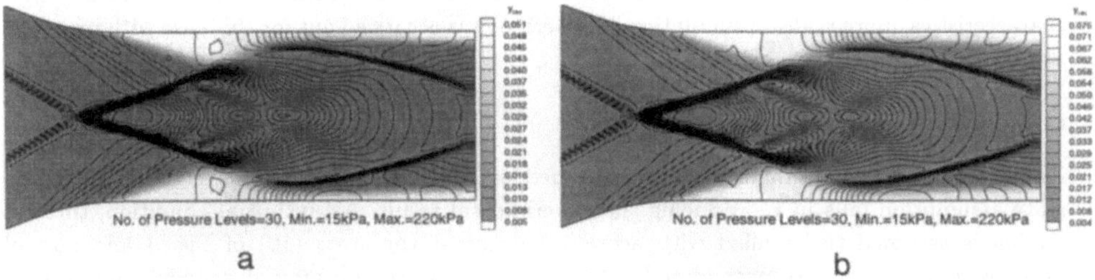

Fig. 5. Overlaid temperature and OH mass fraction distributions from the numerical simulation of the model ram accelerator configuration with steady state assumption; a, $2H_2+O_2+17N_2$; b, $2H_2+O_2+12N_2$

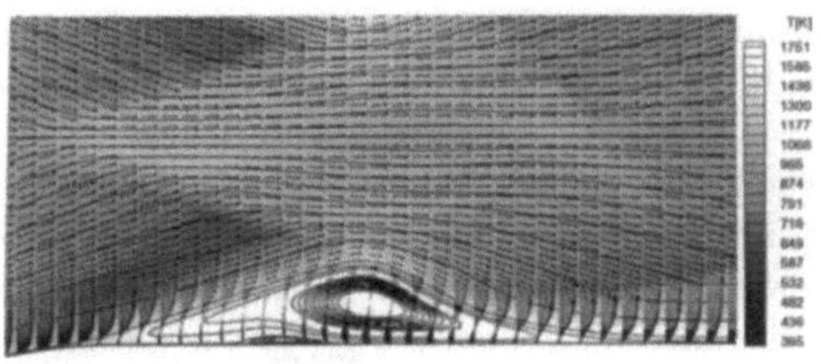

Fig. 6. Overlaid plot of temperature distribution, velocity vector field and stream lines in the case of $2H_2+O_2+17N_2$ mixture

5. Results of steady state simulation

Numerical analyses were performed for the experimental cases above with steady state assumption. All the flow variables were set to the values of inflow initially. Figure 5 is the resulting overlaid plots of temperature contour and OH mass fraction distribution for both cases. The analysis result of case (a) $2H_2+O_2+17N_2$ mixture shows the shock train in the internal flow field of ram accelerator and the combustion region around the wall. The combustion gas region exactly matches with the recirculation region originated from shock wave/boundary layer interaction as shown in Fig. 6. Thus, aerodynamic heating in this stagnation flow region is considered as an ignition mechanism of mixture flow in this region. Choi et al. (1998a, 1998b) have found the similar mechanism of viscous effect in the unsteady simulation of ram accelerator.

In case of (b) $2H_2+O_2+12N_2$ mixture that is more energetic than the case of (a), the numerical result shows that the combustion region is also formed at both walls only. This result is very similar to case (a), but different to experimental result. In this numerical analysis, contrary to experimental result, there is no combustion region in the central region of test section that is formed by shock heating in the back of shock intersection point. Even though this numerical result corresponds to the theoretical estimation by Morris et al. (1996a), it fails in the reproduction of experimental result. So to find the cause of the above disagreement in the unsteady flow characteristics, more realistic simulation of experiment was carried out for this case with unsteady assumption

6. Results of unsteady simulation

An unsteady numerical simulation was performed for $2H_2+O_2+17N_2$ mixture in which the steady state assumption fails in reproducing the experimental result. As an initial condition, the test section is assumed to be filled with acceleration gas at the state (10) of Fig. 1 and normal shock wave is assumed to propagate from the inlet to the computational domain, that is, post shock condition is applied initially at the inlet. Although the helium is used in the experiment, hydrogen is used as an acceleration gas for the efficiency of numerical simulation. Because the initial conditions at the state (10) are not described in the experimental result and different acceleration gas is assumed in this study, the initial conditions at state (10) and the post shock conditions at state (20) are calculated from the normal shock theory and contact surface conditions with the assumed temperature of 300K at state (10). The calculated conditions are summarized in Fig. 7.

In actual situation the experiment, the moving Mach number M_{20} of shock wave S2 is too small to form an attached oblique shock wave at the body surface. Thus, a bow shock is reflected from the body and normal shock wave moves forward to inlet. Consequently, this backward flow will collide with contact surface C2 in front of the inlet. Therefore, a considerable distance in front of test model should be included in computational domain in order to capture this colliding flow. Actually, the time needed for the passage of the acceleration gas is about 300 μs, (Morris et al. 1996b) and this flow collision will be occur at the end of the acceleration gas time. The inclusion of this physical acceleration gas time will be a great burden of computation due to the computational time need for the simulation of the physical time and to the large computational domain need to capture the colliding flow which occurs far ahead from the test section. Therefore, 50 μs of acceleration gas time is assumed. Although this value is so small in comparison with the actual operation time of acceleration gas (about 300 μs), it is considered to be a proper value in regards of the load of time required for computation.

Figure 8 is the computational grid for the unsteady numerical simulation obtained by considering the 50 μs of acceleration gas time. The computational domain is composed by 300×201 grid system that are equally spaced in flow direction and clustered to both walls. No-slip and adiabatic boundary condition is applied at the body surface and supersonic boundary condition is applied at the exit. The physical time step used in the whole computational process is 7.27 ns which corresponds to CFL number of 2 for minimum grid spacing. Because operating fluid is hydrogen only during the acceleration gas time of 50 μsec, computation was performed excluding the chemistry source term for the computational efficiency. After this time acceleration gas time, the flow conditions of mixture at the back of contact surface C2 (state (5)) were used as inflow boundary condition and computation was performed including source term by chemical reactions. For reacting flow simulation, the solution stability needs 4 sub-iterations and entropy fixing parameter is set to 0.4 even though 2 sub-iterations and entropy fixing parameter of 0.01 seemed to be sufficient for the non-reacting flow calculation of acceleration gas.

The time marching progress of flow field of acceleration gas is plotted in Fig. 9. As the normal shock S2 rushes into model ram accelerator, a typical detached bow shock is established by the shock reflection at the body surface. As time goes on, two bow shock waves from upper and lower walls intersect with each other. And then, the bow shock waves reflect at the walls and form

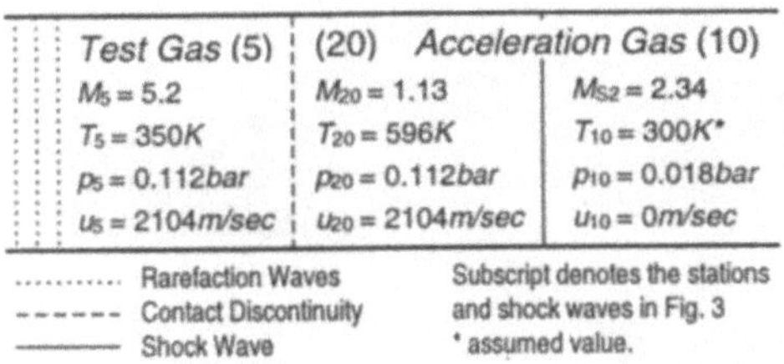

Fig. 7. Deduced flow conditions of the acceleration gas with the conditions of the test gas and the acceleration gas temperature of 300K

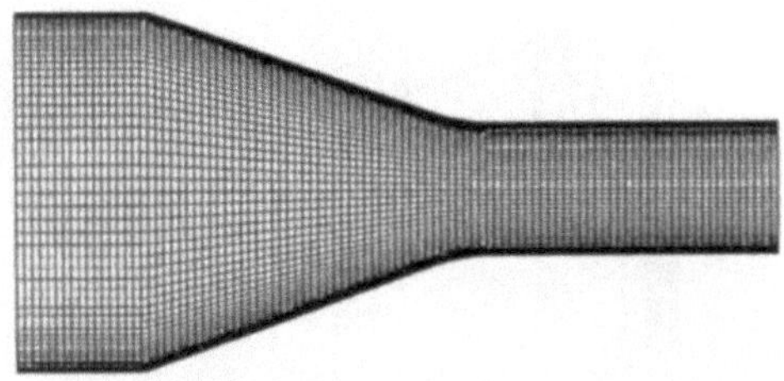

Fig. 8. 300×201 computational grid for the unsteady numerical simulation of the expansion tube experiment. Every third point is plotted for clarity

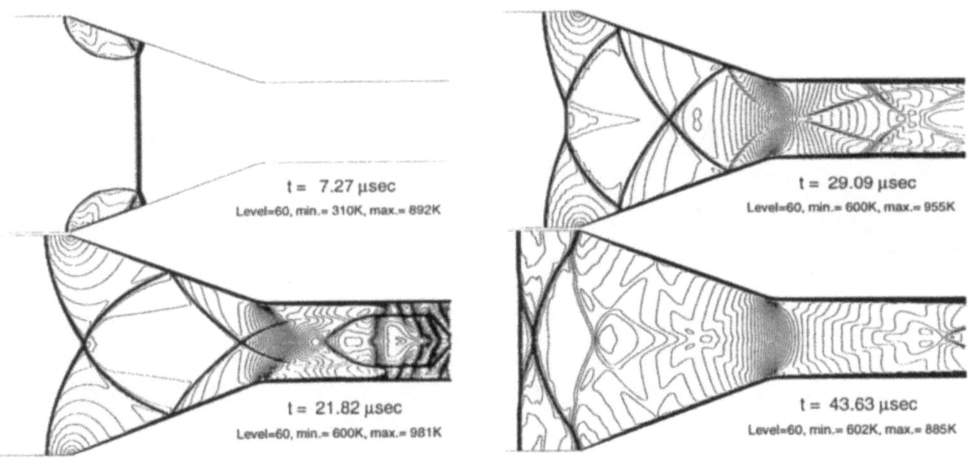

Fig. 9. Resulting temperature distributions from the unsteady simulation for the acceleration gas. The final flow field data are used as an initial condition for the unsteady simulation for the test gas

complex intersected shock train. In addition, boundary layers develop at the both walls. After the several times of intersection/reflection process, a normal shock wave is formed ahead of the test section and runs forward to the inlet. At this stage, the intensity of shock train becomes very weak and the gradient of flow field is mitigated. Finally at 50 μsec, normal shock moves up to inflow boundary, and from that on, the reacting flow calculation is started with application of the post-contact surface mixture condition at the inflow boundary.

The resulting sequential temperature distributions of reactive flow calculation are plotted in Fig. 10. At t=58.17 μsec, 8.17 μsec after the beginning of reacting flow calculation, a fast and

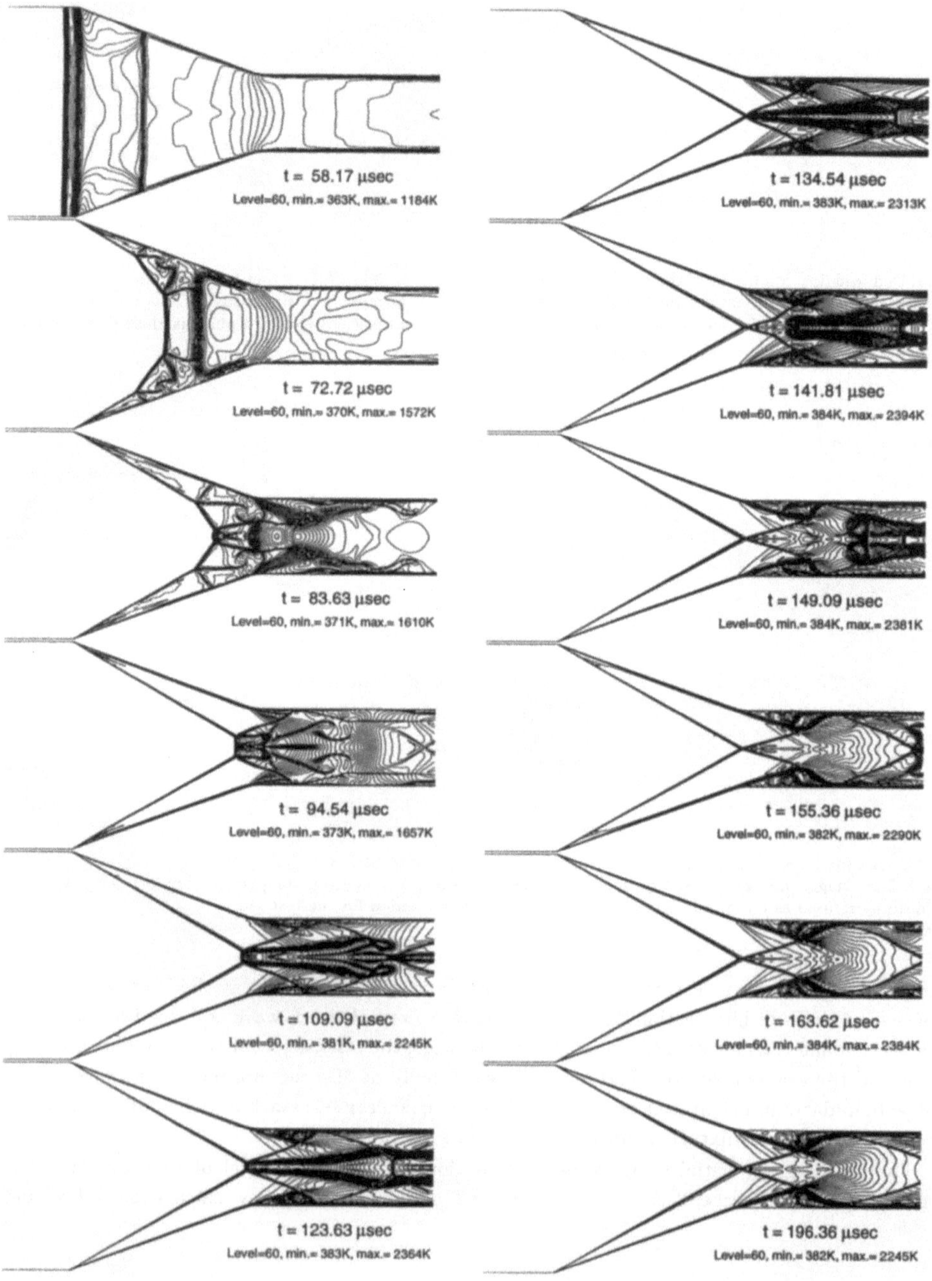

Fig. 10. Resulting temperature distributions from the unsteady simulatin for the test gas

weak normal shock wave, a contact surface and a slow and strong normal shock wave are observed as a result of intersection between the contact surface C2 and the forward running normal shock wave. As the contact surface and strong normal shock rush into the test section, dual angled oblique shock waves are formed at the nose of the model ram accelerator. Between the oblique shock waves of different angles, the small angled one would be the conventional weak oblique shock wave. Mach waves and slip lines are formed at the intersection point. As the time goes on, the cross section area of the normal shock wave is getting smaller and two strong oblique shock waves intersect with each other around 80.00 μsec.

The intersection of oblique shock waves forms a new very strong at the center of the test section. The strength of the new normal shock would be greater than the previous normal shock, and the self-ignition of the mixture flow is initiated by the shock heating at the center point of test section. Therefore, the new normal shock wave transits to a normal detonation wave. Due to the present of strong normal detonation, the large value of entropy fixing parameter was needed in the computation, since the small value of it results in carbuncle and even-odd de-coupling phenomena. After the ignition, burned gas flows downstream and the flows are being stabilized after 100 μsec. Complex wave intersections are noticed during this process but those would not be important in the viewpoint of major combustion phenomena. The normal detonation is maintained with a configuration of triple point interaction mechanism for a long time after disappearance of strong oblique shock waves. Besides, an oblique shock wave is reflected at the triple interaction point to the wall. A contact surface (slip line) is formed parallel to the wall and act as flame boundary. The flame boundary is represented as thick temperature gradient across which the burned gas and the unburned gas are segregated. The combustion initiated by the normal detonation wave forms a long burned gas core in the center of the test section. The burned gas flows downstream and extends to the exit at t=123.63 μsec. This transient result can be compared with experimental result of the case (b) in Fig. 4.

The burned gas core is maintained for sufficiently long time as long as the normal detonation exists. However, the cross section of the normal detonation is getting smaller and disappears finally after t=134.54 μs. After the disappearance of the normal detonation, two oblique shock waves intersect with each other at the center of the test section. Therefore, the experimentally obtained schilieren/OH PLIF image of case (b) in Fig. 4 is not considered as an image of stabilized flow but as an intermediate one at time around 120 μs.

As the disappearance of the normal detonation wave, the ignition source in the core of test section, the burned gas region detaches at the intersection point of oblique shock wave and flashes downstream after t=138.18 μs. Even though the oblique shock waves in the rear part of the test section enhance the combustion, it is not sufficient to be a flame holding mechanism such as oblique detonation. Thus, the central burned gas region completely disappears. The flow field shows nearly stabilized manner after this situation though there is some disturbances in the boundary layer.

On the other hand, after t=95.54 μs, combustion progresses in the separated flow region that originates from the adverse pressure gradient owing to shock wave/boundary layer interaction. The separated flow region expands with the progress of combustion, and is bounded by the tail of expansion fan, reflected oblique shock wave wall and the wall. The combustion of this separated flow region is ignited by the aerodynamic heating of stagnated flow and is maintained in the recirculation flow to the end of computation. Owing to these inviscid and viscous combustion characteristics the final solution of unsteady simulation agrees with the result of steady state simulation in Fig 5. Even though there could be a skeptical question about the structure of vortex in the separated flow region due to the assumption of laminar flow, it would not be an important problem in the viewpoint of combustion mechanism. Because the separated flow region is bounded by inviscid flow characteristics and does not have a great influence on the global flow features. In addition, the computational result is considered as showing reasonable agreement with the experimental OH PLIF image in Fig. 4. From the result of numerical simulation, the test time required for flow stabilization is estimated about 150 μs, that seems to be somewhat large value in comparison with 100 $\sim$ 200 μs test time of expansion tube used in the experiment (Morris et al. 1996a).

7. Conclusion

The present results of steady and unsteady numerical simulations of model ram accelerator flow field shows the reasonable reproduction of experiments and presents a solution to the previous questions. The experimentally observed combustion region near wall surfaces agrees with the separated flow region originated by shock wave/boundary layer interaction. The region is bounded by the tail of expansion fan, reflected oblique shock wave and projectile walls and the combustion in the separated flow region is initiated by the aerodynamic heating in stagnated flow. In the case of $2H_2+O_2+12N_2$ mixture, the combustion region at the center of the test section observed in experiment would be a result of unsteady wave interaction phenomena in the developing stage of the flow field. That is, the intersection of strong oblique shock waves that exist at the initial stage makes the strong normal wave in the center of the test section. The combustion initiated by normal shock results a transition from the normal shock wave to a detonation wave. The detonation wave is sustained for relatively long time due to the heat addition behind the wave. However, the detonation wave disappears and the combustion region flash out downstream accordingly. Thus, the final stabilized combustion is formed along the projectile surface.

From the result of present simulations, we can understand that there could be two kinds of ignition mechanism in ram accelerator flow field. The one is the shock heating and the other is viscous heating. The ignition by shock heating seems to be explosive but somewhat unstable, and the ignition by viscous heating is stable but restricted to solid surface.

Acknowledgement

This study is supported by the Academic Research Promotion Fund for Mechanical Engineering of Korean Ministry of Education with contract number ME97-G-05.

References

Bird RB (1960) Transport Phenomena, John Wiley & Sons, New York

Bogdanoff DW (1992) Ram accelerator direct space launch system: New concepts. J Prop Power 8:481-490

Choi JY, Jeung IS, Yoon Y (1997) Structure of stabilized oblique detonation wave in ram-accelerator. ASPACC 97, pp 452-455

Choi JY, Jeung IS, Yoon Y (1998a) Numerical study of scram accelerator starting characteristics. AIAA J 36, in press

Choi JY, Jeung IS, Yoon Y (1998) Comparisons of numerical methods for the analysis of unsteady shock-induced combustion. In: Takayama K, Sasoh A (eds) Ram Accelerators, Springer-Verlag, Heidelberg, pp 243-254

Gardiner WC (1984) Combustion Chemistry. Springer-Verlag, New York

Hertzberg AB, Bruckner AP, Bogdanoff DW (1988) Ram accelerator: A new chemical method for accelerating projectiles to ultrahigh velocities. AIAA J 26:195-203

Hirsch C (1990) Numerical Computation of Internal and External Flows, John Wiley & Sons, Chichester

Jachimowski CJ (1988) An analytical study of the hydrogen-air reaction mechanism with application to scramjet combustion. NASA TP-2791

Kaneshige MJ, Shepherd JE (1996) Oblique detonation stabilized on a hypervelocity projectile. Proc 26th Symp (Int) Comb, pp 3015-3022

Li C, Kailasanath K, Oran ES (1996) Stability of projectiles in thermally choked ram accelerators. J Prop Power 12:807-809

Morris CI, Kamel MR, Hanson RK (1996a) Expansion tube investigation of ram-accelerator projectile flow fields. AIAA paper 96-2680

Morris CI, Kamel MR, Stouklov IG, Hanson RK (1996b) PLIF imaging of supersonic reactive flows around projectiles in an expansion tube. AIAA paper 96-085

Nusca MJ, Kruczynski DL (1996) Reacting flow simulation for a large-scale ram accelerato. J Prop Power 12:61-69

Shapiro AH (1954) The Dynamics and Thermodynamics of Compressible Fluid Flow, John Wiley & Sons

Shuen S, Yoon S (1989) Numerical study of chemically reacting flows using a lower-upper symmetric successive overrelaxation scheme. AIAA J 27:1752-1760

Srulijes J, Smeets G, Seiler F, George A, Mathieu G, Resweber R (1992) Shock tube validation experiments for simulation of ram-accelrator-related combustion and gasdynamic problems. Shock Waves, Proc 18th Int Symp Shock Waves, pp 611-616

Viguier C, Guerraud C, Desbordes D (1994) H_2-air and CH_4-air detonations and combustion behind oblique shock waves. Proc 25th Symp (Int) Comb, pp 53-59

Yungster S (1992) Numerical study of shock-wave/ boundary-layer interactions in premixed combustible gases. AIAA J 30:2379-2387

CFD computations of steady and non-reactive flow around fin-guided ram projectiles

M. Henner, M. Giraud, J.F. Legendre, C. Berner
ISL, French-German Research Institute of Saint-Louis, BP 34, F-68301, Saint-Louis Cedex, France

Abstract. The numerical simulation of the flow around the body has been conducted with the Navier-Stokes code TASCflow, used in a non-reactive, steady and three-dimensional version. The aim of this computational work is to contribute to the projectile shape optimization under pure aerodynamical conditions, and therefore, to localize shock waves and their interactions, recirculation zone and high-temperature areas. Results presented in this paper show the effects of the fins on the flow configuration around the body and at the base (recirculation zone). Parameters such as the profile of leading edges and the number and size of fins are taken into account to compare the different flow fields. The results of computation, related to experiments performed in inert gas with different projectiles, have been used to highlight the influence of these parameters on the efficiency of the diffuser formed by the projectile afterbody. The pressure and temperature in the flow around the projectile when entering the ram tube and just before the initiation of the combustion are given. The present work demonstrates that the computer code TASCflow can provide a valuable tool for the optimization of the projectile geometry.

Key words: CFD, Projectile shape, Starting process

1. Introduction

The projectile used in a ramac launcher has a biconical shape, and the reacting mixture flowing around the body is compressed by the nose and expands at the afterbody. The projectile must be correctly guided along the ram tube. This should be done without a projectile incidence that can modify the symmetry of the flow and increase the drag (Giraud et al 1992). Such a disturbance in the projectile position can create undesirable conditions of Mach, pressure and temperature distribution, which, in turn, can provoke an unstart.

Therefore, the fins are used to guide the projectile, and they are preferably located behind the throat. Several experiments have been carried out to optimize their number, size and shape (Imrich 1995), in order to control the aerothermal conditions necessary for the mixture initiation at the rear part of the projectile, and to preserve the projectile from thermomechanical strains (Henner et al 1996).

The flow must be compressed and heated in order to control the combustion (Hertzberg et al 1986). This is obtained by both the variation of the cross-sectional area around the projectile and non-isentropic phenomena such as shock waves on the nose and on the fins. The compression and heating of the flow must be adjusted to an appropriate ratio in order to avoid any unexpected ignition. An earlier combustion occurring before the throat would lead to an unstart, which would also happen with a deflagration-detonation transition (Giraud et al 1995a). These phenomena depend on the projectile shape and Mach number, and also on the mixture composition, i.e. its sensitivity to ignition and its tendency to detonate (Legendre 1996). Finally, aerothermal conditions must be adjusted in order to trigger and control the combustion and of course, to deliver the proper amount of energy that will accelerate the projectile.

Shadowgraphs of ram projectiles in free flight in ambient air can be observed to illustrate the complex structure of the flow around the projectile. In Fig. 1, two 30-mm-caliber projectiles are presented at Mach number 3.4 under atmospheric conditions. The first one is equipped with 4 fins of constant 1/6 caliber thickness, the leading edge has the same inclination of 12.5° as the

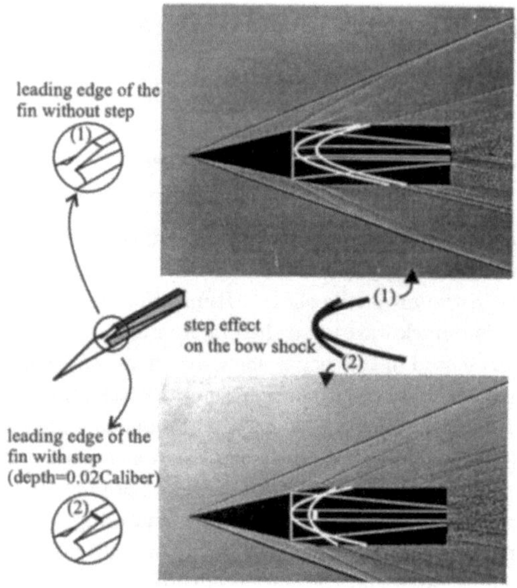

Fig. 1. Ramac projectile in free flight (4 fins of constant thickness=0.17 caliber, Mach=3.4)

forebody. One can easily observe the conical shock attached to the nose, and two shock waves attached to the base and the extremity of the leading edge. These latter are highlighted on the figure with white curves. The second shadowgraph shows the same projectile geometry, except for the step on the leading edge. The shock at the extremity of the leading edge is more important when the fin is equipped with a step, as can be seen on the upper and lower figures. The two shocks with white curves are stronger than the previous ones.

These differences show that a slight change in the projectile geometry can modify the structure of the flow, and therefore, conditions inside a ram tube can vary into large ranges of pressure, temperature and Mach number (Henner 1996).

Because of all these complex phenomena, the understanding of ram experiments requires some knowledge concerning flow fields around the ram projectile. As this knowledge cannot be obtained experimentally in a ram tube, some investigations have been conducted with numerical means in order to obtain some characteristics of the flow. The aim is to detect geometrical parameters that cause strong modifications in the flow, especially in terms of Mach number, pressure and temperature. This can change the initiation conditions of the mixture. Finally, this numerical study will provide some information on the influences of the projectile geometry and the Mach number for a given geometry.

2. Numerical study

The numerical study has been conducted with the 3D, steady and Navier-Stokes code TASCflow, which is a commercial code used in its non-reactive version (ASC Ltd 1995). The turbulence is modelized through a simple (k-ϵ) model, and computation has been done for ideal gas flowing around and behind the projectile.

Results have been obtained from an assumption of steady flow, and it must be admitted that the turbulence model is lacking in precision. However, the code has previously shown its ability to estimate pressure levels with a good accuracy (Henner 1996). Temperature can be used for comparison between different projectile shapes at different Mach numbers. Each computation has been done with the same conditions of supersonic inflow, with assumptions of a smooth and

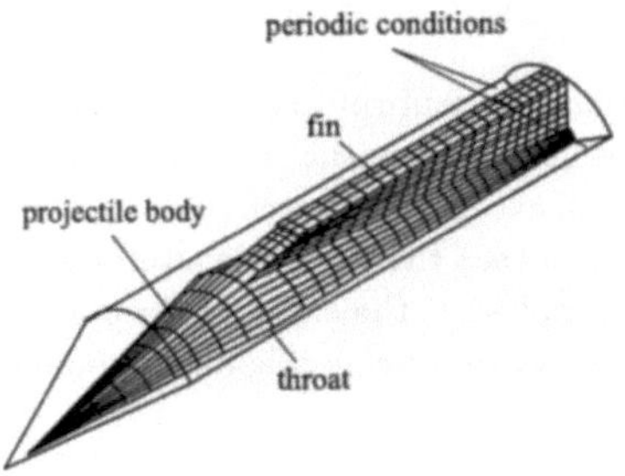

Fig. 2. computation domain

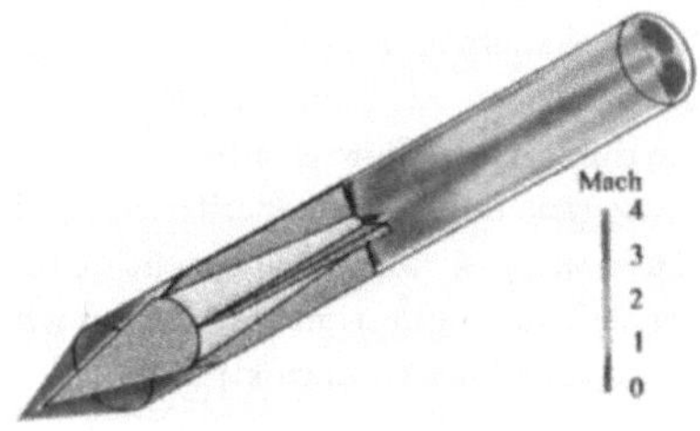

Fig. 3. Mach number distribution (P_0=4.5MPa, T_0=298K, 4 fins of various thickness 0.02 to 0.17 caliber

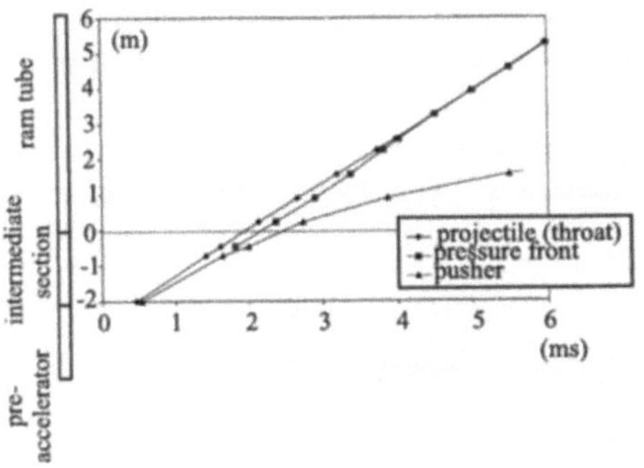

Fig. 4. position versus time (30mm caliber projectile with 4 fins of constant thickness 0.17 caliber, P_0=2MPa, $3CH_4+2O_2+10N_2$)

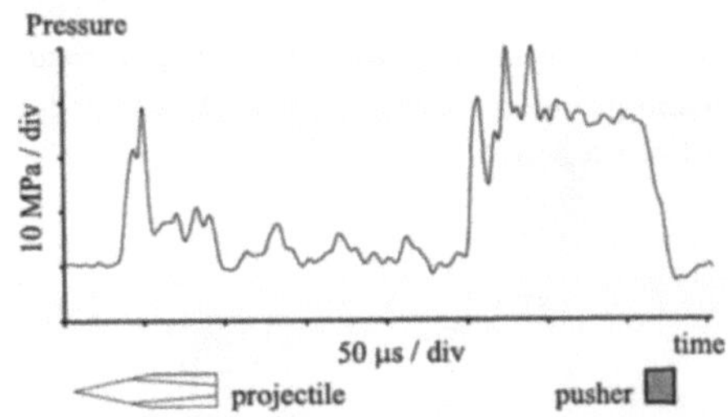

Fig. 5. wall pressure at the ramac entrance (30mm caliber projectile with 4 fins of constant thickness 0.17 caliber, P_0=2MPa, $3CH_4+2O_2+10N_2$)

adiabatic tube wall. The temperature of the projectile wall is limited to the melting temperature of the aluminum or magnesium alloy (about 850 K) used at ISL in subdetonative mode.

In order to take into account the real conditions of sliding between the gas flow and the tube wall, the latter moves numerically at the supersonic speed of the gas at the ramac entrance.

The computation domain around the projectile shown in figure 2 includes about 10^5 nodes. Only a quarter or a third of the volume around the projectile is meshed, for a 4- and 3-fin projectile, respectively. Periodic conditions are applied to both sections, on each side of the fin.

3. Numerical and experimental results

Figure 3 shows the Mach number distribution on the nose and in the wake of the projectile. It clearly indicates that the flow is three-dimensional and strongly influenced by the fins. In the wake, variations of the Mach number can be clearly noticed in the cross-section of the tube.

Although high pressures and high temperatures are assessed around the projectile body, especially on the leading edge of the fins, several experiments have demonstrated that these conditions are not able to ignite the mixture that is currently used in sub-detonative combustion at ISL, i.e. $3CH_4+2O_2+10N_2$ (Henner et al 1997). For instance, the experimental result in 30-mm-caliber presented in Fig. 4 shows that the mixture ignition occurs after the projectile entrance in the ram tube, somewhere between the projectile and its pusher. A combustion front is observed. It catches up with the projectile and accelerates it after several meters.

It has been observed that the pusher exiting from the pre-accelerator after the projectile creates a high-pressure zone with a shock that moves upstream (Schultz 1996). Pressure sensors signals are presented in Fig. 5. Two pressure fronts can be noticed, originating from the projectile and the pusher, respectively.

This phenomenon can be numerically simulated with a high-pressure outlet condition, similar to the one that occurs when a piston enters the ram tube. When an expansion is correctly started in the diffuser, the flow is supersonic throughout the domain, except in the recirculation zone at the base (the diffuser is constituted by the expansion zone between the projectile afterbody and the tube wall). A numerical condition of supersonic flow is applied to the outlet of the domain.

Another configuration is obtained with a choked diffuser, a shock being originated in a high-pressure zone located downstream.

4. Numerical simulation of the flow in the diffuser

Axisymmetric flows inside the diffuser have been simulated and compared for different outlet conditions of pressure (Fig. 6). Results obtained for Mach number 4 have demonstrated that the highest pressure zone is located at the projectile throat in the case of a supersonic flow at the outlet. This pressure resulting from the conical shock attached to the nose and reflecting on the tube wall reaches 5 times the initial pressure. The highest temperatures are located in the wake of the projectile, about 1000 K, but in a limited low-pressure area (1 MPa, i.e. about 1/5 of the initial pressure).

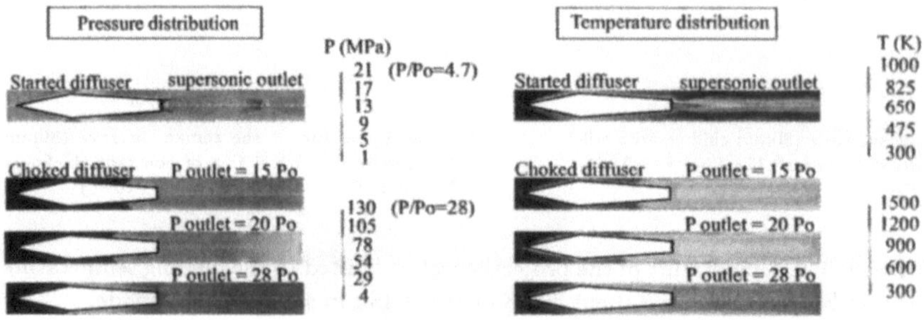

Fig. 6. conditions of pressure and temperature in the diffuser ($Mach_0=4$, $P_0=4.5MPa$, $T_0=298K$)

A high-pressure outlet condition at one projectile length is used to simulate a choked diffuser. Increased values of pressure show that the pressure front moves from the back of the projectile to its throat, which is reached with an outlet pressure of 28 times the initial pressure. The high-pressure zone fills all the space in the diffuser and in the wake, and the temperature rises to 1500 K.

Experiments in which the diffuser was not started have demonstrated that these conditions of pressure and temperature were able to ignite the $3CH_4+2O_2+10N_2$ mixture. If the pressure front is situated along the nose, an immediate unstart occurs. Under ideal conditions, the initiation in the diffuser would take place upon the projectile entrance in the ram tube. The diffuser must be started, and an igniting device can be used to trigger the combustion at the base, or at least in the projectile wake. This device can be the sabot itself, or another technical piece situated between the projectile and the sabot, such as the one patented by ISL (Giraud et al 1995b).

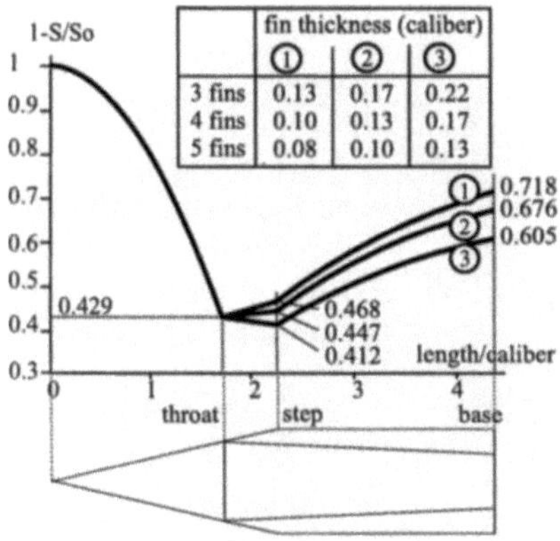

Fig. 7. cross-sectional area of the diffuser for various thickness and number of rectangular fins

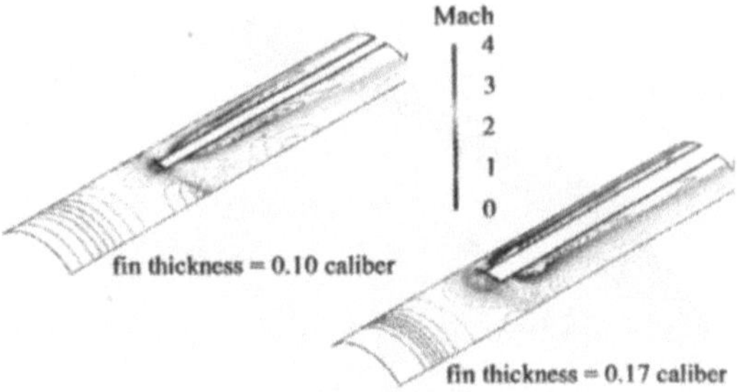

Fig. 8. distribution of Mach on each side of a fin ($Mach_0$=4, P_0=4.5MPa)

5. Effects of the fins in the diffuser starting

Preliminary experiments in RAMAC 30 and in RAMAC 90 have shown that the entrance conditions of the projectile in the ram tube are particularly important to achieve the starting of the diffuser. Given a projectile geometry, the parameters that govern this starting are:

- the distance between the pusher and the projectile,
- the projectile Mach number,
- the filling pressure of the ram tube.

Investigations concerning the Mach number and the projectile geometry have been carried out with our numerical means. They have been undertaken to explain why a projectile with 3 fins (1/6-caliber-thickness) accelerates, or generates an unstart when equipped with 4 identical fins. These different projectile behaviors have been attributed to the various abilities of the diffuser to be started, depending on its geometry.

5.1 Thickness and number of fins. First computations have been done for a projectile propagating at Mach number 4 in a 4.5 MPa mixture. Projectiles equipped with 3, 4 or 5 fins are compared with equivalent cross-sectional areas along the projectile body. The diffuser area (A) is calculated at a given position with the projectile cross-section (S) and the tube cross section (S_0): $A = 1\text{-}S/S_0$. This area, plotted on Fig. 7, describes the size of the diffuser along the nose and the projectile afterbody, based on the thickness and number of fins. The diameter of the projectile is 0.77 caliber at the throat and 0.33 caliber at the base. One can also observe that comparisons are made with different fin thicknesses for a given distribution of the cross-sectional area of the diffuser.

Computational results have shown that both compression and heating ratios depend on the diffuser area, and also on the geometry given by the number and thickness of fins when flowing sections are equivalent.

For instance, Fig. 8 shows the effect of the fins thickness on the distribution of the Mach numbers in the flow. Mach numbers are presented in a plane near the tube wall, on each side of one fin. It can be observed that the thicker the leading edge, the stronger the shock attached to the fin, and the greater the shock angle. This shock has a strong influence on the flow downstream from the throat, because it interacts with the shock coming from another fin in a zone between the afterbody and the tube wall. This interaction rises pressure and temperature. These parameters

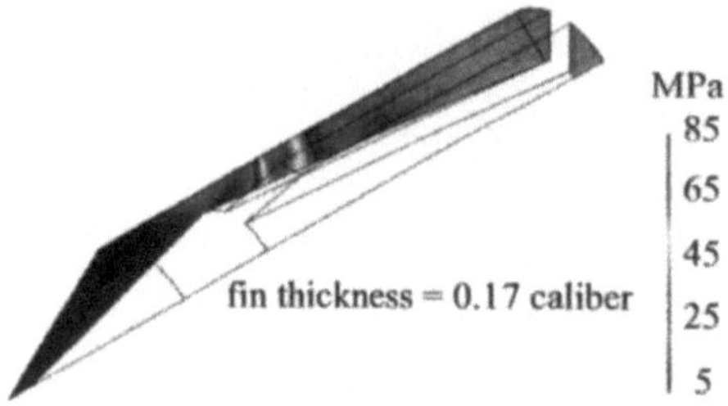

Fig. 9. pressure distribution in a plane between two fins (4-fin projectile, $Mach_0$=4, P_0=4.5MPa)

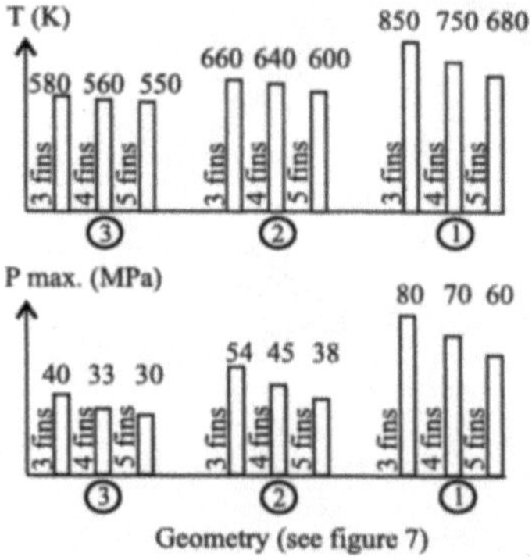

Fig. 10. maximum pressure and temperature at the same location in the plane between two rectangular fins, for given diffuser cross-sectional areas ($Mach_0$=4, P_0=4.5MPa, T_0=298K, started diffuser)

are greatest in the longitudinal plane at an equal distance from one fin to the other. Pressures have been plotted in this plane for different projectiles (see Fig. 9).

Some comparisons of numerical results are presented in Fig. 10, showing the maximum pressure (from the plane shown in Fig. 9) and the temperature at the same location. These values confirm that different flowing sections change both compression and heating ratios of the mixture. Another effect can be observed at a given flowing section: it appears that the intensity of shock is less important for a thin leading edge than for a thick one. The shock is also more inclined in the flow direction and the interaction of the two shocks coming from two fins occurs more downstream in the diffuser, where the flow expands. Therefore, values of pressure and temperature are lower for a given flowing section when a greater number of fins is involved, because the leading edges are thinner.

Both the flowing section variations and the fin thickness effects create various conditions in the flow, when the diffuser is started. It can be considered that high pressure and temperature would enhance the ignition of the mixture, but would also facilitate the choking of the diffuser at the RAMAC entrance or after acceleration at a high Mach number.

Others computations with two shapes of fins, on a 3- or 4-fins projectile have been conducted to study the effect of the initial Mach number.

5.2 Effects of the fins shape at different initial Mach numbers. At first, constant-thickness fins were considered (1/6 caliber). Then, a variable thickness, i.e. from 1/30 caliber at the leading edge to 1/6 caliber at the base, was studied. The different shapes and the number of fins (3 or 4) yield various flowing sections presented in the table of Fig. 11. The numerical study has been conducted for a projectile moving in a 4.5 MPa mixture, at Mach numbers of 3, 4, 5 and 6.

The leading edge is five times thicker with the constant-thickness fin than with the other shape, and the corresponding flowing section is smaller between the throat and the base. Consequently, the shocks and their interactions give higher pressures and temperatures with a thick leading edge, given a number of fins. These results are presented in Fig. 12, where a comparison between 3- or 4-fins projectiles gives increasing values of pressure and temperature as the number of fins increases. Pressures for a projectile with 4 constant-thickness-fins reach 4 times those obtained with a projectile with 3-fins of varying thickness. Likewise, temperatures are twice as high.

It can also be emphasized that increased Mach number will enlarge pressures and temperatures calculated in the flow. Thus, the reacting mixture will be submitted to varying conditions of

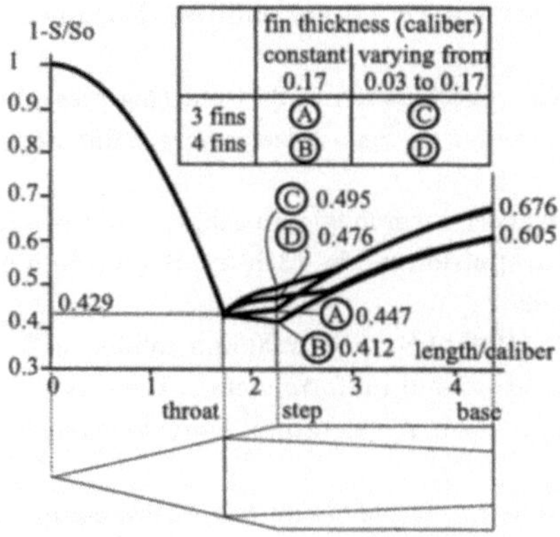

Fig. 11. cross-sectional area of the diffuser for various shapes and numbers of fins

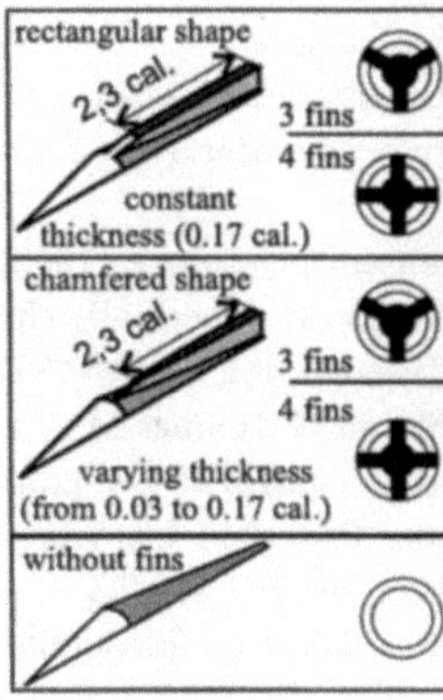

	Mach 4 P_{max} MPa	Mach 4 T (K)	Mach 5 P_{max} MPa	Mach 5 T (K)	Mach 6 P_{max} MPa	Mach 6 T (K)
rectangular shape, constant thickness (0.17 cal.), 3 fins	62	650	85	800	108	1200
rectangular shape, constant thickness (0.17 cal.), 4 fins	85	750	113	950	160	1400
	Maximum values					
chamfered shape, varying thickness (from 0.03 to 0.17 cal.), 3 fins	22	470	28	520	37	590
	Minimum values					
chamfered shape, varying thickness (from 0.03 to 0.17 cal.), 4 fins	24	500	30	550	38	610
without fins	16	470	21	440	30	500

Fig. 12. maximum pressure and temperature at the same location in the longitudinal plane between two fins (see Fig. 9, P_0=4.5MPa, T_0=298K)

ignition during the acceleration. The maximum values of pressure and temperature supported by the mixture before the onset of a detonation will represent one parameter that determine the maximum velocity that can be obtained in subdetonative propulsion mode. A mixture should be preferably located out of the detonable envelope derived from the estimated pattern of a specific projectile flow (Legendre et al 1997).

6. Conclusions

Shadowgraphs of ram projectiles in free flight have shown the complex structure of the flow around the body and the fins. Numerical simulations have been conducted to provide estimations of the aerodynamical conditions inside the ram tube, and to highlight the effect of the fins in the process of compression and heating of the mixture. Computations have been carried out with the 3-dimensional code TASCflow used to study both the projectiles geometries and the flow conditions in the diffuser.

When the diffuser is choked, the conditions of pressure and temperature estimated in the projectile wake are largely above those created by the projectile entering the ram tube at the same initial Mach number. This confirms that the ignition of the reacting mixture in a RAMAC must be obtained when the diffuser is started by local conditions behind the projectile. These conditions must be calibrated either by the pusher or by a specific device such as the one patented by ISL.

The values of pressure and temperature distribution calculated in the flow vary from one projectile geometry to the other. Considering only the shapes and number of fins, the computations provide the values of pressure in the ratio of 1 to 4, and of temperature in the ratio of 1 to 2. Consequently, it must be noticed that the projectile cannot be considered a simple biconical body (Bruckner et al 1993), and the fins play an important role in the flow compression and heating, and thus in the mechanism of ignition.

References

Advanced Scientific Computing Ltd (1995) TASCflow, version 2.4 Theory and User Documentation. Waterloo, Ontario, Canada

Bruckner AP, Hinkey J, Burnham E, Knowlen C (1993) Investigation of 3-D reacting flow phenomena in a 38 mm ram-accelerator. Proc 1st Int Workshop on Ram Accelerators, Saint-Louis, France

Giraud M, Legendre JF, Simon G, Catoire L (1992) Ram accelerator in 90 mm caliber. First results concerning the scale effect in the thermally choked propulsion mode. 13th Int. Symposium on Ballistics, Stockholm, Sweden, ISL report CO 210/92

Giraud M, Legendre JF, Simon G, Henner M, Voisin D (1995a) Ramac in 90 mm caliber or RAMAC 90. Starting process, control of the ignition location and performances in the thermally choked propulsion mode. Proc 2nd Int Workshop on Ram Accelerators, Univ Washington, Seattle, WA, USA, ISL report PU 349/95

Giraud M, Simon H (1995b) Sabot for projectiles of ram accelerators and projectiles equipped with such a sabot. United States Patent Number 5, 394, 805

Henner M, Berner C, Giraud M (1996) Preliminary study in RAMAC 90: non reacting gas flow around the projectile - Some consequences of the guiding fins design. 2nd Int Meeting on Properties of Reactive Fluids and their Application to Propulsion, Poitiers, France, ISL Report PU 343/96

Henner M (1996) Contribution to the design of a new RAMAC projectile. Modelisation and experiments. 47th ARA Meeting, Saint-Louis, France, ISL report PU 361/96

Henner M, Legendre JF, Giraud M, Bauer P (1997) Initiation of reactive mixtures in a ram accelerator. AIAA paper 97-3173, ISL report to be published.

Hertzberg A, Bruckner A, Bogdanoff D (1986) Ram accelerator: A new chemical method of achieving ultra-high velocities. 37th ARA Meeting, Québec, Canada

Imrich TS (1995) The impact of projectile geometry on ram accelerator performance. Master thesis, Aeronautics and Astronautics, Univ Washington, Seattle, WA, USA

Legendre JF (1996) Contribution à l'étude de la sensibilité et des caractéristiques de détonation de mélanges explosifs gazeux denses à base de méthane utilisés pour la propulsion dans les accélérateurs à effet stato. Ph.D dissertation, University of Poitiers, Poitiers, France

Legendre JF, Bauer P, Giraud M (1998) RAMAC 90: detonation initiation of insensitive dense methane-based mixtures by normal shock waves. In: Takayama K, Sasoh A (eds) Ram Accelerators, Springer-Verlag, Heidelberg, pp 223-231

Schultz E (1996) Ignition of the ram accelerator at low projectile entrance velocity. 47th ARA Meeting, Saint-Louis, France

Ignition of a reactive gas by focusing of a shock wave

M. Rose, U. Uphoff and P. Roth
Institut für Verbrennung und Gasdynamik, Gerhard–Mercator–Universität, 47048 Duisburg, Germany

Abstract. An adaptive mesh refinement technique is used to calculate the focussing process of shock waves in a reactive mixture. The fluid flow is described by the reactive Euler equations in two dimensions, which are integrated in time using an operator splitting technique. The convective part is integrated by the Harten–Yee TVD method, and the integration of chemical source terms is performed by CHEMEQ, an explicid Euler method. A one–step reaction mechanism is used to model the chemistry of a stoichiometric H_2/O_2–mixture. Calculations are performed for shock Mach numbers of 1.6 and 1.9. At lower shock strength, the complex interaction of different shock and expansion waves during the focussing process dominate the wave field and chemistry is not significant. The computed results show the same wave patterns as obtained by experiments for a similar set-up. In case of the higher Mach number of 1.9 the calculations lead to the ignition of a spherical detonation wave in the focal region which interacts with the diffracted shock wave and the reflector, resulting in an even more complex wave field.

Key words: Shock waves, Focussing, Detonation, AMR

1. Introduction

A shock wave reflected by a concave reflector generates a region of high energy density near the gasdynamic focus. For weak shocks up to a Mach number of 1.5 the evolution of a complex wave field near the focal region has been investigated in many experimental, analytical, and numerical works. Experimental results for a variety of Mach numbers were obtained by Sturtevant and Kulkarny (1994) and Müller and Grönig (1986). Numerical calculations of the nonlinear wave interaction in shock wave focussing are commonly based on the nonlinear theory of shock dynamics as given by Whitham (1957). Computations have been performed using various numerical techniques. Cates and Sturtevant (1996) developed an improved finite difference formulation for calculations based on geometrical shock dynamics. But the accuracy of this method deteriorates for Mach numbers less than 1.1. A random choise method (RCM) combined with an operator splitting technique was used by Olivier and Grönig (1986). They obtained good agreement between calculations and experiments for low Mach numbers ($M_a \leq 1.3$), but for stronger shock waves, the errors due to the choise of random numbers and due to the splitting technique became unacceptable. The results obtained by the RCM were improved by Sommerfeld (1989), who used a second order extension of Godunov's method to capture discontinuities. Choi and Baek (1996) extended a symmetric finite difference scheme of the total variation diminishing (TVD) type to a cell centered finite volume method to guarantee reasonable accuracy and resolution required for calculations of the complex nonlinear interaction of shock waves. The resolution of spatial discontinuities becomes even more important if strong shock waves ($M_a \geq 1.5$) propagate into a combustible mixture. The local peaks in pressure and temperature generated in the focal region can cause the mixture to ignite.

In the present work the focussing process leading to ignition of a stoichiometric H_2/O_2–mixture is investigated by numerical simulations based on the reactive Euler equations in two dimensions. An operator splitting technique is applied to handle fluid flow and chemistry separately. The convective part is integrated using the Harten–Yee (1987) TVD method and the integration of chemical source terms is performed by an explicid Euler method. A one–step reaction mechanism is used to model the chemistry of the premixed gases and an adaptive mesh refinement technique

(AMR) is applied to get an adequate resolution of steep gradients especially in the ignition and reaction zones of the combustion process. The use of a very fine grid size is crucial for a reliable calculation of the reactive parts of the fluid flow.

2. Modeling

2.1 Governing equations The Euler equations are commonly used to describe the dynamics of a compressible, inviscid fluid. In two space dimensions these equations are of the following form:

$$\frac{\partial \mathbf{U}}{\partial t} + \frac{\partial \mathbf{F}}{\partial x} + \frac{\partial \mathbf{G}}{\partial y} = \mathbf{S} \tag{1}$$

In this equation **U**, **F**, **G** and **S** represent the vectors of the conserved properties, the fluxes, and the source terms, respectively. For a gas mixture containing N species, which are of density ρ_1 to ρ_N, these vectors have the form:

$$\mathbf{U} = \begin{pmatrix} \rho_1 \\ \vdots \\ \rho_N \\ \rho u \\ \rho v \\ \rho E \end{pmatrix}, \quad \mathbf{F} = \begin{pmatrix} \rho_1 u \\ \vdots \\ \rho_N u \\ \rho u^2 + p \\ \rho u v \\ (\rho E + p)u \end{pmatrix}, \quad \mathbf{G} = \begin{pmatrix} \rho_1 v \\ \vdots \\ \rho_N v \\ \rho u v \\ \rho v^2 + p \\ (\rho E + p)v \end{pmatrix}, \quad \mathbf{S} = \begin{pmatrix} \dot\omega_1 \\ \vdots \\ \dot\omega_N \\ 0 \\ 0 \\ 0 \end{pmatrix} \tag{2}$$

The density of the mixture is $\rho = \sum_{i=1}^{N} \rho_i$, its velocity in cartesian co-ordinates is $\mathbf{v} = (u, v)^t$, and its temperature and pressure are denoted with T and p, respectively. The total energy E of the gas phase is calculated as the sum of the internal and kinetic energies:

$$E = \int_{T_{ref}}^{T} c_v dT + \frac{1}{\rho} \sum_{i=1}^{N} \rho_i H_i^{T_{ref}} + \frac{1}{2}\left(u^2 + v^2\right) \tag{3}$$

$H_i^{T_{ref}}$ is the enthalpy of formation of species i at a temperature T_{ref}. It is assumed that the gas obays the ideal gas law:

$$p = \rho R T \tag{4}$$

The formation and destruction of species 1 to N due to homogeneous reactions are included in the source terms $\dot\omega_1, \ldots, \dot\omega_N$. To model the chemistry, a simple one–step reaction mechanism is used, envolving two gaseous species (N=2):

$$A \longrightarrow B \tag{5}$$

It discribes the irreversible conversion of a gas species A into a gas product B. The rate coefficient k is described by a one–step Arrhenius law:

$$k = k_0\, exp(-E_a/RT) \tag{6}$$

The chemical source terms $\dot\omega_A$ and $\dot\omega_B$ become under these simplified assumptions:

$$\dot\omega_A = -\, k\, \rho_A\,, \quad \dot\omega_B = k\, \rho_A \tag{7}$$

2.2 Numerical method The integration of the Euler equations (1) is performed by an operator splitting technique to handle fluid flow and chemistry separately:

$$\mathbf{U}_{t+\Delta t} = \mathcal{L}_{flow}^{\Delta t/2} \quad \mathcal{L}_{chem}^{\Delta t} \quad \mathcal{L}_{flow}^{\Delta t/2} \quad \mathbf{U}_t \tag{8}$$

The convective part is integrated using the Harten–Yee (1987) TVD method represented by

$$\mathcal{L}_{flow}^{\Delta t/2} = \mathcal{L}_{y}^{\Delta t/2} \, \mathcal{L}_{x}^{\Delta t/2} \tag{8}$$

The integration operators in the x– and y–direction on a discretized cartesian grid are of the form:

$$\mathcal{L}_{x}^{\Delta t/2} \, \mathbf{U}_t^{i,j} = \mathbf{U}_t^{i,j} - \frac{\Delta t/2}{\Delta x} \mathbf{J_F}^{-1} \left(\tilde{\mathbf{F}}_t^{i+1/2} - \tilde{\mathbf{F}}_t^{i-1/2} \right) \tag{9}$$

$$\mathcal{L}_{y}^{\Delta t/2} \, \mathbf{U}_t^{i,j} = \mathbf{U}_t^{i,j} - \frac{\Delta t/2}{\Delta y} \mathbf{J_G}^{-1} \left(\tilde{\mathbf{G}}_t^{j+1/2} - \tilde{\mathbf{G}}_t^{j-1/2} \right) \tag{10}$$

$\mathbf{J_F}$ and $\mathbf{J_G}$ are the Jacobians of the flux vectors $\mathbf{J_F} = \partial \mathbf{F}/\partial \mathbf{U}$ and $\mathbf{J_G} = \partial \mathbf{G}/\partial \mathbf{U}$, evaluated somewhere between the states $\mathbf{U}_t^{i,j}$ and $\mathbf{U}_t^{i+1,j+1}$. The numerical fluxes $\tilde{\mathbf{F}}_t^{i+1/2}$ and $\tilde{\mathbf{G}}_t^{j+1/2}$ at the center of computational cell (i, j) are calculated from the physical flux terms **F** and **G** in equation (1) by:

$$\tilde{\mathbf{F}}_t^{i+1/2} = \frac{1}{2} \left[\mathbf{F}\left(\mathbf{U}_t^{i,j}\right) + \mathbf{F}\left(\mathbf{U}_t^{i+1,j}\right) + \mathbf{R}_\mathbf{F}^{i+1/2} \Phi_\mathbf{F}^{i+1/2} \right] \tag{11}$$

$$\tilde{\mathbf{G}}_t^{j+1/2} = \frac{1}{2} \left[\mathbf{G}\left(\mathbf{U}_t^{i,j}\right) + \mathbf{G}\left(\mathbf{U}_t^{i,j+1}\right) + \mathbf{R}_\mathbf{G}^{j+1/2} \Phi_\mathbf{G}^{j+1/2} \right] \tag{12}$$

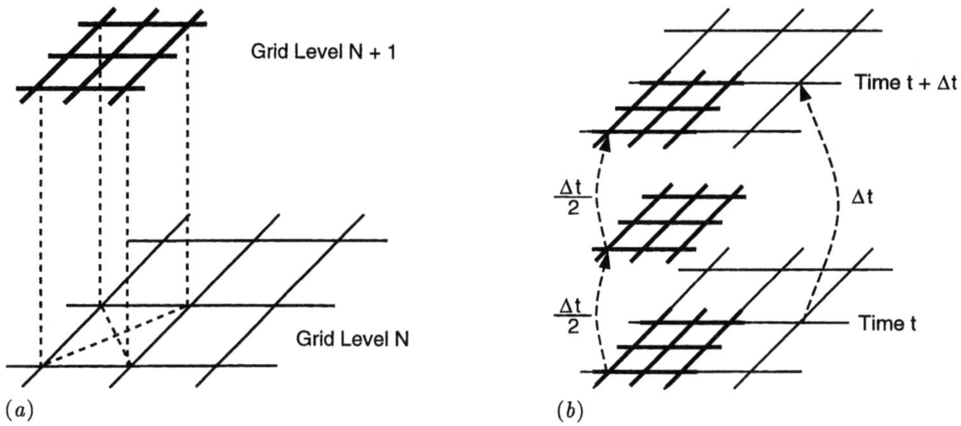

Fig. 1. Illustration of the mesh refinement algorithm: a, Cell on grid level N flagged for refinement by a factor of two in both coordinate directions; b, Cell refined by a factor of two is integrated in time two times by half the timestep used for the course mesh

The matrices $\mathbf{R}_\mathbf{F}^{i+1/2}$ and $\mathbf{R}_\mathbf{G}^{i+1/2}$ consist of the right eigenvectors of the flux Jacobians $\mathbf{J_F}$ and $\mathbf{J_G}$, respectively, and the vectors $\Phi_\mathbf{F}^{i+1/2}$ and $\Phi_\mathbf{G}^{j+1/2}$ depend on the eigenvalues of the corresponding Jacobians. By using the antidiffusive flux terms $\mathbf{R}_\mathbf{F}^{i+1/2}\Phi_\mathbf{F}^{i+1/2}$ and $\mathbf{R}_\mathbf{G}^{j+1/2}\Phi_\mathbf{G}^{j+1/2}$ the large dissipation of first order numerical schemes is corrected in a nonlinear way and the method becomes higher order. The operator by which the integration of chemical source terms is performed, is represented by $\mathcal{L}_{chem}$ in equation (3). The scheme is based on an explicid Euler method implemented in the code CHEMEQ described by Young (1977).

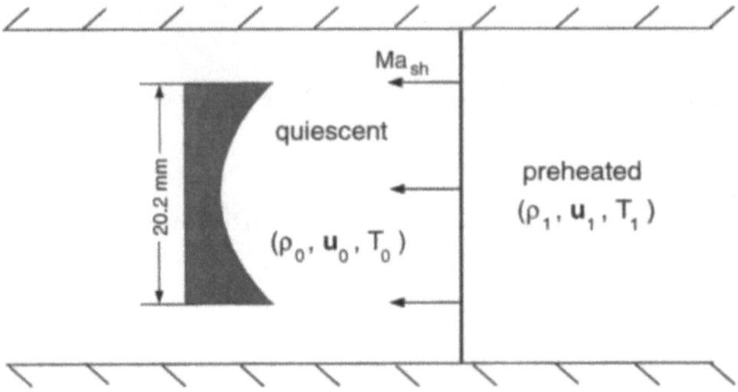

Fig. 2. Schematics of a shock wave in a combustible gas mixture propagating towards a concave reflector

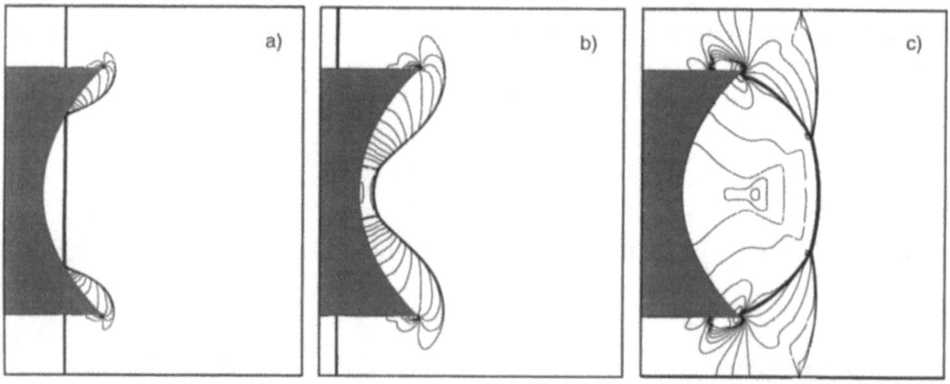

Fig. 3. Isopycnics for an initial shock strength of M_a=1.6

To get the desired resolution of regions with strong gradients and intensive chemical activity, the computational mesh is locally refined by superimposing grids of finer meshes. The mesh refinement algorithm is illustrated in Fig. 1.

One cell on grid level N has been flagged for refinement according to some criterion, i.e. a strong gradient in pressure or density, and is then divided into four smaller cells on the next higher grid level shown in Fig. 1a). After interpolating appropriate boundary conditions from the grids on level N to the new grids on level $N+1$, the integration in time can be performed as illustrated in Fig. 1b). To guarantee the CFL-condition, the finer grid is integrated in time two times by half the timestep used for the course mesh. The mesh refinement technique is described in detail by Quirk (1991).

3. Results

The described numerical method is applied to investigate the wave fields that develop when a shock wave propagates into a quiescent stoichiometric H_2/O_2–mixture with initial conditions p^0 = 10 kPa and T^0 = 300 K. The properties used in the Arrhenius expression (2) are $k_0 = 1.5 \times 10^9$ s^{-1} and E_a = 12000 J mol^{-1}, respectively. A schematical drawing of the set-up is given in Fig. 2.

Three levels of refinement are used on a curvilinear cartesian grid. The criteria according to which cells are flagged for refinement is based on the local variation of density. The refinement factor 2 was used from the basic grid level to level one, and from level one to level two, whereas a refinement factor of 4 was used from level two to level three. The spatial resolution on the finest grid is 5×10^{-6} m. Calculations were performed for Mach numbers of 1.6 and 1.9. The density

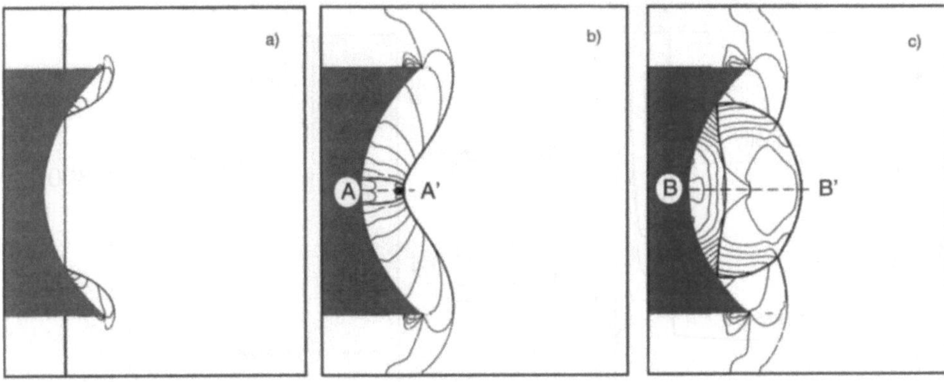

Fig. 4. Isopycnic for an initial shock strength of $M_a = 1.9$

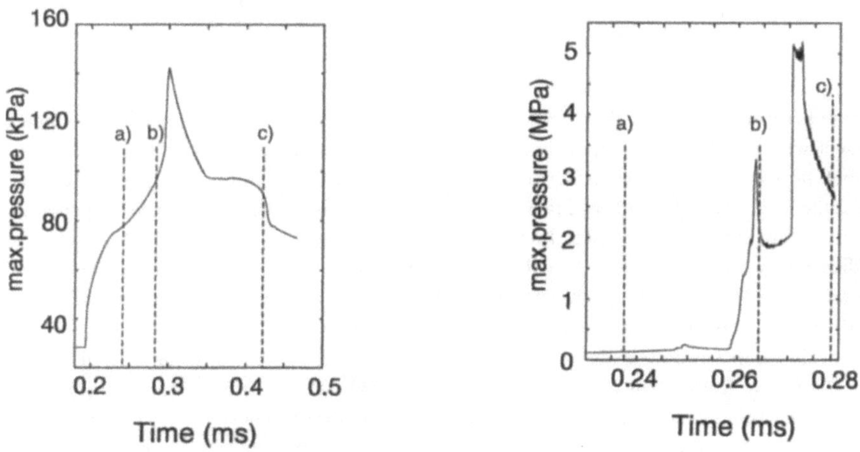

Fig. 5. Evolution of the maximum pressure during the focussing process for an initial shock strength of M_a=1.6 (left part) and M_a = 1.9 (right part)

fields illustrating the focussing process of the reflected waves are presented in Figs. 3 and 4 and the development of maximum pressure in the volume is given in Fig. 5.

In Fig. 3 the conditions are such that ignition and reaction do not occur whereas in Fig. 4 a detonation wave is formed. According to Figs. 3a and 4a the incoming shocks are diffracted at the solid surface and the reflected shocks are accompanied by an expansion wave.

The shock strength is continuously amplified during the reflection process. After total reflection (Fig. 3b), the inner part of the reflected shock has a concave shape and converges towards the focus. The concave portions of the diffracted wavefront become stronger while the convex parts are weakened leading to a non–uniform shock strength along the front. Due to this variation of shock strength, expansion waves travel into regions of high pressure behind the concave front and compression waves of small amplitude propagate into regions of lower pressure behind the convex fronts. By the theory of geometrical acoustics for weak shock waves (Keller 1954), the convergence of the concave parts would lead to an infinite amplitude in pressure in the focal regions. But due to the high strength of the initial shock wave, the focussing process is determined by nonlinear gasdynamic effects, leading to different velocities of the shockfront. This results in an acceleration of the more concave parts relative to the others as described by Sturtevant and Kulkarny (1994).

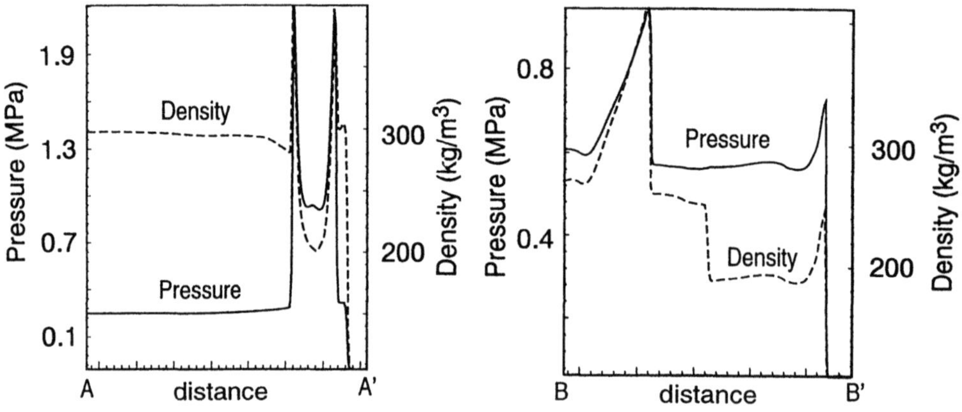

Fig. 6. Pressure and density distribution along lines $\overline{AA'}$ and $\overline{BB'}$ in Fig. 4

The maximum pressure during the total focussing process is obtained when the converging shock-fronts intersect with the diffracted expansion wave at the axis of symmetrie. For a shock of initial Mach number of 1.6 a stem shock is formed by Mach reflection of the converging shocks as shown in Fig. 3c. For the higher initial Mach number of 1.9 in Fig. 4, the local peaks in pressure and temperature are sufficient to ignite the mixture. As shown in Fig. 4b) a detonation wave is formed, indicated by the first pressure peak in the right part of Fig. 5. The combustion wave then propagates into the preheated unburned gas. The pressure and density distributions along the line $\overline{AA'}$ (Fig. 4b) are shown in the left of Fig. 6.

The detonation fronts propagating to the right and to the left become obvious as well as the reflected shock wave in front of the detonation wave. At the time shown in Fig. 4c, the detonation wave has overtaken the diffracted shock thereby forming a contact discontinuity, which becomes visible in the right of Fig. 6 by the jump in density, whereas the pressure remains constant. The rear part of the ignition induced pressure wave is reflected again, forming a secondary shock of higher amplitude. While not being driven by chemical heat release, this shock does not further contribute to the propagation of the combustion wave.

Fig. 7 illustrates the result of the adaptive calculations for a time instant shortly after the reflection of the detonation wave.

Both the isopycnics in the upper part and the corresponding computational mesh with all levels of refinement in the lower part are shown. The clustering of very fine meshes close to the reflected shock wave and at the position of the detonation wave is obvious.

4. Conclusions

The reactive Euler equations were solved in two dimensions by an adaptive mesh refinement technique to investigate the focussing behaviour of a diffracted shock wave in a combustible mixture. For low Mach number shock waves of about 1.6, the wave field in the focal region shows nearly the same characteristics as observed by experiments. For a higher shock Mach number of about 1.9, the local peaks in pressure and temperature in the focal point are sufficient to ignite the mixture, giving rise to a detonation wave, which partly interacts with the diffracted shock wave. This interaction of a hydrodynamic shock with a combustion wave results in multifacetted wave patterns. The combination of an adaptive method with high resolution, shock capturing techniques yields a good representation of the diffracted shock as well as the detonation wave.

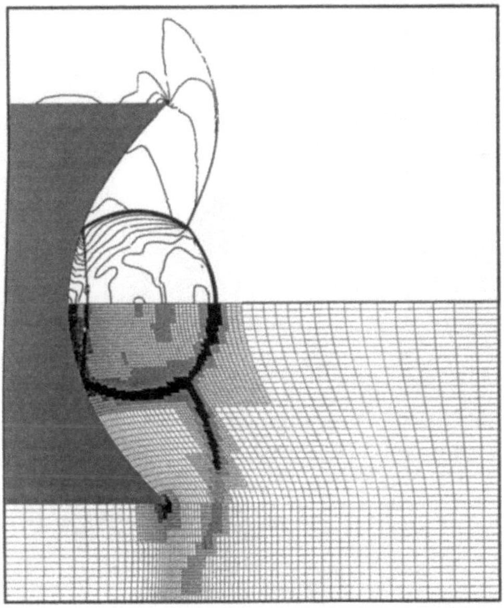

Fig. 7. Upper part: isopycnics at a time instant illustrating the wave pattern developed after reflection of an initial shock strength of $M_a = 1.9$. Lower part; corresponding computational grid distribution

Due to the use of a locally refined grid close to the reaction zone of the combustion wave this region of high chemical activity is resolved by reasonable computational costs.

References

Cates J, Sturtevant B (1996) Calculation of focussing of weak shocks using geometrical shock dynamics. In: Sturtevant B et al (eds) Proc 20th Int Symp on Shock Waves, Vol I, pp 423-428

Choi HS and Baek JH (1996) Computations of nonlinear wave interaction in shock wave focussing. Comp Fluids 25:509-525

Keller JB (1954) Geometrical acoustics I, the theory of weak shock waves. J Appl Phys 25:938ff

Müller H, Grönig H (1986) Experimental investigations on shock wave focussing in water. Proc 12th Int Congress in Acoustics III H3

Olivier H, Grönig H (1986) The random choice method applied to two-dimensional shock focussing and diffraction. J Comp Phys 63:85-106

Quirk JJ (1991) An adaptive grid algorithm for computational shock hydrodynamics. College Aero, Cranfield Inst Tech

Sommerfeld M (1989) Numerical prediction of shock wave focussing phenomena in air with experimental verification. Nonlinear Hyperbolic Equations–Theory, Computation Methods, and Applications, Notes on Numerical Fluid Mechanics 24:562-573

Sturtevant B, Kulkarny VA (1994) The focussing of weak shock waves. J Fluid Mech 73:651-671

Whitham GB (1957) A new approach to problems of shock dynamics I. J Fluid Mech 2:145-171

Yee HC (1987) Upwind and symmetric shock capturing schemes. NASA TM 89464

Young TR (1977) CHEMEQ–subroutine for solving stiff ordinary differential equations associated with the chemical kinetics of reactive flow problems. J Phys Chem 81:2424-2427

Author index

Key word index

MIX
Papier aus verantwortungsvollen Quellen
Paper from responsible sources
FSC® C105338

If you have any concerns about our products,
you can contact us on
ProductSafety@springernature.com

In case Publisher is established outside the EU,
the EU authorized representative is:
Springer Nature Customer Service Center GmbH
Europaplatz 3, 69115 Heidelberg, Germany

Printed by Libri Plureos GmbH
in Hamburg, Germany